海洋生物基因组学概论

Introduction to Marine Genomics

石　琼　孙　颖◎主　编

游欣欣　白　洁◎副主编

编　委：（排名不分先后）

陈洁明　刘珊珊　杨成业　彭　超

汪金兔　卞　超　张新辉　邱　樱

高　鹏　陆　君　黄　玉　金　桃

刘　欢　李文浩　穆　茜　阮志强

范明君　徐军民　于　辉　吴海辉

潘　莹　黄萌理　王　吉　黄小芳

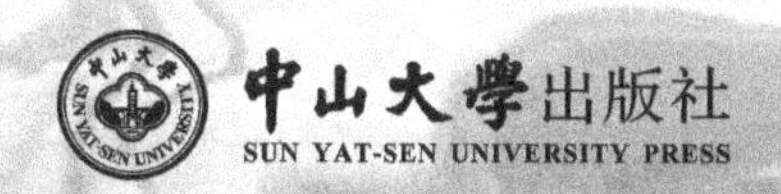

·广州·

图书在版编目（CIP）数据

海洋生物基因组学概论/石琼，孙颖主编；游欣欣，白洁副主编．— 广州：中山大学出版社，2015.8

ISBN 978-7-306-05357-2

Ⅰ.①海…　Ⅱ.①石…②孙…③游…④白…　Ⅲ.①海洋生物—基因组—研究　Ⅳ.①Q178.53

中国版本图书馆 CIP 数据核字（2015）第 161960 号

原书名：Introduction to Marine Genomics

原作者：杰·马可·科克　克莉丝汀·太斯马-莱柏　凯瑟琳·博延　弗德雷克·维亚德

出 版 人：徐　劲

策划编辑：李　文

责任编辑：李　文

封面设计：曾　斌

责任校对：江克清

责任技编：何雅涛

出版发行：中山大学出版社

电　　话：编辑部 020-84111996，84113349，84111997，84110779
　　　　　发行部 020-84111998，84111981，84111160

地　　址：广州市新港西路 135 号

邮　　编：510275　　传真：020-84036565

网　　址：http://www.zsup.com.cn　　E-mail:zdcbs@mail.sysu.edu.cn

印 刷 者：虎彩印艺股份有限公司

规　　格：787mm×960mm　1/16　26.5 印张　723 千字

版次印次：2015 年 8 月第 1 版　2015 年 8 月第 1 次印刷

定　　价：70.00 元

如发现本书因印装质量影响阅读，请与出版社发行部联系调换

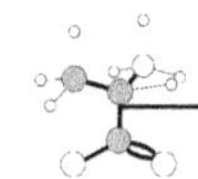

中文译序

本书是2010年Springer出版社出版的Introduction to Marine Genomics一书的中文译本，宗旨是推动我国海洋生物学研究的快速发展，希翼我国科学家能在基因组学背后的大科学、大数据和大产业上占领制高点，在国际上拥有海洋科技和海洋经济的领先地位与话语权。

自从2011年5月海洋平台成立以来，我们在华大基因这个国际领先的平台上与国内外同行们开展了广泛的科技合作。由我们主持、深圳野生动物保护中心和厦门大学等共同参与的弹涂鱼基因组文章已在Nature Communications上发表（2014，5：5594）；我们参与海洋三所和浙江大学牵头的大黄鱼基因组工作，文章已在PLOS Genetics上发表（2015，11（4）：e1005118）；参与国家基因库牵头的企鹅基因组，文章在Gigascience上发表（2014，3：27）。我们与中科院昆明动物所合作主持的金线鲃基因组、参与广东海洋大学主持的马氏珠母贝基因组、参与武汉大学主持的黄鳝基因组等研究论文皆已投稿，与中科院南海海洋研究所合作的海马基因组、与水科院珠江所等合作的金龙鱼基因组、与中山大学合作的石斑鱼基因组等研究项目论文，预期在2015年投稿；我们还积极参与华中农业大学牵头的武昌鱼基因组、与水科院淡水中心合作主持的刀鱼基因组、与大连海洋大学和上海海洋大学等联合开展的中华绒螯蟹基因组等工作。

我们也与解放军防化研究院合作开展南海芋螺的全基因组、转录组与蛋白质组研究，与华侨大学联合开展海蚯蚓类海洋生物的基因组与转录组研究，与深圳大学生科院联合实施硒蛋白、抗菌肽和吸附重金属细菌等研究，并且牵头实施国际合作的“千种鱼类转录

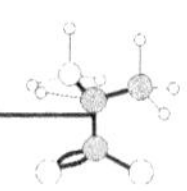

组计划”。我们还主持启动国际水生哺乳动物基因组计划，由海洋三所和汕头大学等联合牵头的中华白海豚基因组和由南京师范大学主持的江豚基因组等项目就是该计划的开路先锋。

海洋平台只是我国众多开展海洋生物基因组学研究单位中的一个小小单元。我们希望与同行们协同推进海洋生物科技的同时，也能大力发展海洋产业的研发。2012 年，我们成立了深圳华大水产科技有限公司，致力于高端水产的育种与养殖示范。2014 年后陆续在江苏镇江、广东广州、河南长垣、湖北恩施、广西北海、贵州兴义、海南海口、广东电白等地布局和建设基地。

本书的出版得到深圳市海洋生物基因组学重点实验室（主任石琼)、深圳国家基因库（执行主任张勇)、深圳华大基因学院（院长杨焕明院士）和广东省海洋经济动物分子育种重点实验室（主任石琼）的经费资助。初稿已在深圳华大基因学院和中国科学院华大教育中心试用，得到同行们的热情鼓励。深圳华大基因研究院海洋平台、深圳国家基因库海洋分库、深圳华大水产科技有限公司和深圳华大基因科技服务有限公司的不少同事们参与了本书的翻译和校对，在此一并致以诚挚的谢意。还要特别感谢中山大学出版社李文主任和 Springer 北京办事处陈青经理，感谢他们的努力协调和细致审校。本书的出版还得到中山大学林浩然院士、大连海洋大学副校长宋林生教授、华大基因董事长汪建和深圳华大基因研究院院长王俊等领导的关怀与支持，在此一并致以诚挚的谢意。

期盼本书能发展成为高校的教材，为我国的海洋生物学教育奉献绵薄之力。

由于译者的水平有限，仓促之中存在诸多错漏之外，还请大家批评指正。

深圳华大基因研究院海洋平台负责人
深圳华大水产科技有限公司科技副总
石　琼

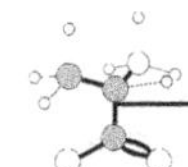

目　录

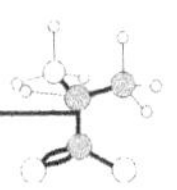

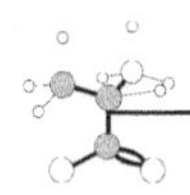

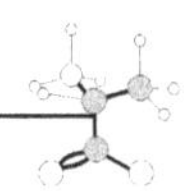

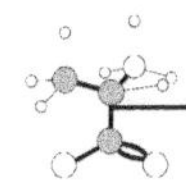

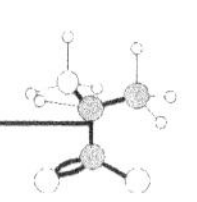

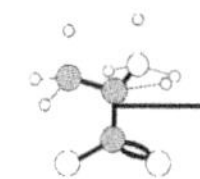

前　言

基因组学（genomics）可定义为研究基因组（genome）结构、功能与多样性的科学，而基因组则是某一特定生物包含的所有遗传信息的总称。基因组学方法与传统生物学方法的主要差别在于研究的规模，因为基因组学的目标在于广泛分析大量的基因，甚至可能涉及组成基因组的全套基因，而不再局限于一个或少量基因。因此，很难定义传统生物学方法终止和基因组学方法起始的界限，甚至可能暗示这两种方法之间不存在本质的差别。这是一种极端的看法，然而从本书阐明的例子来看，对适用于基因集分析的高通量研究方法的发展诉求，极大地促进技术进步，也导致探索生物学新方法的开发。

基因组学是一门新学科，起始于最初尝试通过测定基因组 DNA 序列或大量的 cDNA 序列来获得单个物种的大规模测序数据。早期的测序过程花费昂贵，因此研究者主要将精力集中在包括大肠杆菌（*Escherichia coli*）、酵母（*Saccharomyces cerevisiae*）、线虫（*Caenorhabditis elegans*）、果蝇（*Drosophila melanogaster*）和拟南芥（*Arabidopsis thaliana*）等在内的模式生物上。这些生物大量序列数据的获得促进了其他基因组学研究工具的发展，譬如微阵列系统就可用来分析基因的表达和序列标识的突变体等。

这些早期基因组研究的模式生物并不代表海洋生物，因此对海洋生物的生物学、生态学和演化历史等的研究远远落后于基因组学方法在几种陆地模式生物中的应用程度。近些年来这种状况有所改观，主要基于 DNA 测序费用的大幅下降以及对大数据分析能力的提高。测序花费的下降不仅使得基因组学方法应用于更多更广的生物种类，而且也开启了包括应用创新性测序方法的宏基因组学和宏转

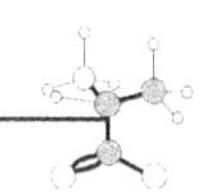

录组学在内的诸多全新的基因组学研究领域。海洋生物学一直处在这些新应用领域的前沿。

基因组学研究方法正广泛应用于解决海洋生物学的众多难题，诸如通过探索海洋生物众多的系统发育多样性来研究发育过程的进化，探讨在全球的地球化学循环中发挥重要作用的海洋生态系统，研发新的生物大分子以及理解重要海洋生态系统之间的生态关联作用。这些领域在探索全球气候变化时发挥越来越重要的作用。

本书的目标是概述基因组学方法应用于海洋生物学研究领域中应用的最新进展。每章包含一个特定领域，对该领域做一个介绍，而且对基因组学方法如何适应该领域研究的特殊需求进行详细说明。最后一章探讨基因组学技术在实际应用中的问题，重点探讨生物信息学方面面临的挑战。本书多个章所涉及的研究领域还将在“海洋生物基因组学进展”系列丛书（the Advances in Marine Genomics book series）中开展进一步的探讨。

第1章探讨基因组学技术如何应用于探索和检测海洋环境中的生物多样性。考虑到全球气候加速变化，全球气温变暖越来越影响诸多重要的海洋生态系统，本章将重点讨论海洋中生物多样性是如何构成生态系统，以及这些不同等级层次的生态系统之间相互影响的重要性。在研究生物多样性时，基因组学方法将从分类层次、基因水平和功能水平三个层面来描述。

第2章探讨快速发展的宏基因组（或称环境测序），其序列分析方法应用于几个或多个物种构成的群落而非单个孤立的个体。宏基因组学将作为一门新学科来描述，重点提及在海洋生态系统中的重要作用。本章重点阐明宏基因组学方法的创新特性，特别是能对生态系统提供全球视野，并可提供描述不能在实验室中培养的生物方法。对近来研发的宏转录组学和宏蛋白质组学等方法也展开了讨论。本章给新近进入这个领域的研究人员提供了一些有用的建议。

第3章讨论基因组学方法在群体水平上研究海洋生物的最新应用。从DNA和RNA水平来探讨，前者可阐述诸如生物多样性的测度、物种疆界的勘察和物种漂移的评估等问题，后者可提供信息以说明通过基因表达的选择过程或调控来适应特殊的栖息地。一个物

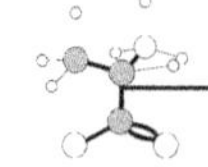

种对环境干扰的适应能力信息将对预测气候变化或其他人为因素的效应至关重要。本章先对可用于在DNA和RNA水平上进行分析的可行性方法做一综述，接着讨论这些工具对研究问题的相对优点，最后通过一些例子来说明基因组学方法如何用来解答海洋环境中群体结构和动态变化与生态系统等问题。

接下来的两章将集中探讨基因组学对多细胞生物研究的影响。第4章讲述基因组学方法已用于促进对多细胞生物系统发生的理解。早期对多细胞生物进化历史的推演，只是基于形态学与发育特征，但是由于曾经复杂的世系出现形态简化，以及用于计算和评估这种特征的方法存在内在难度等原因，因此，推演很容易出现错误。随着基于基因的系统进化分析方法的引入，这个研究领域发生了翻天覆地的变革，从而形成了所谓的动物系统发育的“新视野”，但多细胞生物的诸多关系方面还是未知或相互矛盾的。现在不少科学家正在应用基于更全面的基因组规模的大数据系统基因组学方法，来阐述这些科学问题。本章描述基因组信息是如何用于系统基因组学研究，并举了一些由这些方法产生重要进展的例子（譬如近来对脊索动物进化关系的重新评估）。

第5章重点探讨动物世系的复杂演变。对模式生物和更多物种（包括诸多海洋生物）开展广泛的工作，为研究多细胞生物的物种进化（尤其是潜在的分子遗传进程）提供了更完整的视野。这些研究对后代比祖先复杂的假设提出挑战，也表明简单化确实存在于几个动物世系中。通过解析不同进化阶段的分子进程，为几个进化过渡问题（例如多细胞性过渡或胚层起源）提供重要线索。

第6章集中探索基因组学方法在海洋藻类研究中的应用。藻类是一个变化多样的分类类群的总称，不过它们都有一个共同的特点，就是能够进行光合作用。这些生物在地球化学循环中发挥重要作用，而且是新的生物大分子和生物过程的丰富来源。本章对基因组测序和宏基因组学分析在加快近来海洋藻类生物学研究中的重要贡献也有所讨论，还特别强调近来从数种主要藻类类群中研发模式生物的重要性。

水产养殖是全球快速扩张的领域，但有关的基因组学资源开发

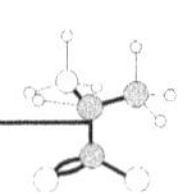

程度还比较滞后。第7章讨论基因组学方法在鱼类和贝类研究中的应用，同时探讨产生的数据如何应用于阐述生殖、生长和营养、产品质量与安全以及病菌等问题。本章还对种群多样性的评估和基因组学数据在选择程序中的应用进行探讨，也讨论了基因组学方法在研究和监测自然界中鱼类和贝类种群变化、阐明在生态系统的相互关系等领域中的应用。

随后的第8章探讨海洋生物工程领域的快速发展。本章强调就系统发育多样性而言在大海中可以发现的众多生物，而且探讨全基因组测序和大规模宏基因组项目正在如何快速拓展这个星球上可用于海洋生物工程研发的遗传资源。同其他的领域一样，海洋生物工程也没有发展到与陆地生物相应的研究高度，这可能是由于海洋环境难以接近的缘故。基因组学方法的出现正在改变这种状况，我们可以预期在不久的将来将出现令人振奋的进展。本章对可获得的基因组学资源进行描述，而且对众多海洋生物（包含病毒、古菌、细菌、藻类、真菌和海洋动物）来源的生物工程产品和工艺展开举例说明。

正如前面提及，本书的最后一章（第9章）专注于海洋生物基因组学的一些实际应用方面。本章涉及目前DNA测序所用的不同方法，并就数据管理问题展开讨论。近十多年来，随着新测序技术产生海量数据的极速攀升，人们对数据管理问题的关注程度越来越高。本章探讨基因组学数据的生物信息处理，涉及EST聚类、基因组组装、基因预测、基因功能分配和全基因组学注释。同时，也概述了转录组数据分析，特别是基于微阵列杂交技术的数据分析。

贡献者（Contributors）

共55位。名录及联系信息请参阅英文原著第xi－xiv页，此处省略。

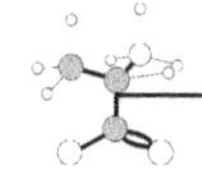

第1章　基因组学在发现与检测海洋生物多样性中的应用

摘要：海洋生物多样性包含遗传、物种、生态系统和功能多样性，不同层面的相互作用最终决定了海洋生物的分布与丰度、进化潜力和恢复能力。环境和生态系统被破坏的程度日益加剧，因此，调查和监测生物多样性结构与功能组成的变化变得越来越迫切。传统的海洋生态与保护研究主要集中在物种和群落层面，如今的研究方向已转变为上述生物多样性各层面之间的关系，尤其是在生态系统中的重要作用，如对全球养分循环和气候的影响。这里我们强调基因组学和遗传规律在阐明生物多样性的不同生物学水平（从基因水平到细胞水平，再到群体水平和生态系统水平的过程）相互作用的重要性。基因组方法在检测未知种类（如DNA条形码），遗传（如454测序）和功能（如基因表达、代谢水平分析）多样性研究等方面极其有效，当然，也可用于新物种和代谢通路鉴定。

1.1　全球视野下的海洋生物多样性与基因组学

1.1.1　海洋生物多样性：结构与功能成分

海洋生物多样性存在以下几个不争的事实：海洋拥有非常丰富的生物多样性；海洋生物遭受的威胁日益严重；生物多样性在生态系统功能中极其重要。

（1）海洋中生物多样性是非常丰富的。海洋占据地球约70%的表面，为28个门的动物提供了各种各样的栖息环境，其中13个门的动物是海洋领域特有的，而11个门的陆生动物中只有1个门是特有的（Angel 1992）。海洋物种多样性高，与之相应的是这些生物以各自不同的方式生活在差异巨大的环境中，从浮游到潜游，或忍受空气暴露栖居在潮间带，甚至可以在超过3500米深的深海热液口附近生存。而且，生命起源于海洋，因此海洋生物类群的出

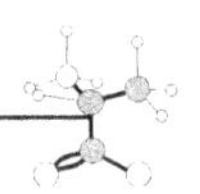

现远远早于陆地相似生物（时间差异可达27亿年）。几乎所有现存门类在海洋中都有代表物种，这与只有将近一半现存门类具有陆地代表生物形成鲜明的对比（Ray 1991）。此外，随着先进分类方法的出现（Savolainen et al. 2005），以及新技术使人类可以探索海洋中更多的处女地，越来越多的海洋新物种被发现（例如Santelli et al. 2008）。其中有用显微镜才能观察到的动植物或微生物（Venter et al. 2004，Gomez et al. 2007），也有人们更熟悉的大型生物，诸如鱼、甲壳动物、珊瑚和软体动物等（Bouchet 2005）。例如，曾经以为海生苔藓虫 *Celleporella hyalina* 在全球只有一个种，而DNA条形码和交配实验证明地理隔离形成了超过20个分布在不同深海水域的遗传世系，但彼此之间存在着生殖隔离（Gomez et al. 2007；见图1.1）。而且，如此形态类似的世系却存在生殖隔离，其形成过程到现在依然是未知的。这种隐藏的多样性最近被澳大利亚外海的新发现所证实，科学家们在水下山脉和塔斯马尼亚峡谷发现了包括鱼、古生珊瑚、软体动物、甲壳动物和海绵等在内的270个新种（http://www.csiro.au/science/SeamountBiodivisity.html）。这种现象也在一些生物地理区域之间（如路西塔尼亚和寒温带之间）的海洋过渡区域得到证实（Maggs et al. 2008）。除了一种世界性分布的硅藻以外，其他目前鉴定为同一种的硅藻实际上也是由多个种组成（Medlin 2007）。

（2）虽然现存生物的多样性水平很高，但是海洋系统遭受来自环境改变和人类活动造成的威胁日益严重，污染、过度开发、富营养化、物种入侵和气候变化导致物种分布和丰度的变化（Worm et al. 2006），还会导致局部性物种灭绝。为了更好的应对这样的状况，我们不仅需要了解变化产生的机制和影响，更要创造有利的环境来促进物种丰度的恢复，减少环境干扰的影响（Palumbi et al. 2008a）。

（3）海洋生物多样性支撑着生态系统功能的范围和力度。海洋生物群在全球营养循环和气候调控等方面发挥着重要作用，给人类提供了大量的资源，自然环境的生命过程和各种产物服务于全球生态系统，包括碳蓄积、大气层的气体调节、废物处理、食物供应和原材料。事实上，海洋藻类贡献了全球将近40%的光能合成产物。据估计，全球远洋生态系统每年提供的经济价值超过8.4万亿美元，近海生态系统的这一数值则超过了12.6万亿美元（Costanza et al. 1997）。深海领域在全球的生态与生物地球化学循环过程中发挥重要的调控作用。保护海洋生物多样性是保证世界海洋可持续性发展的前提。

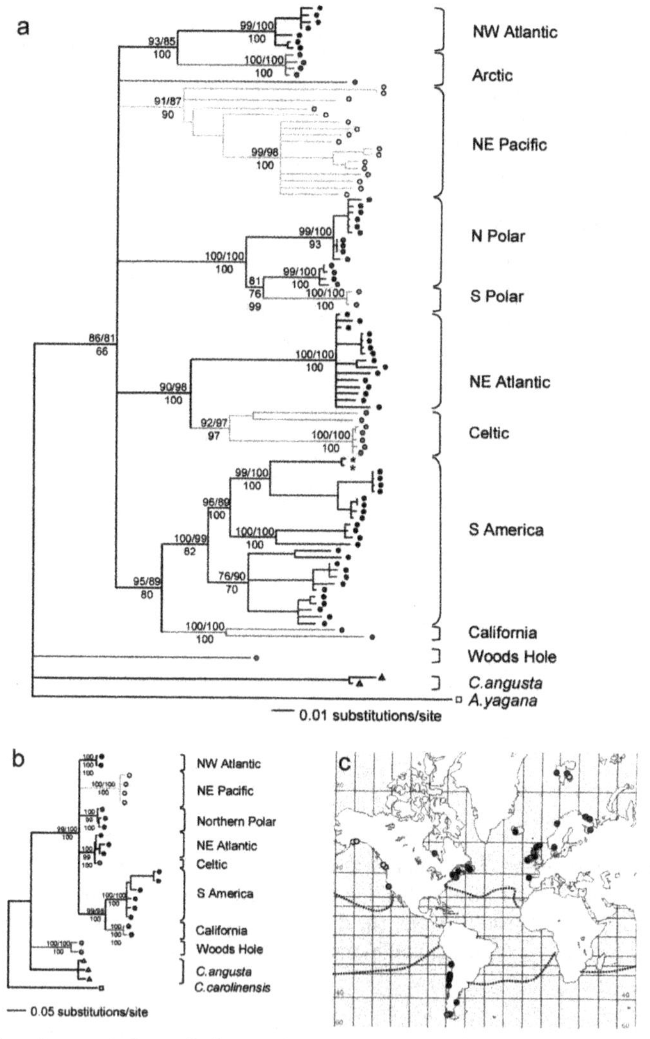

Fig. 1.1 Cryptic speciation of the marine bryozoan, *Celleporalla hyalina*. (After Gómez et al. 2007). (**a**) Maximum-likelihood tree of haplotype data from the barcoding gene, COI. (**b**) Maximum-likelihood tree of the nuclear gene, elongation factor, EF-1a haplotype data. Individuals traditionally described as *C. hyalina* are marked with circles coloured according to the geographical region listed to the right. (**c**) Map of the sample locations included in the genetic analysis. Coloured *circles* indicate the major lineage according to the phylogenetic analysis. *Dotted lines* indicate the limits of the temperate oceans (20C isotherm)

图 1.1　海洋苔藓虫 *Celleporalla hyalina* 的隐种（Gomez et al. 2007）。(a) 以条形码基因 COI 单体型数据构建的进化树；(b) 以核基因延伸因子 EF-1a 单体型数据构建的进化树；传统上描述为苔藓虫的个体按右边地理区划标成不同颜色的圆圈；(c) 用于遗传分析的取样点分布地图。彩色圆圈显示根据系统发生区分的主要世系，虚线表示温带海洋的界限（20 摄氏度等温线）

图中英文注释：NW Atlantic，西北大西洋；Arcti，北冰洋；NE Pacific，东北太平洋；N Polar，北极；S Polar，南极；NE Atlantic，东北大西洋；Celtic，凯尔特；S America，南美洲；California，(美国) 加州；Woods Hole，(美国) 伍兹霍尔

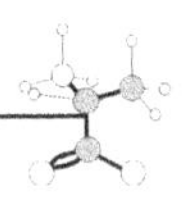

（4）越来越多的证据表明，可持续的生态系统功能依赖于多种多样的生物（参阅综述 Palumbi et al. 2008a）。例如，用几个独立的生态系统功能和效率参数，从全球 116 个深海位点获得的研究结果证实，生态系统的功能与深海生物多样性呈指数正相关（Danovaro et al. 2008，图 1.2）。类似的研究（Palumbi et al. 2008b）进一步证明，丰富的生物多样性支撑着高速有效的生态系统过程，其中碳合成和循环效率与关键类群的生物量成正相关。因此，生物多样性的减少极有可能会导致生态系统功能的显著下降。

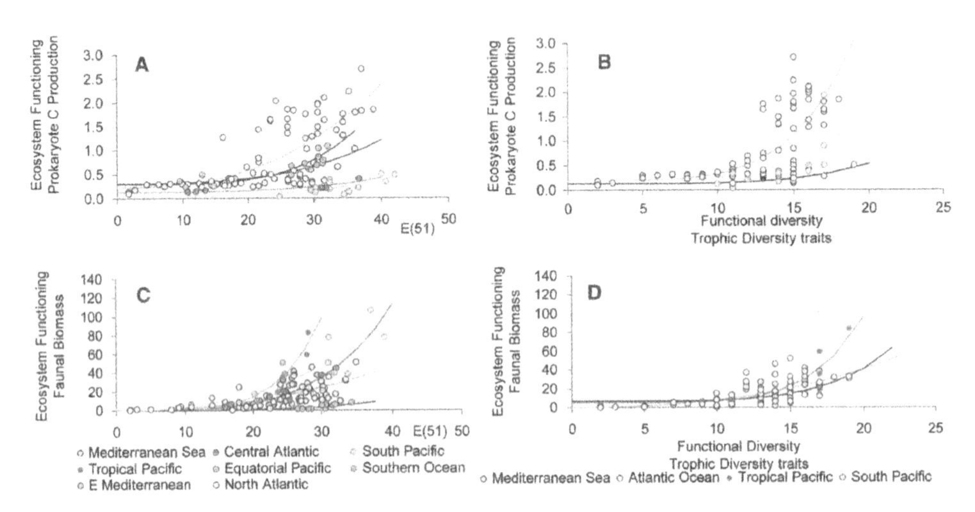

Fig. 1.2 Relationship between biodiversity and ecosystem function (after Danovaro et al. 2008). Data show that as the biodiversity of benthic meiofauna increases (estimated geographically, (**a**, **c**) and from diversity of trophic traits (**b**, **d**)), so do various proxies of ecosystem function (e.g. prokaryote C production (**a**, **b**), faunal biomass (**c**, **d**))

图 1.2　生物多样性和生态系统功能的关系（Danovaro et al. 2008）。结果显示，随着深海底栖小型生物多样性的增加（a 和 c 地理学差异，b 和 d 营养性状多样性），生态系统功能的参数（譬如 a 和 b 原核生物碳合成量，c 和 d 动物生物量）也增加。

图中英文注释：Ecosystem Functioning，生态系统功能；Prokaryote C Production，原核生物碳合成量；Functional diversity，功能多样性；Trophic Diversity traits，营养多样性特征；Faunal Biomass，动物生物量；Mediterranean Sea，地中海；Central Atlantic，中大西洋；South Pacific，南太平洋；Tropical Pacific，热带太平洋；Equatorial Pacific，赤道太平洋；Southern Ocean，南大西洋；E. Mediterranean，东地中海；North Atlantic，北大西洋；Atlantic Ocean，大西洋

我们的海洋具有丰富且不断变化的物种多样性，外加全球相互依存的生物多样性、能量流动和物质循环体系，如一个巨大的宝藏吸引和促使我们加快鉴定新物种的速度。近年来基因组学技术和相关遗传学理论迅速发展，为

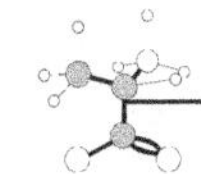

描述和保护海洋生物多样性提供了经验和思路。我们将在本章着重介绍它们是如何发挥作用的。最关键的是，我们强调了基因组学可以用来阐述不同生物学水平（从基因和细胞水平，到群落和生态系统水平）上多样性间的相互

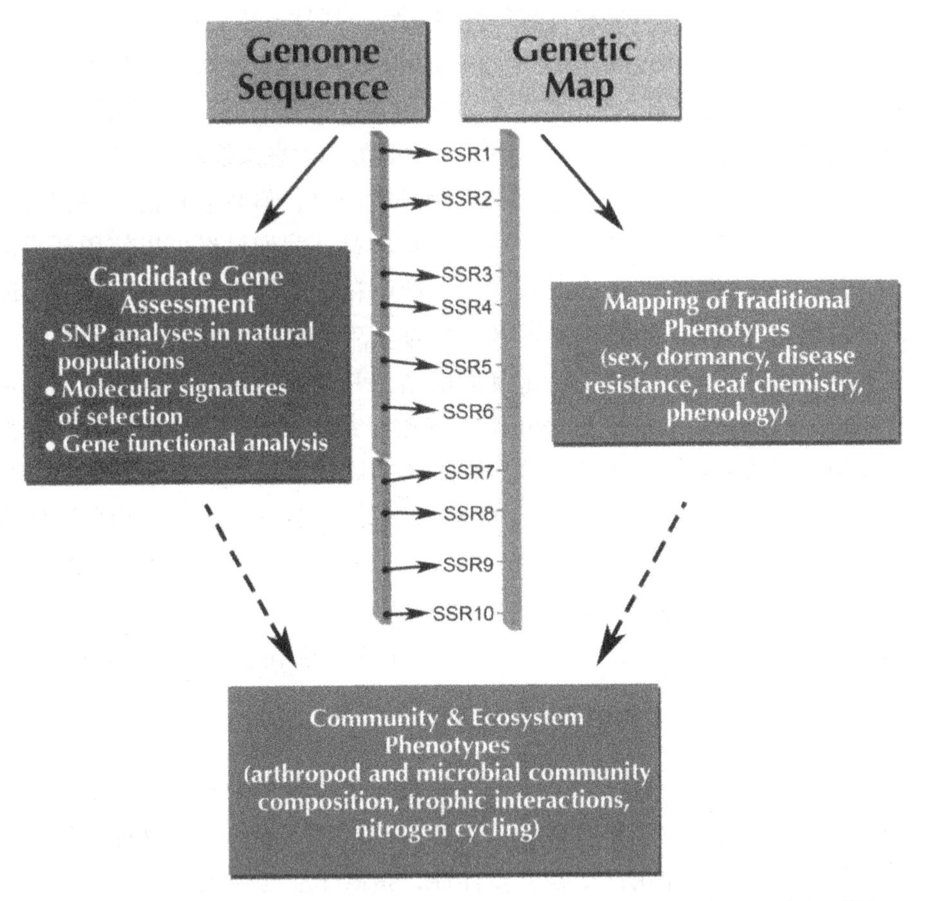

Fig. 1.3 Exploring linkages between genomic information and ecosystem function (after Witham et al. 2008). See text for further information. SSR, simple sequence repeats

图 1.3　探索基因组信息和生态系统功能之间联系的流程图（Witham et al. 2008）。详细信息请参阅文本。SSR，简单重复序列

图中英文注释：Genome Sequence，基因组序列；Candidate Gene Assessment，候选基因评估；SNP analyses，单核苷酸多态性分析；Natural Populations，自然种群；Molecular Signature，分子标记；Selection，选择；Gene Functional Analyses，基因功能分析；Genetic Map，遗传图谱；Mapping of Traditional Phenotypes，传统表型的映射；Sex，性别；Dormancy，休眠；Disease Resistance，疾病抵抗力；Leaf Chemistry，叶片化学；Phenology，物候现象；Community & Ecosystem Phenotypes，群落和生态系统特征；Arthropod，节肢动物；Microbial Community Composition，微生物群落组成；Trophic Interaction，营养关系；Nitrogen Cycling，氮循环

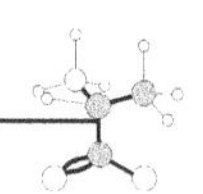

联系（图 1.3）。此外，还可以使用数量性状基因座（QTL，Witham et al. 2008）的方法，来定位与特定表型相关的基因组区域。首先，根据 DNA 序列信息来筛选候选基因，并将它们定位在有序列标签（如简单重复序列 SSR）的遗传图谱上。其次，通过在自然种群中筛查选择性清除、局部连锁不平衡、同义和非同义多态性的比率，来进一步选择候选基因。再次，分析被选择的这些基因的功能，如使用基因芯片和定量 PCR 方法来分析基因表达。最后，通过田间实验和野生种群调查的方法将其与群落和生态系统联系起来。

虽然将基因组学应用于管理海洋生物时机尚早，但我们相信一些相关的遗传和进化原理还是大有用处的。令人高兴的是，我们已经可以使用新手段（如高通量 DNA 条形码，Hebert et al. 2003）和宏基因组学方法（Allen and Banfield 2005，DeLong 2005）来研究生物的生活史，以及鉴定传统分类学无法识别的新物种。此外，研究特定基因的结构和功能还可能发现新的代谢途径（Venter et al. 2004），有助于我们理解和探究海洋生物、生态过程和新型海洋产物之间的关系。这种关系的研究方法与维护和管理生态系统的方法相一致（Pikitch et al. 2004），其研究重点都从单个组分（如种群、物种）转变为它们在群落和营养水平上的关系。接下来，我们将先介绍一下海洋生物多样性的性质，然后仔细讲解海洋生物多样性动力学和分布的研究进展。本章节只是简要地解释其研究方法及应用，在接下来的章节将会展开来详细阐述。

1.1.2　海洋生物多样性的本质

生物多样性是一个概括性的术语，用来描述各种生物及它们的生境，包含四个方面的内容：①遗传多样性，指种内的遗传变异，对于物种和种群抵抗环境扰动起决定性作用；②物种多样性，指某个生态系统中的物种或其他生物分类群的丰度，是决定栖息地的复杂度和恢复能力的关键指标；③生态系统多样性，指各个生物群落及其与周围环境相互依存、相互作用的关系和动力学。生态系统多样性与前两个多样性是有明显区别的，因为它既包括了生物因子，还包括非生物因子；④功能多样性，指某个特定生态系统的一系列生物过程、功能或特征。一些人认为，功能多样性可能是评估生物多样性最有意义的方面，因为它不需要编目生态系统中所有的物种，从而成为保护海洋生态系统的一种便捷方法。虽然基因组学方法很好地支持这一观点，然而它的应用主要局限在日常研究中，即探讨不同空间范围的生物多样性和功能的联系（Bulling et al. 2006，Naeem 2006），因为到目前为止许多物种和它们的功能都还不明确。

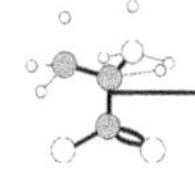

1.1.3 实践和理论的进步

当前所提到的基因组学是一门学科，它主要研究生物基因组中的基因和产物的结构、功能及多样性，从而了解生物体和周围生物、非生物因子之间的关系。不同生物学水平的发展已经彻底改变了我们对基因组结构和功能的分析能力（Wilson et al. 2005）：① DNA 水平，高通量测序技术和下一代测序技术的发展（Rothberg and Leamon 2008）；②基因表达水平（转录组学），基因芯片或数字基因表达谱的运用；③蛋白水平（蛋白质组学），串联质谱技术对蛋白分析的促进；④代谢水平（代谢组学），多种高端分析技术（如热分解气相色谱法和红外法），可以详细分析低分子量的细胞成分。总之，每个水平技术的进步都有助于获得大量的数据，我们将会在第 3 章中进行详细阐述。除了分子生物学方面的发展，基因组学的革命还与另外三个技术的进步息息相关，它们分别是发生于 20 世纪 90 年代的显微技术、计算技术和通信技术（Van Straalen and Roelofs 2006）。

显微技术：使用新的激光技术来研究只有几个微米大小的分子，对基因芯片的部署至关重要。

计算技术：高通量测序产生大量的 DNA 数据、RNA 数据和蛋白数据，它们都需要强大的计算能力来进行生物信息学分析；而高速计算机和大容量数据存储方法的出现，推动了基因组学技术的进步。

通信技术：使用互联网实时访问和集成全球数据库是绘制和解读基因组数据的基本条件，特别是当人们越来越关注非模式生物时（Cock et al. 2005）。

利用基因组学方法探索生物多样性，其目的是为了能够同时分析多个生物体或基因，从而更好地在各个生物学水平上将遗传和表型的多样性联系起来。

从概念上讲，在使用基因组学分析生物多样性时有三个要点需要特别注意。首先，尽管基因组学最初是基于模式生物（如面包酵母、果蝇、线虫、拟南芥以及近期常用的小白鼠）发展起来的，但现在随着技术的进步，人们对具有显著生态学特征的物种开始感兴趣并有所涉足（Vera et al. 2008，Witham et al. 2008）。实验室模式生物确实在研究生物体生长和发育的主要调控过程方面起了很大作用，但在预测生物体对环境变化的反应以及它们在生态系统中的作用时往往意义不大。

近年来 DNA 序列数据快速增加，包括硅藻、海胆、水螅、鱼、虾和褐藻等在内的海洋生物（Wilson et al. 2005，Cock et al. 2005），连同表达序列标签库（EST）也被迅速扩充，都显著增强了海洋类群在整个生态系统中的代表

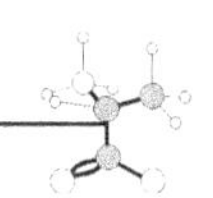

性（见第3和第7章）。此外，在整个进化树上，一些所谓的“关键种”被看作系统发育过程相似生物的模型，针对它们进行基因组学研究（Cock et al. 2005），将进一步扩大我们研究生物体生活方式、适应性和栖息地的范围。其次，直接从环境中分离和测序的大片段DNA，即“宏基因组学”方法，使得对微生物群落的分析研究激增（第2章）。此外，在这些过程中还可以发现新的生化途径（如Peers and Price 2006）和新物种（Massana et al. 2006）。例如，使用“全基因组鸟枪测序”法对从马尾藻海采集的微生物种群进行研究，根据序列关联性鉴定了至少1800个物种，其中包括148个尚未报导的细菌种类（Venter et al. 2004）。这项研究还鉴定了超过120万个未知基因，包括至少782个新的视紫红质基因，揭示了海洋微生物在种类和功能水平上前所未有的多样性。这种方法进一步提高了直接分析野外样品的能力，并将大大提升我们对生物体和生态系统中功能基因的认识。最后，我们已经意识到生物多样性是生态系统恢复和抗干扰能力的基础（参见综述Palumbi et al. 2008a），在环境管理中已将重点从个体、种群和物种转变到群落和生态系统上来，尤其是对生物多样性和生态系统之间功能关系的关注。该功能关系在基因组学研究中是非常重要，因为它代表了细胞、代谢和生态水平上的基因结构、功能和表型多样性之间的联系，反映了生产力、养分和能量流动模式上的多变性。

以上概述了海洋生物多样性研究中基因组学的发展，接下来我们将探讨一些鉴定海洋生物和区分海洋群落功能的分子生物学方法。

1.2 海洋生物多样性的分子生物学鉴定

随着人们越来越多地关注生物多样性和生态系统功能之间的关系（Chapin et al. 1997，Duffy and Stachowicz 2006），包括元素循环、产物和营养转移过程等，正确地评估物种多样性和准确量化生态功能变得非常重要。海洋环境中物种多样性的评估还面临着很大的挑战，因为目前我们还只能探索非常有限的海洋领域及这些生境中的生物群落。因此，最新的基因组学方法成为评估以前未发现的生物多样性的强有力的研究工具。

物种鉴定是研究海洋生物的重要出发点。基于表型性状的传统鉴定方法简单直接，然而很多情况下，这种方法可能会效果不好甚至失败，譬如一些同形种，本来就很难分开的类群，或在分类学上难以区分的卵和幼虫（Kochzius et al. 2008）。随着新的海洋生境和物种不断被发现，环境变化和栖

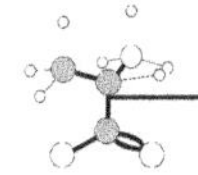

息地受到的干扰对海洋物种的威胁越来越大，发展快速和强大的方法来描述和编目海洋生物多样性显得更加重要。广泛使用的分子生物学工具，如最近提出的 DNA 条形码，指某个通用基因中严格标准化的、分子量较小且序列较短的 DNA 片段，存放于序列数据库中，和来源、现状等信息一起附在凭证标本上，可以打破传统鉴定方法的局限性，为准确鉴定个体，乃至卵、幼虫或肢体碎片提供了一个简单、有效的方法。这个系统必须依靠一系列以同源基因分析为基础的分子生物学方法来界定物种，并使用相关的“基因组筛选”法来识别其功能，这类同源基因通常包括如细胞色素 c 氧化酶亚基 I 基因（COI）、16S、18S 和 ITS。

Hebert 等人在 2003 年提出 DNA 条形码的概念，并提出了一个关于物种鉴定的整体性方案，这将有利于突破传统分类学方法所面临的很多限制。这个方法通过分析单基因中一个较短片段如 COI 的序列，能够准确鉴定很多动物种类。因此，DNA 条形码提供了一个标准化方法，可以快速、简单、准确地鉴定物种。无论选取基因组中哪一段区域，DNA 条形码的基本原理和方法在本质上都是相同的。其基本前提是，每个已知种都可以与 DNA 条形码一一对应。由此产生的“对应假设”在原理上支持了新的分子生物学鉴定系统，该系统同时还扩展和包含了经典的林奈系统，而不是将其替代（Costa and Carvalho 2007）。此外，根据生物物种的概念，种间与种内较小的差异代表了物种间具有生殖隔离，也符合林奈的双命名系统（Goméz et al. 2007）。经过专家鉴定的物种的条形码凭证标本都被存放在博物馆，既可用于日后检查，也可用于长期研究。一旦这个参考数据库完善起来，就可以用于鉴定物种了。

虽然以 COI 基因为基础的条形码鉴定技术有较好的发展势头，但它并不是对所有的真核生物都有效，而在细菌、古细菌和病毒的鉴定中也通常用其他的标记。这主要是因为在一些真核生物（如线虫）中，COI 基因具有独特的分子进化速率和过程。在线虫中，COI 基因具有较高的突变率，富含 A + T，且替代模式具有偏好性（Blouin et al. 1998，Blouin 2000），不适用于 DNA 分类（Hajibabaei et al. 2007）。因此对某些类群需要使用其他标记，而我们正努力尝试建立一系列的标记以适应 DNA 条形码的发展。目前已经有一些使用其他标记的例子，在过去几十年里曾使用核糖体基因鉴定了大量的微生物，在 20 世纪 60 年代，首次报道了核糖体基因（rDNA）和它的基因产物（rRNA），用于微生物的分类学鉴定（Doi and Igarashi 1965，Dubnau et al. 1965，Pace and Campbell 1971a，b）。

在过去的几十年里，比对分析同源基因序列已成为研究微生物系统发育

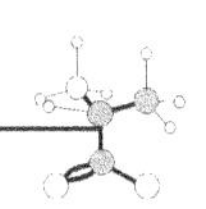

和多样性的必要方法（Díez et al. 2001，Evans et al. 2007）。用于编码 rRNA 的基因特别适用于系统发育分析，因为所有的细胞生物都有这些基因；它们分子量比较大；同时含有高度保守区和可变区，没有明显的横向基因转移（Woese 1987）。核糖体基因序列的数目正在不断增加，目前已有大约550 000条序列（Pruesse et al. 2007），一些文献还详细描述了核糖体基因序列在鉴定原核和真核微生物方面的作用（Amann et al. 1990，Simon et al. 2000，Groben et al. 2004）。对于环境样本的核糖体小亚基 DNA（18S rDNA）的直接克隆和测序，使我们对超微型浮游生物群落结构和组成的分析更加全面（Giovannoni et al. 1990，López-Garcia et al. 2001，Medlin et al. 2006）。

1.2.1 对微生物群落多样性和功能的分析

据推测，海洋生境中超过 99% 的生物目前是无法培养的，因此，海洋蕴藏着地球上最大的未知的基因组多样性（Beja 2004）。事实上，平均每毫升海水中就有超过 100 万个病毒，由于它们在宿主体外没有办法繁殖，因此海洋病毒也许代表了地球上最少被采样研究的生物。迄今，一些以 PCR 为基础的研究方法通过分析特定的同源基因（如核糖体 16S 或 18S 基因）来鉴定物种或种系型，包括克隆文库、变性梯度凝胶电泳（DGGE）（Muyzer 1999）和限制性片段长度多态性（RFLP）分析等方法，使我们对海洋微生物（古细菌、原核生物、真核生物）巨大的多样性有了初步的认识（Kemp and Aller 2004）。DGGE 可以快速分析和比较微生物群落。DGGE 胶图中每个条带在原则上代表了一个微生物种系型，因此我们可以直观地看到群落组成的多样性（Tzeneva et al. 2008）。对每个条带进行回收和测序，可以得到它所代表的类群的信息，也可用于系统发育分析。由于病毒必须寄生在细胞内，它们缺少核糖体基因编码区，因此，有必要对每一个类群寻找特定标记（如 RNA 聚合酶和 DNA 聚合酶）来研究它们的生物多样性。

尽管这些研究都强调了自然界中绝对丰富的生物多样性，可是并没有证明群落具有的功能意义，因为这些用于评估生物多样性的分子标记的基因组功能十分保守。多年来，生物多样性研究的最终目标已不仅仅是探究多样性，而是评估生物多样性的功能，亦即确定多样性是如何帮助生态系统发挥作用的。现在科学家们已尝试使用多种手段来研究功能多样性，如功能和环境基因组学、转录组学、蛋白质组学和代谢组学。这些手段通过重点研究已知的代谢通路，为研究自然界生物多样性的功能提供了重要的线索。

最初海洋功能基因组学研究最大的成功，也许是实现了在细菌人工染色体 BAC（bacterial artificial chromosome）和具有 F1 因子的柯氏质粒载体 F 粘粒中插入多达 300 kb 的 DNA 片段，以在大肠杆菌（*E. coli*）中进一步研究。利用多重 PCR 技术对 BAC 上携带的 rRNA 基因不同片段同时进行扩增，以此鉴定该克隆是来源于哪个生物类群。对 BAC 进行完整的测序能够进一步识别特定的代谢基因，以便推断某个生物体或类群的代谢能力。1996 年，这些技术首次由 Stein et al. 在研究一个沿海水域的海洋古细菌种群时使用，随后在超微型浮游生物的研究中也用到过（Suzuki et al. 2004）。用于海洋环境的 BAC 文库已经非常多，但在构建时需要采集大量的海水，如果无法采集数千升海水，也至少需要数百升海水。后来出现了一些新的改进技术，如使用末端限制性片段长度多态性（TRFLP）和内转录间隔区长度多态性 PCR（ITS-LH-PCR）来筛选 rRNA 基因（Suzuki et al. 2004，Babcock et al. 2007）。使用 BAC 技术最大的成就可能是识别了一种光驱动光子泵（现在称作细菌视紫红质），这是通过构建未经培养的一种 γ 变形菌 SAR86 的 BAC 文库实现的（Beja et al. 2000）。从那时起，BAC 文库已被广泛应用于自然界海洋生物群落和某些海洋菌株中，如变形菌纲 *Congregibacter litoralis*（Fuchs et al. 2007）、东欧和太平洋牡蛎 *Crassostrea virginica* 和 *C. gigas*（Cunningham et al. 2006）以及原索动物海鞘 *Botryllus schlosseri*（de Tomaso and Weissman 2003）。

除了传统的大规模测序方法，随着焦磷酸测序技术的发展，现在可以直接用鸟枪法对环境样品（Blow 2008）进行大规模的测序（Huse et al. 2007，Huber et al. 2007，Mou et al. 2008）。这种宏基因组学方法（Handelsmann 2004）通过研究大量未知和未经培养的生物，对研究海洋环境产生了重大影响。迄今，链终止和焦磷酸测序方法已被综合应用于研究不同的海洋环境中真核生物，细菌与病毒的多样性，样本来源于沿海、开放海域、海洋表面、底层、深海、珊瑚等（Breitbart et al. 2004，Venter et al. 2004，Angly et al. 2006，Culley et al. 2006，Sogin et al. 2006，Bench et al. 2007，Biddle et al. 2008，Dinsdale et al. 2008，Quaiser et al. 2008，Williamson et al. 2008）。但仍然需要注意的是，如此大量信息也有它的局限性：目前绝大多数识别出的基因功能仍然未知，且它们的生物来源完全是个谜。有时还可能会找到与 DNA 片段关联的系统发育标记，但更多时候基因组序列的来源是未知的。不过，随着更新的测序技术的到来，这样的困难将会得到解决。

并非研究所有生物多样性的功能都需要庞大的测序工作。例如，可以使用一种病毒的基因芯片来检测同一科的其他病毒的基因（Allen et al. 2007）。这种技术可以提供生物功能方面的重要信息，它主要通过识别基因组区域和

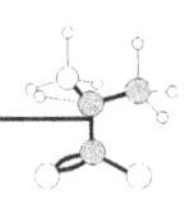

单个基因来实现，基因芯片与被检测样本之间序列差异的大小决定了是否能进行杂交。一旦通过芯片杂交获取了目标基因足够的信息，就能简单总结出该物种代谢功能和通路方面的特征。

1.2.2　介于微生物和后生动物之间的生物：真核原生生物

海洋生态系统的功能是以各种微小生物为基础的，这些微生物大小不同，包括超微型、微型和小型浮游生物，主要由细菌和真核原生生物组成，如浮游植物或异养型的原生动物就属于真核原生生物。相比于海洋环境中原核微生物重要的生态作用，我们对真核原生生物的生物多样性方面的认识还非常有限。评估该类群的生物多样性面临许多挑战。首先，生物体的大小，尤其是超微型浮游生物，使用显微镜来观察和鉴定非常困难，甚至有时是不可能的。海洋中真核超微型浮游生物的多样性或生态学信息非常有限，但它们却有着非常重要的生态学作用。譬如在开放的贫营养海域，超微型浮游生物的生物量高达80%，是异养型鞭毛虫的重要食物来源（Ishizaka et al. 1997，Caron et al. 1999）。硅藻是另一个例子，它是具有重要生态价值的真核微生物，关于它的生物多样性和生态学信息却非常少。它们每年通过光合作用固定的碳源至少占全球碳源的20%（Mann，1999）。然而，对这个物种来说生物多样性评估的限制因素并不是生物个体大小，而是很难用形态学差异来鉴定亲缘关系接近的种，因为差异非常小而且存在不确定性，全球性分布的硅藻通常还有隐种（Evans et al. 2007，Medlin 2007）。隐种会导致我们低估物种的数量，而相反同种二形又会导致对生物多样性的高估。二形性物种通常在生活周期的不同阶段表现出不同的形态特征，因此容易被误认为不同的物种。在微藻的隐藻纲（Hoef-Emden 2003，Melkonian 2003）和定鞭金藻纲中曾出现过这种情况（参阅综述 Billard 1994）。

1.2.2.1　核糖体探针

在核糖体数据库项目（RDP，Maidak et al. 2001）中，藻类有效的18S rDNA 序列不断地在增加，因此在系统进化分析中，根据这些序列可设计针对18S rDNA 的探针，其特异性可从物种水平到更高的分类层次（Groben et al. 2004）。分子探针在研究微生物多样性时发挥了强大的功能，特别是对超微型生物。核糖体探针可以与多种杂交方法结合使用，如荧光原位杂交（Fluorescence in situ hybridization，FISH）（Eller et al. 2007），基于 RNA 的核酸生物传感器（Metfies et al. 2005）和 DNA 芯片（Metfies et al. 2006）。FISH 广泛用于定量检测微生物，但是荧光显微镜对样本进行处理和定量分析时速度较慢，

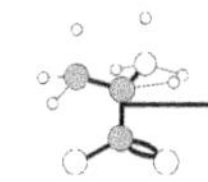

因为荧光染料的限制，每次只能处理一个探针，因此耗时较长。然而，结合了分子探针和 FISH 技术的流式细胞技术大大增加了鉴定微生物的速度和准确性。使用基于芯片技术的杂交手段，如核酸生物传感器或 DNA 芯片，可以在一次实验中同时对多个类群进行鉴定和定量分析，因此对微生物群落的检测就更加精准了（Biegala et al. 2003）。分子探针被固定在传感器芯片的表面，当它们与目标物种的 rRNA 或 rDNA 杂交时，便可以鉴定物种（Metfies et al. 2006）。使用基因芯片（图 1.4）来评估超微型浮游生物群落组成的可行性已经在超微型真核生物 *prasinophytes* 中得到证明（Gescher et al. 2008）。

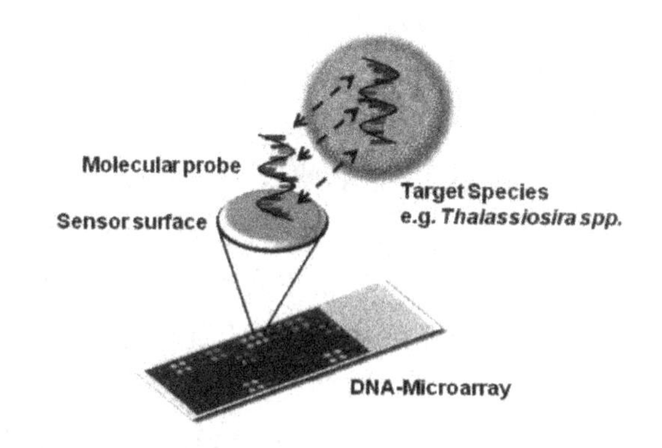

Fig. 1.4 Microbial species identification with DNA-Microarrays (PHYLOCHIPS). A PHYLOCHIP contains an ordered set of species specific molecular probes immobilized to the chip-surface. The identification of a microbial species is based on a specific interaction between the immobilized molecular probe and the complementary nucleic acid of the target species

图 1.4　使用基因芯片鉴定微生物物种的示意图。一个基因芯片包含一套特异性分子探针，它们按序排列后固定在芯片表面。这些分子探针可以与目标物种的核酸进行互补配对，以此鉴定微生物种类

图中英文注释：Molecular probe，分子探针；Sensor surface，传感器表面；Target Species，目标物种；DNA-Microarray，基因芯片

1.2.2.2　在亚种水平上的生物多样性评估

如果群落的物种多样性信息还需要补充，尤其是对以下水平上信息的补充，可以通过一些指纹图谱的方法，如限制性长度多态性（RFLP）、扩增片段长度多态性（AFLP）或微卫星。这些成熟的方法可用于判断某些种群的生物多样性程度，许多文献也证明了使用指纹图谱方法来评估微生物的生物多样性的可行性（Adachi et al. 2003，John et al. 2004，Iglesias-Rodrguez et al . 2006，Medlin 2007）。迄今，所有的研究都证明海洋中微生物种群具有独特的组成，即便地理位置很近的两个海域之间的基因流动都会受到限制。因此，

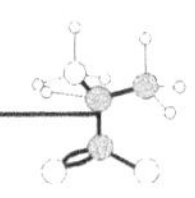

如果根据群落组成和基因流，海洋可以被细分为很多的小区域，而几十年前我们并没有这样的认识。

1.2.3 小型底栖动物群落的多样性和生态分析

小型软体底栖动物无处不在，在群落中具有非常高的丰度；个体大小介于45~500 μm之间的小型底栖动物包含了50%浅水和90%深水的线虫，在海洋生态系统中发挥了至关重要的作用。小型底栖动物在远洋的养分循环、水体流动、污染物分布、次级生产和稳定沉积物等方面有着重要贡献（Snelgrove et al. 1997，Smith et al. 2000，Snelgrove et al. 2000）。尽管它们如此重要（Danovaro et al. 2008），但由于我们对全球海洋小型动物的分类学和多样性缺乏了解，我们对生物多样性与生态系统功能之间联系的研究受到严重阻碍。例如，目前只能靠推测来估计全球线虫约100万种（Lambshead and Boucher 2003）。其中只有大约2万种被描述过，其中约4000种是海洋种（Platt and Warwick 1983），现在研究样本中发现的线虫有30%到40%属于新种（Lambshead and Boucher 2003）。这种状况无疑是由于它们个体较小而形态又非常保守所导致的，非专业人员对它们的鉴定更加困难。事实上，即使是专家，因为许多雌雄个体形态特征存在种内差异，也只能依靠传统光学和电子显微镜，并通过查阅相关专业文献才能来鉴定成熟的个体（Floyd et al. 2002，DeLey et al. 2005）。如果说付出1倍的努力就可以根据形态特征成功鉴定所有的脊椎动物，那么付出120倍以上的努力科学家才只能鉴定10%的线虫（Lawton et al. 1998）。在38 cm^3的温带底栖细砂样品中就可以找到50多种线虫，个体总量多达5000多个，这个例子说明了小型底栖动物学家在分析环境样本时所面临的巨大挑战。

在过去的几年里除了使用种系型方法来鉴定微生物以外，还建立了DNA条形码鉴定较大型后生动物的系统（Hebert et al. 2003）。这个方法有一个缺点，即为了从生物体内提取DNA用于后期的PCR实验，不得不牺牲掉整个个体，破坏了凭证标本。这个问题的有效解决方案是使用具备视频采集和编辑功能（VCE）的显微镜（DeLey et al. 2005），它能够从多角度通过视频记录下分类学相关的“3D”显微照片，充当“电子凭证标本”。这些图像可以公开存放在合适的数据库里，如 http://nematol. unh. edu/，以方便研究机构查询使用。

通常生物多样性问题一般都出现在小型生物群落中，而不是单个个体。条形码之外的另一个方法，类似于研究原核生物时所采用的方法，是大量测

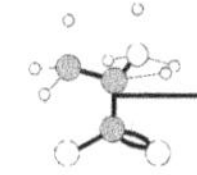

序环境样品，最近的研究也强调了使用18S rDNA 分子操作分类单位（molecular operational taxonomic unit，MOTU）处理环境样品的效果（Floyd et al. 2002，Blaxter and Floyd 2003，Bhadury et al. 2006）。虽然 MOTUs 和公开的物种描述没有任何正式的关系，但是我们可以根据现有的数据库来来鉴定物种，或者是使用一种被称为"反向分类法"的未来分类方法（Markmann and Tautz 2005）。迄今为止，基于 MOTU 的研究已应用于检测多样性，它主要是对生物个体进行测序，或者是对几百个环境样品的 PCR 产物进行克隆和测序（Floyd et al. 2002，Blaxter et al. 2005）。这些数据都包含了大量信息，但分子多样性累积曲线却没有达到渐近线，这表示需要采集更多的样本以获得能够代表小型底栖生物的生物群体（Markmann and Tautz 2005）。和以前的同源基因分析、宏基因组分析一样，更先进的测序技术和基于芯片的下游技术可能可以直接定性地评估分子多样性（Creer 2008），类似于衡量物种丰度的方法。此外，使用 VCE 有一个显著优势，即可以参考凭证标本的营养方式、个体大小和生活史策略，通过功能多样性反过来推断环境群落代表生物（Moens and Vincx 1997，Schratzberger et al. 2007）

1.2.4 DNA 条形码和渔业

我们将以 DNA 条形码在海洋渔业中的应用为例来说明它在全球海洋资源管理中的作用。2004 年 5 月，一个国际组织联盟，生命条码联盟（CBOL；http://barcoding.si.edu/），倡导在全球范围内实施 DNA 条形码技术，从而推出大规模的水平基因组项目，即不同的物种都需要提供同一个基因的序列。全球第一个 DNA 条形码计划开始实施，包括鱼类条形码（FISH-BOL；http://barcoding.si.edu/）和鸟类条形码（ABBI；http://www.barcodingbirds.org/），旨在创建一个包含所有鱼类和鸟类 DNA 条形码的参考数据库。FISH-BOL 预计在 2010 年完成世界大部分已知的鱼类（估计共有29 112种），至今约 5600 种已具备了 COI 基因条形码，占全球鱼类的 19%。CBOL 在全球范围内协调和推动 DNA 条形码技术，并将 DNA 条形码数据公开。无论是生命条形码数据库 BOLD（Ratnasingham and Hebert 2007），还是现有的公共基因库，包括美国国家生物技术信息中心的数据库 NCBI、欧洲分子生物学实验室的数据库 EMBL 和日本 DNA 数据库 DDBJ，都免费提供 DNA 条形码数据。

由于种类繁多，具有较高的经济价值以及面临环境污染的威胁，鱼类是非常合适在全球范围内实施 DNA 条形码技术的模型（Costa and Carvalho 2007，也可参阅第七章）。鱼类和渔业资源是研究的主要目标类群，DNA 条形

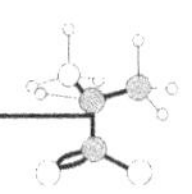

码技术预计将会带来更大更直接的收益（Lleonart et al. 2006）。这个系统将提供一个简单，同时也会越来越快、越越来越便宜的方法，不仅准确鉴定整条鱼，还能鉴定鱼卵、幼鱼、鱼碎片、鱼片和加工过的鱼。这种功能在管理和维护渔业资源方面具有重要作用，也会产生更多更严格的数据，以便于在开发渔业资源、探究渔业的生态和地理范围时进行参照，增加育苗场和产卵场所需的相关知识，以及厘清在分类学上的紊乱（Rock et al. 2008 年）。例如，DNA 条形码在鱼卵和幼鱼阶段鉴定鱼的种类效果是显著的，因为在鱼类生活史的早期通过表型来鉴定尤其困难（Pegg et al. 2006）。采用有效的分子标记技术鉴定鱼卵后发现，根据表型而错误鉴定的鱼卵超过 60%（Fox et al. 2005）。有调查显示，黑线鳕和牙鳕的卵曾被错误报道为鳕鱼卵，而该报导可能是爱尔兰海鳕鱼股票上涨的原因。此外，在爱尔兰海曾发现过早期黑线鳕卵，表明该物种曾在这里有过群体性产卵（Fox et al. 2005）。在环境不断变化的大背景下（如全球气候变暖），从生活史的各个阶段来准确鉴定鱼的种类对于评估鱼类分布范围、产卵场和育苗区的转变是非常有用的（Fox et al. 2008）。

DNA 条形码技术的另一个作用是可以用于鉴定物种的食性，并且可直接根据捕食者消化道的内容物来鉴定。这种研究可以提供水体中关于食物链更为详细的信息，以揭示哪种鱼被其他鱼（Sigler et al. 2006）或海鸟（Phillips et al. 1999）所捕食。生态学信息也会包含在模型中，为管理和保护工作提供新的资料。

鱼类 DNA 条形码在法律上的应用潜力主要有：监测渔业配额和副渔获物、检查渔业市场和产品、控制濒危物种贸易及提高鱼类产品的可追踪性（Ogden 2008）。例如在澳大利亚水域，为了得到鱼翅，鲨鱼常常被非法捕捉。品质好的鱼翅在香港可以卖到 6000~8000 美元/公斤。据估计，全球每年有超过 100 万条鲨鱼被捕杀。鲨鱼由于生长缓慢、寿命较长、孕期较长且繁殖力较低，特别容易受到过度捕杀的影响。许多鲨鱼种类在形态上非常相似，而且很多种类都受到了保护（RD Ward，私人通信）。如果某个工具能够在渔船到餐厅的汤里面，通过鱼翅来准确识别鲨鱼的种类，对于执法和保护濒危物种具有重大意义（Chan et al. 2003）。DNA 条形码技术也可以被用在鱼市和食品检验中，用于筛选假冒物种，这项应用也引起了消费者的广泛关注（Wong and Hamer 2008）。一个鲜明的例子是红鲷鱼，墨西哥湾最具有经济价值的鱼类之一，因为存量枯竭的问题一直受到严格的捕鱼限制。Marko 等人在 2004 年使用线粒体细胞色素 b 的基因序列（一种与 DNA 条形码非常类似的方法），发现在美国市场上高达 77% 的红鲷鱼鱼片被贴错标签。这种高程度的错贴标签现象可能会对估计鱼类储量产生不利影响，并在消费者和生产企业中造成假象，即错误地认为该鱼的供应能够满足需求。

因此，DNA 条形码技术提供了一个标准化的工具来描述和监测鱼的物种多样性。它不仅能在野外运用，也能在整个食品供应链中使用，有利于执法和消费者权益保护（参见第 7 章）。此外，一个全球性标准化的 DNA 条形码数据库意味着非专业人士也可以利用其中的信息来确定物种，更重要的是提倡广泛合作，在物种的分布范围内记录生物多样性。

1.2.5 海洋系统中的幼虫

很早之前就有研究认为，幼虫是种群动力学和种群连通性的主要影响因素，尤其是在成体附着生长的情况下（Underwood and Fairweather 1989）。最近有研究发现，幼虫自身的行为和停留能力在它们的分布中非常重要，从而质疑了传统模式的观点，即幼虫是被动散播的（Shanks and Brink 2005，Kinlan et al. 2005，Marta-Almeida et al. 2006）。事实上，准确估计幼虫的分布、数量、产生和散播模式，是可持续管理海洋生物资源（如 Taylor et al. 2002，Fox et al. 2005，Kochzius et al. 2008）和修复人类活动引起的生态系统退化（如珊瑚礁）的先决条件（Bellwood et al. 2004，Shearer and Coffroth 2006）。不管是对未探索过的新领域如深海（Grassle and Maciolek 1992），还是对已经有很多记录的沿海水域（Harding 1999，Lindley and Batten 2002），估算包括幼虫在内的海洋物种的丰度都是至关重要的。

幼虫，尤其是那些无脊椎动物的幼体，在海洋动物名录中通常是一个较为神秘的部分（Mariani et al. 2003）。它们的生态学意义仍然知之甚少，且直到最近才开始强调它们在海洋食物链中的重要性（Bullard et al. 1999，Rosel and Kocher 2002，Metaxas and Burdett-Coutts 2006）。有几个因素造成了这种现象：它们体积较小，缺乏形态鉴定特征以至于物种间的界限较为模糊，又分为多个幼虫阶段，且表型多变，需要大量的时间和专业知识来进行准确识别。即使一些鉴定特征存在，也仅仅在一定的地理区域中有效（见综述，Roger 2001）。培养捕获的幼虫进行反向鉴定是幼虫分类学的另一大困难。然而，幼虫代表了海洋生态系统的一个基本组成部分，是许多海洋生物在生命周期中的关键阶段。

许多分子生物学方法已经应用到幼虫生物学领域，尤其是在物种鉴定方面（表 1.1）。但是它们大部分以高通量的方式，且都是以 PCR 为基础，往往具有破坏性，很少能维持其形态特征。因此尽管已经有大量的技术方法，在生态学研究中却很少能使用。

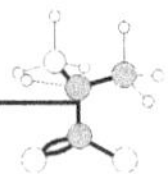

表 1.1　近期用于鉴定海洋幼虫的分子生物学进展

Table 1.1 Recent molecular advances towards the identification of marine larvae

	Molecular approach	Organism targeted	Tested on larvae?	Used in ecological studies?
Hansen and Larsen (2005)	Single step nested multiplex PCR	*Mytilus edulis*/*Musculus marmoratus*, *Ensis* sp., *Myoida* spp., *Cariids*, *Spisula* spp., *Macoma*/*Abra* spp. (bivalves)	Yes	Yes
Patil et al. (2005)	COI nested PCR	*Crassostrea gigas* (Pacific oyster)	Yes	No
Noell et al. (2001)	Control region PCR	*Hyporhamphus melanochir*, *H. regularis* (Garfish)	Yes	No
Santaclara et al. (2007)	18S multiplex PCR and RFLP	*Xenostrobus secures*, *Mytillus galloprovincialis* (mussels)	Yes	No
Comtet et al. (2000)	ITS2 PCR and RFLP	*Bathymodiolus azoricus* (deep-sea vent bivalve)	Yes	No
Shearer and Coffroth (2006)	COI PCR and RFLP	*Agaricia agaricites*, *Porites astreoides* (scleractinian corals)	Yes	Yes
Karaiskou et al. (2007)	Cyt b PCR and RFLP	*Trachurus trachurus*, *T. mediterraneus*, *T. picturatus* (European horse mackerel species)	Yes	No
Hosoi et al. (2004)	D1/D2/D3 PCR and RFLP	*Barbatia* (*Abarbatia*) *virescens*, *Mytilus galloprovincialis*, *Pinctada martensii*, *Pinna bicolor*, *Chlamys* (*Azumapecten*) *farreri nipponensis*, *Anomia chinensis*, *Crassostrea gigas*, *Fuluvia mutica*, *Lasaea undulata*, *Moerella jedoensis*, *Theora fragilis*, *Ruditapes philippinarum*, *Paphia undulata* (bivalves)	Yes	No
von der Heyden et al. (2007)	Control region PCR and sequencing	*Merluccius capensis*, *M. paradoxus* (Cape hakes)	Yes	No
Richardson et al. (2007)	Cyt b PCR and sequencing	*Istiophorus platypterus*, *Makaira nigricans*, *Tetrapturus albidus*, *T. pfluegeri* (billfish), *Thunnus atlanticus*, *T. albacares*, *T. thynnus*, *T. obsesus*, *T. alalunga* (tuna)	Yes	Yes
Webb et al. (2006)	COI, 16S and 18S PCR and sequencing	Solasteridae, Echinidae, Ophiuridae, Echinidae, Syllida, Serpulidae, Nudibranchia, Gastropoda, Nemertea, Urochordata, Ophiuroidea, Terebellidea, *Adamussium*	Yes	No

续上表

Table 1.1 (continued)

	Molecular approach	Organism targeted	Tested on larvae?	Used in ecological studies?
		colbecki, (Antarctic scallop) *Parborlasia corrugatus*, (Antarctic nemertean)		
Kirby and Lindley (2005)	16 S nested PCR and sequencing	*Echinocardium cordatum*, *Marthasterias glacialis*, *Spatangus purpureus*, *Amphiura filiformis* (echinoderms)	Yes	Yes
Barber and Boyce (2006)	COI PCR and sequencing	Stomatopod larvae (scleractinian corals)	Yes	Yes
Goffredi et al. (2006)	18S PCR and sandwich hybridisation	Thoracica, *Balanus glandula* (barnacles)	Yes	Yes
Jones et al. (2008)	18S PCR and sandwich hybridisation	*Carcinus maenas* (green crab), *Mytilus edulis* (native blue mussel), *Balanus* sp. (barnacle), *Osedax* sp. and *Ophelia* sp. (polychaetes)	Yes	Yes
Deagle et al. (2003)	PCR and DGGE	*Asterias amurensis* (northern Pacific seastar)	Yes	No
Fox et al. (2005)	TaqMan PCR	*Gadus morhua* (cod)	Yes	Yes
Vadopalas et al. (2006)	Real-time PCR	*Haliotis kamtschatkana* (pinto abalone)	Yes	No
Morgan and Rogers (2001)	Microsatellites	*Ostrea edulis* (flat oyster)	Yes	No
Zhan et al. (2008)	Microsatellites	*Chlamys farreri* (Zhikong scallop)	Yes	No
Barki et al. (2000)	AFLP	*Parerythropodium fluvum fluvum* (soft coral)	Yes	No
Arnold et al. (2005)	Blot hybridization with18S rRNA targeted probes	*Mercenaria* (bivalve)	Yes	Yes
Kochzius et al. (2008)	Microarrays	*Boops boops*, *Engraulis encrasicolus*, *Helicolenus dactylopterus*, *Lophius budegassa*, *Pagellus acarne*, *Scomber scombrus*, *Scophthalmus rhombus*, *Serranus cabrilla*, *Sparus aurata*, *Trachurus* sp., Triglidae (fish)	No	No
Pradillon et al. (2007)	Whole larvae in situ hybridization with 18S rRNA targeted probes	*Riftia pachyptila*, *Tevnia jerichonana* (hydrothermal vent polychaetes), *Crassostrea gigas* (Pacific oyster)	Yes	No

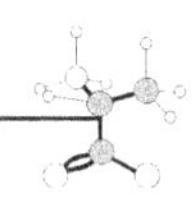

续上表

Table 1.1 (continued)

	Molecular approach	Organism targeted	Tested on larvae?	Used in ecological studies?
Le Goff-Vitry et al. (2007)	Whole larvae in situ hybridization with 18S rRNA targeted probes	*Ostrea edulis*, *Cerastoderma edule*, *Macoma balthica*, *Mytilus* sp., *Nucula* sp., *Glycymeris* sp., Pectinidae, Veneridae (bivalves)	Yes	No

表中英文注释：Molecular Approach，分子生物学方法；Organism Targeted，目标生物；Tested on Larvae? 能否检验幼虫？Used in Ecological Studies? 是否用于生态学研究？PCR，聚合酶链式反应；RFLP，限制性片段长度多态性；ITS2，内转录间隔区 2；DGGE，变性梯度凝胶电泳；Microarrays，基因芯片；Hybridization，杂交；COI，细胞色素 c 氧化酶亚基 1；Cyt b，细胞色素 b；Sequencing，测序

使用 DNA 条形码鉴别幼虫面临着技术性难题，如分子标记的普适性（Shearer and Cofforth 2008）、数据库中数量有限的参考序列（Barber and Boyce 2006，Webb et al. 2006，Richardson et al. 2007）生物分类解析的缺乏（Webb et al. 2006）。分子生物学方法解决了之前的幼虫生态学研究的难题，然而还是有几个困难限制了它在环境研究中的应用，包括与基因芯片平台相关的定量和灵敏度问题（Kochzius et al. 2008），此外还有因幼虫 DNA 总量有限而带来的对 PCR 检测和优化的影响（Patil et al. 2005）。因此，形态学数据仍然是幼虫生态学研究的基础（Minagawa et al. 2004，Richardson and Cowen 2004，Shanks and Brinks 2005）。未来的重点是寻找一个更加综合的方法，结合形态学、生态学和分子生物学数据，以了解海洋生物生命周期的复杂性和动力学（Will and Rubinoff 2004，Janzen et al. 2005，Dasmahapatra and Mallet 2006）。

1.3 海洋生物多样性和生态系统功能

1.3.1 新环境中的微生物

最近对于微生物多样性的研究过程中发现了许多未知的微生物，其中很多物种对海洋变化过程有重大影响。例如深海热液喷口和冷泉含有旺盛的化能合成型微生物群落；海洋的水层中含有丰富的古菌；大量的超微型浮游生物是碳固定和氮循环的主要推动力。如果我们可以得到全球海洋微生物的基因组，就可以获得微生物世界关于选择和进化的基本信息。通过大量采样，

对海洋微生物进行完整的普查，将有助于了解整个代谢过程（Scherer-lorenzen 2005）。所有的栖息地中，多样性变化对生物量、养分循环和生态系统的稳定性有着深刻的影响（Hughes et al. 2006）。较高的生物遗传多样性可以对生态环境的变化提供一定程度的生态防护（Hughes et al. 2006）。

1.3.2　生态系统进程中微生物之间的关系

了解种群中存在哪些生物以及群落结构是如何对环境变化做出反应这些知识，是探讨生物系统如何推动海洋功能的唯一方法。同时，还需要精确衡量微生物和代谢的多样性，以及这种多样性是如何与生物地球化学和物理过程联系在一起的，以进一步探索生物种群动力学、基因组多样性和生物地球化学过程的代谢基础，特别是在有着大量未知类群的微生物领域。所有学科的科学家必须相互合作，设计一个具有预测功能的模型体系，来理解复杂的微生物群体与海洋生物地球化学之间的相互作用。这种预测甚至会挑战最先进的遗传学技术和进化理论，因为海洋生态系统的演变和多样性与陆地生境有很大的不同，而我们目前使用的模型是根据陆地生境来设计的。例如，海洋环境每天都发生一次波动，这意味着那些可以在陆地上生存发展的顶级群落很少能在海洋里生长。DeLong and Karl（2005）使用基因组学技术寻找环境样本中尚未鉴定的微生物的基因组。既然微生物在生态系统功能中起到核心作用，生态基因组学对于了解海洋微生物群落的多样性具有巨大的影响作用（(Eisen 2007)，同时如果我们能认清微生物的组合和功能，并映射到它们的地理位置上，生态基因组学也可以为保护生态系统提供重要线索（Hoffman and Gaines 2008）。

1.3.3　环境变化与微生物多样性

全球海洋正面临着日益严重的威胁，全球生态系统的物种组成正经历着巨大的改变，这都是由人类活动的干扰所造成的。自然群落以资源的方式影响着物理环境，并与其他物种相互影响，这都表明生物多样性是生态系统发挥功能或保持可持续性的必要条件。它们的改变通常会导致物种多样性的减少。物种组成、丰度或功能类型的变化，会影响生态系统中资源利用的效率，这意味着生物多样性减少会导致生态系统的生物地球化学功能受损。然而，现在仍然不知道物种丰度和功能类群是如何影响生态系统的。之前研究物种丰度和生态系统功能之间关系的实验主要局限于陆地系统，但最近海洋系统也常常被涉及，虽然实际的观测结果会和模拟的结论有一些冲突。最近几年，

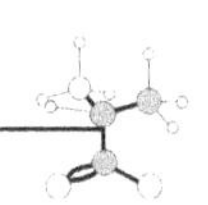

许多模型都试图说明物种丰度变化是如何影响生态系统功能的。总之，一个特定的生态系统功能应包括以下几个方面：①生物多样性和其中生物体的功能特性；②相关的生物地球化学过程；③非生物环境。“恢复力”这一概念，在生态系统领域是指生态系统在面临新的干扰时维护其原有功能的能力。这个概念与“稳定性”相关，但它强调的是功能的维护和抵抗新的干扰。因此，“恢复力”还包含了人类社会对生态系统服务具有依赖性和人类行为造成的干扰越来越多这两层意思（Webb 2007）。从而生态学家可以通过复杂适应系统（complex adaptive systems，CAS）或社会—生态系统（social-ecological systems，SES）来研究生态系统的恢复力，他们认为人类因素和生态因素在生态系统的功能和维护中是同等重要（Webb 2007）。

虽然海洋环境中的生物很少会全球灭绝，但是局部灭绝和丰度发生巨大变化的现象还是很普遍的。由于这些灭绝威胁到了生态系统为社会提供产品和服务的能力，它们产生的原因及其对生态系统的功能和稳定性的影响成为当前研究的热点。迄今，很多研究存在着争议，大家对于是多样性在发挥作用，还是单个物种或功能类群在发挥作用意见不一致。海洋环境的物种和功能多样性都比陆地系统的要高。现存在几种假说，有的观点认为多样性不影响生态系统功能，有的观点认为多样性有助于生态系统功能，且必须通过较高的多样性来应对环境剧烈变化，有的观点认为多样性的减少给生态系统带来的改变也许不会被预测到。探究生物多样性如何影响生态系统功能，也就是探究生物群落是如何随着生物多样性变化的。

很少有数据来评估种内（已证明具有重要功能的物种）遗传多样性在生态系统水平上的重要性。不过，Hughes and Stachowicz（2004）指出，对于栖息地的海草 *Zostera marina*，群落的基因型多样性越高，其抵抗鹅啃食所造成的干扰的能力就越强。此外，随着大叶藻基因多样性的增加，群落恢复到干扰前的密度所需的时间也越少。一个以微藻为对象的研究，采集了南极洲周围所有主要的大陆环流中的海藻 *Phaeocystis antarctica*，并利用微卫星技术来研究种群结构，由于环流间的基因流动较少，使得每个大陆环流都具有较高的基因多样性（Gaebler and Medlin，未发表）。同时，每个环流中分离的微藻在生理特性上是不同的。例如，每年都结冰的环流里分离的微藻能够适应的盐度范围是18‰～70‰，只是生长过程有一点迟缓，而终年不结冰的环流里的微藻只能在盐度为33‰的海水里生存。环境变化似乎会引起生物体表型上的反应，因此如果气候发生变化，当地物种灭绝，种群的重建将取决于其他地区种群间的基因流和分散度。每个环流里较高的多样性表明，海藻已经占据

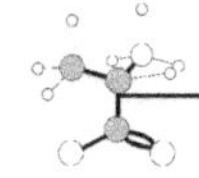

了每个环流里可用的生态位，以便形成一个更加稳健的种群来应对环流里所有可能产生的环境条件。

1.4 结束语

气候变化导致生物分布和丰度的变化（Umina et al. 2005，Bradshaw and Holzapfel 2006），表明了环境正在快速改变。生物的承受能力是由生物体的遗传结构和生理适应能力所决定的。当环境发生极端变化时，生物体会产生四种反应：迁移到一个适宜生存的地区；发生生理变化、扩大代谢范围以便生存和繁殖；选择性死亡以适应变化的环境；当地种群灭绝。

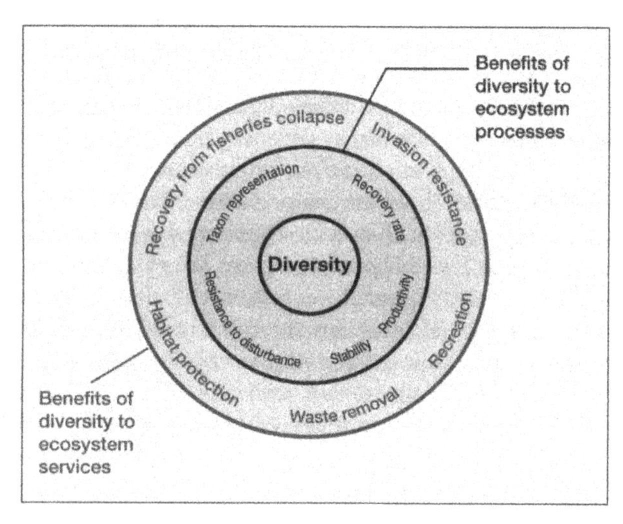

Fig. 1.5 Schematic representation of ecosystem benefits of marine biodiversity (after Palumbi et al. 2008b). Biodiversity (*red portion*) at the various biological levels (genetic, species, ecosystem and functional) enhances a variety of ecological processes (*blue portion*). Ecological processes enhance the benefits that ecosystems cab provide in terms of recovery, resistance, protection, recycling etc. (*green portion*)

图 1.5 海洋生物多样性的生态效益示意图（Palumbi et al. 2008 b）。不同生物学水平（基因、物种、生态系统和功能）上的生物多样性（红色部分）增强各种生态过程（蓝色部分），生态过程提高生态系统在复苏、反抗、保护和回收等方面（绿色部分）的效益

图中英文注释：Diversity，多样性；Ecosystem service，生态系统服务；Ecosystem process，生态系统过程；Fisheries collapse，渔业崩溃；Habitat protection，栖息地保护；Waste removal，废物处理；Recreation，再创造；Invasion resistance，入侵抵抗；Taxon representation，分类群代表；Recovery tate，恢复速率；Resistence to disturbance，抗干扰性；Stability，稳定性；Productivity，生产力

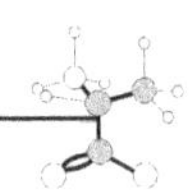

最重要的是，遗传标记可以协助监测环境变化对生物体基因、蛋白或代谢产物水平上的影响，同时可以评估环境变化所引起选择的本质，并估计种群对自然和人类活动所引起变化的适应潜力（Hoffmann and Willi 2008）。除了已经确定的中性分子标记的使用外（Carvalho et al. 2002，Avise 2004），对表型变化相关基因的研究已发生很大转变。虽然许多新物种的发现令人兴奋（Venter et al. 2004，Santelli et al. 2008），然而过度开采所引发持续的种群崩溃，海洋濒危物种日益增多，生境和生态系统的服务能力逐步退化等情况，都不容乐观（Palumbi et al. 2008 b）。利用基因技术，并结合适当的遗传学和生态保护策略，为阻止这些状况提供了一个有效的方法。

上述对于海洋生物多样性的介绍包括描述和检测的一系列分级水平，以及这些水平之间的内在联系。在许多情况下，如确定管理单位开发物种时，鉴定和监测单独的物种比较实用，然而当研究温度对底栖生物的一些过程（如碳同化作用）的影响时，侧重于基因和代谢途径可能会更加合适。总之，生态系统稳定性的组成部分，即恢复力、抵抗力和可逆性（Palumbi et al. 2008 a），是衡量其面对环境变化时的弹性和强度的指标。因此需要在不同的生物、空间和时间尺度上对该目标进行研究。正如最近所强调那样（Palumbi et al. 2008 b），在未来管理海洋生物多样性时，至少有两个方法是可以确定优先考虑的：首先，阐明生物多样性和生态系统功能之间的联系；其次，开展调查以证明增加生物多样性的同时可以增强和稳定生态系统的功能（Worm et al. 2006，Danovaro et al. 2008）。我们坚信，基因组学将在这两个方面发挥至关重要的作用。

参考文献

Adachi M, Kanno T, Okamoto R, Itakura S, Yamaguchi M, Nishijima T (2003) Population structure of Alexandrium (Dinophyceae) cyst formation-promoting bacteria in Hiroshima Bay, Japan. Appl Environ Microbiol 69:6560 – 6568

Allen EE, Banfield JF (2005) Community genomics and microbial ecology and evolution. Nat Rev Genet 3:489 – 498

Allen MJ, Martinez-Martinez J et al (2007) Use of microarrays to assess viral diversity: from genotype to phenotype. Environ Microbiol Apr 9(4):971 – 982

Allen MJ, Wilson WH (2008) Aquatic virus diversity accessed through omic techniques: a route map to function. Curr Opin Microbiol 11(3):226 – 232

Amann RI, Binder, BJ, Olson, RJ et al (1990) Combination of 16S rRNA-targeted oligonucle-

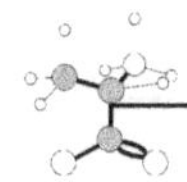

otide probes with flow cytometry for analysing mixed microbial populations. Appl Environ Microbiol 56:1919 – 1925

Angel M (1992) Managing biodiversity in the oceans. In: Peterson M (ed) Diversity of oceanic life. CSIS, Washington

Angly FE, Felts B et al (2006) The marine viromes of four oceanic regions. PLoS Biol 4 (11):e368

Arnold WS, Hitchcock GL, Frischer ME, Wanninkhof R, Sheng P (2005) Dispersal of an induced larval cohort in a coastal lagoon. Limnol Oceanogr 50:587 – 597

Arise JC (2004) Molecular Markers, Natural History, and Evolution 2nd ed. Sinauer Assoc Inc, Sunderland, Massachusetts

Babcock DA, Wawrik B et al (2007) Rapid screening of a large insert BAC library for specific 16S rRNA genes using TRFLP. J Microbiol Methods 71(2):156 – 161

Barber P, Boyce SL (2006) Estimating diversity of Indo-Pacific coral reef stomatopods through DNA barcoding of stomatopod larvae. Proc R Soc B 273:2053 – 2061

Barki Y, Douek J, Graur D, Gateno D, Rinkevich B (2000) Polymorphism in soft coral larvae revealed by amplified fragment-length polymorphism (AFLP) markers. Mar Biol 136:37 – 41

Beja O (2004) To BAC or not to BAC: marine ecogenomics. Curr Opin Biotechnol 15(3): 187 – 190

Beja O, Aravind L et al (2000) Bacterial rhodopsin: evidence for a new type of phototrophy in the sea. Science 289(5486):1902 – 1906

Bellwood DR, Hughes TP, Folke C, Nystro M (2004) Confronting the coral reef crisis. Nature 429:827 – 833

Bench SR, Hanson TE et al (2007) Metagenomic characterization of Chesapeake Bay virio-plank-ton. Appl Environ Microbiol 73(23):7629 – 7641

Bhadury P, Austen MC et al (2006) Development and evaluation of a DNA-barcoding approach for the rapid identification of nematodes. Mar Ecol Prog Ser 320:1 – 9

Biddle JF, Fitz-Gibbon S et al (2008) Metagenomic signatures of the Peru Margin subseafloor biosphere show a genetically distinct environment. Proc Natl Acad Sci USA 105 (30): 10583 – 10588

Biegala IC, Not F, Vaulot D, Simon N (2003) Quantitative assessment of picoeukaryotes in the natural environment by using taxon-specific oligonucleotide probes in association with tyra-mide signal amplification, fluorescence-in-situ-hybridization and flow cytometry. Appl Environ Microbiol 69:5519 – 5529

Billard C (1994) Life cycles. In: Green JC, Leadbeater BSC (eds). *The Haptophyte Algae*, The Systematics Association Special, Vol. 51. Clarendon Press, Oxford, pp 167 – 186

Blaxter M, Floyd R (2003) Molecular taxonomics for biodiversity surveys: already a reality.

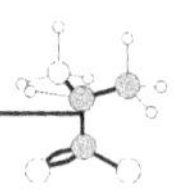

Trends Ecol Evol 18(6):268 – 269

Blaxter M, Mann J et al (2005) Defining operational taxonomic units using DNA barcode data. Phil Trans R Soc London B 360. doi:10.1098/rstb.2005.1725

Blouin MS (2000) Neutrality tests on mtDNA: unusual results from nematodes. J Hered 91 (2):156 – 158

Blouin MS, Yowell CA, Courtney CH, Dame JB (1998) Substitution bias, rapid saturation, and the use of mtDNA for nematode systematics. Mol Biol Evol 15(12):1719 – 1727

Blow N (2008) Exploring unseen communities. Nature 453:687 – 690

Bouchet P (2005) The mangitide of marine biodiversity. In: Duarte CM (ed) The exploration of marine biodiversity. Fundavión BBVA, Paris

Bradshaw, WE and Holzapfel, M (2006) Evolutionary response to rapid climate change. Science 312:1477 – 1478

Breitbart M, Felts B et al (2004) Diversity and population structure of a near-shore marine-sediment viral community. Proc Biol Sci 271(1539):565 – 574

Bullard SG, Lindquist N, Hay ME (1999) Susceptibility of invertebrate larvae to predators: how common are post-capture larval defenses?. Mar Ecol Prog Ser 191:153 – 161

Bulling MT, White PCL, Rafaelli D et al (2006) Using model systems to address the biodiversity-ecosystem functioning process. Mar Ecol Prog Ser 311:295 – 309

Caron DA, Peele ER, Lim EL, Dennett MR (1999) Picoplankton and nanoplankton and their trophic coupling in surface waters of the Sargasso Sea south of Bermuda. Limnol Oceanogr 44:259 – 272

Carvalho GR, van Oosterhout C, Hauser, L et al (2002) Measuring genetic variation in wild popu-lations: from molecular markers to adaptive traits, In: Behringer, J Hails, RS, Godfrey C (eds) Genes in the Environment. Blackwell Science, pp. 91 – 111

Chan RWK, Dixon PI, Peperrell JG et al (2003) Application of DNA-based techniques for the identification of whaler sharks (*Carcharhinus* spp.) caught in protective beach meshing and by recreational fisheries off the coast of New South Wales. Fish Bull 101:910 – 914

Chapin FS, Walker BH et al (1997) Biotic control over the functioning of ecosystems. Science 277(5325):500 – 504

Cock JM, Scornet D, Coelho S et al (2005) Marine Genomics and the exploration of marine biodiversity. Biodiversity Conservation in the Coastal Zone. "Exploring Marine Biodiversity: Scientific and Technological Challenges", Madrid, 29th November 2005.

Comtet T, Jollivet D, Khripounoff A et al (2000) Molecular and morphological identification of settlement-stage vent mussel larvae, Bathymodiolus azoricus (Bivalvia: Mytilidae), preserved in situ at active vent fields on the Mid-Atlantic Ridge. Limnol Oceanogr 45:1655 – 1661

Costa FO, Carvalho GR (2007) The barcode of life initiative: synopsis and prospective societal

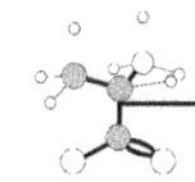

impacts of DNA barcoding of Fish. Genom Soc Pol 3:29 – 40

Costanza R, d'Arge R, de Groot R et al (1997) The value of the world's ecosystem services and natural capital. Nature 387:253 – 260

Creer S (2008) New technologies. In: Lambshead PJD, Packer M (eds) Manual for the molecular barcoding of deep-sea nematodes. International Seabed Authority, United Nations, Kingston
Culley AI, Lang AS et al (2006) Metagenomic analysis of coastal RNA virus communities. Science 312(5781):1795 – 1798

Cunningham C, Hikima J et al (2006) New resources for marine genomics: bacterial artificial chromosome libraries for the Eastern and Pacific oysters (*Crassostrea virginica* and *C. gigas*). Mar Biotechnol (NY) 8(5):521 – 533

Danovaro R, Gambi C et al (2008) Exponential decline of deep-sea ecosystem functioning linked to benthic biodiversity loss. Curr Biol 18:1 – 8

Danovaro R, Gambi C, Dell'Anno A et al (2008) Exponential decline of deep-sea ecosystem functioning linked to benthic biodiversity loss. Curr Biol 18:1 – 8

Dasmahapatra KK, Mallet J (2006) DNA barcodes: recent successes and future prospects. Heredity 97:254 – 255

Deagle BE, Bax N, Hewitt CL, Patil JG (2003) Development and evaluation of a PCR-based test for detection of Asterias (Echinodermata: Asteroidea) larvae in Australian plankton samples from ballast water. Mar Freshwat Res 54:709 – 719

DeLey P, DeLey IT et al (2005) An integrated approach to fast and informative morphological vouchering of nematodes for applications in molecular barcoding. Phil Trans R Soc London B 360: 1945 – 1958

DeLong EF (2005) Microbial community genomics in the ocean. Nat Rev Microbiol 3:459 – 469

DeLong EF and Karl DM (2005) Genomic perspectives in microbial oceanography. Nature 437:336 – 342

de Tomaso AW, Weissman IL (2003) Construction and characterization of large-insert genomic libraries (BAC and F 粘粒) from the Ascidian Botryllus schlosseri and initial physical mapping of a histocompatibility locus. Mar Biotechnol (NY) 5(2):103 – 115

Díez B, Pedrós-Alío C, Massana R (2001) Study of genetic diversity of eukaryotic picoplankton in different oceanic regions by small-subunit rRNA gene cloning and sequencing. Appl. Environ Microbial 67:2932 – 2941

Dinsdale EA, Pantos O et al (2008) Microbial ecology of four coral atolls in the Northern Line Islands. PLoS ONE 3(2):e1584

Doi RH, Igarashi RT (1965) Conservation of ribosomal and messenger ribonucleic acid cistrons in Bacillus species. J Bacteriol 90:384 – 390

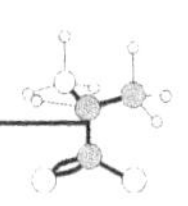

Dubnau D, Smith I, Morell P, Marmur J (1965) Gene conservation in Bacillus species. I. Conserved genetic and nucleic acid base sequence homologies. Proc Natl Acad Sci USA 54(2): 491-498

Duffy JE, Stachowicz JJ (2006) Why biodiversity is important to oceanography: potential roles of genetic, species, and trophic diversity in pelagic ecosystem processes. Mar Ecol Prog Ser 311: 179-189

Eisen JA (2007) Environmental shotgun sequencing: Its potential and challenges for studying the hidden world of microbes. PLoS Biol 5: e82

Eller G, Töbe K, Medlin LK (2007) A set of hierarchical FISH probes for the Haptophyta and a division level probe for the Heterokonta. J Plank Res 29: 629-640

Evans KM, Wortley AH, Mann DG (2007) An assessment of potential diatom "Barcode" genes (cox1, rbcL, 18S and ITS rDNA) and their effectiveness in determining relationships in sellaphora (Bacillariophyta). Protist 158: 349-364

Floyd R, Abebe E et al (2002) Molecular barcodes for soil nematode identification. Mol Ecol 11(4): 839-850

Fox CJ, Taylor MI et al (2008) Mapping the spawning grounds of Noth Sea cod (*Gadus morhua*) by direct and indirect means. Proc R Soc Lond 275: 1543-1548

Fox CJ, Taylor CJ, Pereyra R, Willasana MI, Rico C (2005) TaqMan DNA technology confirms likely overestimation of cod (*Gadus morhua* L.) egg abundance in the Irish Sea: Implications for the assessment of the cod stock and mapping of spawning areas using egg-based methods. Mol Ecol 14: 879-884

Fuchs BM, Spring S et al (2007) Characterization of a marine gammaproteobacterium capable of aerobic anoxygenic photosynthesis. Proc Natl Acad Sci U S A 104(8): 2891-2896

Gescher C, Metfies K, Frickenhaus S, Knefelkamp B, Wiltshire K, Medlin LK (2008) Feasibility of assessing the community composition of prasinophytes at the helgoland roads sampling site with a DNA microarray. Appl Environ Microbiol 74: 5305-5316

Giovannoni SJ, Britschgi TB, Moyer CL et al (1990) Genetic diversity in sargasso sea bacterloplankton. Nature (London) 345: 60-63

Goffredi SK, Jones WJ, Scholin CA, Marin R, Vrijenhoek RC (2006) Molecular detection of marine invertebrate larvae. Mar Biotech 8: 149-160

Gómez A, Wright PJ, Lunt DH et al (2007) Mating trials validate the use of DNA barcoding to reveal cryptic speciation of a marine bryozoan taxon. Proc R Soc B 274: 199-207

Grassle FJ, Maciolek NJ (1992) Deep-sea species richness: regional and local diversity estimates from quantitative bottom samples. Am Nat 139: 313-341

Groben R, John U, Eller G, Lange M, Medlin LK (2004) Using fluorescently-labelled rRNA probes for hierarchical estimation of phytoplankton diversity. Nova Hedwigia 79: 313-320

Hajibabaei M, Singer GA, Hebert PD, Hickey DA (2007) DNA barcoding: how it complements taxonomy, molecular phylogenetics and population genetics. Trends Genet 23(4): 167 – 172

Handelsmann J (2004) Metagenomics: application of genomics to uncultured microorganisms. Microbiol Mol Biol Rev 68: 669 – 685

Hansen BW, Larsen JB (2005) Spatial distribution of velichoncha larvae (Bivalvia) identified by SSNM-PCR. J Shell Res 24: 561 – 565

Harding JM (1999) Selective feeding behavior of larval naked gobies (Gobiosoma bosc) and blennies (Chasmodes bosquianus and Hypsoblennius hentzi): preferences for bivalve veligers. Mar Ecol Prog Ser 179: 145 – 153

Hebert PDN, Cywinska A, Ball SL, de Waard JR (2003) Biological identifications through DNA barcodes. Proc R Soc Lond B 270: 313 – 321

Hoef-Emden K, Melkonian M (2003) Revision of the genus Cryptomonas (Cryptophyceae): a combination of molecular phylogeny and morphology provides insights in a long-hidden dimorphism. Protist 154: 371 – 409

Hoffmann AA, Willi Y (2008) Detecting genetic responses to environment change. Nature Rev Genet 9: 421 – 432

Hofmann GE, Gaines SE (2008) New tools to meet new challenges: Emerging technologies for managing marine ecosystems for resilience. BioScience 58(1): 43 – 52

Hosoi M, Hosoi-Tanabe S, Sawada H et al (2004) Sequence and polymerase chain reaction-restriction fragment length polymorphism analysis of the large subunit rRNA gene of bivalve: Simple and widely applicable technique for multiple species identification of bivalve larva. Fish Sci 70: 629 – 637

Huber JA, Welch DBM et al (2007) Microbial population structures in the deep marine biosphere. Science 318: 97 – 100

Hughes TP, Bellwood DR, Folke C, Steneck RS, Wilson J (2006) New paradigms for supporting the resilience of marine ecosystems. Trends Ecol Evol 20: 380 – 386

Hughes AR, Stachowicz JJ (2004) Genetic diversity enhances the resistance of a seagrass ecosystem to disturbance. Proc Natl Acad Sci 101: 8898 – 9002

Huse SM, Huber JA, Morrison HG, Sogin ML, Welch DM (2007) Accuracy and quality of massively parallel DNA pyrosequencing. Genome Biol 8(7): R143

Igelsia-Rodriguez D, Schofield OM, Batley PJ, Medlin LK, Hayes PK (2006) Extensive intraspe-cific genetic diversity in the marine coccolithophorid Emiliania huxleyi: the use of microsatellite analysis in marine phytoplankton populations studies. J Phycol 42: 526 – 536

Ishizaka J, Harada K, Ishikawa K, Kiyosawa H, Furusawa H, Watanabe Y, Ishida H, Suzuki K, Handa N, Takahash M (1997) Size and taxonomic plankton community structure and carbon flow at the equator, 175°E during 1990 – 1994. Deep Sea Res II 44: 1927 – 1949

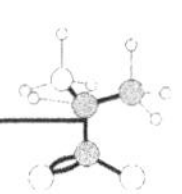

Janzen DH, Hajibabaei M, Burns JM et al (2005) Wedding biodiversity inventory of a large and complex Lepidoptera fauna with DNA barcoding. Phil Trans R Soc B 360:1835 - 1845

John U, Groben R, Beszteri B, Medlin L (2004) Utility of Amplified Fragment Length Polymorphisms (AFLP) to analyse genetic structures within the Alexandrium tamarense species complex. Protist Jun 155(2):169 - 179

Jones WJ, Preston CM, Marin R, Scholin CA, Vrijenhoek RC (2008) A robotic molecular method for in situ detection of marine invertebrate larvae. Mol Ecol Res 8:540 - 550

Jonston AW, Li Y, Ogilvie L (2005) Metagenomic marine nitrogen fixation-feast or famine? Trends Microbiol 13:416 - 420

Karaiskou N, Triantafyllidis A, Alvarez P et al (2007) Horse mackerel egg identification using DNA methodology. Mar Ecol 28:429 - 434

Kemp PF, Aller JY (2004) Bacterial diversity in aquatic and other environments: what 16S rDNA libraries can tell us. FEMS Microbiol Ecol 47:161 - 177

Kinlan BP, Gaines SD, Lester SE (2005) Propagule dispersal and the scales of marine community process. Div Distrib 11:139 - 148

Kirby RR, Lindley JA (2005) Molecular analysis of continuous plankton recorder samples, an examination of echinoderm larvae in the North Sea. J Mar Biol Ass UK 85:451 - 459

Kochzius M, Nölte M, Weber H et al (2008) Microarrays for identifying fishes. Mar Biotechnol 10:207 - 217

Lambshead PJD, Boucher G (2003) Marine nematode deep-sea biodiversity-hyperdiverse or hype?. J Biogeography 30(4):475 - 485

Lawton JHD, Bignell E et al (1998) Biodiversity inventories, indicator taxa and effects of habitat modification in tropical forest. Nature 391:72 - 76

Le Goff-Vitry MC, Chipman AD, Comtet T (2007) In situ hybridization on whole larvae: a novel method for monitoring bivalve larvae. Mar Ecol Prog Ser 343:161 - 172

Lindley JA, Batten SD (2002) Long-term variability in the diversity of North Sea zooplankton. J Mar Biol Ass UK 82:31 - 40

Lleonart J, Taconet M, Lamboeuf M (2006) Integrating information on marine species identifica-tion for fishery purposes. Mar Ecol Prog Ser 316:231 - 238

López-García P, Rodríguez-Valera F, Pedrós-Alió C et al (2001) Unexpected diversity of small eukaryotes in deep sea Antarctic plankton. Nature 409:603 - 607

Maggs CA, Castilho R, Foltz D, Henzler C, Jolly MT, Kelly J, Olsen J, Perez KE, Stam W, Väinölä R, Viard F, Wares J (2008) Evaluating signatures of glacial refugia for North Atlantic marine organisms?. Ecology 89:S108 - S122

Maidak BL, Cole JR, Lilburn TG, Parker CTJ, Saxman PR, Farris RJ, Garrity GM, Olson GJ, Schmidt TM, Tiedje JM (2001) The RDP-II (Ribosomal Database Project). Nucleic Acids Res

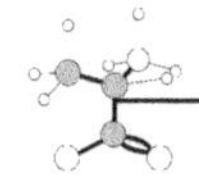

29:173 – 174

Mann DG (1999) The species concept in diatoms. Phycologia 38:437 – 495

Mariani S, Uriz MJ, Turon X (2003) Methodological bias in the estimations of important meroplanktonic components from near-shore bottoms. Mar Ecol Prog Ser 253:67 – 75

Markmann M, Tautz D (2005) Reverse taxonomy: an approach towards determining the diversity of meiobenthic organisms based on ribosomal RNA signature sequences. Phil Trans R Soc London B 360:1917 – 1924

Marko B, Lee SC, Rice AM et al (2004) Mislabelling of a depleted reef fish. Nature 430:309 – 310

Marta-Almeida M, Dubert J, Peliz A, Queiroga H (2006) Influence of vertical migration pattern on retention of crab larvae in a seasonal upwelling system. Mar Ecol Prog Ser 307:1 – 19

Massana R, Terrado R, Forn I et al (2006) Distribution and abundance of unclutured heterotrophic flagellates in the world oceans. Env Microbiol 8:1515 – 1522

Medlin LK (2007) If everything is everywhere, do they share a common gene pool?. Gene 405:180 – 183

Medlin LK, Metfies K, Mehl H, Wiltshire K, Valentin K (2006) Picoplankton Diversity at the Helgoland Time Series Site as assessed by three molecular methods. Micro Ecol 167:1432 – 1451

Metaxas A, Burdett-Coutts V (2006) Response of invertebrate larvae to the presence of the ctenophore Bolinopsis infundibulum, a potential predator. J Exp Mar Biol Ecol 334:187 – 195

Metfies K, Hujic S, Lange M, Medlin L (2005) Electrochemical detection of the toxic dinoflagellate *A. ostenfeldii* with a DNA Biosensor. Biosen Bioselec 20:1349 – 1357

Metfies K, Töbe K, Scholin C, Medlin LK (2006) Laboratory and field applications of ribosomal RNA probes to aid the detection and monitoring of harmful algae. In: Granéli E, Turner J (eds) Ecology of harmful algae. Springer, New York, pp 311 – 326

Minagawa G, Miller MJ, Aoyama J, Wouthuyzen S, Tsukamoto K (2004) Contrasting assemblages of leptocephali in the western Pacific. Mar Ecol Prog Ser 271:245 – 259

Moens T, Vincx M (1997) Observations on the feeding ecology of estuarine nematodes. J Mar Biol Assoc UK 77(1):211 – 227

Morgan TS, Rogers AD (2001) Specificity and sensitivity of microsatellite markers for the identification of larvae. Mar Biol 139:967 – 973

Mou X, Sun S et al (2008) Bacterial carbon processing by generalist species in the coastal ocean. Nature 451:708 – 711

Muyzer G (1999) DGGE/TGGE a method for identifying genes from natural ecosystems. Curr Opin Microbiol 2:317 – 322

Naeem S (2006) Expanding scales in biodiversity-based research: challenges and solutions for marine systems. Mar Ecol Prog Ser 311:273 – 283

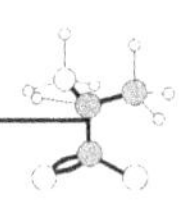

Noell CJ, Donnellan S, Foster R, Haigh L (2001) Molecular discrimination of garfish Hyporhamphus (Beloniformes) larvae in southern australian waters. Mar Biotechnol 3:509 – 514

Ogden R (2008) Fisheries Forensics: the use of DNA tools for improving compliance, traceability and enforcement in the fishing industry. Fish Fisheries 9:462 – 472

Pace B, Campbell LL (1971a) Homology of ribosomal ribonucleic acid of Desulfovibrio species with Desulfovibrio vulgaris. J Bacteriol 106(3):717 – 719

Pace B, Campbell LL (1971b) Homology of ribosomal ribonucleic acid diverse bacterial species with Escherichia coli and Bacillus stearothermophilus. J Bacteriol 107(2):543 – 547

Palumbi SR, McLeod KL, Grunbaum D (2008a) Ecosystems in action: lessons from marine ecology about recovery, resistance and reversibility. Bioscience 58:33 – 42

Palumbi SR, Sandifer PA, Allan JD (2008b) Managing for ocean biodiversity to sustain marine ecosystem services. Front Ecol Environ. doi:10. 1890/070135

Patil JG, Gunasekera RM, Deagle BE, Bax NJ (2005) Specific detection of pacific oyster (Crassostrea gigas) larvae in plankton samples using nested polymerase chain reaction. Mar Biotechnol 7:11 – 20

Peers G, Price NM (2006) Copper-containing plastocyanin used for electron transport by an oceanic diatom. Nature 44:341 – 344

Pegg GC, Snclair B, Briskey L et al (2006) MtDNA barcode identification of fish larvae in the southern Great Barrier Reef, Australia. Sci Mar 70:7 – 12

Phillips RA, Petersen MK, Lilliendahl K et al (1999) Diet of the northern fulmar *Fulmarus glacialis*: reliance on commercial fisheries?. Mar Biol 135:159 – 170

Pikitch EK, Santora C, Babcock EA, Bakun A, Bonfil R, Conover DO, Dayton P, Doukakis P, Fluharty D, Heneman B, Houde ED, Link J, Livingston PA, Mangel M, McAllister MK, Pope J, Sainsbury KJ (2004) Ecosystem-Based Fisheries Management. Science 305:346 – 347

Platt HM, Warwick RM (1983) Free-living marine nematodes. Part I British Enoplids. Cambridge University Press, Cambridge

Pradillon F, Schmidt A, Peplies J, Dubilier N (2007) Species identification of marine invertebrate early stages by whole-larvae in situ hybridisation of 18S ribosomal RNA. Mar Ecol Prog Ser 333:103 – 116

Pruesse E, Quast C, Knittel K, Fuchs BM, Ludwig W, Peplies J, Glöckner FO (2007) SILVA: a comprehensive online resource for quality checked and aligned ribosomal RNA sequence data compatible with ARB. Nucleic Acids Res 35(21):7188 – 7196

Quaiser A, Lopez-Garcia P et al (2008) Comparative analysis of genome fragments of Acidobacteria from deep Mediterranean plankton. Environ Microbiol 10:2704 – 2717

Ratnasingham S, Hebert PDN (2007) BOLD: The barcode of life data system (http://www.barcodinglife.org). Mol Ecol Notes 7:355 – 364

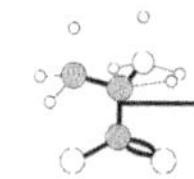

Ray GC (1991) Coastal-zone biodiversity patterns. BioScience 41:490 – 498

Richardson DE, Cowen RK (2004) Diversity of leptocephalus larvae around the island of Barbados (West Indies): relevance to regional distributions. Mar Ecol Prog Ser 282:271 – 284

Richardson DE, Vanwye JD, Exum AM, Cowen RK, Crawford DL (2007) High-throughput species identification: from DNA isolation to bioinformatics. Mol Ecol Notes 7:199 – 207

Rock J, Costa FO, Walker DI et al (2008) DNA barcodes for fish of the Antarctic Scotia Sea indicate priority groups for taxonomic and systematic focus. Antarctic Sci 20:253 – 262

Rogers AD (2001) Molecular ecology and identification of marine invertebrate larvae. In: Atkinson D, Thorndyke M (eds) Environment and animal development: genes, life histories and plasticity. BIOS Scientific Publishers Ltd., Oxford, pp 29 – 69

Rosel PE, Kocher TD (2002) DNA-based identification of larval cod in stomach contents of predatory fishes. J Exp Mar Biol Ecol 267:75 – 88

Rothberg JM, Leamon JH (2008) The development and impact of 454 sequencing. Nat Biotech 26:1117 – 1124

Santaclara FJ, Espineira M, Vieites JM (2007) Molecular detection of Xenostrobus securis and Mytillus galloprovincialis larvae in galician coast (Spain). Mar Biotechnol 9:722 – 732

Santelli CM, Orcutt BN, Banning E et al (2008) Abundance and diversity of microbial life in ocean crust. Nature 453:653 – 656

Savolainen R, Cowan RS, Vogler AP (2005) DNA barcoding of life. Phil Trans R Soc Lond B 360:1805 – 1980

Scherer-Lorenzen M (2005) Biodiversity and ecosystem functioning: basic principles. In: Wilhelm Barthlott, K Eduard L, Stefan P (eds) Biodiversity: structure and function. Encyclopedia of Life Support Systems (EOLSS), Developed under the Auspices of the UNESCO, Eolss Publishers, Oxford. [http://www.eolss.net]

Schratzberger M, Warr K et al (2007) Functional diversity of nematode communities in the southwestern North Sea. Mar Environ Res 63(4):368 – 389

Shanks AL, Brink L (2005) Upwelling, downwelling, and cross-shelf transport of bivalve larvae: test of a hypothesis. Mar Ecol Prog Ser 302:1 – 12

Shearer TL, Coffroth MA (2006) Genetic identification of Caribbean scleractinian coral recruits at the flower garden banks and the florida keys. Mar Ecol Prog Ser 306:133 – 142

Shearer TL, Coffroth MA (2008) Barcoding corals: limited by interspecific divergence, not intraspecific variation. Mol Ecol Res 8:247 – 255

Sigler MF, Hulbert LB, Lunsford CR et al (2006) Diet of Pacific sleeper shark, a potential Steller sea lion predator, in the north-east Pacific Ocean. J Fish Biol 69:392 – 405

Simon N, Campbell L, Ornolfsdottir E et al (2000) Oligonucleotide probes for the identification of three algal groups by dot biot and fluorescent whole-cell hybridization. J EuR Microbial 47:

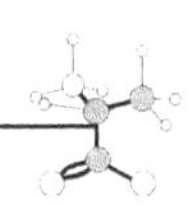

76 – 84

Smith CR, Austen MC et al (2000) Global change and biodiversity linkages across the sediment-water interface. Bioscience 50(12):1108 – 1120

Snelgrove PVR, Austen MC et al (2000) Linking biodiversity above and below the marine sediment-water interface. BioScience 50(12):1076 – 1088

Snelgrove P, Blackburn TH et al (1997) The importance of marine sediment biodiversity in ecosystem processes. Ambio 26:578 – 583

Sogin ML, Morrison HG et al (2006) Microbial diversity in the deep sea and the underexplored "rare biosphere". Proc Natl Acad Sci USA 103:12115 – 12120

Stein JL, Marsh TL et al (1996) Characterization of uncultivated prokaryotes: isolation and anal-ysis of a 40-kilobase-pair genome fragment from a planktonic marine archaeon. J Bacteriol 178 (3):591 – 599

Suzuki MT, Preston CM et al (2004) Phylogenetic screening of ribosomal RNA gene-containing clones in Bacterial Artificial Chromosome (BAC) libraries from different depths in Monterey Bay. Microb Ecol 48(4):473 – 488

Taylor MI, Fox C, Rico I, Rico C (2002) Species-specific TaqMan probes for simultaneous identification of (Gadus morhua L.), haddock (Melanogrammus aeglefinus L.) and whiting (Merlangius merlangus L. Mol Ecol Notes 2:599 – 601

Tzeneva VA, Heilig HG, van Vliet WA, Akkermans AD, de Vos WM, Smidt H (2008) 16S rRNA targeted DGGE fingerprinting of microbial communities. Methods Mol Biol 410:335 – 349

Umina PA, Weeks AR, Kearney MR et al (2005) A rapid shift in classic clinical pattern in *Drosophila* reflecting climate change. Science 308:691 – 693

Underwood AJ, Fairweather PG (1989) Supply-side ecology and marine benthic assemblages. Trends Ecol Evol 4:16 – 19

Vadopalas B, Bouma JV, Jackels CR, Friedman CS (2006) Application of real-time PCR for simultaneous identification and quantification of larval abalone. J Exp Mar Biol Ecol 334:219 – 228

Van Straalen NM, Roelofs D (2006) An introduction to ecological genomics. Oxford University Press, Oxford

Venter JC, Remington K et al (2004) Environmental genome sequencing of the Sargasso Sea. Science 304:66 – 74

Vera JC, Wheat CW, Fescemyer HW et al (2008) Rapid transcriptome characterisation for a non-model organism using 454 pyrosequencing. Mol Ecol 17:1636 – 1647

Von der Heyden S, Lipnski MR, Matthee CA (2007) Species-specific genetic markers for identifica-tion of early life-history stages of Cape hakes, Merluccius capensis and Merluccius paradoxus in the southern Benguela. Curr J Fish Biol 70:262 – 268

Webb CT (2007) What is the role of ecology in understanding ecosystem resilience?. BioSci-

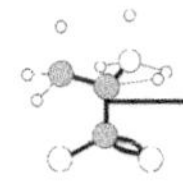

ence 5:470 – 471

Webb KE, Barnes DKA, Clark MS, Bowden DA (2006) DNA barcoding: A molecular tool to identify Antarctic marine larvae. Deep-Sea Res II 53:1053 – 1060

Will KW, Rubinoff D (2004) Myth of the molecule: DNA barcodes for species cannot replace morphology for identification and classification. Cladistics 20:47 – 55

Williamson SJ, Rusch DB et al (2008) The Sorcerer II global ocean sampling expedition: metage-nomic characterization of viruses within aquatic microbial samples. PLoS ONE 3(1):e1456

Wilson K, Thorndyke M, Nilsen F et al (2005) Marine systems: moving into the genomics era. Mar Ecol 26:3 – 16

Witham TG, DiFazio SP, Schweitzer JA et al (2008) Extending genomics to natural communities and ecosystems. Science 320:492 – 495

Woese CR (1987) Bacterial evolution. Microbiol Rev 51:221 – 271

Wong EH-K, Hanner RH (2008) DNA barcoding detects market substitution in North American seafood. Food Res Int 41:828 – 837

Worm B, Barbier EB, Baumont N et al (2006) Impacts of biodiversity loss on ocean ecosystem services. Science 314:787 – 790

Zhan A, Bao Z, Hu X et al (2008) Accurate methods of DNA extraction and PCR-based genotyping for single scallop embryos/larvae long preserved in ethanol. Mol Ecol Res 8:790 – 795

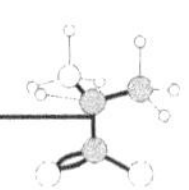

第2章　宏基因组分析

摘要：宏基因组学（metagenomics）代表着分子生物学与生物信息学工具的结合，用于评估一个群落中的遗传信息，而不需要经过任何微生物培养。这种方法，对于那些包含大量不可培养微生物的样本研究非常有意义。整个过程只需要提取一个环境样本或目标微生物富集后的基因组DNA，然后进行测序和序列分析。宏基因组学领域发展迅速，这应归功于高通量测序技术和日益增长的计算能力。宏转录组和宏蛋白质组也结合到宏基因组学中，不仅可以评价微生物群落的遗传信息，而且可以进一步研究在特定环境中的基因表达情况。本章将介绍宏基因组学早期研究概况和应用方向，讨论宏基因组分析在分子生物学和生物信息学上面临的挑战，提供亚群落富集、基因组提取和纯化的方法，阐述大片段与小片段文库的构建和筛选的方法，以及利用高通量测序技术直接测序的分析方法。生物信息学部分介绍组装和合并的工具、基因预测程序和注释的系统，同时指出宏基因组学在研究复杂群落时遇到的问题，以及讨论基于数据集内分析系统发育和功能多样性的方法。本章目的是提供宏基因组学的分子和生物信息学方面的基础知识，以便为从事相关研究的读者提供有用的方法和策略。

2.1　前言

地球上的微生物种类繁多，且生物量巨大，在氮固定与有机质矿化过程中起了关键的酶催化作用。70%的地球表面是由海洋覆盖，全面了解海洋环境中微生物介导的元素循环，对了解这些物质的全球循环至关重要。按照现在对全球变暖的论述，海洋环境中温室气体的形成、贮存、排放以及降解备受关注。尽管调研海洋的生物化学地球循环的重要性已经受到普遍认可，但是对海洋环境微生物群落以及它们对生物化学地球循环的贡献、个别物种的生态生理学研究等仍处于萌芽状态。宏基因组学不依赖培养法即可评估生物体的遗传信息，为环境研究提供了新的思路。本章主要叙述宏基因组学从实

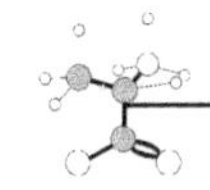

验室技术到生物信息分析方面取得的研究进展。

列文虎克在 17 世纪首次用显微镜证明了微生物的存在（综述：Hall 1989）。几乎过了两个世纪，微生物学家如 Louis Pasteur（综述：Schwartz 2001）、Robert Koch（综述：Kaufmann and Schaible 2005）、Martinus Beijerinck（综述：Chung and Ferris 1996）等，才建立起培养和富集微生物的方法，开始进一步描述微生物。在 19 世纪和 20 世纪早期对微生物的研究集中在医学领域，但是有些先驱者如 Ferdinand Cohn 就开展了藻类和光合细菌的研究，他同时还描述了硫细菌中的贝氏硫菌属 *Beggiatoa*（综述：Drews 2000）。Sergei Winogradsky 首次提出了化能营养这个概念，揭示了微生物在生物化学地球循环中的重要作用。而且，他首次分离和描述了固氮细菌和硝化细菌（综述：Schlegel 1996），但是这些微生物的研究局限于实验室分离和鉴定。直到分子生物学被引进这个领域，这种僵局才被打破，16S rRNA 基因可以用作系统发生的标记，描述不可培养微生物的多样性（Olsen et al. 1994）。早期这种方法报道了环境中大量的微生物（Torsvik et al. 1990），并且随着特定 rRNA 探针的设计和应用，实现了原位观察微生物群落的组成（Stahl and Amann 1991，Amann and Fuchs 2008）。

用标准的培养法大概仅能发现生物圈中约 1% 的微生物（Amann et al. 1995，Curtis et al. 2002）。而为了得到更多的微生物，研究者们也尝试了各种新的培养方法。（Connon，Giovannoni 2002，Rappe et al. 2002，Zengler et al. 2002）。然而，随着分子生物学方法的应用，大量的微生物物种被鉴定，目前的培养法已经无法与之并驾齐驱。这些分子生物学工具包括基于 PCR 的直接扩增、克隆，以及基于环境中 rRNA 基因的分析方法（Pace et al. 1985，Olsen et al. 1986，Giovannoni et al. 1990，Ward et al. 1990）。最近几年，DNA 扩增方法和测序技术的日新月异为 16S RNA 基因多样性分析提供了新视角，可对环境样品中不同微生物来源的 16S RNA 基因可变区进行平行的测序（Sogin et al. 2006）。

仅仅基于 rRNA 基因的系统发育关系，很难对未培养微生物进行生态生理学鉴定。因此还需要研究关于适应性代谢、抗性和防御等功能基因，它们保证了微生物在环境中的正常生存。宏基因组学技术的出现，为深入分析不可培养微生物的代谢打开了新的窗口，并在多样性和功能之间搭建了桥梁。

2.2 宏基因组学的历史和应用

“宏基因组分析”（metagenome analysis）由 Jo Handelsman 于 1998 年首次

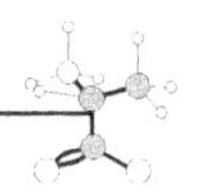

提出。这个术语源于 meta 分析中的统计学概念，是将单独分析与基因组学中微生物遗传信息的复杂分析综合在一起的统计过程。具体包括取得某个环境中或目标细胞富集后的基因组 DNA 片段，然后进行测序或功能分析。宏基因组学分析通常还被称为环境基因组学（environmental genomics）、生态基因组学（ecogenomics）和群落基因组学（community genomics）。

实际上，宏基因组分析在“宏基因组学”提出前已经开展。Thomas M. Schmidt、Edward F. DeLong 和 Norman R. Pace 首次在海洋生态系统中建立不依赖 PCR 无偏向性的宏基因组文库，对宏基因组文库进行筛选。研究者们提取了北太平洋中部 8000 升海水中的超微型浮游生物的 DNA，并将其克隆到 λ 噬菌体载体，总的文库达到 10^7 个单克隆，其中一部分插入片段为 10 ～ 20 kbp，约 3.2×10^4 个。克隆用探针对 16S rRNA 基因进行杂交，鉴定得到 16 个特异克隆。这个研究小组已经注意到了宏基因组学还能同时获取与 16S rRNA 基因关联的其他序列信息（Schmidt et al. 1991）。随后 DeLong 和他的团队用了五年时间，对一种不可培养的浮游类古细菌的基因组构建了 F 粘粒文库并进行测序（Stein et al. 1996）。凭借测序技术和生物信息工具，宏基因组学在上世纪 90 年代末得到越来越广泛的应用。

大片段宏基因组 DNA 被克隆至高容量载体，如 cosmid 粘粒（Collins and Hohn 1978）、F 粘粒（Kim et al. 1992）和细菌人工染色体 BACs（Shizuya et al. 1992）（参见本章 2.3.5.2）。从这些克隆的大片段（>30 kbp）中有机会得到完整的操纵子，在同一个片段上找到进化标记，推测微生物的系统发生起源。一个很好的例子就是借助大片段插入文库发现变形菌视紫质，该结果源于提取海洋浮游细菌的 DNA 构建的 130 kbp BAC 文库，其中一个克隆包含了类视紫质基因和 16S rDNA 基因，属于 γ 变形杆菌 SAR86 基因簇，至今这仍是宏基因组研究中最突出的成果之一（Beja et al. 2000，2001）。随后的研究使我们充分了解到海洋中的好氧不产氧光养细菌及其与全球碳和能量平衡的关系（Bryant and Frigaard 2006），这种菌是一类能够产生菌绿素 a，但不依赖光能而只将其作为辅助能量来源，且生活在有氧环境中（译者注）。最近，细菌的完整基因组可以基于宏基因组 F 粘粒文库克隆来实现组装，该方法曾用在水稻 cluster I 细菌的分析：对该菌进行富集培养后提取 DNA 构建文库，总共约3700个 F 粘粒克隆，用来进行全基因组组装（Erkel et al. 2006）。此外 F 粘粒末端测序还用来分析不同深度的浮游生物群体的碳和能量代谢趋势（DeLong et al. 2006）。

随着高通量测序的成本降低和技术改善，基于小片段插入文库（约1.5～

3 kbp)的宏基因组分析开始成为主流。这种方法特别适用于多样性较低的环境样本，如研究酸性矿山排水系统中的微生物群落（Tyson et al. 2004）。还有对海洋蠕虫 *Olavius algarvensis* 的共生菌的研究，结合利用了小片段插入文库鸟枪法测序和 F 粘粒文库测序（Woyke et al. 2006）。这些方法为研究微生物群落遗传特征、微生物之间以及它们与宿主之间的相互作用提供了新视角。迄今为止，最大的小片段宏基因组文库测序是最近一项名为“全球海洋取样考察”（Global Ocean Sampling expedition，GOS）的研究计划（Rusch et al. 2007，Yooseph et al. 2007）。该研究是马尾藻海鸟枪测序计划的延续（Venter et al. 2004），在之前的这项研究中就已经鉴定得到 148 个新的类群和69901个新基因，包括 782 个变形菌视紫质基因。

之后，一种新的焦磷酸测序方法被开发和使用，此技术不需要克隆，因此没有克隆偏好性（Margulies et al. 2005）。焦磷酸测序目前可产生 400 bp 长度的序列，此技术初期的读长仅 100 bp，但已经证实对群落多样性研究（Leininger et al. 2006，Sogin et al. 2006）和宏基因组分析极有价值（Edwards et al. 2006）。

除了 DNA 测序技术革新外，应用噬菌体 ϕ29 聚合酶进行等温多重置换扩增（multiple displacement amplification，MDA）产生的大量结果都使分子生态学家非常感兴趣。该技术可从微量环境样品中扩增基因组 DNA，从而有助于深入了解微生物群落样品，Abulencia 等人把这项技术应用在土壤微生物的研究中（Abulencia et al. 2006）。MDA 同样用于单细胞的扩增，尽管此技术存在缺陷（Hutchison and Venter 2006），但用来获取单个未培养细菌的基因组概况还是非常有效的（Marcy et al. 2007）。单细胞策略是为了从环境中分离单个未培养微生物，随后描述单个微生物的遗传特征。单细胞基因组学与宏基因组学是评估未培养物的互补方案，下文中介绍的一些技术可同时应用在这两方面。

2.3 宏基因组分析的技术挑战

2.3.1 宏基因组研究策略

宏基因组文库的构建与分析可遵循一个常规流程（图 2. 1）。基因组 DNA 可直接从环境样品或富集后提取，为了防止酚类或金属离子污染、干扰后续的酶反应，DNA 通常需要进一步纯化。如果 DNA 的量不足以进行后续的操作，可用 MDA 法增加 DNA 总量（Binga et al. 2008）。随后，DNA 被克隆到不

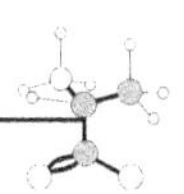

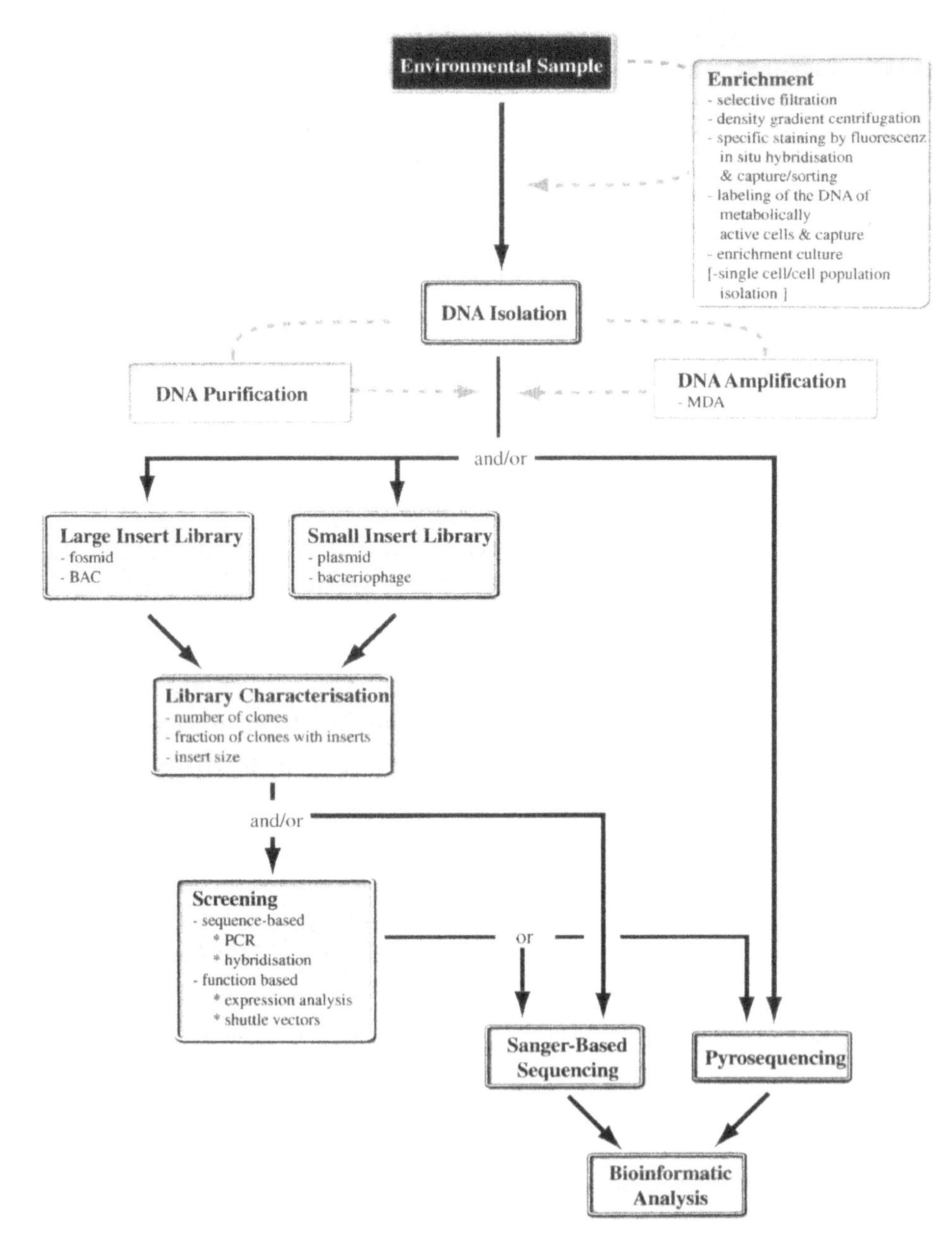

Fig. 2.1 Metagenomic workflow. Steps indicated by dashed lines are optional

图2.1　宏基因组学研究流程图（虚线所示为可选择的步骤）。从环境样品（或经过富集）分离DNA，可以经过DNA纯化或MDA基因组扩增之后构建大片段或者小片段插入文库，经过鉴定、筛选进行sanger法测序，也可以不建库直接进行焦磷酸测序，最后进行生物信息分析

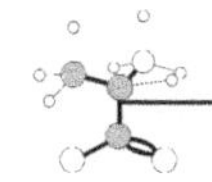

英文注释：Enrichment，富集；DNA Isolation，DNA 提取；DNA Purification，DNA 纯化；DNA Amplification，DNA 扩增；Library Characterisation，文库描述；Bioinformatic Analysis，生物信息分析

同容量的载体，大片段载体 BAC、F 粘粒、Cosmid 或小片段载体 plamid，也可不构建文库而直接测序，如焦磷酸测序法。如果构建了文库，通常建议做单克隆的微阵列，以及进行文库特征鉴定，包括重组克隆数，平均插入片段大小等。文库可直接用来进行插入片段的末端测序或筛选目的 DNA 序列，也可以表达蛋白质。如果不同的克隆携带同一个目的基因，可将其全部测序，随后进行生物信息分析，包括对同种或同类的序列进行合并、contig 组装、基因的功能注释以及特定代谢功能的分析。

有多种微生物学和分子生物学工具可供生态学家应用于宏基因组分析，工具的选择取决于要解决的问题。一般来说需要重点关注的是所研究的栖息地中微生物群落的多样性。同时还需要确定的是关注整个群落，还是其中一个群体的遗传特性，以此决定是否进行细胞或 DNA 富集。此外，为达到研究目标，序列的质量要求、测序长度和数据产量等都需要确定。

有多种方法可以来评估一个复杂微生物群落的遗传组成。通过鸟枪法测序、小片段插入文库（Venter et al. 2004，Tringe et al. 2005，Rusch et al. 2007）或焦磷酸测序（Angly et al. 2006，Edwards et al. 2006，Biddle et al. 2008，Dinsdale et al. 2008）即可获得简要的遗传信息，并不包括基因在基因组上的顺序和系统发育关系。如果需要进一步了解这些信息，可选用末端测序或筛选大片段插入文库的方法（DeLong et al. 2006），进一步对携带目的基因的克隆子进行测序。

在研究复杂群落中特定物种的遗传功能时，可采取对目标细胞富集的策略以减少筛选与测序的工作量。富集成功后，再采用鸟枪法或焦磷酸法测序和组装。如果仅仅只有少量高纯度的目标细胞，可引入 DNA 扩增步骤（Mussmann et al. 2007，Podar et al. 2007）。如果没有合适的方法进行富集，可尝试构建大片段插入文库，以便可靠地评估序列的系统发育关系，从中筛选出目标微生物的序列（Beja et al. 2000，Krüger et al. 2003，Teeling et al. 2004，Bryant et al. 2007）。综上所述，有效结合各种分析工具，将大大地节约时间和金钱成本。

2.3.2 富集策略

在开展宏基因组分析之前，可以用多种方法来富集微生物群落的亚群体。

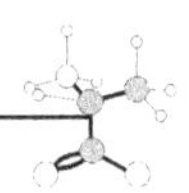

浮游生物样品可通过不同孔径的滤器进行连续过滤将样品分级分离。真核浮游生物和颗粒可通过20～0.8 μm孔径的网筛或滤膜去除（Schmidt et al. 1991，Stein et al. 1996，Beja et al. 2000，de la Torre et al. 2003，Lopez-Garcia et al. 2004，Venter et al. 2004，Angly et al. 2006，Culley et al. 2006，DeLong et al. 2006，Martín-Cuadrado et al. 2007，Rusch et al. 2007）。滤液可进一步细分（Rusch et al. 2007），或用0.22 μm孔径的滤膜来收集原核生物，或用离心法分离（Schmidt et al. 1991，Beja et al. 2000）。针对病毒宏基因组学研究，原核细胞可用0.22 μm孔径的滤膜去除（Angly et al. 2006，Culley et al. 2006，Rusch et al. 2007）。寄生于海洋无脊椎动物的原核细胞可通过密度梯度离心从宿主细胞里分离出来（Hughes et al. 1997，Schirmer et al. 2005，Woyke et al. 2006，Robidart et al. 2008）。海底微生物通过超声破碎法从沉积物中脱离，再经过密度梯度离心分开，富集后过滤获得微生物细胞（Schleper et al. 1998，Hallam et al. 2003）。除此之外，我们甚至可能手动分离丝状细胞，此法已用在多细胞贝氏硫细菌（*Beggiatoa*）的细胞分离（Mussmann et al. 2007）。如果目标DNA与大部分DNA的GC含量有显著不同，根据目标细胞的GC含量富集DNA也是可行的方法之一。如果必须从一个群落中富集亚群体，还可采用根据细胞壁成分差异或生化特性不同来选择性地富集细胞。

细胞群体亦可以用16S rRNA探针进行原位杂交。探针可被标记（如生物素），杂交之后目标细胞被链霉亲和素包被的磁珠捕获（Stoffels et al. 1999）。另一种微孔板富集方法也曾被使用（Zwirglmaier et al. 2004）。一种被称为磁力荧光原位杂交（Magneto-FISH）的改进方法，结合了催化受体沉积荧光原位杂交（catalysed-reporter deposition uorescence in situ hybridisation，CARD-FISH）与磁珠捕获，这两种方法最近被用来富集沉积物中嗜甲烷菌类微生物（Pernthaler et al. 2008）。细胞群体被FISH染色后，由流式细胞仪分选（Podar et al. 2007）。

除了基于物理学与分子生物学特征的方法富集细胞外，细胞群体也可以根据代谢特征进行分离。微生物群落中的活性组分DNA可经过短时间孵育被5-溴-2-脱氧尿苷（BrdU）标记。BrdU是一类胸苷类似物，被加入到新复制细胞的DNA中，标记的细胞可在免疫染色和流式细胞分选的时候显示出来。标记DNA也可被磁珠偶联的BrdU特异的单克隆抗体捕获（Urbach et al. 1999，Mou et al. 2008），此方法可标记对特定刺激做出响应的生长细胞（Borneman 1999）。微生物群落还可用稳定同位素标记的底物培养（Dumont，

Murrell 2005，Dumont et al. 2006，Neufeld et al. 2008）。在一项研究中，海洋表层水用^{13}C标记的甲醇培养，从而使能够代谢甲醇的微生物中含有^{13}C。分子量较大的DNA与较小的DNA通过氯化铯密度梯度分离开来，可用于利用甲醇微生物的遗传功能宏基因组分析（Neufeld et al. 2008）。

另一种增加目标微生物基因片段的方法是经典微生物学技术。产甲烷水稻cluster I细菌的环状基因组（Erkel et al. 2006）与不可培养的厌氧氨氧化菌*Kuenenia stuttgartiensis*（Strous et al. 2006）的全基因组都用了经典方法从宏基因组数据中分离得到，结合使用了鸟枪法测序，F粘粒和BAC文库构建、筛选和测序等方法。

研究未培养原核生物的趋势为单细胞基因组学。在一项研究中，海洋浮游的单细胞是由高速荧光激活细胞分选法（fluorescence activated cell sorting，FACS）通过96孔板分选出来（Stepanauskas and Sieracki 2007）。此外，人们开发了微流控设备通过稀释（Ottesen et al. 2006）或单个细胞的标记（Marcy et al. 2007）进行单细胞分离。显微操作同样适用于分离FISH染色的细胞（Ishoy et al. 2006，Kvist et al. 2007）。应用激光钳（Huber et al. 1995）和激光捕获显微解剖（Gloess et al. 2008，Thornhill et al. 2008）技术进行微操作，也是从环境样品中分离单细胞的可选工具。

2.3.3 基因组DNA的分离纯化

分离宏基因组DNA的方法很大程度上取决于下游的分析策略。在不需要文库构建或用鸟枪法构建小片段/大片段文库时，可采取“基于液相”的方法提取宏基因组DNA，即环境样品直接与裂解缓冲液混合，能得到最大长度的150～200 bp的片段。如果需要构建平均插入长度超过50 kbp（Rondon et al. 2000，MacNeil et al. 2001）的BAC文库，此方法并不适合。为了获取大片段DNA进行文库构建，环境样品需在裂解之前嵌入到一种基质中（如琼脂），用来避免对DNA的过度剪切。

液相提取方法是最直接的DNA分离技术，几乎适用于每种环境样品。有商业DNA分离试剂盒可直接使用。然而如果用基于二氧化硅的分离试剂盒，可导致平均片段长度变小、所获得的DNA量较少。如果需要DNA大片段来构建大片段文库，可尝试由Zhou et al.（1996）发表的一种方法，细胞在含有蛋白酶K、CTAB、SDS的高浓度盐缓冲液中裂解，革兰氏阳性菌可通过几次反复冻融促进裂解。从浮游生物中提取的DNA可直接用于下游测序或文库构建，一些其他样品的DNA（如沉积物）则需要进一步纯化才可用于下游的酶

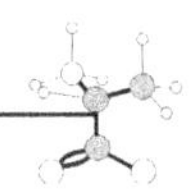

反应操作。纯化可用凝胶电泳（Rondon et al. 2000，Quaiser et al. 2003）、离子交换层析（Krüger et al. 2003）、密度梯度离心（MacNeil et al. 2001，Courtois et al. 2003）等方法或用商业试剂盒完成。

提取片段大于200 kbp 的高质量 DNA 会更加困难。100 kbp 以上的片段容易在移液或样品混匀时被剪切，因此需要把样本包埋在琼脂中，随后用酶和去污剂在不同缓冲液下裂解，去除细胞中的蛋白质和脂类，仅剩下 DNA（Green et al. 1997，Sambrook and Russel 2001）。经过这些处理，浮游生物样本的 DNA 可直接用来构建 BAC 文库（Beja et al. 2000）。但这种方法用在其他污染严重的环境样品时（如沉积物样品），会导致琼脂中带入部分降解的基因组和酶抑制剂，无法在裂解过程中去除。这时需要进一步的纯化，如常规琼脂糖电泳或聚乙烯吡咯烷酮双向琼脂糖电泳（Quaiser et al. 2002），或者在含甲酰胺的高盐浓度中透析（Liles et al. 2008）。然而，上述流程常常导致 DNA 量的减少和部分剪切，有可能导致后续构建大片段 BAC 文库的失败。

2.3.4 基因组 DNA 的扩增

为了使结果具有代表性，宏基因组分析通常需要微克量级的基因组 DNA（见下文）。理论上，如果群落多态性低或者采取减少多样性的策略来富集目的细胞，较少量的 DNA 也足够开展宏基因组分析。微克以下的 DNA 不足以完成标准的宏基因组文库构建与测序。受条件所限有些情况下无法获得足量 DNA，如深海样本（Webster et al. 2003，Biddle et al. 2008），这些微量的起始 DNA 在分析之前需要被扩增。为解决这个问题科学家们开发了不同的基因组 DNA 扩增方法（Telenius et al. 1992，Zhang et al. 1992，Breitbart et al. 2002，Breitbart，Rohwer 2005，Pinard et al. 2006），其中采用噬菌体 ϕ29 聚合酶进行的多重置换扩增法（MDA）得到了广泛应用。

噬菌体 ϕ29DNA 聚合酶是从枯草芽孢杆菌中发现的，是这个19285 bp 的噬菌体基因组复制时所需的酶（Blanco，Salas 1985b）。除此之外，这个66520 Da 的单肽（Blanco，Salas 1984，Watabe et al. 1984）具备 3′→5′核酸外切酶活性（Blanco and Salas 1985a）。ϕ29DNA 聚合酶的错误率为 10^{-5}～10^{-7}，比 Taq DNA 聚合酶低约 10～100 倍，而且持续合成能力很高（Eckert and Kunkel 1990）。由蛋白质引发的噬菌体 ϕ29 基因组 DNA 复制过程需要8 分钟（Blanco et al. 1989）。ϕ29 聚合酶同样能够延伸 DNA 引发的反应，这归功于 ϕ29 聚合酶的链置换能力，单链环状 M13 基因组有7250个碱基，在等温条件下完成复

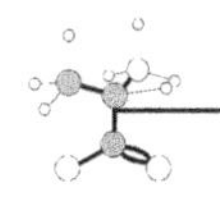

制需要 5 分钟，40 分钟后可产生大于 70 kbp 的长链（Blanco et al. 1989）。凭借 φ29 聚合酶的链置换能力，新合成 DNA 的 5′端可脱离模板，进入滚环复制（rolling circle ampli cation，RCA）（Fire and Xu 1995，Liu et al. 1996）。

在宏基因组分析中，φ29 聚合酶以六碱基随机引物（N6）进行基因组扩增，同时进行 3′末端修饰，防止被外切酶作用，扩增反应呈指数增长，最后可获得的 DNA 是起始核酸量的10000倍（Dean et al. 2001）甚至更多（Dean et al. 2002）。N6 引物结合到模板上某些位点，由 φ29 聚合酶延伸，如果反应链到达下一个六聚体结合的 5′端，那么之前合成的 DNA 链将被替换，被替换的单链又重新结合到六聚体上进行延伸反应，最终形成扩增产物（图 2.2）。该方法缺点在于嵌合体的形成、扩增偏好性和非特异性 DNA 的扩增。MDA 反应时嵌合体形成伴随着每 10 – 22 kb 的 MDA 产物发生一次重排的频率（Zhang et al. 2006，Lasken and Stockwell 2007），其中 85% 的上述情况可观察到反向序列（Lasken and Stockwell 2007）。在 MDA 文库中，嵌合体产生的频率可通过以下措施的结合使用来降低：φ29 聚合酶去分支，S1 核酸酶水解单链区域，用 DNA 聚合酶 I 进行缺口翻译。然而，在 3 kbp 左右的小片段文库中始终会存在 6%～8% 的嵌合体插入，增加测序量和反复多次组装可在一定程度上解决这个问题（Zhang et al. 2006）。第二个问题是在 MDA 反应中观察到扩增偏好性（Dean et al. 2002，Hosono et al. 2003，Abulencia et al. 2006，Yokouchi et al. 2006），在扩增几种微生物或单细胞时尤为严重（Raghunathan et al. 2005，Zhang et al. 2006，Kvist et al. 2007，Podar et al. 2007）。因为这个偏好的产生不依赖于后续的测序，所以深度测序以及将相同样品的多个 MDA 产物结合起来分析可能会解决该问题（Raghunathan et al. 2005，Abulencia et al. 2006，Zhang et al. 2006）。在样品量较少时容易产生非特异扩增，可能是引物二聚体或者痕量 DNA 污染的存在（Raghunathan et al. 2005，Zhang et al. 2006）。在一些研究中通过从环境样品中分离单细胞，将 MDA 用于单个未培养微生物的全基因组扩增（WGA；Zhang et al. 2006，Kvist et al. 2007，Marcy et al. 2007，Stepanauskas and Sieracki 2007）。

对一份含有 8 种细菌且基因组大小与 GC 含量均不同的混合样品进行 MDA 扩增，结果显示对特定的菌株有明显的偏好（Abulencia et al. 2006）。在一项污染土壤中低生物量样品的研究中，通过 MDA 法处理群落 DNA 后构建 16S RNA 文库，与未经 MDA 得到的结果相比，得到的微生物类群只有微小的差异，一些物种的相对丰度有所增加（Abulencia et al. 2006）。

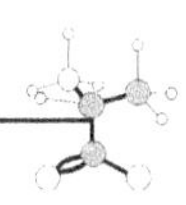

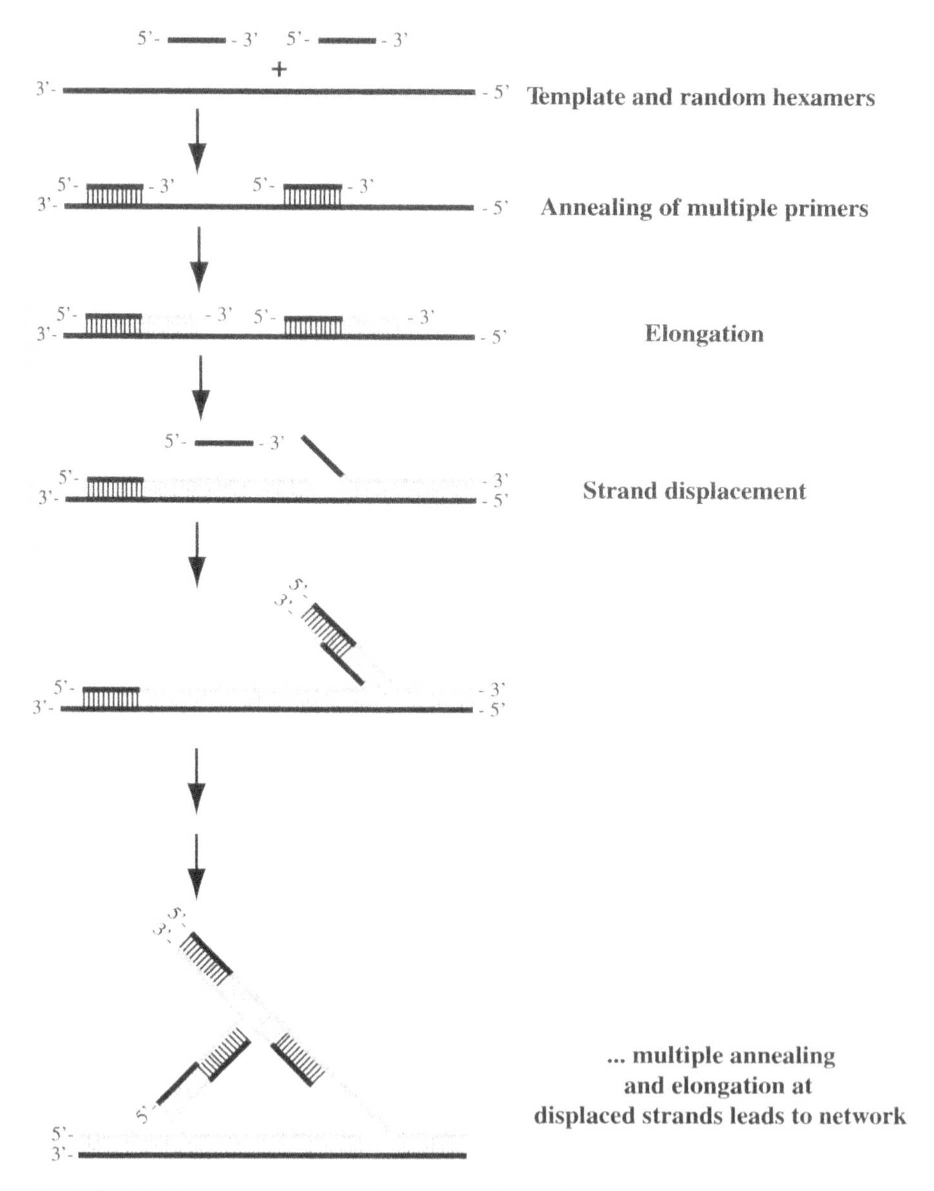

Fig. 2.2 Multiple displacement amplification of DNA. Template and random hexamers are indicated in black. Newly synthesised DNA is coloured in grey (after Binga et al. 2008)

图 2.2 DNA 的多重置换扩增。黑色：模板和随机六碱基引物；灰色：新合成的 DNA（从 Binga et al. 2008）。模板与六聚体经过多重引物，延长和链的置换，在置换链上进行多步退火和延伸，从而形成链的网络

英文注释：Template，模板；Random hexamers，随机六碱基引物；Annealing，退火；Elongation，延长

另一项深海微生物群落调查研究结果显示，经过 MDA 扩增与未经扩增的样品在系统发生和功能类别分布的结果相似。然而，16S rRNA 基因和核糖体蛋白质的系统发生分布在该研究中差异显著，原因可能是数据库中的可参考基因数太少（Biddle et al. 2008）。用变性梯度凝胶电泳（DGGE）来评价 MDA 的偏好性，结果显示，当 MDA 反应的起始 DNA 量小于 1 ng 时可产生轻度的扩增偏好（Neufeld et al. 2008）。MDA 法扩增 DNA 已成功地将稳定同位素探针实验得到的 DNA 用在 F 粘粒文库构建。在插入片段末端和中间观察到嵌合体时，有必要对其基因顺序进行 PCR 验证。尽管如此，这种方法标志着我们在评估遗传信息（包括海洋表层水域未培养活性微生物的完整操纵子结构）方面，前进了重要的一步（Neufeld et al. 2008）。

在某些其他研究中，宏基因组 DNA 的扩增是研究未培养海洋微生物的先决条件，包括海洋病毒研究（Angly et al. 2006）、两种未培养海洋大型的细胞丝状硫细菌的遗传信息分析（Mussmann et al. 2007）以及由原位杂交和磁珠捕获法分离的海洋嗜甲烷菌群落的分析（Pernthaler et al. 2008）。

2.3.5 宏基因组文库的构建和分析

2.3.5.1 小片段插入的宏基因组文库

以质粒或 λ 噬菌体为载体的小片段插入文库（通常 1.5 ～ 3 kbp 的片段）是宏基因组的信息资源库，可以通过测序或功能实验筛选。小片段文库（Cottrell et al. 1999，Henne et al. 1999）的早期研究集中在未培养群落中的单个代谢相关基因。带有了报告基因的小片段文库可用于底物诱导的基因表达筛选时鉴定新的分解代谢相关的操纵子（Uchiyama et al. 2005）。小文库构建是多数研究中基于 Sanger 法进行宏基因组测序的先决条件。小片段文库的高通量测序是获得基因组框架（例如 Tyson et al. 2004，Venter et al. 2004，Woyke et al. 2006）和评估复杂环境中系统发生与代谢多样性的强大工具（Venter et al. 2004，Rusch et al. 2007）。小片段文库的构建流程已经明确，然而在基于序列的宏基因组学研究时，还存在一些大的弊端。首先，一个系统发生标记可以对隶属于特定系统发生类群的片段进行可靠评估，但是在小片段文库中的标记常常丢失。如果数据量不足或样品多样性高（如土壤；Handelsman et al. 2002，Tringe et al. 2005），那么对特定物种的基因组框架图的重构建，甚至遗传功能的评估几乎是不可能。并且较大片段的电脑模拟组装存在嵌合体的风险，尤其重组装的重叠片段太短或存在重复序列时。

2.3.5.2 大片段插入的宏基因组文库

宏基因组大片段文库（ >30 kbp）的优势在于通常可以归类到特定物种，

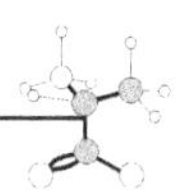

可根据在相同DNA片段上系统发生标记（如16S rRNA基因）进行物种的分选（Schleper et al. 1997，Beja et al. 2002）。并且可尝试通过核酸频率（Krüger et al. 2003，Teeling et al. 2004）将大片段组装到含有系统发生标记的参考序列上。大片段文库可得到完整的操纵子（Krüger et al. 2003）或基因岛（Schübbe et al. 2003，Muβmann et al. 2005），形成嵌合体风险低，且测序量适中合理，因此大片段文库在早期研究中受到青睐。改良的单细胞测序技术降低每个碱基的测序成本，转移了近些年来的关注焦点。但大片段插入文库在未来的宏基因组研究中是否不可替代，仍然是个未知数。

大片段文库可克隆到不同载体中（Green et al. 1997，Tao and Zhang 1998），其中有三种载体常用来构建宏基因组文库：cosmid、F粘粒 和细菌人工染色体（BAC）。Cosmid载体为传统质粒，含有1至2个噬菌体λ *cos*整合位点（Collins，Hohn 1978），这使λ噬菌体进行体外DNA包装成为可能。克隆到cosmid质粒的DNA片段平均长度为30～45 kbp之间，具体取决于载体的大小。载体加上插入片段的大小需为平均λ噬菌体基因组的78%～105%，便于DNA包装进入噬菌体头部（Sambrook and Russel 2001）。一旦进入宿主细胞，cosmids便呈现高拷贝，这方便了进一步的文库筛选（Collins，Hohn 1978，Sambrook，Russel 2001）。Cosmids已经用于有些宏基因组文库的构建（Entcheva et al. 2001，Piel 2002，Courtois et al. 2003，Schmeisser et al. 2003，Sebat et al. 2003，Lopez-Garcia et al. 2004），然而这个系统存在两个严重不足：第一，可造成嵌合体、重排或缺失（Monaco and Larin 1994）；第二，插入片段大小基本上是一致的，但是相对于一个完整基因组来说略小。

Fosmid载体（Kim et al. 1992）在目前宏基因组分析中应用最广泛（例如Stein et al. 1996，Schleper et al. 1997，1998，Beja et al. 2002，Quaiser et al. 2002，2003，Meyerdierks et al. 2005，DeLong et al. 2006，Neufeld et al. 2008）。Fosmid克隆同样具有λ噬菌体体外包装的优势，插入片段大小与cosmid载体相似。然而相对于cosmid载体来说，F粘粒载体来源为*E. coli*的致育因子（F因子），携带了F因子复制和分区相关的序列。结果每个细胞中仅有1～2个F粘粒拷贝，并且而在一个细胞里无法维持2个不同的F粘粒。因为拷贝数较少，嵌合体的频率也较低，因此可导致宿主死亡的有毒基因表达量会减少。为了便于筛选，携带了第二个诱导复制子的改良F粘粒载体被构建了（Wild et al. 2002）。这使文库维持在低拷贝，同时又能选择性诱导至每个细胞10～50个拷贝，以利于进一步分析。这些改良的F粘粒克隆使筛选更清晰明确，这也许是这些载体成功用于宏基因组分析的主要原因。

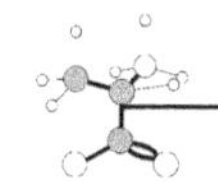

BAC 载体（Shizuya et al. 1992）克服了 F 粘粒固有的 λ 噬菌体包装系统插入片段大小的限制。BAC 载体与 F 粘粒同样携带大肠杆菌的 F 因子与分割序列，而且可用于二次诱导复制（Handelsman et al. 2002，Wild et al. 2002）。两种克隆系统的主要区别在于诱导重组 DNA 进入宿主的方法。相对于 F 粘粒克隆，BAC 文库构建应用电穿孔技术，避免 λ 噬菌体包装系统对片段大小的限制。特定的大肠杆菌用于转化这种带有长 DNA 插入片段的载体时效果很好（Sheng et al. 1995）。大于 300 kbp 的 DNA 片段可以克隆到 BAC 载体，并在大肠杆菌中稳定存在（Shizuya et al. 1992，Kim et al. 1996，Zimmer，Verrinder 1997）。例如，Bryant 等用系统发生标记基因和光营养基因分析了一个 271 kbp 的光营养群落宏基因组 BAC 克隆（Bryant et al. 2007）。然而 BAC 克隆比 F 粘粒克隆的效率低 100 ～ 1000 倍，原因是电穿孔比体外包装的转化效率低，并且 BAC 文库的平均长度与产生的 BAC 克隆数成负相关（Leonardo and Sedivy 1990，Woo et al. 1994，Sheng et al. 1995，Zimmer and Verrinder 1997）。对于宏基因组工作，尤其是研究土壤和海洋沉积物时，BAC 克隆仍然存在挑战，有时构建的宏基因组 BAC 文库的平均长度并不高于 F 粘粒文库（例如 Hughes et al. 1997，Beja et al. 2000，Rondon et al. 2000，MacNeil et al. 2001，de la Torre et al. 2003，Dumont et al. 2006）。最近发表的关于土壤样品 BAC 文库构建的文章指出："……大片段的插入构建在某种程度上讲是一个特殊的过程……"（Liles et al. 2008）。这清楚地指出了 BAC 克隆可能遭遇的困难。然而，插入片段越大，在染色体步移法分析时越省力，尤其对于复杂样品。

对大片段文库构建的技术环节已经被详细描述，例如 Sambrook and Russel（2001）和 Green et al.（1997），以及相关公司克隆试剂盒的操作说明。构建流程如图 2.3 所示。近年来大部分改良方法已被发表，由于篇幅限制在这里不再赘述。但是要强调一个关键步骤，即克隆之前插入 DNA 片段的长度选择，它大大影响了文库的大小和质量：在构建 F 粘粒文库时选择理想的长度，避免了在同一载体上连接两个或多个片段产生的嵌合体。并且，克隆的效率同样受连接产物的影响，片段太长或太短都会导致难以包装到噬菌体头部。在构建 BAC 文库时，小的质粒在电穿孔时会优先导入宿主细胞，这样可能会导致文库的平均长度变小（Osoegawa et al. 1998）。BAC 文库的片段大小选择由脉冲场电泳（PFGE）完成，而不是常规琼脂糖电泳。在常规琼脂糖电泳中电流是恒定，15 ～ 25 kbp 分子的迁移率相差不多。比较而言，PFGE 允许 5 Mbp 以下的 DNA 分子基于大小分离（Sambrook and Russel 2001）。如果凝胶中待分离的 DNA 量大，实施两轮的大小选择更加合适（Osoegawa et al. 1998，

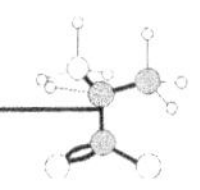

Rondon et al. 2000）。

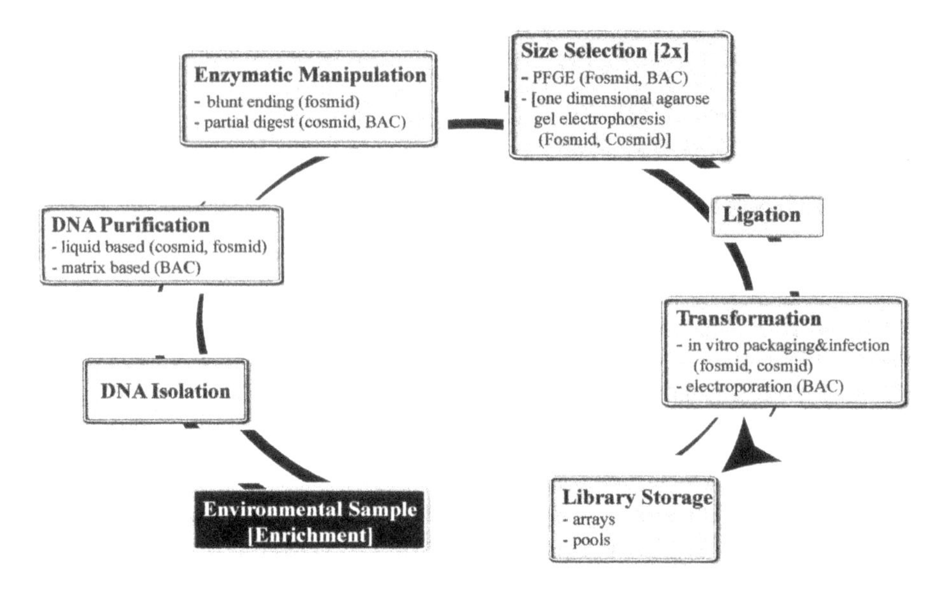

Fig. 2.3 Overview of the construction of large insert metagenomic libraries

图 2.3 大片段宏基因组插入文库构建概述。收集环境样品（或者经过富集）之后，进行 DNA 分离、纯化、酶处理、片段选择、连接和转化等步骤，最后将文库贮存

英文注释：Isolation，分离；Purification，纯化；Enzymatic Manipulation，酶处理；Ligation，连接；Transformation，转化；library，文库；Storage，贮存

2.3.5.3 宏基因组文库的大小

对于纯培养群落的文库大小，它覆盖基因组的概率定义为 $N = [\ln(1-P)]/[(\ln(1-f))]$，其中 P 为期望概率，f 是插入片段与基因组长度之比（即平均插入长度/基因组长度；Sambrook and Russel 2001）。

在宏基因组分析中，需要考虑一些额外因素。宏基因组文库的大小需要代表一个环境样品，或者从统计学的角度来讲，一个克隆携带一个目的标记基因，而文库的大小计算是很困难的。同时需要考虑的还包括群落的多样性、组成以及样品中不同微生物的基因组大小。通过对 NCBI 主页（September 2009；http://www.ncbi.nlm.nih.gov/）列出的 815 种细菌和古菌进行统计，得出微生物平均基因组大小为 3.2 Mbp。事实上基因组大小变化很大，当前两个极端的例子，一个是 *Candidatus* Carsonella ruddii（159 662 bp），属于内共生的 γ 变形菌门，另外一个是 *Sorangium cellulosum* "So ce 56"（13 Mbp），是土壤中的粘细菌。此外，复杂样品中的细胞不完全裂解会导致偏好性产生；重

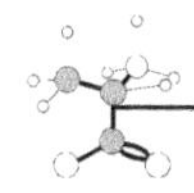

复序列与异常 GC 含量（Abulencia et al. 2006）以及存在有毒基因产物时（如噬菌体基因；Edwards and Rohwer 2005）都会导致偏好性增高。

2.3.5.4 宏基因组文库的保存

宏基因组文库可在不同阶段贮存。如果文库不能立即涂平板，在连接或包装阶段的噬菌体头部，可贮存于 -80℃。在深度冷冻之前，噬菌体溶液需存于 DMSO（Sambrook and Russel 2001，Promega Corporation 2007）或甘油（Epicentre Biotechnologies 2007）中。重组细胞转化后，也要保存于甘油中并立即 -80℃保存。

用来长期保存涂板文库的方法取决于许多因素，诸如样品的均一性、文库的大小、以及下一步筛选方案的选择。一般推荐保存多个文库的拷贝。文库可以用含有低温防冻剂（例如 5%～7% 甘油）的培养基从琼脂板上洗脱，以重组克隆的混合体的方式存放在 -80℃。如果将来有测序插入末端序列的计划，一般会实行单克隆的挑选和阵列式保存。

2.3.5.5 宏基因组文库的筛选

宏基因组文库可以基于碱基序列或功能进行筛选。

基于序列筛选的策略包括 PCR 筛选、杂交和插入片段测序。当有 DNA 库或重组克隆库时，PCR 筛选是最快捷的文库筛选方法。已经有文章发表了相关的方法（Kim et al. 1996，Asakawa et al. 1997）。一些策略可以避免由引物交叉反应导致的宿主染色体 DNA 的扩增。有研究指出，在 PCR 筛选之前将宿主细胞通过 ATP 依赖的 DNase 处理可将宿主染色体 DNA 选择性水解（Bega et al. 2000，Liles et al. 2003），低宿主专一性且末端修饰的寡核苷酸也加入到 PCR 反应中（Goodman，Liles 2001，Liles et al. 2003）。另外一项研究结合了 PCR 和限制性片段长度多态性（RFLP）分析，目的是挑选阳性克隆（Liles et al. 2003）。杂交筛选通常选用的方法为群落印迹（Asakawa et al. 1997，Osoegawa et al. 2000））和高密度 DNA 阵列（Rondon et al. 1999）与微阵列（Park et al. 2008）。最后，通过插入末端测序可以找到特定的基因（Kube et al. 2005）或染色体步移法的重叠克隆（Meyerdierks et al. 2005）也可以描述和比较微生物群落（DeLong et al. 2006）。

功能筛选是检测克隆基因在宿主细胞中的异源表达。这种方法主要用于鉴定具有生物技术用途的酶类。如果宿主与供体菌株的转录和翻译机制不兼容，也可用穿梭载体将文库转移到一个不同宿主菌株中（Handelsman et al. 2002，Riesenfeld et al. 2004）。

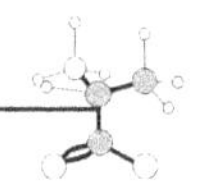

2.3.6 不依赖文库的宏基因组分析

基于双脱氧核苷酸的DNA测序方法是数十年来DNA测序的标准方法。此方法由Sanger（1977）等首先提出，后来被进一步改进，包括以荧光代替放射性标记的双脱氧核苷酸作为链合成终止物（Prober et al. 1987）。此方法进行宏基因组测序之前需要构建鸟枪法文库。对小片段鸟枪文库进行批量Sanger测序的最成功案例是马尾藻海的表层水宏基因组测序（Venter et al. 2004）和随之的全球海洋调查（Rusch et al. 2007，Yooseph et al. 2007）。

21世纪初，一些新的测序方法变得可行。这些技术是高通量、快速、经济的DNA测序，并且不需要构建鸟枪文库（Mardis 2008）。这些技术中，由454 Life Science开发的焦磷酸测序技术（Margulies et al. 2005），在宏基因组分析中应用最广泛。焦磷酸测序基于边测序边合成技术。片段化DNA（300～800 bp）为平末端，有两种接头，其中一个生物当标记的接头被连接到片段上。两侧通过带有相同接头的片段在链霉亲和素与生物素下移除。变性后的单链DNA片段被稀释，一个片段只结合一个DNA捕获磁珠。随后，DNA片段采用接头区的特异引物，通过微乳液PCR使各反应平行独立进行。有40万（GS FLX）到100万（GS FLX Titanium）的磁珠携带扩增片段放入PTP（Pico Titer Plate）板中，一个孔只含一个磁珠。此时按照测序仪的流体子系统按固定的顺序加入测序酶。反应体系把单核苷酸加入孔里包括测序引物、DNA模板、DNA聚合酶、ATP硫酸化酶、荧光素酶和腺苷三磷酸双磷酸酶、底物、腺苷5′-磷酸化硫酸（APS）和荧光素。通过DNA聚合酶催化，引物结合dNTP后延长，同时释放等摩尔的焦磷酸（PPi）。如果目的DNA上有连续相同碱基，可能会整合不止一个相同核苷酸，所以等摩尔的PPi释放非常重要。在APS存在时，ATP硫酸化酶将PPi转移到ATP上，在荧光素酶作用下，ATP使荧光素氧化成为氧化荧光素，因此产生与ATP合成成正比的可见光。采用CCD摄像机检测到光信号，表现为pyrogram™的峰，每个光信号与加入的核苷酸成正比例。在下一个dNTP加入之前，每个循环通过腺苷三磷酸双磷酸酶来降解未结合的dNTP和过剩的ATP。

焦磷酸测序有许多优势，高通量测序将产生海量的数据信息。此方法比Sanger法快100倍（Rogers and Venter 2005），每个碱基的成本也较低（Wheeler et al. 2008），而且没有克隆偏好性，对常见于基因组中的可造成测序终止的高GC序列不敏感（Goldberg et al. 2006，Wheeler et al. 2008）。与Sanger法相比，454测序仪一个明显的劣势是读序列的长度比较短。然而第一代454 GS20测序仪的100 bp读长已经足够支持基因组测序项目（Goldberg et al. 2006）和重测序项目（Margulies et al. 2005）。454测序仪已经成功应用于一个未培养的TM7细胞（Marcy et al. 2007）和大约600个未培养海洋贝氏硫细菌 *Beggiatoa* 细胞丝（Mussmann et al. 2007）的新陈代谢组分析研究。焦磷酸测序使微生物群落的代谢分析前进了一大步，同时加速了不同环境中的微生物群落差异研究（Angly et al. 2006，Edwards et al. 2006，Biddle et al. 2008，Dinsdale et al. 2008）。然而，由于读长短，在鉴定基因的系统发育关系、合并相同或相近微生物种类序列、分析群落中特定微生物的功能时，存在一定的困难（Frias-Lopez et al. 2008）。随着454 GS FLX系统的读长提升至200～300 bp，新的454 GS FLX Titanium系统的读长可超过400 bp。另外的一个技术问题是，由于连续相同碱基在模板上延伸造成了高错误率（Margulies et al. 2005，Huse et al. 2007，Wheeler et al. 2008）。但是，从数据中严格地过滤低质量序列可以很大程度上弥补上述不足（Margulies et al. 2005，Huse et al. 2007，Wheeler et al. 2008）。

2.4 生物信息学在宏基因组分析时面临的挑战

对于完整或框架基因组测序，宏基因组数据分析的流程都已建立（Glöckner and Meyerdierks 2006，Stothard and Wishart 2006）。无论是大片段还是小片段的克隆文库，在得到原始序列后，组装通常是数据处理的第一步。如果长的contig（连续的序列）或者scaffold（一些包含了测序缺口的contig）能够成功地确定，基因的查找和注释可深入了解样本的系统发育和功能的多样性。为了得到样品的功能和代谢能力的整体信息，可以在Subsystems（数据系统）、KEGG（Kanehisa et al. 2004）和COG（Tatusov et al. 1997）数据库中搜索那些蛋白质的编码基因的信息。比较宏基因组（Tringe et al. 2005，DeLong et al. 2006）和功能宏基因组技术，如宏转录组学（Poretsky et al. 2005，Frias-Lopez et al. 2008）和宏蛋白质组学（Ram et al. 2005，Wilmes and Bond 2006），都是目前新兴的技术，用于动态地了解差异和生物体对环境的适应性。

尽管宏基因组原始序列数据的处理看起来比较简单，但实际操作时并非如此，这个过程类似于用上百万具有相似颜色和材质的碎片构建一个拼图。在研究多样性高的环境（如海洋生态系统），尤其数据产生覆盖度低、序列读长短时，都会给分析带来困难。由于宏基因组数据分析还处于起步发展阶段，

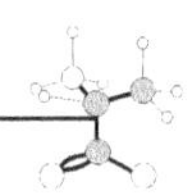

所以会遇到这类问题。为了解决这些问题，宏基因组数据的生物信息分析可两种情况，分析组装后的宏基因数据成直接分析高通量测序产生的短序列。

2.4.1 数据组装与合并

为了将片段组装成长度几千个碱基的 scaffold 和 contig，测序大片段插入文库 BAC 或 F 粘粒是最佳选择，这些文库通常用前文所述的鸟枪法进行测序(Sambrook and Russel 2001)。一个 40 kbp 的 F 粘粒克隆要得到覆盖度为 8 次的数据量，就需要 400 个反应。因为所有序列读长相近，因此易于用标准化程序组装，如 Phrap（www.phrap.com)、JAZZ（Aparicio et al. 2002)、Arachne 2（Jaffe et al. 2003）或 Celera Assembler（http://sourceforge.net/projects/wgs-assembler)。如果不能进行大量测序，那么就需要使用系统发育或代谢标记基因进行筛选，或通过末端随机测序来确定 BAC 或 F 粘粒文库中的目标克隆再进行测序。

如果只用鸟枪法来做宏基因组测序，情况会立即变得更复杂。此分析方法成功与否依赖于样品的复杂度。在低多样性环境中，仅含有少量优势种（5～10 个)，可能获得高质量数据集，在合适的测序量时，scaffolds 长度有可能达到几 Mb（Tyson et al. 2004，Martin et al. 2006，Woyke et al. 2006)。然而，高复杂度的样本会带来一些问题，即使是含有相近的物种，另外，如转座子、病毒、插入噬菌体等普遍存在的基因组元件也会导致组装的困难。Mavromatis et al.(2007）在最近的一项研究中指出了尤其对于 8 kbp 以下的测序片段嵌合 contig 形成的几率增加。总的来说，目前没有一个组装程序是专为宏基因组而设计的。特别是越来越多的改进比对程序如 Phrap 或 JAZZ，试图合并尽可能多的序列，产生大量的 contig，这是单基因组组装的主要目的。对于宏基因组，则应该使用比较相对保守的方法如 Arachne assembler，以降低组装错误。

为了在 contig 数量与嵌合体形成几率两者之间保持一个最优的平衡，可采用“框并 - 组装”的多次循环。

框并法（binning）所描述的是不同长度的测序片段聚类为“框”（bin）的过程。每个 bin 可能代表一个单独的生物。近几年提出一些合并的方法，最简单的是基于 BLAST 的最好结果，只将现有的系统发育标记基因，或者所有 contig 上的基因比对到相应的物种上（Treusch et al. 2004，Huson et al. 2007)。比较复杂的方法是利用基因组自身的特征，基于寡核苷酸频率来分析片段相关性，甚至系统发育聚类。许多研究表明，DNA 序列中寡核苷酸的频

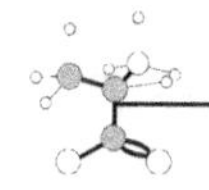

率呈现出物种特异性模式（Karlin et al. 1998）；四核苷酸的频率甚至代表了一种固有的但是比较微弱的系统发生信号（Pride et al. 2003）。目前可用的技术可分为分类法和聚类法，分类法是基于监督的方法，而聚类则是无监督（McHardy and Rigoutsos 2007）。这里监督的意思是测序片段基于自身特征划分到一种或多种类群，而这些类群是先前有研究过的模式生物（例如所有细菌基因组）。最近的例子有 Naive Bayesian Classifier（Sandberg et al. 2001）和 Phylophytia（McHardy et al. 2007）。这些方法的主要优势是，将这些宏基因组片段序列加入训练集分析时，可直接得到相关的物种分类信息。

基于监督的分类其主要缺点是训练集中有明确分类信息的序列是有限的。只有在训练集中包含近缘物种的基因组信息时，才能得到较好的分类结果（Mavromatis et al. 2007，McHardy et al. 2007，Warnecke et al. 2007）。而在分析环境样本时，通常并非如此，所以会存在一些问题。

无监督的聚类，如 TETRA（Teeling et al. 2004）或 Self Organizing Maps（SOM）（Abe et al. 2005，Abe et al. 2006）不需要训练集，不依据系统发育关系进行聚类。在后续的步骤中，在每个聚类里，至少有一个以上的片段携带了系统发育标记基因，以此把这个聚类比对到相应的物种。这类方法一个明显的优势在于它参考了更多的序列特征（如 GC 含量和序列深度），甚至考虑到寡核苷酸的频率和分布。因此能将宏基因组来自相同物种的序列聚类到一起，哪怕缺乏样本中这些物种的信息甚至没有样品中普遍存在的有机体的相关信息。这类技术的能力最近已经在一些研究中得到证明，如具有甲烷厌氧氧化能力的微生物群落研究（Krüger et al. 2003，Meyerdierks et al. 2005）、贝氏硫细菌单细胞基因组分析（Mussmann et al. 2007），以及海洋寡毛纲蠕虫 *Olavius sp.* 的四种共生微生物的基因组组装（Woyke et al. 2006）。

然而，以上方法都是基于测序数据的统计分析，所以如果序列长度小于 8 kbp 时很难进行。对于组装的数据，通常都没有问题，因为目前标准的组装方法很少有小于 8 kbp 的 Contig 的要求（Lontigs Mavromatis et al. 2007）。对于复杂的环境样本，通常会得到大量的单条序列，可以经过最初一轮的保守组装，生成的 scaffold 与 contig 再用无监督的方法进行聚类，得到初始的结果作为种子进行进一步分析：如把单条序列或双末端配对后的序列进行关联或聚类，或使用训练集如 Phylophytia 来对已得到的聚类进行物种分析。这个过程能有效进行的条件是测序深度足够得到原始 contig，或还有大片段插入的序列。

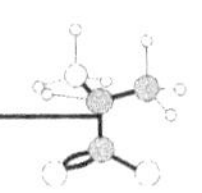

2.4.2 基因预测

经典的基因预测软件如ZCurve（Guo et al. 2003）、Glimmer（Delcher et al. 2007）或者与一些基因查询工具的结合（Glöckner et al. 2003），它们在鸟枪法和大片段文库生成的高质量contig的前提下可以正常工作。如果只有少量contig要分析，会产生10%～20%的高估，这是可接受的并在随后的手工校对中会被修正。基因预测采用策略如在已有的基因组或宏基因组数据库中进行基于相似性的搜索，记录编码区（Badger and Olsen 1999，Krause et al. 2006），但这种做法比较耗时，在数据库中没有同源信息时容易丢失基因。对于数据量大的宏基因组分析，处理时间和潜在的信息损失是较为严重的限制因素。除此之外，宏基因组基因预测需要处理的还有：①片段化的基因；②导致移码的低质量序列；③高系统多样性，这就限制了基因查询的起始训练步骤。MetaGene，一种最新开发的宏基因组片段的基因预测软件，能够弥补上述大部分的不足。MetaGene利用对已知序列GC含量的评估而得到的双密码子频率。这项评估结合了ORF的长度分布、最左端起始密码子的距离、方向以及到邻近ORF的距离，通过分析约130个细菌和古菌的基因组中得到这些参数，用来基因预测（Noguchi et al. 2006）。这个系统很快，而且据我们的经验，通过对个原核生物宏基因组进行测试，得到的结果都是最好的。

2.4.3 功能注释

在宏基因组分析中，功能注释被看作是最重要的步骤。正是在这个过程中，研究者可获得样品中大量遗传信息。注释过程需要谨慎，因为注释效果不好会出现滚雪球效应，在公共数据库中将会出现连续的错误传导。通常情况下，错误可通过以下原因引入：在不同的甚至单个的基因组中不一致的功能注释，或是由于过分简单化的给基因分配潜在功能。不幸的是，现在宏基因组注释并没有可用的“金标准”，注释软件也没有可遵循的约束规则（Raes et al. 2007）。为了对可用的数据资源进行功能预测，需要全面的软件系统来对数据进行存贮、分析和可视化数据，通过不同的基于序列分析的工具提供信息来支持处理流程。一个高水平的数据分析流程需要运用基因查找，基于相似性、模式和物种谱图的搜索，以及预测信号肽、跨膜蛋白螺旋、转移RNA和其他稳定RNA的标准生物信息分析工具。此外，全局和部分的GC含量分析和密码子，以及其他统计参数有助于将编码区与非编码区分开。完备的注释系统还应包括蛋白编码区的自动注释，通过适合用户的各种注释软件

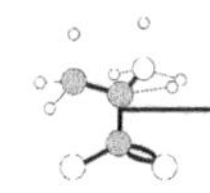

进行手动注释来修正分析过程（细节详见 Stothard and Wishart 2006）。最后代谢途径和网络的重构也有助于将序列信息转换成生物学知识。

2.4.4 基于网络的注释流程

现在，创始者如 Joint Genome Institute（JGI）的联合测序计划、Gordon and Betty Moore 基金会资助的微生物基因组测序计划，或者 Genoscope 的工作，皆促进世界范围内的研究者对感兴趣的基因组或宏基因组进行测序。并且这些机构在初期资助了一部分项目，解决了重要的经费问题。不幸的是，研究者获得序列后，由于在安装、维护和运行基因组注释软件方面缺乏经验，处理序列时可能面临很多问题。首先，测序中心通过在线系统提供生物信息支持，如 IMG 和 IMG/M（Markowitz et al. 2006，2008）、Magnifying Genomes（Vallenet et al. 2006）或 CAMERA（Seshadri et al. 2007）。随后建立的在线系统如 BASys 和 PUMA2 可接收原始宏基因组数据并在处理后提供基于网络的可视化结果，这两个系统，支持功能分配和代谢通路重建，（Van Domselaar et al. 2005，Maltsev et al. 2006）。近来，用于快速注释的 RAST 服务器已经公布，它运用了子系统技术（Aziz et al. 2008）和比较基因组学（Overbeek et al. 2005，Ye et al. 2005）。除了基因预测和注释，这个系统利用注释信息去构建研究样品的功能代谢网络，且所有数据都可以被下载。最近，一个专门为宏基因组分析而设计的实验系统正在开发中，可通过 http://metagenomics.ics.nmpdr.org 访问。

2.4.5 注释系统的本地安装

注释系统本地化安装取决于能否提高操作和可视化的灵活性（Gans and Wolinsky 2007），在应用网络注释系统对数据进行第一轮筛选之后，用户一般会对数据进行个性化分析。为了满足这个需求，需要对数据库和工具具有完全访问权限。自从 1990 年基因组学问世，有些注释系统可以使用本地安装。其中最杰出的是 MAGPIE（Gaasterland and Sensen 1996）、PEDANT Pro（Frishman et al. 2001）、WIT/ERGO（Overbeek et al. 1999，2000）和 ARTEMIS（Rutherford et al. 2000）。现在，最先进的是 GenDB system（Meyer et al. 2003），它被“欧洲海洋基因组学”认定为标准数据处理注释平台。

关于不同注释系统的利弊此处不再详细叙述（可参阅 Stothard and Wishart 2006），我们认为 GenDB 的数据模式和可视化最能满足宏基因组分析的要求。从基因预测开始，GenDB 中有提供几个选项：①运行单基因预测，如 Glimmer

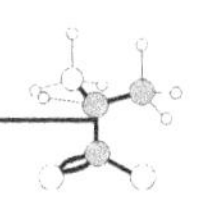

(Delcher et al. 2007) 和 Critica (Badger and Olsen 1999); ②运用整合了两种基因查找方法的软件 Reganor ((McHardy et al. 2004)。在用本地安装后可很容易地在预配置的基因预测软件之间互换，如 MetaGene 或联合交换系统 (Bauer et al. 2006)。可为预测基因提供功能注释结果的标准工具和数据集如表 2.1 所示。蛋白质定位的信息可由软件 signalP 预测信号肽 (Bendtsen et al. 2004) 和软件 TMHMM 预测跨膜螺旋 (Krogh et al. 2001), tRNAscan-SE (Lowe and Eddy 1997) 可用来查找和分选序列中的 tRNA。

表 2.1　蛋白质功能相关数据库及比对工具

Table 2.1 Standard tools and databases providing functional observations

Tool	Databases	References
BLASTn	GenBank	http://www.ncbi.nlm.nih.gov
	EMBL	http://www.ebi.ac.uk/
BLASTp or BLASTx	GenBank	http://www.ncbi.nlm.nih.gov
	UniProt	Apweiler et al. (2004) and
	Swiss-Prot	Boeckmann et al. (2003)
HMMER	Pfam	Bateman et al. (2004)
InterProscan[a]	InterPro	Mulder et al. (2003)

[a]InterPro itself is a metadatabase that provides access to commonly used signature database such as Prosite, Prints, Pfam, ProDom, SMART, TIGR-fams, SCOP, Cath and MSD see http://www.ebi.ac.uk/interpro/.

[a]InterPro 自身是一个宏数据库 (http://www.ebi.ac.uk/interpro/), 为特征数据库如 Prosite、Prints、Pfam、ProDom、SMART、TIGR-fams、SCOP、Cath 和 MSD 提供使用入口。

一旦完成计算，自动注释系统如 GenDB 提供的 Metanor，或者 MicHanThi (Quast 2006) 将为各自工具分析预测的所有基因自动生成注释信息。该系统也支持通过其他的信息手动注释哺乳动物基因组。MicHanThi 系统现在可用量化的方式来描述和注释假定基因和保守的假定基因。对于在一级和二级数据库中得到匹配度高的基因，UniProt 或 Pfam 系统始终可以一致地将其注释到正确的功能类别。在随后的手动注释过程中，每个基因都应检查在数据库中高度匹配的情况。首先匹配到 Swiss-Prot，综合考虑 Pfam 和 InterPro 的结果，整合附加的信息，最后确定基因功能。GenDB 应用的历史系统跟踪了所有的注释变化，因此不同研究者或自动注释系统可对同一个基因做平行的注释。在完成注释过程之后，可以通过数据集构建代谢通路，最简单的方法是自动比对 EC 号到 GenDB 系统提供的相应的 KEGG 通路图上 (Kanehisa et al. 2004)。

基于网络可视化的 GenDB 系统运行比较缓慢，它最近被 JCoast (一个用来数据筛选和原核宏基因组比对的软件工具) 弥补 (Richter et al. 2008)。JCoast 提供了一个灵活的用户绘图界面 (GUI) 以及应用程序界面 (API),

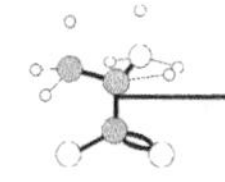

方便了后期数据的直接访问权限。此系统提供了单个基因组、跨基因组和宏基因组分析，协助生物学家开发大而复杂的宏基因组数据集。只要有合适的后期数据集，这个系统就能够独立于 GenDB 运作。

2.4.6 高多样性环境的浅度测序和短读长技术

宏基因组信息分析，随着环境多样性的变高而变得越来越困难，尤其是测序深度不够的情况下。这个问题由最近发表的全球海洋取样调查项目提出（Venter et al. 2004，Seshadri et al. 2007，Yooseph et al. 2007）。尽管获得了近10亿碱基，但只有组装得到两个微生物的基因组，而且必须是群落中的优势种才可能组装得到较长的 scaffold。尽管如此，公共数据库中数量惊人的基因信息，也使科学家们进行广泛的后续研究，主要用于调研海洋生态系统中微生物的多样性和功能（DeLong 2005，DeLong and Karl 2005，Harrington et al. 2007，Sabehi et al. 2007，Yutin et al. 2007，Kagan et al. 2008）。

当组装失败时，基因预测需要额外处理片段化问题。在这种情况下，用环境基因组标签（EGT；Tringe and Rubin 2005）在 UniProt 或 Swiss - Prot 中进行 BLASTx 搜索，至少能够提供一些可用的功能信息。两个方案可用来弥补这些不足：①将多个样品的测序数据合并在一起可用来增加覆盖度；②构建一个参考基因组数据集，作为组装和比对的模板（详见 Gordon and Betty Moore Marine Microbiology Project）。数据集已采用了合并不同样本序列的策略（Rusch et al. 2007），尽管这附带地给生态学家带来新的问题。建立“海洋群落基因组”导致了一些数据丢失，包括对海洋领域特异性适应以及这些适应与栖息地参数之间的联系。在用了短读长技术如焦磷酸测序时，这个问题变得更加复杂。最近一项关于读长对功能预测影响的研究指出，根据 Sargasso、AMD（Tyson et al. page38 2004）、Chesapeake bay datasets（Bench et al. 2007，Wommack et al. 2008）等数据库的长序列（～750 bp）BLASTx 的结果，如果只用 100 bp 的片段来做分析，会丢失平均 72% 的结果。然而，值得一提的是 Mou et al.（2008）在一项关于有机碳源代谢的微生物群落研究中，指出只要应用符合目标的研究策略，就可实现从短序列中获得有价值的信息。

2.4.7 比较宏基因组学结果展示

系统发育与功能多样性的展示可用来更好地理解所调研的宏基因组，尤其适用于比较生物多样性高的环境样品的宏基因组。

2.4.7.1 系统发生的多样性

用来描述系统发育多样性的方法包括：①前面所描述的片段聚类和分类；

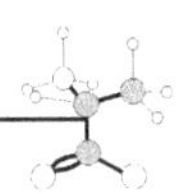

②RNA 基因的系统发育分析；③最优 BLAST 匹配比对；④单拷贝和等拷贝的基因分析。rRNA 基因的系统发育分析是最简单的，可将序列比对到核糖体 RNA 数据库 SILVA（Pruesse et al. 2007）或 RDP II（Cole et al. 2007）。SILVA 的优势在于它可以为真核生物提供质量检查和序列比对，也可提供 16S/18S、23S/28S rRNA 数据库下载（www. arb-silva. de）。对 SILVA 兼容的 ARB 软件可用于精细的系统发育树构建，并提供高级比对、建树和可视化工具（Ludwig et al. 2004）。

为展示宏基因组的分类构成，一个常用的选择就是把 contig 或基因比对到 UniProt 或 nr 数据库上获得最相近的物种信息（Treusch et al. 2004, DeLong et al. 2006, Turnbaugh et al. 2006）。由于在数据库中错误的基因分类，基因组数据集有时会不如 GenBank 提供的 genomesDB 那么准确和精细，当分类出现问题时，可以使用 JCoast 系统（Richter et al. 2008）来做一些补充分析。最近报导的一种方法是把单拷贝或等拷贝基因比对到一个参考物种来进行系统发育分析（Ciccarelli et al. 2006, von Mering et al. 2007）。与分析 rRNA 序列时只有两个标记相比，这个系统的优势是有 31 个系统标记基因可用。值得注意的是，与 80 多万个已公布的 rRNA 基因相比，系统发育蛋白标记数据库的信息要少得多，仅为 600 个/基因。然而这两种方法提供了不同的数据分析视角，彼此可相互补充（Raes et al. 2007）。

2.4.7.2　功能多样性

如前所述，功能多样性主要包括经典的注释和代谢途径分析，通过手动操作综合处理。然而，缺乏统一的标准意味着手动操作会导致不同的人处理和阐述数据的质量并不一致，因此影响了比较宏基因组分析（Raes et al. 2007）。而且，宏基因组测序产生海量数据常常致使分析耗时且不切实际，导致分析常局限于常规统计描述，并对基因（Tringe et al. 2005），或微生物群落丰度、物种组成、群落结构的估计过高或不足（Schloss and Handelsman 2008）。在比较群体的功能适应时，通常只能采用这样的一些分析方法，尤其在一些样品只有浅测序和短读长数据时。此类分析的成功例子是对北太平洋副热带环流圈中分层微生物的调查（DeLong et al. 2006）、DMSP 修正的实验生态系统（Mou et al. 2008）和对一些病毒群落的调查（Edwards and Rohwer 2005, Angly et al. 2006, Culley et al. 2006）。

总之，“以基因为中心”的高多样环境催生了上述提及的众多分析和结果展示的工具，加上功能分类的差异，这些方法对生态学和微生物群落功能的研究十分有用。在低多态性群落中，仅存在少数优势物种时，这种经典的

"以基因为中心"的分析方法更适合，原因是它提供了更多详细信息。把组装的 contig 匹配到物种的 bins，结合随后的注释评估，往往能够重建个体代谢特征，推断出有机体如何分享他们的资源和能量的（Tyson et al. 2004，Meyer-dierks et al. 2005，Martin et al. 2006，Woyke et al. 2006）。现在更具挑战的是扩展生物信息分析工具，使其能够处理复杂环境，并对海洋生态系统的功能提供一个综合解释。

2.5 展望

在当前全球变化的环境下，对地球上的主要成员及其生命调节过程的广泛了解是极其重要。海洋生态系统覆盖了超过70%地球表面，代表着这个行星的主要生物量，为全球的物质和能量循环做出了巨大贡献。众所周知，微生物是这些过程的"看门者"，调查它们的生活方式和适应性将会提升我们对监测、模拟和预测未来变化的能力。对自然环境中未经培养的生物的 DNA 进行测序，为在分子水平调研微生物多样性和功能提供了前所未有的机会。从长远来看，这些研究能帮助我们解决海洋生态学的中心问题：①环境中有哪些微生物；②它们的丰度如何；③它们具备什么功能；④它们在不同环境条件下的活性和适应性如何。

最近宏基因组学在解答一些问题时发挥了重要作用。Dinsdale et al.（2008）揭示了功能差异可以用来区分 9 个独立的群落；Gianoulis et al.（2009）通过结合相关的环境变量，把代谢通路的权重作为代谢指纹，使研究向前迈出了一大步。他们甚至建议在没有可测量的环境因子可用时，这些指纹可用作环境指示物。

尽管宏基因组学被看作是强大的工具，但仍然有些问题尚待提出和解决。一个主要的问题是无法把具体的功能对应到个体，尤其对多样性高的样品。主流的系统发育和功能多样性结果无法回答下列问题，例如"谁做了什么"和"他们如何一起工作和交换能量与营养物"。基于物理分离的单细胞基因组学和随后的全基因组多重置换扩增（MDA；Lasken 2007）是一个新兴的技术。此方法为了解硫化物氧化者如贝氏硫细菌（Mussmann et al. 2007）和一组从海湾浮游细菌中分离的生物（Stepanauskas and Sieracki 2007）的遗传信息提供了极有价值的帮助。

以测序分析为主的宏基因组学面临的另一个问题是数据积累的速度与产生知识的速度极不相符。为了加快形成稳定的基因目录，需要关于基因表达

的信息，尤其是来自不同采样点和环境条件的基因。尽管宏转录组学（Poretsky et al. 2005，Bailly et al. 2007）和宏蛋白组学（Ram et al. 2005，Wilmes and Bond 2006）还处于初期阶段，但它们逐渐被人们所认识，并且对微生物天然群落的基因动力学和调控的研究非常有帮助。在最近的一项研究中，Frias－Lopez et al.（2008）通过 cDNA 焦磷酸测序来研究海洋表层水的基因表达情况，除了鉴定关键代谢通路中的基因表达外，此研究还检测到了大量高表达的假定基因。这些基因最可能与这些物种的特异性适应有关，应该是功能评估的重要研究对象。

为了将现今大量宏基因组数据转化为生物学知识，需要将数据生成和存贮标准化，以便进一步处理序列、基因组、基因和预测代谢功能，并将其与正在分析的环境相关联（DeLong and Karl 2005，Lombardot et al. 2006，Markowitz 2007）。近来成立的“基因组标准化联合会”，对我们收集到的所有基因组和宏基因组数据提出了要求（Field et al. 2007，2008）。宏基因级序列的最小信息（MIMS）最初是为了将数据采集标准化，研究者需要在时空范围里通过提供最小量信息包括 GPS 坐标、深度或高度和采样时间，来确定宏基因组信息（详见 http：//gensc. org）。样品的理化学信息也需要提供，这样基因组特征就可与其栖息地特征联系起来。同时也鉴定生物的特异适应性，以及评估它在环境中的作用和影响。为了弥补丢失的环境信息和获得所调研的生态系统稳定性的动力学图像，我们可以通过全球数据和原位测量相互补充。目前有一种针对宏基因组信息整合全球海洋数据层补充的栖息地参数的方法可以使用（详见 www. megx. net；Lombardot et al. 2006）。如果此法进一步开发，还可能做到在物理、化学和生物数据上附加功能、代谢通路和系统发育的多样性，从而将各海洋区域上标记微生物特征。

另一方面，在数据处理时，不同宏基因组研究标准之间的比较也至关重要。海洋领域中计算框架结构（如 CAMERA）对于处理一些重要数据有所帮助。Raes et al.（2007）提出了 MINIMESS，即最小宏基因组序列分析标准，需要对每一个宏基因组记录下基因的组装参数、物种和功能组成与覆盖度，以及生物和技术因素。这些工作需要相融贯通，让数据提供者和生物学家清楚易懂。这将是一个团体的工作，共同致力于我们逐渐增多的宏基因组资源的完整与品质，以提高我们对海洋生态系统如何行使功能的理解。从长远看，这些数据对于环境中复杂的相互作用网络模型化可能会是一个起点，将产生生态系统生物学这一门新的学科。

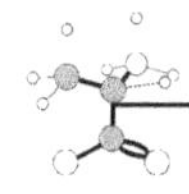

参考文献

Abe T, Sugawara H, Kanaya S et al (2006) A novel bioinformatics tool for phylogenetic classifica-tion of genomic sequence fragments derived from mixed genomes of uncultured environmental microbes. Polar Biosci 20:103 – 112

Abe T, Sugawara H, Kinouchi M et al (2005) Novel phylogenetic studies of genomic sequence fragments derived from uncultured microbe mixtures in environmental and clinical samples. DNA Res 12:281 – 290

Abulencia CB, Wyborski DL, Garcia JA et al (2006) Environmental whole-genome amplification to access microbial populations in contaminated sediments. Appl Environ Microbiol 72:3291 – 3301

Amann R, Fuchs BM (2008) Single-cell identification in microbial communities by improved fluorescence in situ hybridization techniques. Nat Rev Microbiol 6:339 – 348

Amann RI, Ludwig W, Schleifer KH (1995) Phylogenetic identification and in situ detection of individual microbial cells without cultivation. Microbiol Rev 59:143 – 169

Angly FE, Felts B, Breitbart M et al (2006) The marine viromes of four oceanic regions. PLoS Biol 4:2121 – 2131

Aparicio S, Chapman J, Stupka E et al (2002) Whole-genome shotgun assembly and analysis of the genome of *Fugu rubripes*. Science 297:1301 – 1310

Apweiler R, Bairoch A, Wu CH et al (2004) UniProt: the Universal Protein knowledgebase. Nucleic Acids Res 32:D115 – D119

Asakawa S, Abe I, Kudoh Y et al (1997) Human BAC library: construction and rapid screening. Gene 191:69 – 79

Aziz RK, Bartels D, Best AA et al (2008) The RAST server: rapid annotations using subsystems technology. BMC Genomics 9:75

Badger JH, Olsen GJ (1999) CRITICA: coding region identification tool invoking comparative analysis. Mol Biol Evol 16:512 – 524

Bailly J, Fraissinet-Tachet L, Verner MC et al (2007) Soil eukaryotic functional diversity, a metatranscriptomic approach. ISME J 1:632 – 642

Bateman A, Coin L, Durbin R et al (2004) The PFAM protein families database. Nucleic Acids Res 32:D138 – D141

Bauer M, Kube M, Teeling H et al (2006) Whole genome analysis of the marine Bacteroidetes 'Gramella forsetii' reveals adaptations to degradation of polymeric organic matter. Environ Microbiol 8:2201 – 2213

Beja O, Aravind L, Koonin EV et al (2000) Bacterial rhodopsin: evidence for a new type of

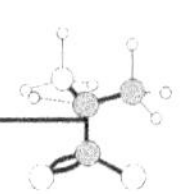

phototrophy in the sea. Science 289:1902 – 1906

Beja O, Koonin EV, Aravind L et al (2002) Comparative genomic analysis of archaeal genotypic variants in a single population and in two different oceanic provinces. Appl Environ Microbiol 68:335 – 345

Beja O, Spudich EN, Spudich JL et al (2001) Proteorhodopsin phototrophy in the ocean. Nature 411:786 – 789

Bench SR, Hanson TE, Williamson KE et al (2007) Metagenomic characterization of Chesapeake bay virioplankton. Appl Environ Microbiol 73:7629 – 7641

Bendtsen JD, Nielsen H, von Heijne G et al (2004) Improved prediction of signal peptides: SignalP 3. 0. J Mol Biol 340:783 – 795

Biddle JF, Fitz-Gibbon S, Schuster SC et al (2008) Metagenomic signatures of the Peru Margin subseafloor biosphere show a genetically distinct environment. Proc Natl Acad Sci U S A 105: 10583 – 10588

Binga EK, Lasken RS, Neufeld JD (2008) Something from (almost) nothing: the impact of multiple displacement amplification on microbial ecology. ISME J 2:233 – 241

Blanco L, Bernad A, Lazaro JM et al (1989) Highly efficient DNA-synthesis by the phage phi-29 DNA-Polymerase-symmetrical mode of DNA-replication. J Biol Chem 264:8935 – 8940

Blanco L, Salas M (1984) Characterization and purification of a phage phi-29-encoded DNA-polymerase required for the initiation of replication. Proc Natl Acad Sci USA-Biol Sci 81:5325 – 5329

Blanco L, Salas M (1985a) Characterization of a 3 – 5 exonuclease activity in the phage phi-29-encoded DNA-polymerase. Nucleic Acids Res 13:1239 – 1249

Blanco L, Salas M (1985b) Replication of phage phi-29 DNA with purified terminal protein and DNA-polymerase-synthesis of full-length phi-29 DNA. Proc Natl Acad Sci U S A 82:6404 – 6408

Boeckmann B, Bairoch A, Apweiler R et al (2003) The SWISS-PROT protein knowledgebase and its supplement TrEMBL in 2003. Nucleic Acids Res 31:365 – 370

Borneman J (1999) Culture-independent identification of microorganisms that respond to specified stimuli. Appl Environ Microbiol 65:3398 – 3400

Breitbart M, Rohwer F (2005) Method for discovering novel DNA viruses in blood using viral particle selection and shotgun sequencing. Biotechniques 39:729 – 736

Breitbart M, Salamon P, Andresen B et al (2002) Genomic analysis of uncultured marine viral communities. Proc Natl Acad Sci U S A 99:14250 – 14255

Bryant DA, Costas AMG, Maresca JA et al (2007) *Candidatus* Chloracidobacterium thermophilum: an aerobic phototrophic acidobacterium. Science 317:523 – 526

Bryant DA, Frigaard NU (2006) Prokaryotic photosynthesis and phototrophy illuminated.

Trends Microbiol 14:488 – 496

Chung KT, Ferris DH (1996) Martinus Willem Beijerinck (1851 – 1931)-Pioneer of general microbiology. ASM News 62:539 – 543

Ciccarelli FD, Doerks T, von Mering C et al (2006) Toward automatic reconstruction of a highly resolved tree of life. Science 311:1283 – 1287

Cole JR, Chai B, Farris RJ et al (2007) The ribosomal database project (RDP-II): introducing myRDP space and quality controlled public data. Nucleic Acids Res 35:D169 – D172

Collins J, Hohn B (1978) Cosmids-type of plasmid gene-cloning vector that is packageable in vitro in bacteriophage lambda-heads. Proc Natl Acad Sci U S A 75:4242 – 4246

Connon SA, Giovannoni SJ (2002) High-throughput methods for culturing microorganisms in very-low-nutrient media yield diverse new marine isolates. Appl Environ Microbiol 68:3878 – 3885

Cottrell MT, Moore JA, Kirchman DL (1999) Chitinases from uncultured marine microorganisms. Appl Environ Microbiol 65:2553 – 2557

Courtois S, Cappellano CM, Ball M et al (2003) Recombinant environmental libraries provide access to microbial diversity for drug discovery from natural products. Appl Environ Microbiol 69: 49 – 55

Culley AI, Lang AS, Suttle CA (2006) Metagenomic analysis of coastal RNA virus communities. Science 312:1795 – 1798

Curtis TP, Sloan WT, Scannell JW (2002) Estimating prokaryotic diversity and its limits. Proc Natl Acad Sci U S A 99:10494 – 10499

de la Torre JR, Christianson LM, Beja O et al (2003) Proteorhodopsin genes are distributed among divergent marine bacterial taxa. Proc Natl Acad Sci U S A 100:12830 – 12835

DeLong EE (2005) Microbial community genomics in the ocean. Nat Rev Microbiol 3:459 – 469

DeLong EF, Karl DM (2005) Genomic perspectives in microbial oceanography. Nature 437: 336 – 342

DeLong EF, Preston CM, Mincer T et al (2006) Community genomics among stratified microbial assemblages in the ocean's interior. Science 311:496 – 503

Dean FB, Hosono S, Fang LH et al (2002) Comprehensive human genome amplification using multiple displacement amplification. Proc Natl Acad Sci U S A 99:5261 – 5266

Dean FB, Nelson JR, Giesler TL et al (2001) Rapid amplification of plasmid and phage DNA using phi29 DNA polymerase and multiply-primed rolling circle amplification. Genome Res 11: 1095 – 1099

Delcher AL, Bratke KA, Powers EC et al (2007) Identifying bacterial genes and endosymbiont DNA with Glimmer. Bioinformatics 23:673 – 679

Dinsdale EA, Edwards RA, Hall D et al (2008) Functional metagenomic profiling of nine bi-

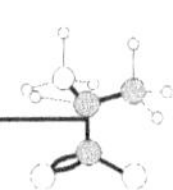

omes. Nature 452:629 – U632

Drews G (2000) The roots of microbiology and the influence of Ferdinand Cohn on microbiology of the 19th century. FEMS Microbiol Rev 24:225 – 249

Dumont MG, Murrell JC (2005) Stable isotope probing-linking microbial identity to function. Nat Rev Microbiol 3:499 – 504

Dumont MG, Radajewski SM, Miguez CB et al (2006) Identification of a complete methane monooxygenase operon from soil by combining stable isotope probing and metagenomic analysis. Environ Microbiol 8:1240 – 1250

Eckert KA, Kunkel TA (1990) High fidelity DNA-synthesis by the *Thermus aquaticus* DNA-polymerase. Nucleic Acids Res 18:3739 – 3744

Edwards RA, Rodriguez-Brito B, Wegley L et al (2006) Using pyrosequencing to shed light on deep mine microbial ecology. BMC Genomics 7:57

Edwards RA, Rohwer F (2005) Viral metagenomics. Nat Rev Microbiol 3:504 – 510

Entcheva P, Liebl W, Johann A et al (2001) Direct cloning from enrichment cultures, a reliable strategy for isolation of complete operons and genes from microbial consortia. Appl Environ Microbiol 67:89 – 99

Epicentre Biotechnologies (2007) CopyControl™ Fosmid Library Production Kit

Erkel C, Kube M, Reinhardt R et al (2006) Genome of Rice Cluster I archaea-the key methane producers in the rice rhizosphere. Science 313:370 – 372

Esteban JA, Salas M, Blanco L (1993) Fidelity of phi-29 DNA-polymerase-comparison between protein-primed initiation and DNA polymerization. J Biol Chem 268:2719 – 2726

Field D, Garrity G, Gray T et al (2007) eGenomics: cataloguing our complete genome collection III. Comp Funct Genom 2007:1 – 7

Field D, Garrity G, Selengut J et al (2008) Towards a richer description of our complete collection of genomes and metagenomes: the "Minimum Information about a Genome Sequence" (MIGS) specification. Nat Biotechnol 26:541 – 547

Fire A, Xu SQ (1995) Rolling replication of short DNA circles. Proc Natl Acad Sci U S A 92:4641 – 4645

Frias-Lopez J, Shi Y, Tyson GW et al (2008) Microbial community gene expression in ocean surface waters. Proc Natl Acad Sci U S A 105:3805 – 3810

Frishman D, Albermann K, Hani J et al (2001) Functional and structural genomics using PEDANT. Bioinformatics 17:44 – 57

Gaasterland T, Sensen CW (1996) MAGPIE: automated genome interpretation. Trends Genet 12:76 – 78

Gans JD, Wolinsky M (2007) Genomorama: genome visualization and analysis. BMC Bioinformatics 8:204

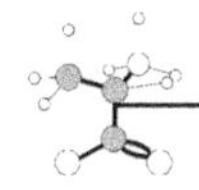

Gianoulis TA et al (2009) Quantifying environmental adaptation of metabolic pathways in metagenomics. Proc Natl Acad Sci U S A 106:1374 – 1379

Giovannoni SJ, Britschgi TB, Moyer CL et al (1990) Genetic diversity in Sargasso Sea bacterio-plankton. Nature 345:60 – 63

Glöckner FO, Kube M, Bauer M et al (2003) Complete genome sequence of the marine planctomycete *Pirellula* sp. strain 1. Proc Natl Acad Sci U S A 100:8298 – 8303

Glöckner FO, Meyerdierks A (2006) Metagenome analysis. In: Stackebrandt E (ed) Molecular identification, systematics, and population structure of prokaryotes. Springer-Verlag, Heidelberg

Gloess S, Grossart HP, Allgaier M et al (2008) Use of laser microdissection for phylogenetic char-acterization of polyphosphate-accumulating bacteria. Appl Environ Microbiol 74:4231 – 4235

Goldberg SMD, Johnson J, Busam D et al (2006) A Sanger/pyrosequencing hybrid approach for

the generation of high-quality draft assemblies of marine microbial genomes. Proc Natl Acad Sci U S A 103:11240 – 11245

Goodman RM, Liles M (2001) Template specific termination in a polymerase chain reaction. US Patent 6,248,567

Green ED, Birren B, Klapholz S et al (1997) Genome analysis: a laboratory manual, 1st edn. Cold Spring Harbor Laboratory Press, New York

Guo FB, Ou HY, Zhang CT (2003) ZCURVE: a new system for recognizing protein-coding genes in bacterial and archaeal genomes. Nucleic Acids Res 31:1780 – 1789

Hall AR (1989) The Leeuwenhoek Lecture, 1988-Antoni van Leeuwenhoek 1632 – 1723. Notes Rec R Soc Lond 43:249 – 273

Hallam SJ, Girguis PR, Preston CM et al (2003) Identification of methyl coenzyme M reductase A (*mcr*A) genes associated with methane-oxidizing archaea. Appl Environ Microbiol 69:5483 – 5491

Handelsman JRM (1998) Molecular biological access to the chemistry of unknown soil microbes-a new frontier for natural products. Chem Biol 5:R245 – R249

Handelsman J, Liles M, Mann D et al (2002) Cloning the metagenome: culture-independent access to the diversity and functions of the uncultivated microbial world. In: Functional microbial genomics. Academic Press Inc., San Diego, pp 241 – 255

Harrington ED, Singh AH, Doerks T et al (2007) Quantitative assessment of protein func-tion prediction from metagenomics shotgun sequences. Proc Natl Acad Sci U S A 104:13913 – 13918

Henne A, Daniel R, Schmitz RA et al (1999) Construction of environmental DNA libraries in *Escherichia coli* and screening for the presence of genes conferring utilization of 4-hydroxybutyrate. Appl Environ Microbiol 65:3901 – 3907

Hosono S, Faruqi AF, Dean FB et al (2003) Unbiased whole-genome amplification directly

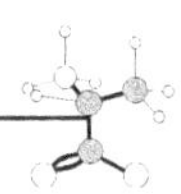

from clinical samples. Genome Res 13:954 - 964

Huber R, Burggraf S, Mayer T et al (1995) Isolation of a hyperthermophilic archaeum predicted by in situ RNA Analysis. Nature 376:57 - 58

Hughes DS, Felbeck H, Stein JL (1997) A histidine protein kinase homolog from the endosymbiont of the hydrothermal vent tubeworm Riftia pachyptila. Appl Environ Microbiol 63:3494 - 3498

Huse SM, Huber JA, Morrison HG et al (2007) Accuracy and quality of massively parallel DNA pyrosequencing. Genome Biol 8:9

Huson DH, Auch AF, Qi J et al (2007) MEGAN analysis of metagenomic data. Genome Res 17:377 - 386

Hutchison CA, Venter JC (2006) Single-cell genomics. Nat Biotechnol 24:657 - 658

Ishoy T, Kvist T, Westermann P et al (2006) An improved method for single cell isolation of prokaryotes from meso-, thermo-and hyperthermophilic environments using micromanipula-tion. Appl Microbiol Biotechnol 69:510 - 514

Jaffe DB, Butler J, Gnerre S et al (2003) Whole-genome sequence assembly for mammalian genomes: Arachne 2. Genome Res 13:91 - 96

Johnson PLF, Slatkin M (2006) Inference of population genetic parameters in metagenomics: a clean look at messy data. Genome Res 16:1320 - 1327

Kagan J, Sharon I, Beja O et al (2008) The tryptophan pathway genes of the Sargasso Sea metagenome: new operon structures and the prevalence of non-operon organization. Genome Biol 9:R20

Kanehisa M, Goto S, Kawashima S et al (2004) The KEGG resource for deciphering the genome. Nucleic Acids Res 32:D277 - D280

Karlin S, Campbell AM, Mrazek J (1998) Comparative DNA analysis across diverse genomes. Ann Rev Genet 32:185 - 225

Kaufmann SHE, Schaible UE (2005) 100th anniversary of Robert Koch's Nobel Prize for the discovery of the tubercle bacillus. Trends Microbiol 13:469 - 475

Kim UJ, Birren BW, Slepak T et al (1996) Construction and characterization of a human bacterial artificial chromosome library. Genomics 34:213 - 218

Kim UJ, Shizuya H, Dejong PJ et al (1992) Stable propagation of cosmid sized human DNA inserts in an F-factor based vector. Nucleic Acids Res 20:1083 - 1085

Krause L, Diaz NN, Bartels D et al (2006) Finding novel genes in bacterial communities isolated from the environment. Bioinformatics 22:E281 - E289

Krogh A, Larsson B, von Heijne G et al (2001) Predicting transmembrane protein topology with a hidden Markov model: application to complete genomes. J Mol Biol 305:567 - 580

Krüger M, Meyerdierks A, Glockner FO et al (2003) A conspicuous nickel protein in microbial mats that oxidize methane anaerobically. Nature 426:878 - 881

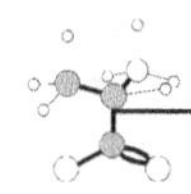

Kube M, Beck A, Meyerdierks A et al (2005) A catabolic gene cluster for anaerobic benzoate degradation in methanotrophic microbial Black Sea mats. Syst Appl Microbiol 28:287 – 294

Kvist T, Ahring BK, Lasken RS et al (2007) Specific single-cell isolation and genomic amplification of uncultured microorganisms. Appl Microbiol Biotechnol 74:926 – 935

Lasken RS (2007) Single-cell genomic sequencing using Multiple Displacement Amplification. Curr Opin Microbiol 10:510 – 516

Lasken RS, Stockwell TB (2007) Mechanism of chimera formation during the Multiple Displacement Amplification reaction. BMC Biotechnol 7:19

Leininger S, Urich T, Schloter M et al (2006) Archaea predominate among ammonia-oxidizing prokaryotes in soils. Nature 442:806 – 809

Leonardo ED, Sedivy JM (1990) A new vector for cloning large eukaryotic DNA segments in *Escherichia coli*. Bio-Technology 8:841 – 844

Liles MR, Manske BF, Bintrim SB et al (2003) A census of rRNA genes and linked genomic sequences within a soil metagenomic library. Appl Environ Microbiol 69:2684 – 2691

Liles MR, Williamson LL, Rodbumrer J et al (2008) Recovery, purification, and cloning of high-molecular-weight DNA from soil microorganisms. Appl Environ Microbiol 74:3302 – 3305

Liu DY, Daubendiek SL, Zillman MA et al (1996) Rolling circle DNA synthesis: small circular oligonucleotides as efficient templates for DNA polymerases. J Am Chem Soc 118:1587 – 1594

Lombardot T, Kottmann R, Pfeffer H et al (2006) Megx. net-database resource for marine ecological genomics. Nucleic Acids Res 34:D390 – D393

Lopez-Garcia P, Brochier C, Moreira D et al (2004) Comparative analysis of a genome fragment of an uncultivated mesopelagic crenarchaeote reveals multiple horizontal gene transfers. Environ Microbiol 6:19 – 34

Lowe TM, Eddy SR (1997) tRNAscan-SE: a program for improved detection of transfer RNA genes in genomic sequence. Nucleic Acids Res 25:955 – 964

Ludwig W, Strunk O, Westram R et al (2004) ARB: a software environment for sequence data. Nucleic Acids Res 32:1363 – 1371

MacNeil IA, Tiong CL, Minor C et al (2001) Expression and isolation of antimicrobial small molecules from soil DNA libraries. J Mol Microbiol Biotechnol 3:301 – 308

Maltsev N, Glass E, Sulakhe D et al (2006) PUMA2-grid-based high-throughput analysis of genomes and metabolic pathways. Nucleic Acids Res 34:D369 – D372

Marcy Y, Ouverney C, Bik EM et al (2007) Dissecting biological "dark matter" with single-cell genetic analysis of rare and uncultivated TM7 microbes from the human mouth. Proc Natl Acad Sci U S A 104:11889 – 11894

Mardis ER (2008) Next-generation DNA sequencing methods. Annu Rev Genomics Hum Genet 9:387 – 402

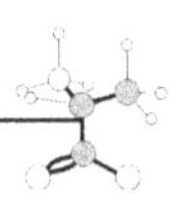

Margulies M, Egholm M, Altman WE et al (2005) Genome sequencing in microfabricated high-density picolitre reactors. Nature 437:376 – 380

Markowitz VM (2007) Microbial genome data resources. Curr Opin Biotechnol 18:267 – 272

Markowitz VM, Ivanova NN, Szeto E et al (2008) IMG/M: a data management and analysis system for metagenomes. Nucleic Acids Res 36:D534 – D538

Markowitz VM, Korzeniewski F, Palaniappan K et al (2006) The integrated microbial genomes (IMG) system. Nucleic Acids Res 34:D344 – D348

Martin HG, Ivanova N, Kunin V et al (2006) Metagenomic analysis of two enhanced biological phosphorus removal (EBPR) sludge communities. Nat Biotechnol 24:1263 – 1269

Martín-Cuadrado AB, López-García P, Alba JC et al (2007) Metagenomics of the deep Mediterranean, a warm bathypelagic habitat. PLoS One 2:e914

Mavromatis K, Ivanova N, Barry K et al (2007) Use of simulated data sets to evaluate the fidelity of metagenomic processing methods. Nat Methods 4:495 – 500

McHardy AC, Goesmann A, Pühler A et al (2004) Development of joint application strategies for two microbial gene finders. Bioinformatics 20:1622 – 1631

McHardy AC, Martin HG, Tsirigos A et al (2007) Accurate phylogenetic classification of variable-length DNA fragments. Nat Methods 4:63 – 72

McHardy AC, Rigoutsos I (2007) What's in the mix: phylogenetic classification of metagenome sequence samples. Curr Opin Microbiol 10:499 – 503

Meyer F, Goesmann A, McHardy AC et al (2003) GenDB-an open source genome annotation system for prokaryote genomes. Nucleic Acids Res 31:2187 – 2195

Meyerdierks A, Kube M, Lombardot T et al (2005) Insights into the genomes of archaea mediating the anaerobic oxidation of methane. Environ Microbiol 7:1937 – 1951

Monaco AP, Larin Z (1994) YACs, BACs, PACs and MACs-artificial chromosomes as research tools. Trends Biotechnol 12:280 – 286

Mou XZ, Sun SL, Edwards RA et al (2008) Bacterial carbon processing by generalist species in the coastal ocean. Nature 451:708 – U711

Mulder NJ, Apweiler R, Attwood TK et al (2003) The InterPro Database, 2003 brings increased coverage and new features. Nucleic Acids Res 31:315 – 318

Mussmann M, Hu FZ, Richter M et al (2007) Insights into the genome of large sulfur bacteria revealed by analysis of single filaments. PLoS Biol 5:1923 – 1937

Muβmann M, Richter M, Lombardot T et al (2005) Clustered genes related to sulfate respiration in uncultured prokaryotes support the theory of their concomitant horizontal transfer. J Bacteriol 187:7126 – 7137

Nelson JR, Cai YC, Giesler TL et al (2002) TempliPhi, phi 29 DNA polymerase based rolling circle amplification of templates for DNA sequencing. Biotechniques, 44 – 47

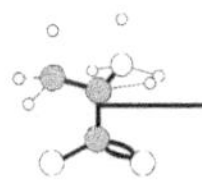

Neufeld JD, Chen Y, Dumont MG et al (2008) Marine methylotrophs revealed by stable-isotope probing, multiple displacement amplification and metagenomics. Environ Microbiol 10:1526 – 1535

Noguchi H, Park J, Takagi T (2006) MetaGene: prokaryotic gene finding from environmental genome shotgun sequences. Nucleic Acids Res 34:5623 – 5630

Olsen GJ, Lane DJ, Giovannoni SJ et al (1986) Microbial ecology and evolution-a ribosomal-RNA Approach. Ann Review of Microbiol 40:337 – 365

Olsen GJ, Woese CR, Overbeek R (1994) The winds of (evolutionary) change-breathing new life into microbiology. J Bacteriol 176:1 – 6

Osoegawa K, Tateno M, Woon PY et al (2000) Bacterial artificial chromosome libraries for mouse sequencing and functional analysis. Genome Res 10:116 – 128

Osoegawa K, Woon PY, Zhao B et al (1998) An improved approach for construction of bacterial artificial chromosome libraries. Genomics 52:1 – 8

Ottesen EA, Hong JW, Quake SR et al (2006) Microfluidic digital PCR enables multigene analysis of individual environmental bacteria. Science 314:1464 – 1467

Overbeek R, Begley T, Butler RM et al (2005) The subsystems approach to genome annotation and its use in the project to annotate 1000 genomes. Nucleic Acids Res 33:5691 – 5702

Overbeek R, Fonstein M, D'Souza M et al (1999) The use of gene clusters to infer functional coupling. Proc Natl Acad Sci U S A 96:2896 – 2901

Overbeek R, Larsen N, Pusch GD et al (2000) WIT: integrated system for high-throughput genome sequence analysis and metabolic reconstruction. Nucleic Acids Res 28:123 – 125

Pace NR, Stahl DA, Olsen GJ et al (1985) Analyzing natural microbial populations by rRNA sequences. ASM News 51:4 – 12

Park SJ, Kang CH, Chae JC et al (2008) Metagenome microarray for screening of F 粘粒 clones containing specific genes. FEMS Microbiol Lett 284:28 – 34

Pernthaler A, Dekas AE, Brown CT et al (2008) Diverse syntrophic partnerships from deep-sea methane vents revealed by direct cell capture and metagenomics. Proc Natl Acad Sci U S A 105:7052 – 7057

Piel J (2002) A polyketide synthase-peptide synthetase gene cluster from an uncultured bacterial symbiont of *Paederus* beetles. Proc Natl Acad Sci U S A 99:14002 – 14007

Pinard R, de Winter A, Sarkis GJ et al (2006) Assessment of whole genome amplification-induced bias through high-throughput, massively parallel whole genome sequencing. BMC Genomics 7:21

Podar M, Abulencia CB, Walcher M et al (2007) Targeted access to the genomes of low-abundance organisms in complex microbial communities. Appl Environ Microbiol 73:3205 – 3214

Poretsky RS, Bano N, Buchan A et al (2005) Analysis of microbial gene transcripts in envi-

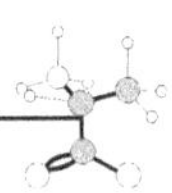

ron-mental samples. Appl Environ Microbiol 71:4121 – 4126

Pride DT, Meinersmann RJ, Wassenaar TM et al (2003) Evolutionary implications of microbial genome tetranucleotide frequency biases. Genome Res 13:145 – 158

Prober JM, Trainor GL, Dam RJ et al (1987) A system for rapid DNA sequencing with fluorescent chain-terminating dideoxynucleotides. Science 238:336 – 341

Promega Corporation (2007) Technical Bulletin: Packagene R Lambda DNA Packaging System.

Pruesse E, Quast C, Knittel K et al (2007) SILVA: a comprehensive online resource for quality checked and aligned ribosomal RNA sequence data compatible with ARB. Nucleic Acids Res 35:7188 – 7196

Quaiser A, Ochsenreiter T, Klenk HP et al (2002) First insight into the genome of an uncultivated crenarchaeote from soil. Environ Microbiol 4:603 – 611

Quaiser A, Ochsenreiter T, Lanz C et al (2003) Acidobacteria form a coherent but highly diverse group within the bacterial domain: evidence from environmental genomics. Mol Microbiol 50: 563 – 575

Quast C (2006) MicHanThi-design and implementation of a system for the prediction of gene functions in genome annotation projects. Diploma thesis. Department of Computer Science and Microbial Genomics Group. University Bremen and Max Planck Institute for Marine Microbiology, Bremen

Raes J, Foerstner KU, Bork P (2007) Get the most out of your metagenome: computational analysis of environmental sequence data. Curr Opin Microbiol 10:490 – 498

Raghunathan A, Ferguson HR, Bornarth CJ et al (2005) Genomic DNA amplification from a single bacterium. Appl Environ Microbiol 71:3342 – 3347

Ram RJ, VerBerkmoes NC, Thelen MP et al (2005) Community proteomics of a natural microbial biofilm. Science 308:1915 – 1920

Rappe MS, Connon SA, Vergin KL et al (2002) Cultivation of the ubiquitous SAR11 marine bacterioplankton clade. Nature 418:630 – 633

Richter M, Lombardot T, Kostadinov I et al (2008) JCoast-A biologist-centric software tool for data mining and comparison of prokaryotic (meta) genomes. BMC Bioinformatics 9:177

Riesenfeld CS, Schloss PD, Handelsman J (2004) Metagenomics: genomic analysis of microbial communities. Ann Rev Genet 38:525 – 552

Robidart JC, Bench SR, Feldman RA et al (2008) Metabolic versatility of the *Riftia pachyptila* endosymbiont revealed through metagenomics. Environ Microbiol 10:727 – 737

Rogers YH, Venter JC (2005) Genomics-Massively parallel sequencing. Nature 437: 326 – 327

Rondon MR, August PR, Bettermann AD et al (2000) Cloning the soil metagenome: a strategy for accessing the genetic and functional diversity of uncultured microorganisms. Appl Environ Microbiol 66:2541 – 2547

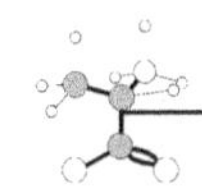

Rondon MR, Raffel SJ, Goodman RM et al (1999) Toward functional genomics in bacteria: anal-ysis of gene expression in *Escherichia coli* from a bacterial artificial chromosome library of *Bacillus cereus*. Proc Natl Acad Sci U S A 96:6451 – 6455

Rusch DB, Halpern AL, Sutton G et al (2007) The Sorcerer II Global Ocean Sampling expedition: Northwest Atlantic through Eastern Tropical Pacific. PLoS Biol 5:398 – 431

Rutherford K, Parkhill J, Crook J et al (2000) Artemis: sequence visualization and annotation. Bioinformatics 16:944 – 945

Sabehi G, Kirkup BC, Rozenberg M et al (2007) Adaptation and spectral tuning in divergent marine proteorhodopsins from the eastern Mediterranean and the Sargasso Seas. ISME J 1:48 – 55

Salzberg SL, Yorke JA (2005) Beware of mis – assembled genomes. Bioinformatics 1:4320 – 4321

Sambrook J, Russel DW (2001) Molecular cloning: a laboratory manual, 3rd edn. Cold Spring Harbor Laboratory Press, Cold Spring Harbor, New York

Sandberg R, Winberg G, Branden CI et al (2001) Capturing whole-genome characteristics in short sequences using a naive Bayesian classifier. Genome Res 11:1404 – 1409

Sanger F, Nicklen S, Coulson AR (1977) DNA sequencing with chain-terminating inhibitors. Proc Natl Acad Sci U S A 74:5463 – 5467

Schirmer A, Gadkari R, Reeves CD et al (2005) Metagenomic analysis reveals diverse polyketide synthase gene clusters in microorganisms associated with the marine sponge *Discodermia dissoluta*. Appl Environ Microbiol 71:4840 – 4849

Schlegel HG (1996) Winogradsky discovered a new Modus vivendi. Anaerobe 2:129 – 136

Schleper C, Delong EF, Preston CM et al (1998) Genomic analysis reveals chromosomal variation in natural populations of the uncultured psychrophilic archaeon *Cenarchaeum symbiosum*. J Bacteriol 180:5003 – 5009

Schleper C, Swanson RV, Mathur EJ et al (1997) Characterization of a DNA polymerase from the uncultivated psychrophilic archaeon *Cenarchaeum symbiosum*. J Bacteriol 179:7803 – 7811

Schloss PD, Handelsman J (2008) A statistical toolbox for metagenomics: assessing functional diversity in microbial communities. BMC Bioinformatics 9:34

Schmeisser C, Stockigt C, Raasch C et al (2003) Metagenome survey of biofilms in drinking-water networks. Appl Environ Microbiol 69:7298 – 7309

Schmidt TM, Delong EF, Pace NR (1991) Analysis of a marine picoplankton community by 16s ribosomal-RNA gene cloning and sequencing. J Bacteriol 173:4371 – 4378

Schwartz M (2001) The life and works of Louis Pasteur. J Appl Microbiol 91:597 – 601

Schübbe S, Kube M, Scheffel A et al (2003) Characterization of a spontaneous nonmagnetic mutant of *Magnetospirillum gryphiswaldense* reveals a large deletion comprising a putative magnetosome island. J Bacteriol 185:5779 – 5790

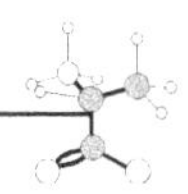

Sebat JL, Colwell FS, Crawford RL (2003) Metagenomic profiling: microarray analysis of an environmental genomic library. Appl Environ Microbiol 69:4927 – 4934

Seshadri R, Kravitz SA, Smarr L et al (2007) CAMERA: a Community Resource for Metagenomics. PLoS Biol 5:e75

Sheng Y, Mancino V, Birren B (1995) Transformation of *Escherichia coli* with large DNA molecules by electroporation. Nucleic Acids Res 23:1990 – 1996

Shizuya H, Birren B, Kim UJ et al (1992) Cloning and stable maintenance of 300-kilobase-pair fragments of human DNA in *Escherichia coli* using an F-factor-based vector. Proc Natl Acad Sci U S A 89:8794 – 8797

Sogin ML, Morrison HG, Huber JA et al (2006) Microbial diversity in the deep sea and the underexplored "rare biosphere". Proc Natl Acad Sci U S A 103:12115 – 12120

Stahl DA, Amann R (1991) Development and application of nucleic acid probes. In: Stackebrandt E, Goodfellow M (eds) Nucleic acid techniques in bacterial systematics. John Wiley & Sons Ltd., Chichester, UK, pp 205 – 248

Stein JL, Marsh TL, Wu KY et al (1996) Characterization of uncultivated prokaryotes: isolation and analysis of a 40-kilobase-pair genome fragment front a planktonic marine archaeon. J Bacteriol 178:591 – 599

Stepanauskas R, Sieracki ME (2007) Matching phylogeny and metabolism in the uncultured marine bacteria, one cell at a time. Proc Natl Acad Sci U S A 104:9052 – 9057

Stoffels M, Ludwig W, Schleifer KH (1999) rRNA probe-based cell fishing of bacteria. Environ Microbiol 1:259 – 271

Stothard P, Wishart DS (2006) Automated bacterial genome analysis and annotation. Curr Opin Microbiol 9:505 – 510

Strous M, Pelletier E, Mangenot S et al (2006) Deciphering the evolution and metabolism of an anammox bacterium from a community genome. Nature 440:790 – 794

Tao Q, Zhang HB (1998) Cloning and stable maintenance of DNA fragments over 300 kb in *Escherichia coli* with conventional plasmid-based vectors. Nucleic Acids Res 26:4901 – 4909

Tatusov RL, Koonin EV, Lipman DJ (1997) A genomic perspective on protein families. Science 278:631 – 637

Teeling H, Meyerdierks A, Bauer M et al (2004) Application of tetranucleotide frequencies for the assignment of genomic fragments. Environ Microbiol 6:938 – 947

Telenius H, Carter NP, Bebb CE et al (1992) Degenerate oligonucleotide-primed PCR-General amplification of target DNA by a single degenerate primer. Genomics 13:718 – 725

Thornhill DJ, Wiley AA, Campbell AL et al (2008) Endosymbionts of *Siboglinum fiordicum* and the phylogeny of bacterial endosymbionts in *Siboglinidae* (Annelida). Biol Bull 214:135 – 144

Torsvik V, Goksoyr J, Daae FL (1990) High diversity in DNA of soil bacteria. Appl Environ

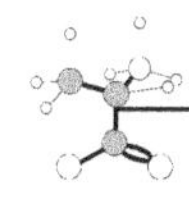

Microbiol 56:782 – 787

Treusch AH, Kletzin A, Raddatz G et al (2004) Characterization of large-insert DNA libraries from soil for environmental genomic studies of Archaea. Environ Microbiol 6:970 – 980

Tringe SG, Rubin EM (2005) Metagenomics: DNA sequencing of environmental samples. Nat Rev Genet 6:805 – 814

Tringe SG, von Mering C, Kobayashi A et al (2005) Comparative metagenomics of microbial communities. Science 308:554 – 557

Turnbaugh PJ, Ley RE, Mahowald MA et al (2006) An obesity-associated gut microbiome with increased capacity for energy harvest. Nature 444:1027 – 1031

Tyson GW, Chapman J, Hugenholtz P et al (2004) Community structure and metabolism through reconstruction of microbial genomes from the environment. Nature 428:37 – 43

Uchiyama T, Abe T, Ikemura T et al (2005) Substrate-induced gene-expression screening of environmental metagenome libraries for isolation of catabolic genes. Nat Biotechnol 23:88 – 93

Urbach E, Vergin KL, Giovannoni SJ (1999) Immunochemical detection and isolation of DNA from metabolically active bacteria. Appl Environ Microbiol 65:1207 – 1213

Vallenet D, Labarre L, Rouy Z et al (2006) MaGe: a microbial genome annotation system supported by synteny results. Nucleic Acids Res 34:53 – 65

Van Domselaar GH, Stothard P, Shrivastava S et al (2005) BASys: a web server for automated bacterial genome annotation. Nucleic Acids Res 33: W455 – W459

Venter JC, Remington K, Heidelberg JF et al (2004) Environmental genome shotgun sequencing of the Sargasso Sea. Science 304:66 – 74

von Mering C, Hugenholtz P, Raes J et al (2007) Quantitative phylogenetic assessment of microbial communities in diverse environments. Science 315:1126 – 1130

Ward DM, Weller R, Bateson MM (1990) 16s ribosomal-RNA sequences reveal numerous uncultured microorganisms in a natural community. Nature 345:63 – 65

Warnecke F, Luginbuhl P, Ivanova N et al (2007) Metagenomic and functional analysis of hindgut microbiota of a wood-feeding higher termite. Nature 450: U560 – U565

Watabe K, Leusch M, Ito J (1984) Replication of bacteriophage phi-29 DNA in vitro-the roles of terminal protein and DNA-polymerase. Proc Natl Acad Sci USA-Biol Sci 81:5374 – 5378

Webster G, Newberry CJ, Fry JC et al (2003) Assessment of bacterial community structure in the deep sub-seafloor biosphere by 16S rDNA-based techniques: a cautionary tale. J Microbiol Meth 55:155 – 164

Wheeler DA, Srinivasan M, Egholm M et al (2008) The complete genome of an individual by massively parallel DNA sequencing. Nature 452: U872 – U875

Wild J, Hradecna Z, Szybalski W (2002) Conditionally amplifiable BACs: switching from single-copy to high-copy vectors and genomic clones. Genome Res 12:1434 – 1444

Wilmes P, Bond PL (2006) Metaproteomics: studying functional gene expression in microbial ecosystems. Trends Microbiol 14:92 – 97

Wommack KE, Bhavsar J, Ravel J (2008) Metagenomics: read length matters. Appl Environ Microbiol 74:1453 – 1463

Woo SS, Jiang JM, Gill BS et al (1994) Construction and characterization of a bacterial artificial chromosome library of *Sorghum bicolor*. Nucleic Acids Res 22:4922 – 4931

Woyke T, Teeling H, Ivanova NN et al (2006) Symbiosis insights through metagenomic analysis of a microbial consortium. Nature 443:950 – 955

Ye YZ, Osterman A, Overbeek R et al (2005) Automatic detection of subsystem/pathway variants in genome analysis. Bioinformatics 21:I478 – I486

Yokouchi H, Fukuoka Y, Mukoyama D et al (2006) Whole-metagenome amplification of a micro-bial community associated with scleractinian coral by multiple displacement amplification using phi 29 polymerase. Environ Microbiol 8:1155 – 1163

Yooseph S, Sutton G, Rusch DB et al (2007) The Sorcerer II Global Ocean Sampling expedition: expanding the universe of protein families. PLoS Biol 5:432 – 466

Yutin N, Suzuki MT, Teeling H et al (2007) Assessing diversity and biogeography of aerobic anoxygenic phototrophic bacteria in surface waters of the Atlantic and Pacific Oceans using the Global Ocean Sampling expedition metagenomes. Environ Microbiol 9:1464 – 1475

Zengler K, Toledo G, Rappe M et al (2002) Cultivating the uncultured. Proc Natl Acad Sci U S A 99:15681 – 15686

Zhang L, Cui XF, Schmitt K et al (1992) Whole Genome Amplification from a single cell-implications for genetic-analysis. Proc Natl Acad Sci U S A 89:5847 – 5851

Zhang K, Martiny AC, Reppas NB et al (2006) Sequencing genomes from single cells by polymerase cloning. Nat Biotechnol 24:680 – 686

Zhou J, Bruns MA, Tiedje JM (1996) DNA recovery from soils of diverse composition. Appl Environ Microbiol 62:316 – 322

Zimmer R, Verrinder GA (1997) Construction and characterization of a large-fragment chicken bacterial artificial chromosome library. Genomics 42:217 – 226

Zwirglmaier K, Ludwig W, Schleifer KH (2004) Improved method for polynucleotide probe-based cell sorting, using DNA – coated microplates. Appl Environ Microbiol 70:494 – 497

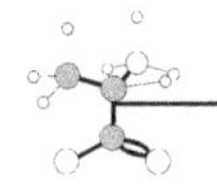

第3章　种群结构和环境适应性研究中的基因组学技术及其研究进展

摘要：随着人类基因治疗技术的发展，生物医学研究从本质上促进了基因组学领域的进展，比如人类基因组计划的实施。这项重大的国际研究项目显著提升了基因组测序能力，极大减少了测序的时间和成本，提高了用于分析的计算机能力。海洋生物学家利用高通量技术的优势解答了许多重要的科学问题，这在5年之前还是难以实现的事情。对于DNA水平的研究，基因组学能够有效地阐明种群遗传结构，并用于物种遗传作图研究和种群遗传漂移概况分析，以及准确度量生物多样性。对于RNA水平的研究，基因组学研究有望能够监测细胞水平的变化。在目前这个急剧改变的大环境下，这些环境变化与物种对特殊生境的适应性息息相关，因此，基因组学研究可预测物种在大环境下的生存能力波动。

本章主要介绍最常见的用于种群结构和环境适应性研究的基因组学技术，它们包括基于DNA水平研究种群的方法学和基于RNA水平研究表达水平的技术。技术的选择取决于研究的物种和可用的资源。在环境相关的研究中，明白以下内容十分重要："资源"不仅仅是用于测序和文库构建的财力，同时也是获取种质资源及在困难条件下成功贮存这些资源的能力。例如，通过巡航去调查一次特别的深海热源喷口，可能一个研究者一生当中仅仅发生一次，因此种质资源是稀缺的、数量是有限的，同时很可能无法在足够低的温度下保存样品以便防止RNA降解，那么基因表达研究就无法实现。同样，物种的可获得性经常是即时存在的，而不是通过计算去决定哪个物种更有研究价值。在其余的研究当中，比如水产品或者入侵物种，必须使用一个针对特殊物种的技术方法，即使获取材料是很棘手问题，原因在于该领域的工作有着明确的需求。所以在进行技术概述后问题是如何去使用它们。

本章主要关注低等脊椎动物（鱼类）和无脊椎动物的物种，研究案例将证明这些技术是如何被应用于阐明海洋环境的重要生态学问题。这些问题包括种群结构分析和基因表达（功能）研究，说明种群如何适应他们的生存环

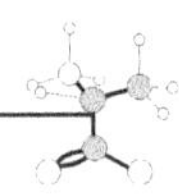

境并与之相互作用。

最后，海洋生物学家所面临的挑战是如何利用好在这些已知海量基因组序列数据的模式物种中开展研究的技术，去开发出一些可用于非模式物种研究的技术。本质上这是从零开始，并不是一个简单的任务。

缩略词：

AFLP	Amplified fragment length polymorphism
BAC	Bacterial artificial chromosome
cDNA	Copy/complementary DNA
COI	Cytochrome C oxidase
dbEST	Database for ESTs：http://www.ncbi.nlm.nih.gov/dbEST/
DNA	Deoxyribonucleic acid
ER	Endoplasmic reticulum
EST	Expressed sequence tag
GH	Growth hormone
GHRH	Growth hormone releasing hormone
GMPD	Glycosylase mediated polymorphism detection
HSP	Heat shock protein
IPCC	Intergovernmental panel on climate change
MPA	Marine protected area
MPSS	Massively parallel sequencing signature
PR	Prolactin
Q-PCR	Quantitative or Real Time PCR
QTL	Quantitative trait locus/loci
RNA	Ribonucleic acid
SNP	Single nucleotide polymorphism
SSR	Simple sequence repeat
SYBR	Green Fluorescent green dye
UTR	Untranslated region

3.1 技术

本节将主要介绍有用的技术，其中一些能同时用在 DNA 和 RNA 研究上；接下来将对它们进行描述。这些技术将在副标题中注明主要是用在 DNA（种

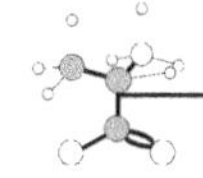

群分析）还是 RNA（功能/表达）研究中。每个技术都会给出简明的描述，同时在当中会有更多详细的内容。

3.1.1 DNA 和 RNA 研究：EST 文库

EST 是表达序列标签（Expressed Sequence Tag）的缩写。一个 EST 对应一个 cDNA 克隆片段的单个测序序列。cDNA 的产生是为了增加稳定性，即与一个基因的表达部分相对应的 RNA 分子转录成 DNA 拷贝的过程。通常会从每个克隆的其中一端开始测序（一般情况下是基因的 5' 端，因为可以避免 polyA 尾巴导致的测序错误）。这种技术能更直接地获取开放阅读框的序列信息，因此可以增加发现潜在基因的机会；这个过程可通过数据库的基因序列相似性搜索匹配来实现。我们将数据输入到一个公共数据库 dbEST（Boguski et al. 1993；http://www.ncbi.nlm.nih.gov/dbEST/）中，不过要注意测序质量可能没有达到 100%，错误随时可能存在。EST 能够从任何细胞、组织和物种当中产生，是一个从非模式物种当中发现和鉴定新基因的好办法。它们也能够用于一些分子标记的发掘，用于对一些特殊 DNA（微卫星和单核酸多态性）的研究。对于基于 DNA 的研究，EST 文库的来源也许不是那么重要，但是对于表达方面的工作来说，EST 文库来源以及使用的文库类型却是极其重要的考虑要素。

制作许多来自同一个特定组织或者细胞类型的 EST 就可建立一个文库。文库可以包含仅有数百条的序列，也可以有成千上万条的序列，具体规模取决于成本以及使用大规模大范围测序设备的便利程度。在本质上，有两个方法常用来建立 EST 文库：

（1）无消减法（Non-subtracted methodologies）。在研究工作中，cDNA 是直接从细胞或组织当中产生的，所以一条序列在文库当中出现的次数是细胞中该基因所对应的 RNA 数量的直接反映。这可能意味着高丰度表达的序列（如肌动蛋白 actin 在肌肉中的表达）可能会掩盖那些少量表达克隆被检测到的可能性。因此，对一个包含较多高丰度表达序列的文库进行测序时，结果通常并不理想，这主要是因为对相同基因进行测序效率低，而且仅仅在评价特异位点的核酸多态性时能起到作用。尤其是当只有少数克隆可以获得并测序时，这个问题尤显重要。如果能够获得同一种 RNA 分子在不同组织中的表达量，对这种研究非常有帮助。

（2）消减法或均一化法（Subtracted andor normalised methodologies）。这种方法涉及到杂交的步骤，用来除去那些在细胞或组织中有最高表达的序列，

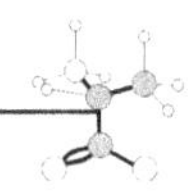

目的是为了增加细胞中那些低丰度表达序列的相对拷贝数量（Suzuki et al. 1997，Carninci et al. 2002，Otsuka et al. 2003）。所以，随着这一新技术的发展，大量的不同序列在各个文库当中被发现（发现基因变得更加高效），但是不能估算基因拷贝的相对数目。这项技术比无消减法要复杂，通常来说它需要大量的RNA，而且在PCR扩增过程中还有扩增偏好性的影响。

两种文库构建技术都各自有优缺点，因此在进行相应的生物信息学分析时要对存在的问题进行系统的分析。EST标签既可以作为数据资源用于DNA标记研究，也可以作为克隆实体用在基因芯片或者目标基因分析等产品上。而在很大程度上，EST文库常被用于前者，也是迄今为止EST文库被运用得最好的地方（见章节3.1.2和3.1.3）。同时，大量文献对EST文库的构建进行了报道（例如Douglas et al. 2007，Govoroun et al. 2006），使其成为未来研究的指导性基因清单或目录。其实EST文库本身并不能提供精确的有关环境适应性方面的信息，任何的推测和发现都必需由深入的功能分析和实验操作来确认。还有，非模式物种缺乏与EST数据相关的功能信息，如许多序列被指定为“未知”或“假定蛋白”，这大大影响了它们在功能和种群研究上的使用（见3.2.1.1）。

3.1.2 DNA研究：微卫星

微卫星也被称为简单重复序列（simple sequence repeats，SSR），是一段短DNA序列，是由少数几个碱基重复串联而成（比如GT重复单元）。因为它们在长度上存在高度多态性，所以自1990年起就被广泛用于绘制遗传连锁图谱，进行种群管理、亲权鉴定和种群扩散等。这项技术的主要缺点是需要在非模式物种中获取大量的具有统计学分析意义的多态性标记（Zane et al. 2002）。

随着越来越多基因组数据的发表，通过在基因组保守区设计通用引物，大大增加了从非模式物种中获得微卫星的可能性。基于EST的SSR数据在不同物种当中的出现频率高度可变。微卫星已经在许多EST文库中被发现，包括许多具有生态或重要经济价值的海洋物种，有无脊椎动物（例如欧洲蛤、紫贻贝、日本牡蛎等，Tanguy et al. 2008；海湾扇贝，Roberts et al. 2005）、鱼类（鳕鱼、大西洋鳕、大比目鱼、庸鲽，Douglas et al. 2007，Weiss et al. 2007）和海洋植物（譬如大叶藻，Oetjen and Reusch 2007b）。在10万条来自于深海通风孔的多毛纲庞贝蠕虫的cDNA序列数据集中，发现有大量的EST-SSR分布在cDNA的3'UTR区域（Daguin and Jollivet，unpublished data）。从基

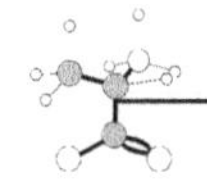

因组 DNA 得到的微卫星和来源于 EST 的微卫星之间有着不同的性能。Oetjen and Reusch（2007b）指出，来源于 EST 的微卫星应该谨慎地用于种群遗传学分析，因为它们很可能与某些受到选择的基因有着强烈的连锁关系。然而，因为这些标记存在于已知基因当中，所以它们可以作为那些使用未知遗传标记方法的可行性选择。它们还具有一些其他优点：

（1）可特异应用于研究自然选择。举个例子，对 GenBank 数据库中的 58,146条大西洋鲑 EST 序列进行分析时，发现 75 个 EST 连锁的微卫星可作为检测大西洋鲑的选择标记（Vasemägi et al. 2005）。

（2）来源于 EST 的微卫星标记通常会在多个物种中相当保守，因此可以在大量非模式物种之间进行比较研究。对于水生动物之间这种比较研究很少报道，但是在植物当中研究比较深入（Ellis and Burke 2007）。跨物种的微卫星扩增存在一个潜在问题，那就是无效的等位基因的增加。这会影响等位基因频率的准确评估，降低表观杂合度，增加近交系数观测值（DeWoody et al. 2006）。然而，相对于未知的微卫星序列，EST－SSRs 侧翼的引物通常都是存在于更为保守的序列当中，所以无效等位基因产生的问题很少。虽然在其他物种当中报道了这种无效等位基因的存在，但是，我们并不知道在水生生物中微卫星标记无效等位基因的具体情况。

（3）通过比较由非编码 DNA（如“传统的”未知微卫星）和编码 DNA（如 EST－SSRs、UTR 和内含子）中所获得微卫星数据，分别计算的种群多样性和分化水平，能使我们更好地对种群遗传参数进行评估，以及更好地评判选择压力对每种标记类型的相对作用强度（Luikart et al. 2003，Oetjen and Reusch 2007b）。

3.1.3 DNA 研究：单核苷酸多态性（SNP）

SNP 被认为是最能反映遗传变异（多态性）的遗传标记类型。在编码区，它们被用来计算同义替换和非同义替换的比率，这是选择进化研究中的主要问题（参阅综述 Ford 2002，Vasemägi and Primmer 2005）。与微卫星类似，在非模式物种中检测 SNPs 比较困难，并且价格昂贵（Kim and Misra 2007）。同样，公共基因组数据库的不断出现和丰富为 SNPs 的研究和开发提供了极大帮助（Kim and Misra 2007，Phillips 2007，Hayes et al. 2007）。例如，通过参考姊妹物种的基因组序列，成功获得了非模式物种太平洋鲑的 SNP 集合（Smith et al. 2005，Campbell and Narum 2008，Ryynanen and Primmer 2006）。同样，从17 056个 EST 序列中分离出了 318 个鳕鱼的 SNP（Moen et al. 2008），这是

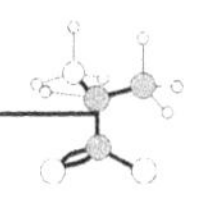

第一个基于 cDNA 文库，从种群水平筛选到的结果（Tanguy et al. 2008，Faure et al. 2007，2008）。SNP 的检测并不是只能依靠 EST 数据库的筛选，也可以通过其他方法获得，详细的 SNP 标记的方法不再赘述（请参阅综述 Kim and Misra 2007，Hudson 2008）。值得一提的是，一项新技术——GMPD 技术（glycosylase mediated polymorphism detection；O'Leary et al. 2006，Vaughan 2000）似乎对于目标特异基因的 SNP 检测特别有效，而且产生的结果可以和使用其他类型标记得到的遗传多样性进行比较。

和所有类型的标记一样，SNP 也有其优缺点。Luikart（2003）曾强调在种群基因组学研究中，使用 SNPs 可能会出现固有偏差，主要是因为选择的标准（如多态性）和检测的方法（如个体数太少）对结果的影响较大。实际上，通过和未知微卫星、同工酶的研究对比，由 SNP 引起的一些偏差已经在切努克鲑鱼当中得到确认（Smith et al. 2007）。虽然没有一种遗传标记是完美的，但最重要的是我们要正确认识到一项技术的优点和局限性，进而有针对性地进行实验和分析（见 3. 2. 1 更深入的讨论）。

3. 1. 4　DNA 研究：扩增片段长度多态性（AFLP）

AFLP 数据相对容易获得，并且具有较高的可靠性。这项技术主要是使用限制性内切酶对基因组 DNA 进行切割，并将接头连接到粘性末端，然后根据接头序列设计引物对酶切片段进行扩增，最后将扩增的不同大小片段通过聚丙烯酰胺凝胶电泳或者毛细管测序仪进行分离。在 AFLP 中只对两种等位基因状态进行统计，就是有和无，通过特异长度的电泳条带来确定一个 AFLP 位点有无某一等位基因。

AFLP 在植物研究中已经成为一项完善成熟的分子技术（Meudt and Clarke 2007）。在非模式物种中获得大于 100 个 AFLP 标记还是相对容易的，相比之下，许多微卫星位点则经常只能获得不足 20 个可用的遗传标记。位点的使用数量对正确评估选择的影响（例如 Beaumont and Nichols 1996）或瓶颈效应（Luikart et al. 1998）是非常重要的。此外，AFLP 还可以被应用于一些其他研究，如近亲繁殖（Dasmahapatra et al. 2008）以及分析具有较弱种群结构的种群特征（Campbell et al. 2003），比如许多海洋生物种群（Ward et al. 1994，DeWoody and Avise 2000）。虽然与陆生动植物相比，海洋无脊椎动物使用较少，但近期的研究表明，AFLP 在研究许多生态问题上是一个非常有用的工具（见表 3. 1）。

表 3. 1　AFLP 技术的应用实例。AFLP 的使用其实并非仅限于此表中的介绍。理论上，大多数的群体遗传学和生态学相关问题都可以用这种方法来解决。

Table 1 Examples of uses of AFLPs. This list is not restrictive and most population genetics and associated ecological issues can theoretically be addressed using such markers

Characteristic	Species	References
Divergence between different ecotypes	*Littorina saxatilis* periwinkle	Oetjen and Reusch (2007a)
Zones of secondary contact	*Crassostrea virginica* oyster	Murray and Hare (2006)
	Anguilla spp. eel species	Albert et al. (2006)
Stock definition	*Cyclina sinensis* clams	Zhao et al. (2007)
	Crangon crangon shrimp	Weetman et al. (2007)
	Solea vulgaris common sole	Garoia et al. (2007)
Anadromous species	*Oncorhynchus keta* chum salmon	Flannery et al. (2007)
Connectivity among populations	*Haliotis rutescens* Californian red abalone	Gruenthal et al. (2007)
	Portunus pelagicus blue swimming crab	Klinbunga et al. (2007)
	Asterina gibbosa sea star	Baus et al. (2005)
	Riftia pachyptila tubeworm	Shank and Halanych (2007)
Forensic issues	Various	Maldini et al. (2006)
Assortative mating	Florida Keys hamlets genus *Hypoplectrus*	Barretto and McCartney (2008)
Reproductive tactics	*Stichopus chloronotus* Holothurian *Heteroxenia fuscescens* coral	Uthicke and Conand (2005) and Fuchs et al. (2006)
Maximising genetic diversity when founding a hatchery population	*Salmo salar* Atlantic salmon	Hayes et al. (2006)

在 AFLP 的应用上，有时会产生一些人为的错误，比如在明显不同的位点上扩增出相同大小条带的可能性（Gort et al. 2006，Pompanon et al. 2005），以及不同个体在等位基因和基因座之间可能出现混淆。这会导致出现同质性（类似于趋同进化的相似性），这是任何一种遗传标记都会遇到的情况（Ellegren 2004）。例如，Wares and Blakeslee（2007）在分析玉黍螺入侵美国北海岸研究时，发现了同质性是采用 AFLP 这一技术的主要不足。因为 AFLP 是显性标记，并没有真正遵从异常基因座的检测规则（参阅文献 Guinand et al. 2004），这些基因座是假定 Hardy-Weinberg 平衡成立的条件下，对等位基因频率做出的评估，然而显性标记并没有通过验证。不过，Bonin et al.（2006）发

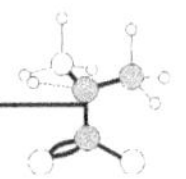

现，由 Hardy-Weinberg 平衡引起的位点特异偏差可能会更大，因此会使异常检测的结果产生不可预知的偏差。最后，AFLP 还有一个主要的缺点，就是它们在非编码区是未知标记。因为它们不能被一一鉴定并且不易与其他基因连锁，因此很难使用它们去填补表型和基因型之间以及观测多样性和适合度之间的缺口。AFLP 最好的用途是促进生态学领域知识的增加，比如更好的库存或物种描述以及表 1 提到的内容。然而，AFLP 并不能很好地与个体的适应性建立可靠的关联，因为在这个问题的研究上使用任何遗传标记都是有争议的（Balloux et al. 2004，Pemberton 2004，Slate et al. 2004，DeWoody and DeWoody 2005）。AFLP 技术现在正和基于 AFLP 片段的焦磷酸测序法共同使用来开发 SNP 和微卫星标记。

3.1.5 DNA 研究：高通量测序

Sanger 测序法已经成为一种常规的实验技术，不过最近新一代测序技术的显著优势，比如大规模平行信号（MPSS，Brenner et al. 2000）和焦磷酸测序法（也称 454 测序；Margulies et al. 2005，Langaee and Ronaghi 2005）已经使测序技术发生了彻底的变革，它们可以同时对数百万个短序列读长进行测序。尽管面临着来自生物信息学方面的挑战，但是这些技术为探索更多生态学与进化问题提供了更多的机会，其中包括对生物多样性的分析（Venter et al. 2004）。此外，二代测序技术是最不可能由于操作不当、缺失、稀有转录及克隆细菌不稳定的原因而产生错误的。技术进步使得这一方法变得不断可靠（Hamady et al. 2008），绝大多数的转录表达（包括那些表达量极少的转录本）都能够被精准地定量（Stolovitzky et al. 2005）。随着序列读长的增加，454 焦磷酸测序法的使用频率将大增，能进一步增加在非模式物种中鉴定基因的概率（Hudson 2008）。由于测序的读长较短，这项技术最初只能用于已测序的模式物种。

短片段测序数据最有效的用途是与其他高质量测序的基因组做比对。如 Vera et al.（2008）利用家蚕做序列比对，从而建立了蛔蝶科有用的 Contigs。家蚕基因组估计有 18 000 个左右的基因，并且假定这是鳞翅目的典型代表，Vera et al.（2008），在这个研究中获得9 000条非冗余序列。这或许表明使用单一焦磷酸测序，至少一半的庆网蛱蝶基因能被识别。如 Ellegren（2008）注意到 Vera et al. 的研究那样。这项研究还表明，454 测序技术为关联比对、QTL、种群遗传研究、候选基因研究等提供了技术支撑（Saastamoinen and Hanski 2008）。虽然庆网蛱蝶的基因组仍然只是草图，但是 Toth et al.（2007）

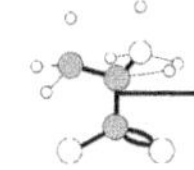

在 Polistes 黄蜂上的研究工作证实了 cDNA 高通量技术测序的确可以用于生态和进化相关物种的研究中。一个更深入的使用高通量技术的案例会在 3. 3. 4 章中给出，该案例主要对杂种优势进行了分析。如前文所述，这项技术会更多地被使用，并且随着成本的降低，会成为基因芯片表达分析的替代工具。

3. 1. 6　DNA 和 RNA 研究：目标基因分析

靶向基因分析或候选基因筛选（Tabor et al. 2002）是指在不同条件下（自然或人工诱导）对一个生物体或者一个群体的特定基因或少数的一系列基因进行深入研究。在非模式物种中执行靶向基因分析是很困难的，因为只有很少或者没有基因数据作为参照。基因序列可以通过 EST 文库获取，然后设计感兴趣基因的特异性引物对基因进行扩增，或者也可以采用随机引物法并使用较低退火温度对其进行 PCR 扩增以获得目标基因。随机引物法可以在不同物种中鉴定同一感兴趣基因，这些物种最好是包含与目标物种同一类别的物种，可以根据蛋白序列设计兼并引物。

在表达研究中，由于基因与环境处理之间的相互作用，基因可能会发生表达水平的变化，那么目标基因分析法正好用于此类研究。目前，研究目标基因的表达分析一般使用 Q-PCR 技术。DNA 研究历来都非常重视系统发育分析以及一些特殊的基因，比如细胞色素 c 氧化酶亚单位 I 的编码基因（COI）和核糖体基因（18s、16s 和 28s；http：// www. barcoding. si. edu/，http：// rdp. cme. msu. edu，http：// bioinformatics. psb. ugent. be/webtools/rRNA/）。COI 基因常用在 DNA 条形码技术上，这已在第一章中进行了详细的描述。然而，目前测序技术的发展使来自同一种群的多个个体之间靶基因的平行重测序成为可能。得到的单倍体信息随后既可用于研究特异等位基因是否和感兴趣的性状相关联，也可以聚合法处理近缘种形成中的选择和杂交过程（Faure et al. 2008）。在这项技术的一个改良技术（ecoTILLING）中，研究者开始使用跟感兴趣的表型相关的基因（往往从基因组当中选择），而不是使用特殊的标记（Comai et al. 2004）。ecoTILLING 方法能够采用 PCR 扩增一个种群的等位基因，快速地检测一个种群基因型的自然基因变异，随后运用 CEL1 或 ENDO1 内切酶来检验错配，以上两个酶会在多态性位点切割 DNA（Comai et al. 2004）。随后，变异等位基因就可以使用分子标记进行跟踪，并与适应性表型进行关联。

3. 1. 7　DNA 研究：条形码技术

测序技术的快速发展使得新出现的测序技术被广泛应用于生物条形码研

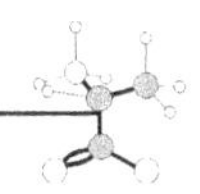

究。条形码技术（Barcoding）是一个形象化的名词，其实就是通过一段合适的DNA片段来分辨不同的物种。这项技术不仅可用于种群和分类学研究，也可用于海洋生物系谱鉴定。这项技术的应用在第一章里已经进行了详细的介绍，因此这里不再赘述，仅在大规模种群分析中会再次提及相关内容。

3.1.8 RNA研究：微阵列或基因芯片

基因芯片代表了对多个基因在同一时间表达情况的分析的方法。成千上万条序列分别独立地附着在一个小玻璃载片上（显微镜载玻片尺寸），每一个点代表着一条基因序列。这些序列可以是cDNA，也可以是通过EST或基因组序列信息设计的寡核苷酸。样品与微阵列芯片的杂交，通过使用单荧光或双荧光染料系统来实现对信号的检测（Gibson 2002）。随后再使用统计学分析手段分析颜色与荧光强度水平，从而确定使用的处理方法是否影响了基因的表达。一般情况下，基因表现出两倍以上的荧光强度增长，就认为是发生了显著变化。这无疑是一个规模化筛选基因表达的强大技术，必须仔细地设计实验，严谨地考虑对照以及规划好重复样本。如果太多的变量被引入实验，那么数据就会变得复杂，不能获得有意义的信息。目前，使用其他转录分析的方法（如Q-PCR）来证实基因芯片的结果很有必要。

3.1.9 RNA研究：Q-PCR

Q-PCR即定量PCR（Quantitative PCR），有时也称为实时PCR（Real-Time PCR）。通过这项技术，可以在对照组和实验组动物中使用荧光标记（通常用SUBR Green）的引物对感兴趣基因进行定量分析，主要是通过在PCR过程中荧光激发来监控这些引物的插入量。荧光的水平能够直接地反映出生成PCR产物的数量。通过对比PCR反应中扩增倍数达到预先设定的、用于比较两个样本区别的对数期时间节点，处理组转录丰度的水平变化便可以被确定。有许多方法可以控制和证实这项技术（参见Pfaffl 2001，Pfaffl et al. 2002，Radonic et al. 2004）。这项技术经常和微阵列分析一起使用，因为在评估表达水平的变化问题上，通过Q-PCR所得的结果比微阵列分析更加精确。但由于设计和检测特异性引物非常耗时，因此微阵列技术更加适用于作最初的、总体的转录活性分析，Q-PCR则应用于分析微阵列研究鉴定出来的候选基因的表达情况。在之前的章节中，对主要用于生态基因组学的研究技术已经有了简要的介绍，需要注意的是，在人类基因组计划催生的技术副产品的刺激下，这样的技术在不断增加。现在重要的是概述它们当中的一些技术如何运

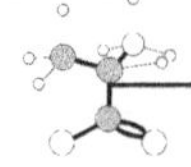

用在海洋环境，并且对我们在种群动态、生物多样性以及环境问题上的认知能做出多大贡献。

3.2 种群基因组学

种群基因组学区别于种群遗传学的地方主要在基因组覆盖度的显著增加以及生物信息学方法的使用来提供分子工具以及分析大数据。种群基因组学把中性和受到选择的位点都作为研究对象，这样不仅可以同时对种群历史（种群大小的改变）和选择过程（环境约束的影响）进行研究，同时又可以对二者进行有效区分。

种群基因组学整合了基因组学发展出来的方法论和种群遗传学的概念框架（Luikart et al. 2003），可被广义地定义为探索进化或生态的问题而对大量DNA片段进行筛选和分析（Black et al. 2001，Luikart et al. 2003，Schlotterer 2003，Feder and Mitchells-Olds 2003）。相关的方法和途径是高度跨学科的。举个例子，为了研究一个表型的遗传基础，可以从个体和群体水平上对基因和环境之间的任何关系进行研究。这样的关系也可以从功能的角度进行分析，比如通过检测个体或群体水平中与这个表型相关的并且与该基因共同表达的调节基因的表达情况（Feder 2007，Crawford and Oleksiak 2007，M arden 2008）。一些学者把环境或生态基因组学定义为种群基因组学和功能基因组学的集合（例如Ungerer et al. 2008），而另外一些学者则将生态基因组学作为独特的领域（Wilson et al. 2005）。我们同意这些不同的观点，显然基因组学广泛而深刻地影响着现今的生物学。我们在这里描述了种群遗传学或数量遗传学的理论框架的用途，它们能帮助我们鉴定和研究那些影响适应性快的遗传基础，这些适应性基因在自然选择过程中起着关键的作用（Ellegren and Sheldon 2008）。这些研究在模式物种当中很常见，现在已经被进一步地应用于生态学、表型可塑性以及基因和环境互作等相关研究（Morin et al. 2004，Nielsen 2005）。

种群基因组学中非常诱人的一个研究领域即是对复杂进化历程的探索，比如动植物对环境的适应模式和速率。通过检验大部分的基因组，种群基因组学可以在理论上解释位点特异效应，比如重组、选择、上位相互作用等等，这些效应影响了一个或一些基因座，从基因组范围内来看也影响到了整个基因组（瓶颈效应、建群、近亲繁殖等；Mitchell-Olds et al. 2007，Stinchcombe and Hoekstra 2008，Ellegren and Sheldon 2008）。由于生物体在进化过程中对环

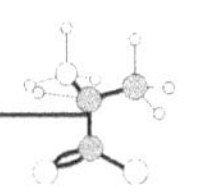

境的长期适应，一些选择过程可能同样会作用于整个基因组（譬如热或压力对蛋白质稳定性的影响）。种群遗传学能应用在探索生态学和环境问题上，比如生物多样性研究、渔业相关的海洋生态系统调查、入侵物种监视和海洋保护区（MPAs）的设立。它们同样提供了探索以下问题的可能性，即在表型水平上所表现出的依赖于基因表达调控和分子进程的权衡现象（Roff 2007）。

3.2.1 分析：选择、局限性和注意事项

3.2.1.1 标记类型

Vasemägi 和 Ryynanen 等综述了几种遗传学标记类型（包括 SNP）在遗传数据分析以及种群参数估计方面的使用情况（Vasemägi and Primmer 2005，Ryynanen et al. 2007）。Vasemägi et al.（2005）在对大西洋鲑鱼的研究中表明了不同遗传标记和可能被选择的基因位点数会使杂合度等位基因多样性，及种群分化水平发生改变在一系列遗传标记上的差异水平、潜在的被选择基因位点数量的变化等问题。他们通过对来自于不同自然环境栖息地（咸水、半咸水和淡水）群体的 95 个基因组和 78 个 EST 微卫星标记（EST - SSR）进行了研究，发现了歧化选择的遗传特征。当把所有调查的群体作为一个整体的来研究时，通过对未知标记或 EST - SSR 标记位点的检测，他们没有发现在杂合度、等位基因多样性或种群分化水平上的显著变化。然而，当按照栖息类群将种群进行划分时，关于遗传和等位基因多样性的显著差异则明显表现出来。他们还发现，EST 中的重复串联标记与未知基因组微卫星标记相比，并没有出现更频繁地偏离中间的期望值。这可能归结于这个研究当中的每个类型标记在数量上的差异，但类似的结果已经在其他生物当中有所报道（譬如 Woodhead et al. 2005）。这个结果可以解释为，有一小部分 EST - SSR 标记（通常与基因连锁）受到了选择。实际上，Vasemägi et al.（2005）最终发现仅有 9 个假定的 EST - SSR（12%）是可能处于选择压力下的。

Moen et al.（2008）使用 SNP 标记对鳕鱼开展了类似的研究。不过鳕鱼的研究并没有使用其他类型的标记去调查基因或等位基因的多样性和种群分化水平。在鳕鱼中，318 个 SNP 中有 48 个（≈15%）遗传分化水平显著，有 29 个是潜在的选择（≈9%）。后者中，有 17 个 SNP 位点和已知功能的基因相关联。

这两项研究表明，当研究中主要使用 EST - SSR 标记或 SNP 位点时，受到选择的标记的百分比要大于 9%。这个数字可以和已使用 AFLP 标记的基因组扫描得出的估值进行比对。这些报道中，1.5%～12.5% 的位点受到了选择

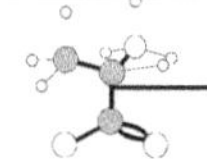

压力的影响，随后，这些位点在特定的生境中不断提高对环境的适应性（Wilding et al. 2001，Campbell and Bernatchez 2004，Bonin et al. 2006，Gruenthal and Burton 2008）。Bonin et al.（2006）认为，报道的 AFLP 的最高值（12.5%，392 个位点中的49 个）是有争议的，因为它取决于用在选择的位点检测的统计方法，以及依赖于数据取样设计（例如在分层数据分析中需要考虑地理跨度和主要生态因子）。对于 AFLP，这个数字很可能是高估的，校正值很可能是2%（Bonin et al. 2006）。需要更多的研究才能对标记类型进行严谨的比较，不过目前的数据显示包含在 EST-SSR 或 SNP 的编码序列是更好的适应性多态性资源。这个数字不可能超过 15%，但也表明建立 EST 文库来覆盖“巨大”数量的潜在选择基因时，需要付出巨大的努力。然而，这并不否定 EST 库是挖掘生物适应位点的一个重要来源，特别是由于它们的开放性以及允许任何人去开发利用。

到目前为止，只有很少的水生物种群研究是单独通过 EST-SSR 标记对遗传变异进行研究的，不过 Vasemägi et al.（2005）的研究揭示了这种方法在非模式物种研究中的一些不足之处。例如，在这项研究中，大量的 EST 序列没有匹配数据库中的任何已知基因。“未知”基因的比例估计在欧洲海鲈（Boutet et al. 2006，Chini et al. 2006）中高达 70%，在大比目鱼中低至约 13%（Douglas et al. 2007）。因此，从任何非模式物种中通过随机 EST 数据库挖掘获得的 EST-SSR 能够得到相似的结果（Vasemägi et al. 2005），在他们的研究中不能直接解释环境适应性的生理基础。如果仅仅使用匹配上了已知基因的 EST-SSR，这种情况可能得到改善。这会提高我们关于生理调节是如何形成特定的表型或形成栖息地分布格局的认识。但是，这也可能极大地减少可用于种群研究的标记，降低基因组学工具在生态和种群研究领域的优势。因此，功能信息不应该成为选择 EST-SSR 标记的一个先决条件。

3.2.1.2 区分选择和种群历史效应

研究适应性遗传基础的时候，碰到一个巨大的挑战是如何区分适应性变异模式是由遗传搭车效应（hitch-hiking）或选择性清除产生的还是由种群历史变化产生的。因此，种群遗传学研究的一个常用的方法是分析大量的非连锁位点（所谓的“基因组扫描”方法，见 Luikart et al. 2003，STORZ 2005），以区分选择作用和种群历史（见 Teshima et al. 2006）。例如 Wenne et al.（2007）指出，基因组扫描中的一些异常位点（即那些在统计学上不遵守中性选择理论的位点）为一些有信息含量标记的产生提供了可能，这些标记适用于解决一些海洋物种中出现的特有问题。根据最早由 Maynard-Smith and Haigh

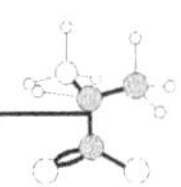

(1974）提出来的理论，异常位点是最有可能的作用位点，但也有可能是与经历或者正在经历选择作用的位点存在物理连锁或连锁不平衡关系，在标记基因位点和涉及生态适应性功能相关的突变之间的连锁不平衡程度，可能因为基因组和研究系统的不同而发生显著变化。连锁不平衡程度受到种群历史、交配系统、重组率、受选择的等位基因的年龄、选择压力的大小等诸多因素影响（Nordborg and Tavare 2002)。一些标记类型能够用于基因组扫描，比如未知的微卫星（两栖类，Bonin et al. 2006)、SNP（人类，Akey et al. 2002)，或者几种遗传标记的结合（鲑鱼，Smith et al. 2007)。由于低廉的成本以及不需要已知基因组信息，显性（Bensch et al. 2002）AFLP（Vos et al. 1995）标记也是一个做这类分析的合适工具。

3.2.1.3 识别适应性状

使用未知的微卫星、EST-SSR 和 AFLP 等遗传分子标记，并不一定是研究所有生态问题的最好方法。因为他们大部分是中性的，而且不能帮助我们了解野生个体中的适应性差异（可以推动它们在给定环境中的适应)。一般来说，并不知道它们是否与某个基因有关。就算已经知道，那么在大量海洋生物里这个基因在新陈代谢或基因网络中所扮演的角色，也是十分模糊的。但是，这些标记的不断发展，对于制作精细连锁图谱还是很关键，而这个连锁图最后将会帮助对适应性性状或基因的识别。

连锁分析是一种传统的分析手段，在模式物种中用来定位染色体区域里的数量性状位点（QTL)。这种方法依靠追踪遗传特征的分离，并试图寻找（共同）遗传特征和大量的遗传标记。因此，对微卫星和 AFLP 等全基因组范围内的分子标记的统计分析，以及对控制杂交的子代表型测量，可以用来定位对某个或一套相关性状的表型分化做出贡献的染色体区域（Mackay 2001，Erickson et al. 2004)。如果能定位这样的区域，那么性状位点就可以被定位于标记位点（诸如微卫星、AFLP 甚至 SNP 等标记物）附近，这时，这些遗传标记就有了自己的“身份”。那么通过研究少数标记的多态性，理论上性状的决定因素可被查明。遗传杂交中的自交系运用是进行 QTL 分析的最有效方法，因为它把标记物与性状位点之间的连锁不平衡最大化了（Lynch and Walsh 1998)。于是，分析就更大地基于谱系了。

连锁分析的效力会随着被研究的减数分裂数目的增加而增强。迄今为止，连锁图谱的构建主要集中在那些能够驯养繁殖的物种中，包括鱼类、昆虫和哺乳动物。在海洋生物中，研究主要集中于水产养殖物种，包括欧洲海鲈（Chistiakhov et al. 2005)、金头鲷（Franch et al. 2006)、欧洲平牡蛎（Lallias et al. 2007a）、太平洋牡蛎（Hubert and Hedgecock 2004，Li and Guo 2004）、美洲牡蛎（Yu and Guo 2003）、海湾扇贝（Qin et al. 2007）、黑鲍（Baranski et al. 2006）和紫贻贝（Lallias et al. 2007b）。同时，适用于水生生物的连锁图谱的数量也在快速增加（Wenne et al. 2007）。目前大部分对 QTL 的研究都集中在与生长相关的性状，只有少部分涉及与抗病相关的性状。尽管生长是适应性的一个重要标志，其他生态相关性状（例如再生能力、稚鱼期以及各种胁迫下的基因表达应答）也应当得到相应的研究。

目前，可用于水生生物的遗传图谱还没有精细到足够定位适应性位点，因为扩散在基因组中的遗传标记很少超过 1000 个。把 QTL 正确定位到染色体群的区段上，需要遗传标记的平均跨度为 20 cM（例如 Rogers et al. 2001，Chistiakhov et al. 2005）。但对于更精细的定位，则需要 1 cM 或更小的标记距离。绘制与感兴趣性状相关联的染色体区域是定位表型遗传变异的第一步。许多基因通常属于同一目标染色体区域，最终鉴定这样的遗传变异需要精确地定位和指定候选基因。到目前为止，在自然种群研究中只有少许这样的例子，如对三刺鱼不同种群的外胚叶发育不全（Eda gene）armour plate 配对基因（Cresko et al. 2004，Colosimo et al. 2005）进行了测绘和鉴定。这个基因座的等位基因变异已经导致了这个淡水生态物种的骨盘减少，这是一个与捕食风险变化相关的过程。

QTL 绘图是研究远交自然种群的一种常见方法，它需要大量的家系样本或大量的姊妹样本以及在这些个体中所测量得到的适应性组成成分，如哺乳动物（Beraldi et al. 2006）。在没有巨大的财力投入和庞大的定向繁殖计划的情况下，这基本上是不可能实现的，因此不是一个切合实际的选择，在没有商业目的的情况下，这种劣势尤为明显。最近的有关这个问题的例子是一项对欧洲海鲈的研究。这个研究报道了了解甚少的谱系个体之间的杂交，并且定位了体型的 QTL（Chatziplis et al. 2007），水产养殖业和野生种群研究都将得益于此。

另一种定位 QTL 的方法是绘制关联或连锁不平衡图谱。它是通过种群样本全基因组扫描得到的连锁不平衡位点与标记之间的关联分析，而不是通过 QTL 系谱分析。这种方法通常比传统的系谱分析具有更高的分辨率，但关联分析效率依赖于用于扫描基因组标记的数量和分布，以及连锁不平衡的程度。这本身是一个参数变化（Jorde 2000）并强烈地依赖于种群或物种的历史（例如 Backström et al. 2006）以及种群结构模型（Yu et al. 2006）。由于全基因组扫描的标记数目低，我们不知道不久的将来是否可以使用这样的方法来研究

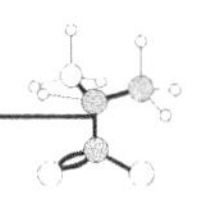

海洋物种序列差异的成因。

承认在当代种群中，QTL对由定向选择驱动的适应表型分化鉴定起到作用，对使用任何上述列举的标记来对全基因组进行扫描都是很重要的。根据定义，QTL可能用于推断导致物种或种群差异的适应性特征的遗传基础，但是它们不依靠本身的作用选择，而可能是相应适应性特征（Hoekstra and Nachman 2003，Rogers and Bernatchez 2005）。基因组扫描同时依赖于检测潜在的选择标记，需要弄清这些标记在特定的表型结构上的功能和角色。然而，如果在种群内部分化偏小或者种群之间分化偏大的群体中扫描所得的选择信号与QTL在染色体的定位相一致，这就表明在这些区域的基因可能存在进化上的适应性（Ellegren and Sheldon 2008）。最近，Rogers and Bernatchez（2005）报道了类似结果。他们结合QTL定位和基因组扫描的方法来区分同一地域的两种鲱形白鲑，验证了一个假说，即矮小和正常的个体在一个生长相关的QTL分化是由选择作用导致的。事实上，他们的研究目的主要包括以下几点：评估生长表型与环境之间的关联，通过QTL定位确定它的遗传基础，筛选自然群体分化的离群水平，最后评估在多种环境下所观察的离群值的分化模式。他们能够决定在种群基因组范围内，离生长速率QTL最近的基因位点是否与表现出较高分化程度的基因位点一致（Campbell and Bernatchez 2004）。他们发现，有8个AFLP位点最接近增长速率的QTL，根据经验确定这个数值位于超过95%的置信范围内，而该置信范围是针对遗传分化、通过440个AFLP位点评估所获得的。从而表明这些位点的分化是由于附近生长速率位点的选择。研究还显示，一个与生长速率QTL相对应的AFLP位点，在中性水平下，正常个体展现出比期望更高的遗传分化水平。这例证了一个正常和矮小个体在不同湖泊中有相似的适应性遗传基础，基因组的某一特定区域是如何与受到多基因控制的表型性状的变化相关联的。因此，这种研究加强了通过QTL和基因组扫描的互补来研究适应性分化的可能性。如Campbell and Bernatchez（2004）的研究，没有使用一幅高密度的连锁图谱（覆盖了40个假定连锁群中的25个），并且“仅仅”用了440个AFLP标记物。这也就是说，像多数海洋生物在缺乏高密度QTL图谱的情况下，我们对姊妹物种或者种群之间的适应性分化遗传基础的理解也会不断增强。

因此，尽管耗资和耗时很大，通过发展连锁图谱、寻找QTL以及将这些信息与基因组信息相结合，代表了一种深入挖掘与适应性相关的基因组信息的最全面的方法（Price 2006；Stinchcombe and Hoekstra 2008）。尽管如此，这个方法仍然有缺点，在棘鱼的例子中可以得到例证。

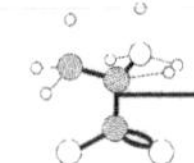

一个基于微卫星的连锁图谱已经定位出基因组中一段长为10M的包含了主效QTL的部分，该QTL与海洋和湖泊种群之间的骨盆形态学适应性遗传变异密切相关（Shapiro et al. 2004）。相关研究包括测序包含棘鱼基因组Pitxl基因鉴定区域的BAC文库（细菌人工染色体），Pitxl基因显示与湖泊种群的骨盆变小相关联的表达差异（Shaporo et al. 2004）。然而，未能够鉴定导致类型改变的确切分子变化机制。这种情况的发生至少引出三个问题：

（1）在没有BAC或相当数量的基因组序列的情况下，类似问题该怎么解决？

（2）如果这个QTL不是一个主效QTL会出现什么情况？研究已经表明，脊椎动物的后肢和骨盆发育过程是由几个主要基因控制的（例如Marcil et al. 2003）。影响着大部分主要生态特性的QTL当然不属于这个范畴，因此不能期望在大多数情况下都能出现这种清晰的基因型与表型的关联。

（3）为什么研究一个基因需要做这么多的努力（比如众多标记物的开发、系谱、连锁图）？有价值吗？为什么一个用来处理一些已知基因的候选基因方法不是来自于生化的、生理学的文献或其他像微阵列的技术呢（见下文，例如Ellegren and Sheldon 2008）？

因此，尽管基因组学和QTL相结合的方法在寻找感兴趣的基因组区域中是有用的，但与单个基因及其周围的顺式与反式作用元件之间的关联确实需要通过深度测序（如BAC）和通过比较基因组学方法对海洋生物中感兴趣基因进行研究和鉴定。

3.3 种群基因组学在海洋环境中的实际应用

上述工具以及种群遗传学研究方法与生态学问题息息相关，并且在这些问题上有诸多应用。

3.3.1 海洋中的扩散：从幼体发育到本地适应和物种形成的过程

了解海洋物种群体间的关联程度十分重要，不仅仅出于物种保护这一目的，也是为了有效地进行渔业管理，更是为了增进我们对海洋生物种群功能和进化过程的了解。大量方法学的存在意味着这一问题可以通过多种方法来得到解决，包括追踪到幼体基因层面的分析。关于特定区域幼体的物种鉴定，详见第1章。同时，我们还将给出三个常见的例子，这些例子都是以基因组

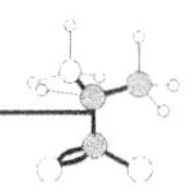

学为基础，将进一步深化对海洋生物扩散的模式及程度的了解。

3.3.1.1 浮游幼体的研究

对于具有 bentho-pelagic 生命周期（海洋无脊椎动物最常见的周期）以及成体之后无法自由移动的海洋生物，幼体就成了他们主要的扩散载体。变态发育物种的幼虫能够推迟自身的变态期来应对各种生物或非生物因素的干扰。这种可滞后的变态发育方式，大大提高了物种的分布潜力及种群的动态。尽管如此，许多研究显示，物种扩散能力与扩散代价之间存在进化上的平衡，比如权衡避免相关个体之间的竞争和在扩散中产生的高死亡率。

尽管阐明海洋无脊椎动物的能力获取时序的机制有极为重要的生物学意义，但关于这方面的研究并不多。虽然代谢通路、幼体发育的抑制和诱导、变态发育等方面的研究已经深入到了多种海洋生物，然而这些生物学进程的分子和遗传机制以及应对选择压力发生的遗传变异方面的研究却鲜有报道。直到最近，基因组学工具才逐步开始应用于非模式物种。正如 Medina（2009）所指出，非模式物种的研究能够为诸如生活史进化等重大问题提供思路。例如，正在进行的一项利用微阵列和定量 PCR 技术对入侵腹足类生物大西洋舟螺的研究揭示了一氧化氮信号通路在其幼体变态发育中的重要作用。另外，Degnan et al.（1997）在对鲍鱼的研究中，通过对包含差异表达基因在内的 cD-NA 片段的鉴定，发掘出了一系列很可能涉及耳鲍变态发育及相关功能的基因。基于 cDNA 微阵列的大量实验结果表明，涉及幼体变态发育的信号通路可能调节了鲍鱼等软体动物新的特异性基因的表达。对这些特异性基因及其多态性的保护与分析，可能会帮助我们了解幼体发育途径中的自然选择过程。

3.3.1.2 高基因流物种适应性分化的遗传学基础

通过幼体扩散的方式使海洋生物往往表现出较高的基因流。这种扩散能力可能降低了缺乏排卵期轮转的相关联群体对多元环境的适应能力。另一方面，高繁殖能力（允许有差别的死亡率）和大规模群体能够在一定程度上抵消基因流的同质化效应。这一问题将决定适应性差异的真实程度，这也解释了在明显不同的环境下存在同一海洋物种的现象；这种适应机制在什么样的条件下产生并维持。

种群基因组学和更具体的基因组扫描分析方法为解决这些问题提供了相关工具。例如，因生境不同而产生差异选择的过程中，选择性清除会使新出现的有益等位基因在群体中固定。接着，与该区域紧密连锁的一些中性等位基因的突变频率将会增加，这就是遗传学意义上的搭车效应。因而，这些与特定位点紧密连锁的中性位点的遗传变异将增大，其变异程度远远大于其他

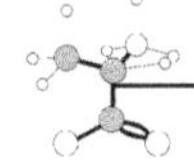

中性位点。基本原理就是利用上述不同的标记方法与技术，先定位外围位点，再定位可能是特定靶标的候选基因。

在对两个区域的紫贻贝的研究发现，11 个基因座中有 1 个正在经历选择性清除，其余无差别（Faure et al. 2008）。Vasemägi et al.（2005）在大西洋鲑中定位了 9 个位点，它们显著偏离了中性期望值，认为它们非常有可能是与环境适应相关的候选基因，值得进一步深入研究。类似的方法还在大叶藻中被应用，利用全基因组扫描技术研究潮间带和潮下带大叶藻群体，结果发现了超过 25 个未知的或基因连锁的微卫星位点受到了选择压力。Oetjenand Reusch（2007a）定位了 3 个严重偏分布位点，其中一个与水运输调控相关的结瘤素基因连锁。这表明在离群位点观察到的遗传分化是有重要的功能意义，它可能反映了栖息地的差异。

3.3.1.3 杂交带以及物种形成的研究

在陆地生态系统中，已经发生分化的种群基因之间存在着足够多的二次接触机会，海洋生态系统也是如此，并由此产生了杂交带，也称为生态过渡带。在这个区域内，既存在由幼体扩散导致的亲本基因组的流入，也包括缺乏适应性的杂交基因组的丢失，二者之间处于平衡状态。正是由于大多数海洋物种存在种群数量大的特征，可以对多种基因型进行精细的筛选，人们期望区带划分精确，在区带里两个分化的基因池（如亚种）表现出栖息地特异性。而生境特异性基因（外部选择）与“物种形成”基因（内部选择）共同导致了局部遗传渐变群的产生。在渐变群中，所有基因连锁不平衡，这比较像是随着环境梯度变化而产生的初级变异。这些渐变群是“冻结”基因组相互作用的结果。所谓“冻结”基因组，就是由于存在这些“物种形成”基因使不同基因组之间无法进行有效的交流。其中一个最好的研究案例就是紫贻贝和地中海贻贝的马赛克镶嵌型结构（Bierne et al. 2002，2003），不过有可能有许多其他未报道的相关案例或者被错误理解的案例。最近，运用多基因联合分析方法对分布在中大西洋海岭和东南太平洋海隆的贻贝群体的调查研究显示，这种伴随杂交的二次接触频繁发生于深海，且以往发生的基因渐渗事件覆盖了大部分海岭区域（Faure et al. 2009）。如果对这些渐变群采用全基因组扫描法进行研究，许多的标记将被分离出来，但仅有很少一部分真正与环境适应相关。在研究许多物种表现出遗传渐变群现象的地点（如波罗的海口；Johannesson and André 2006）时，这一点不应该被低估。今后，生态基因组学在阐明内部及外部选择压力的复杂相互作用等问题上具有巨大潜力，这两种遗传学效应在研究海洋领域的物种形成方面都不可忽视。

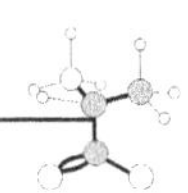

3.3.2 海洋生物入侵：用基因组学方法研究入侵物种

生物入侵是海洋生物多样性和生态系统稳定性的主要威胁之一，这和人类的行为密切相关。尽管不可预见性是生物入侵的主要特性之一，但是已经证实外来物种的入侵与商业航运（Ricciardi and MacIsaac 2000）和水产养殖（Wolff and Reise 2002）有着紧密的联系。例如，英国水域中20%的外来种可能是由于英美之间的航运导致的。同时，沿海水域污染或气候变化（Stachowicz et al. 2002）都可能引起外来物种的入侵（例如在日本，环境污染促使外来物种裂片石莼取代了本土的物种孔石莼；Morand and Briand 1996）。在这种背景下，有几位学者已经强调了在生物入侵的研究中缺少分子手段和基因组相关数据（Holland 2000，Roman and Darling 2007，Sax et al. 2007，Caroll 2008，Darling and Blum 2008）。种群基因组学以及基于DNA的其他分析方法，从不同方面为研究生物入侵的过程提供了重要信息。例如，利用已经测序完成的掌状海带（AJ344328；Oudot et al. 2002）和褐藻线粒体基因组序列（AJ277126；Oudot-LeSecq et al. 2001，Engel et al. 2008）鉴定出了7个多态性基因间隔区，并表明这些间隔区在褐藻纲海带目及部分墨角藻科中是保守的。其中有两个序列曾用于世界范围内约500个裙带菜个体的群体普查，结果显示这种海带利用多种不同的入侵方式成功地分布到了世界各地。

Darling and Blum（2008）近来倡导利用基因组资源并使用基因芯片和定量PCR等方法作为检测微生物和浮游生物多样性的手段。这些生物因其个体较小，且难以鉴定而经常被忽略。入侵物种在新领域内进化的速度和模式是另一个重要且未解决的问题。基于AFLP、SNP、基因组扫描技术或全基因组研究，为分析入侵生物持续的适应过程提供了有效手段。在生物入侵的具体案例中，这些分析可用于检测持续的遗传变异中自然选择的可能性（Barrett and Schluter 2008）。

3.3.3 揭示水产养殖种群中杂种优势的遗传基础

牡蛎有很高的经济价值，研究牡蛎的遗传背景对于维持牡蛎这种经济价值较高的水产生物是至关重要的。Hedgecock et al.（2007）利用MPSS技术，研究了长牡蛎杂种优势和近交衰退的遗传学和生理学机制（这些机制的研究一直是近百年来的难点问题），并提出杂种优势的主要原因有：

- 超显性（影响生物适应性特征基因的杂种优势）。
- 显性（杂交个体显性基因覆盖，即来自一个或其他自交亲本中的显性

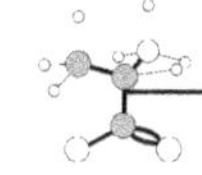

平行基因覆盖杂合体中有害的隐性基因)。

- 上位性(不同位点上等位基因的相互作用)。

以前的研究中,关于蚌类杂种优势并无直接证据。已经建立了杂合子分子标记和适应性特征相关性状的关系。同时,对控制杂交家系中 F2 代杂交群体的研究显示,有害隐性基因突变负荷较高,与显性假说相一致(Launey and Hedgecock 2001)。

尽管对杂交优势中生长性状的生理学机制的研究远不及对遗传机制的研究,但是无论杂交牡蛎的幼体还是成体与近交的个体相比都具有更高的摄食率和饲料转化率(Bayne et al. 1999, Pace et al. 2006)。Hedgecock et al.(2007)的研究有两个主要目标:

- 预测牡蛎杂种优势中相关基因数目。
- 通过全基因组扫描法分析杂种优势的原因。

对与生长性状相关的杂种优势基因的表达模式进行研究:将两个近交家系之间进行正反交,针对产生的两个杂交子代群体($f = 0.375$)中具有生长性状杂种优势的基因表达模式进行研究。克隆 cDNA,获得 4.5Mb 的序列标签。这个序列所包含的23274个不同的标签(即短序列标签)都具有一定程度的表达,中位数为 925 万,众数为 300 万的高度正偏态分布。近一半(57%)的标签表达水平依赖于基因型。结果显示这种现象主要是非加和性效应(在超显性或显性不足的前提下,杂交种偏离近交的平均水平)。超显性在阐明长牡蛎的表达模式中更占优势,与已报道的玉米和果蝇的结果存在一定差异(Swanson-Wagner et al. 2006, Gibson et al. 2004)。

超显性表型的遗传基础可能是由于顺式作用元件或者反式作用元件之间的差异表达。正如 Hedgecock et al.(2007)指出,在下一代的表型和基因型之间的关联分析中,应该可以区分顺式调节和反式调节的表达水平。

- 顺式调节:在候选基因的杂合性和表达水平之间的连锁应保持不变。
- 反式调节:候选基因和反式作用因子(可能引起上位性)之间的重组可以去除或者降低杂合子的超表达。

调节多个基因表达的非加和模式时,反式作用原件的重要性很可能在于它表明了生长相关的杂种优势可能是由于上位基因之间的相互作用产生的。本研究的重要性除了在于探索产生杂种优势的主要原因外,还进一步提出了与杂种优势生长特性相关、并与正反交杂种非加和性表达相一致的约 350 种候选基因。这仅仅代表了约 1.5% 的总转录本,与调节线虫寿命及脂质蓄积生理机能基因的数目大致相同(Ashrafi et al. 2003)。这只是研究杂种优势的一

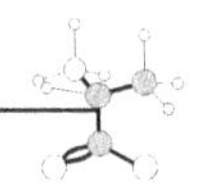

个比较明显的例子，但就成功运用新技术而言，特别在水产养殖业经济利益的驱动下，我们还应该继续在不同的生物中，进行更多的研究。

3.3.4 基因多样性和种群适应性

多次对同一物种不同个体的同一基因进行测序与传统的种群遗传学观念并不相一致（Ellegren and Sheldon 2008，Stinchcombe and Hoekstra 2008）。实际上，这种手段更应该被看做是系统发生学或是一系列的条形码。个别基因的核苷酸变化（基因多态性）能够解释种群内和种群间观测到的表型差异（或种内与种间差异；见 Yang and Bielawski 2002）。

Streelmanand Kocher（2002）报道了尼罗罗非鱼催乳素（*prl*）基因启动子附近的微卫星多态性，表明不同的催乳素表达与不同的微卫星基因型有关。罗非鱼生长性能与盐度的关系，体现或部分体现在盐度梯度下的生长模式（体重数据）的变化。这在海鲈中也一样，栖息地不同时催乳素基因表达也不同（Boutet et al. 2007），表明这个基因座的其他调控机制（可能由于多态性）可能参与比罗非鱼盐度适应更广的适应性调节功能。与前面类似，Almuly et al.（2008）也报道了调节生长激素基因（GH）的核苷酸多态性和小卫星标记的功能。金头鲷的这些位点中，一个生长激素基因型与不同栖息环境（公海与泻湖）有关（L. Chaouiand F. Bonhomme，未发表）。最终，基于谱系分析，Tao and Boulding（2003）报道了红点鲑生长激素释放激素基因座内含子区域的一个 SNP，对早期生长速率有重要影响（≈10%）。他们同样研究了另一些位于 GH 基因启动子上的 SNP，但它们作用并不显著。通过比对基因亲源关系相近物种的核苷酸多样性与物种分化的关系，也能够帮助弄清已知位点中的选择作用。最近 Faure et al.（2007）研究显示，热液喷孔的贻贝中 EF1α 基因的第二内含子受到强烈的负选择压力，而在其他双壳软体动物中呈中性进化。分辨选择性清除、搭车效应，还是种群扩张，只可能通过结合两个不同物种的多态性分析来辨别。

上面的例子都是通过顺式调节区域来对表型的遗传基础进行研究，而不是在编码区。随着海洋生物基因数据的大量产生，种间比较的作用是鉴定出可能参与调控基因表达和表型增多的非编码基因组区域（例如 Lennard Richard et al. 2007）。同时，非编码区的变异推动了表型的改变，这一点已经得到了证实（Kashi et al. 1997，Britten et al. 2003），但是否把非编码区的变异看作是表型变异的是主要根源还存在争议（Hoekstra and Coyne 2007，Wray 2007）。涉及生态学问题的水生生物数据和候选基因编码区多态性的相关报道

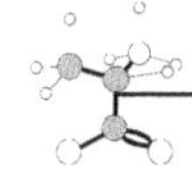

早已有之，各种各样的例子都是关于种间和种内生殖隔离（Palumbi 1999，Moy et al. 2008）和毒性适应的遗传基础研究（Palumbi 1999，Moy et al. 2008）。

3.4 表达研究和环境基因组学

与种群有关的研究主要集中在 DNA 分析，而对 RNA 或表达序列的分析也是一种研究方法。通过 DNA 分析可以检测某一种群是否具有适应性特征，而通过 RNA 研究能够对种群的功能和适应性本身进行分析。举个例子，物种如何适应极端环境及应对环境变化，关键点在于给出未来几年环境变化的预测（IPCC 2007），这在种群中的功能信息分层时称为“环境基因组学”。这种“组学”分支真正联合了实验室的实验与环境观察和样本。这两种方法必须联合起来使用才能产生有意义的功能数据。

例如，南极软体动物热休克蛋白（HSP70）基因的最初克隆和试验生产是在 15℃ 的环境温度下进行的。实际上，这是该基因能在实验室条件下表达的最低温度（Clark et al. 2008b）。如果在该动物中没有检测到 HSP70 的表达，则理论上可以得出如下结论：在这种特定环境条件下该基因是没有活性的，这种基因的特征即是如此，这种方法不仅有效而且很精确。因此，不管怎样，如果在实验室中成功检测到 HSP70，那么从环境中取样的动物标本的应激水平也能被检测到。但是，实际并不是这样的，也表明这些基因控制更为复杂，取决于施加的压力是短期急性的（实验）还是长期慢性的（环境）（Clark et al. 2008d，Clark and Peck 2009）。

虽然上面说的是一个单基因的例子，从本质上来看并非“基因组学”，但其说明了环境基因组学的复杂性以及不同学科（从生态学、生理学到基因表达以及基因组学）之间交叉运用的需求。直到今天，它仍具有非常重要的作用，比如，无脊椎动物大多数环境的应激检测仍是用 HSP70 基因家族（非常好的效果），而很少有利用基因组学方法研究的例子。因此，下面将会介绍一些单基因分析的例子，用来展示用分子生物学方法研究环境问题，然后可以扩展利用基因组学方法。

3.4.1 栖息地范围的定义：生物地理学

“为什么不同的动物会选择不同的栖息环境?”这其实是生态学的一个基本问题。由于群体遗传学可以通过 DNA 的手段来检测种群间存在的遗传差

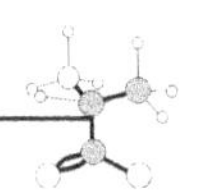

异，即使这两个种群看起来非常相似，因此，通过种群遗传学研究理应能够部分回答一些问题。在这种分析中所用到的 DNA 分子标记一般都是中性的（如某些非编码序列），不会受到自然选择的作用。使用中性 DNA 对动物种群进行研究并不能回答一些基本的问题：为什么物种会选择生存在一些特定的栖息环境？他们需要具备怎样特别的适应能力才能在这些环境中生存？（见 3.2.1.3）。以前我们往往通过形态学、生理学和生物化学对环境适应性进行分类，但是利用分子生物学则能够在细胞水平获得更详尽的信息，所有这些表型的适应性都是在细胞水平调控的。现在已经可以对生活在不同环境条件下的不同种群的蛋白和蛋白组学图谱进行翻译和构建，并且利用这些信息研究潜在的细胞水平上的变化的本质。通过这样的研究，除了能够回答栖息地"选择"的问题，还能预测到当环境改变时，动物面对干扰是如何进行适应的。

这类研究与冷血动物有很大的相关性，海洋环境其实就是一个很好的例子，因为海洋环境的温度与物种分布模式有着密切的关联。这个领域的研究大多集中于潮间带物种，因为这为具有不同生存压力和有机体热量限制且生存在明显不同潮汐区域的物种提供了一个自然的实验室（Roberts et al. 1997，Tomanek and Somero 2000，Tomanek 2002，Halpin et al. 2002，Tomanek 2005）。按照之前的规定，这项工作需要重点集中于监测热休克基因或蛋白的产物，特别是调节 HSP70 的诱导形式，不仅是温度，还包括广泛的应激蛋白。

该研究的发现表明，物种之所以能够占领潮间带区域有部分原因是由于其对热的耐受性。虽然动物能够对环境温度的变化进行适应，相比距离潮间带更近的物种，距离海岸线更远的物种对热休克的应激调节能力要弱得多，因此，离潮间带远的物种最容易受到气候变化的影响。这种现象最初令人非常费解，因为在栖息于最严酷环境中的动物往往被认为会有较强的适应能力。这种现象几乎可以确定是由于细胞能量分配和蛋白的生成所导致的，因此在高海岸线的生物对 HSP 产物增加的需求必须与其他细胞水平上的生物过程相协调。例如潮间带蚌类的相互移植实验导致那些较高海岸线的蚌类生长变慢，它们离海岸线越远生长越慢（Hofmann 2005）。

小规模本地研究和大规模的研究几乎可以得到相同的结果（Hofmann 2005），就是 HSP 产物与生物地理学是相互联系的。相关的研究工作主要集中分布在不同纬度梯度（Halpin et al. 2002，Osovitz and Hofmann 2005）及其边缘范围的物种（Hofmann 2005）。最近，使用微芯片技术大大推进了这项研究。利用微阵列法对一种潮间带加州贻贝的2496条 cDNA 序列进行了研究，

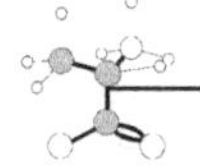

并且对来自沿着北非西海岸横跨17个纬度的四个不同种群的RNA序列进行了分析（Place et al. 2008）。这些分析关注的是基因与环境应激之间的关系（蛋白折叠、蛋白退化和凋亡），在酵母菌的应激反应中，一些蛋白的表达水平大大升高，包括热休克同源的HSC71、β-蛋白酶、elf2-α、一种应激调节反应起始因子和一种完整的膜蛋白。这些基因在后续的工作中成为该生物体应激反应的生物标志物。总而言之，这四种地域特异性的基因表达谱表明栖息地环境的复杂性要远大于纬度。大多数基因表达的改变具有非常重要的意义，基因的表达往往随着采集地点的不同而发生很大的变化，而且这种基因表达差异也表明加州贻贝对物理的以及其他非生物因素的应激具有明显的种群特异性。

这类研究往往是跨纬度或全球采样，而且重要的是这种研究往往针对同物种或亲缘关系很近的物种，那么作为独立的系统发育，遗传变异的适应性效应可以很明确地展示出来。现在，随着非模式物种的有效基因组数据的大幅增加，该研究得到了快速的发展，可分析单基因或利用基因芯片分析数千个基因。因此，在栖息地选择方面的研究需要发展一种更整体化的基因组水平分析方法。

3.4.2 基因芯片：识别与适应性有关的生化通路

这一节探讨复杂通路的识别以及这些通路的变化与扰动与自然环境的循环的关系。值得一提的是基因芯片技术，到目前为止这个领域的研究还十分有限，需要开展大量分子生物学研究，如文库构建、制备基因芯片、杂交芯片和杂交分析。这里面每一项都很重要，需要实验室具备足够的专业技能。用研究模式物种的生物基因芯片去研究非模式物种是可行的。Hogstrand et al.（2002）研究了虹鳟对锌暴露的应答，用的就是来自于日本东方鲀的高密度点阵芯片。虽然这项技术对此类研究十分有效（见Buckley 2007，Kassahn 2008），但随着遗传距离的不断增加，可检测到的基因表达水平的数量级不断降低（Renn et al. 2004），这一点在进行实验设计的时候必须要考虑。还有，非模式物种研究还存在其他问题，如准确鉴定基因的能力有限（见3.2.1.1节），对很多基因功能方面的推断都是根据数据库中已有序列相似性比对得到的，而这些数据库中大多数的已知序列都来自脊椎动物特别是哺乳动物。结果，基因芯片技术对非模式物种的分析几乎仅存在于鱼类，特别是有关水产养殖和生态毒理方面的研究。

尽管如此，现在也开始出现了基于环境因子的基因芯片实验，包括潮间

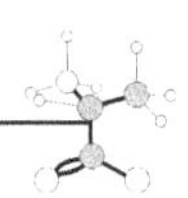

带的脆壳蟹对热胁迫的应答（Teranishi and Stillman 2007）、鲤鱼的冷胁迫应答（Gracey et al. 2004）、日常温度波动对林奈氏澳鳉的影响（Podrabsky and Somero 2004）以及正常环境取样对加州贻贝的影响（Place et al. 2008）。同时，这些基因是随环境改变而改变的，例如在热压力下，基因会显示最容易受影响的。蛋白质折叠、蛋白质降解和蛋白质合成等通路，而这样的基因还是很多的（Teranishi and Stillman 2006）。对每个基因都进行详细的分析是不可能的，这些实验本质上是为以后的研究提供候选基因，如硬脂酰辅酶A脲酶在鲤科鱼类中的冷适应（Gracey et al. 2004）。然而，没有最初全景式筛选成千上万的基因，也不可能识别出这么多候选基因，这将最终帮助我们探寻出环境适应性的本质。

3.4.3 基因组的可塑性与季节波动

要记住的一点是基因组的表达并不是一成不变的。除非开展长期的研究（有许多有差异的种群的样本，且所有的数据都被测量），否则研究结果所都只能反应一个时间点和一系列特殊的环境变量。解释这样的数据不需要生物体大量的背景知识。例如，产卵行为或其他受季节性气温变化（热历史）引起的行为，特定的季节蛋白产物（就像冬天的防冻剂；Buckley et al. 2001，Tomanek 2002，Enevoldsen et al. 2003，Jin and DeVries 2006）能够显著地改变基因表达模式和明显偏离任何的分析研究。我们把根据生物的进化史和环境信号产生的基因表达变化叫做可塑性，它代表一个变化范围或基因（和形成的物种）能够有效调控和适应的表达窗口。

这是另一个受到个体基因水平等因素极大限制的研究领域，虽然这样的研究数据很匮乏，但是它还是表现出这种可塑性应答具有高度的基因依赖性。在大量关于热历史对鱼类肌肉发育影响的分子生物学研究中，肌细胞生成素和FoxK1被确认是温度应答基因（Fernandes et al. 2006，2007），但在所研究的基因之中，至少一半基因都与热刺激无关（Mackenzie2006）。迄今为止，大多数可塑性研究集中在陆生物种，这些物种经历的温度变化比海洋生物（Deere and Chown 2006）或真核代表生物酵母（Stern et al. 2007）大得多。但是，考虑到全球温度升高导致海水温度升高的预测，这对海洋领域来说无疑是一个有重大价值的新兴领域（Peck et al. 2009）。在基因组水平采用基因芯片和新测序技术将大力推动基因适应性的研究。

3.4.4 对极端环境的适应

关于种群基因组学的章节，列举了DNA多态性影响基因表达进而影响物

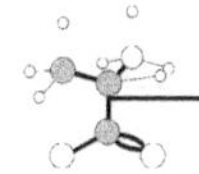

种适应性和物种形成的多个例子（第3.3.4节）。然而，在极端环境中观察到的物种对环境的适应无疑是最明显的。这不仅是一个学术研究领域，也存在潜在商业利益范畴，通过研究新蛋白质的特征来识别在不同情况下能够更好工作的酶变体，可以利用到食品加工或洗涤剂等相关生产上（Clark et al. 2004；Peck et al. 2005）。虽然存在多种类型的极端环境，但本节将主要关注对其中两个最常见的极端自然环境：热液喷口和极地地区，目的在于概述对热和冷这两个相反极端的研究以及动物对人类造成的污染所进行的适应。对热液喷口和极地地区的适应已经存在了数千乃至数百万年，同时生物体却要在仅仅数十、数百年内对污染物进行调整或适应。

3.4.4.1 热液喷口

热液喷口的特点是具有非常特异的物理和化学性质，比如高压（高达420个大气压）；高温和温度骤变（从4℃到400℃）可能在数十厘米空间内同时发生（Piccino et al. 2004），且时间短暂（一分钟内10～50℃；Le Bris et al. 2005）；能够提供共生燃料的高浓度硫化物或甲烷（Childress and Fisher，1992）；化学毒性（重金属和放射性核素；Cherry et al. 1992，Lurther et al. 2001）以及完全黑暗的环境。虽然如此，人们在那些环境中却发现很多生物，例如虾、蛤蜊、贻贝、巨型管状蠕虫、螃蟹和鱼类。这些生物进化出了不同的适应策略，从而确保他们能够从深海热泉流体中摄取生存所必须的营养成分。较多的适应性研究包括：

- 共生现象对缺乏光合作用环境的适应。基于化能自养型细菌所产生的能量和有机分子构建了一条不需要光照的食物链（Minic and Herve 2004，Stewart and Cavanaugh 2006，Duperron et al. 2007）。
- 对高温的适应性（Gaill et al. 1995，Sicot et al. 2000）。
- 对有毒物质的适应性（Company et al. 2004）
- 对低氧或缺氧的适应性（Hourdez and Weber 2005）。

对高温适应性的研究基本上都集中在细菌。它们有一些共同特征，比如带电氨基酸和脯氨酸残基的增加以及精氨酸对赖氨酸的替代导致的氢键的增加（Kumar et al. 2000，Nishio et al. 2003，Robinson et al. 2006）。类似的特征在真核生物中也存在，比如生活在环境比 *P. grasslei* 更热的部分的 *A. pompejana*，带正电氨基酸残基显著增加，蛋白质疏水性也随之增加。一些特定的蛋白质的热稳定性也被系统地研究过，比如胶原蛋白（Sicot et al. 2000）、线粒体（Dahlhoff et al. 1991）和细胞质（Jollivet et al. 1995）、呼吸链蛋白以及血红蛋白（参见 Hourdez and Weberd 2005）。在胶原蛋白中，羟基

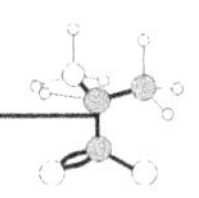

化脯氨酸在分子的热稳定性上似乎发挥着关键的作用，并因而意味着翻译后过程对适应高热状态也是关键的因素。

关于对有毒物质和低氧/缺氧适应过程的相关研究非常少，并且大部分都集中在贻贝属以及重金属或氧化压力对酶活性的影响上（Company et al. 2008），或者金属硫蛋白等特定的基因表达上（Hardivillier et al. 2006）。Pruski and Dixon（2003）的研究也表明，在失压和氧化压力下的DNA链断裂程度和HSP70蛋白表达水平呈正相关关系。对深海热液喷口环节动物的巨型HBL-血红蛋白进行的分子分析就是一个典型的例子，表明了这一动物群是如何适应高毒的硫化物环境的：在这种特殊环境中，呼吸色素能够可逆地把硫化物绑定到A2和B2球蛋白的两个不同的半胱氨酸残基上（Bailly et al. 2002）。Bailly et al.（2003）认为，生活在非还原性栖息地的现代环节动物在进化过程中可能由于达尔文正选择作用已经失去了这种功能。

热液物种的DNA数据极为不足。但是，现在已经进行*Bathymodiolusazoricus*，*Paralvinellagrasslei*，*Alvinellapompejana* 和 *Riftiapachyptyla* 的转录测序（Sanchez et al. 2007，Tanguy et al. 2008，Alvinella联合计划）。第一个cDNA微阵列也已经为 *B. azoricus* 和 *P. grasslei* 建立起来了，目的在于把它们作为模型，进而来研究这些生物是如何应对剧烈的温度变化、重金属影响和共生关系。一个近期的研究已经表明，深海热流喷口贻贝在高于20℃的温度下短期（30～120分钟）曝露会导致基因表达的全局性下调，暗示这些动物似乎更能够适应较冷的温度（Boutet et al. 2008）。

3.4.4.2 极地环境

生活在极寒环境必定会有特殊的遗传适应机制，这里就有一些经典的单基因研究：

- 等离子防冻剂：最初在极地鱼类中被发现（De Vries，1970），但现在已经成为对寒冷海洋环境的标准适应（De Vries，1982）。
- 特定蛋白质的修饰增加分子柔韧性，保证在寒冷环境中正常功能的高效运转（Fields and Somero 1998，Detrich et al. 1989，1992，Fields et al. 2002，Römisch et al. 2003）。
- 在冰鱼中血红蛋白基因的缺失和功能性红细胞的产生能力（Moylan and Sidell 2000，di Prisco et al. 2002）。
- 在很多鱼类和无脊椎物种中经典热休克反应（HSP70热诱导形式的上流调节）的缺失（Hofmann et al. 2000，Clark et al. 2008a，c）。

最近，通过一个试验性的基因芯片技术对南极鱼的适应性进行了研究，

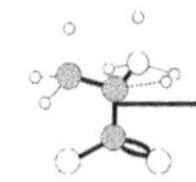

实验中将贝氏肩孔南极鱼的 RNA 杂交到广温性的虾虎鱼生成的阵列芯片上（Buckley and Somero 2009）。这个研究表明南极鱼尽管没有表现出经典的热休克反应，但是与进化保守的细胞应激反应相关的许多基因都增强表达。同时本实验使用了异源性阵列的方法，现在特定生物体阵列对于一些极地的陆地昆虫来说已经成为可能（参见 Purac et al. 2008）。

一般情况下，对鱼类和无脊椎动物冷适应的大多数研究工作已经在南极物种中得到开展，因为极地生物在相对隔离的持续寒冷环境下生存了 1500～2500 万年，进化出了很多适应南极环境的特有物种（参阅 Clarke and Johnston 1996）。与此相反，北极海洋底栖动物群则代表了一个相对年轻的群体，特点是无论来自太平洋还是大西洋的物种都不具有明显地方性（Dunton 1992）。因此，极地物种的比对可以帮助冷适应分离不同的种系发育物种（参见 Verde et al. 2007）。

与极地环境密切相关的是气候变化。目前科学界大都认同的一个观点是，全球正在变暖，而极地变暖的速度又高于世界其他地区。在接下来的 100 年内海洋温度预计将上升 2℃（Murphy and Mitchell 1995），速度比在过去 100 年或上一个冰河更新循环记录的任何时期都要快（Zachos et al. 2001）。但伴随着在 Bellingshausen 海洋表面每 50 年 1℃（Meredith and King 2005）的温度上升，沿南极半岛的地区气候变化也已经加速。但情况远比这复杂的多，二氧化碳排放的影响不仅与温度密切相关，也关系到海洋的酸化和 pH 值的变化。在工业革命开始以来的 250 年里，大气中的二氧化碳水平已经从 280 ppm 上升至 381 ppm（Canadell et al. 2007），并仍在快速上升。

CO_2 被海洋所吸收并形成碳酸，从而降低海水的 pH 值，降低碳酸钙饱和状态（即增加溶解度）。海洋的 pH 值已经下降到平均 8.16～8.05（Caldeira and Wicket 2003），模型预测海洋表面的 pH 值还会持续下降，到 2100 年估计会下降 0.2～0.4 个单位（Caldeira and Wicket 2003，2005，Royal Society 2005，Cao et al. 2007）。在海洋中，如此速度和幅度的 pH 值变化比过去 300 万年所发生的变化还要大，而且要快得多。此外，极地地区还会受到其他因素的影响，因为与当今任何海洋区域相比，南大洋有着最低的碳酸钙饱和率，因此这将是率先达到饱和状态的区域之一。

毫无疑问，判断海水温度和海洋酸化的增加对海洋生物的影响极为重要。虽然所有这些研究都表明生物会受到这些方面的影响，但所受到的影响程度并不相同，很明显地有些生物将适应并生存下来（Dupont et al. 2008）。目前，我们几乎对表型可塑性的遗传机制和基因组基础一无所知。从这种需求来看，

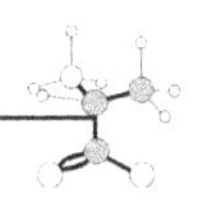

极地物种（特别是南极的物种）非常有价值。例如在针对气候变化的生物标记的开发方面，南极物种的优势是它们对热非常敏感（Peck et al. 2004）。而且反应应答不被污染干扰，因为污染已经影响到这个星球的其他地方了。因此，南极生物为我们提供的气候变化与生物适应性之间的关系是最可靠的，因为没有环境污染因素的影响。极地生物的基因组学和功能研究还处于起步阶段（Clark et al. 2004，Peck et al. 2005），但这样的研究现在正明确指向气候变化及其对功能性生物多样性的影响等有关问题。

3.4.4.3　生态毒理学监测

海洋生态系统，特别是沿海、河口和珊瑚礁，目前正在经历巨大生态危机，因为全球环境变化（重金属、农药、化学品，也包括寄生虫）对生物产生了显著的生理应激压力。了解生态系统的恢复能力和预测环境压力对海洋生物的影响，取决于对这些生物在这些生态系统中的生理状况和可塑性是否了解。与海洋物种有关的传统的生态毒理学研究主要包括“生物标志物”，如酶的活性和应激蛋白定量、免疫学参数评估和生命特征（生长、生殖）评估。基因组和蛋白质组技术的快速发展促使生态毒理学相关研究出现了新的方法，称为生态毒理基因组学。生态毒理基因组学用于研究非模式物种在基因和蛋白表达水平上对环境毒素的应答。DNA 微阵列分析的出现催生了生态毒理基因组学这项技术的出现。微阵列分析能够辅助我们去理解全基因组表达谱、机体的生理状态和传统的毒性终点之间的关系（Irwin et al. 2004，Lee et al. 2003）。

基因组学提供生理多样性和功能等方面的详细信息，因而有助于机械性洞察生物体是如何应对环境压力的。基因组学方法在生态毒理研究中将会：

- 提高对毒物、逆境胁迫或感染机制的认知。
- 发展出一种关于监测生物体内的毒物、逆境胁迫或感染的环境。
- 开发出相关工具来预测毒物、逆境胁迫或感染的传播，进而更好地评估和预测环境健康。

生态毒理基因组学的应用可以通过体外系统推断体内系统，并且可以跨越物种障碍。这将有助于理解特殊的分子事件背后的有毒物行为，因此这些都可以作为生物标志物来确定暴露于环境的压力。

基因表达谱分析可用于表明不同细胞类型在不同有毒物质胁迫下特异基因的表达降低或诱导表达（Troester et al. 2004）。生态毒理基因组学在方法上所面临的主要挑战是如何排除由不同生理状态、年龄、性别和遗传多态性等内在原因所导致的基因表达水平的变化。下面是几个具体需要考虑的方面：

●识别出对毒物环境应答中起到上调作用的保守基因。

●如何将这些基因的表达谱用于诊断毒物胁迫压力。

●鉴定出信息最丰富的基因，并将其整合到更加特异的逆境应答基因阵列中用作监控。

在海洋生物中，大多研究集中在鱼类（Sheader et al. 2006），除此之外，软体动物的相关研究也在不断开展。生态毒理基因组学方法势必会带来很多新的机遇，以促进人们进一步理解生物对环境污染物的毒性反应分子机制（Bradley and Theodorakis 2002，Moore 2001）。

3.5 总结和展望

本章总结了目前海洋生物学家进行海洋生态研究的分子工具。虽然新技术在不断被引进（如 MPSS 和 454 测序），但是其功效还要依赖于对所研究物种的基础生态和生理学方面的理解程度。鉴定具有明显差异表型的同一物种不同种群时，通过表型与基因型之间的差异分析可以发现一系列候选基因，这些基因可以进一步从功能和遗传方面对某一生物学过程进行深入研究。同样，根据已知的设计严密的生理学实验可以避免许多数据解释时存在的问题。因此，全面的生态适应概述只能够通过结合多种方法来实现。这些方法不限于基因组学，还包括生态学和生理学。实际上，在分子生物学技术的辅助下，是这些学科在支配着问题、种群和实验方法的使用。

至于未来对 MPSS 和 454 技术的使用无疑会越来越多，因为这些技术能够使研究人员快速获得大量非模式物种的基因组数据。因此，非模式物种基因资源匮乏这一传统观点在未来的几年内将发生巨大变化。每个领域都应该集中精力产生更多的“新的模式物种”，这对解答其特定领域的生态事件（无论是水产养殖、热液喷口或气候变化和极地物种）极有价值。这些物种应当起源于不同的食性类群，这样，当我们对生物适应性进行研究时才不会限于单一物种。理想的状况应该包括从微生物到高等肉食动物等一系列物种，这样才能让我们对海洋环境有一个从基因到生态系统的全面认识。

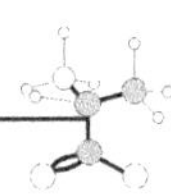

参考文献

Akey JM, Zhang G, Zhang K, Jin L, Shriver MD (2002) Interrogating a high-density SNP map for signatures of natural selection. Genome Res 12:1805 – 1814

Albert V, Jonsson B, Bernatchez L (2006) Natural hybrids in Atlantic eels (*Anguilla anguilla, A. rostrata*): evidence for successful reproduction and fluctuating abundance in space and time. Mol Ecol 15:1903 – 1916

Almuly R, Skopal T, Funkenstein B (2008) Regulatory regions in the promoter and first intron of Sparus aurata growth hormone gene: Repression of gene activity by a polymorphic minisatellite. Comp Biochem Physiol D 3:43 – 50

Ashrafi K, Chang FY, Watts JL, Fraser AG, Kamath RS, Ahringer, Ruvkun G (2003) Genome-wide RNAi analysis of *Caenorhabditis elegans* fat regulatory genes. Nature 421:268 – 272

Backström N, Ovarström A, Gustafsson L, Hellegren H (2006) Levels of linkage disequilibrium in a wild bird population. Biol Lett 2:435 – 438

Bailly X, Jollivet D, Vanin S, Deutsch J, Zal Z, Lallier FH, Toulmond A (2002) Evolution of the sulfide-binding function within the globin multigenic family of the deep-sea hydrothermal vent tubeworm *Riftia pachyptila*. Mol Biol Evol 19:1421 – 1433

Bailly X, Leroy R, Carney S, Collin O, Zal F, Toulmond A, Jollivet D (2003) The loss of the hemoglobin H_2S-binding function reveals molecular adaptation driven by Darwinian positive selection in annelids from sulfide-free habitats. Proc Natl Acad Sci USA 100:5885 – 5890

Balloux F, Amos W, Coulson T (2004) Does heterozygosity estimate inbreeding in real populations?. Mol Ecol 13:3021 – 3031

Baranski M, Loughnan S, Austin CM et al. (2006) A microsatellite linkage map of the blacklip abalone, *Haliotis rubra*. Anim Genet 37:563 – 570

Barrett RD, Schluter D (2008) Adaptation from standing genetic variation. Trends Ecol Evol 23:38 – 44

Barretto FS, McCartney MA (2008) Extraordinary AFLP fingerprint similarity despite strong assortative mating between reef fish color morphospecies. Evolution 62:226 – 233

Baus E, Darrock DJ, Bruford MW (2005) Gene-flow patterns in Atlantic and Mediterranean populations of the Lusitanian sea star *Asterina gibbosa*. Mol Ecol 14:3373 – 3382

Bayne BL, Hedgecock D, McGoldrick D, Rees R (1999) Feeding behaviour and metabolic ef-fi-ciency contribute to may growth heterosis in Pacific oysters [*Crassostrea gigas* (Thunberg)]. J Exp Mar Biol Ecol 233:115 – 130

Beaumont MA, Nichols RA (1996) Evaluating loci for use in the genetic analysis of population structure. Proc R Soc Lond B 263:1619 – 1626

Bensch S, Helbig AJ, Salomon M, Seibold I (2002) Amplified fragment length polymorphism analysis identifies hybrids between two subspecies of warblers. Mol Ecol 11:473 – 481

Beraldi D, McRae AF, Gratten J et al (2006) Development of a linkage map and mapping of phenotypic polymorphisms in a free-living population of Soay sheep (*Ovis aries*). Genetics 173: 1521 – 1537

Bierne N, Borsa P, Daguin C, Jollivet D, Viard F, Bonhomme F, David P (2003) Introgression patterns in the mosaic hybrid zone between *Mytilus edulis* and *M. galloprovincialis*. Mol Ecol 12: 447 – 462

Bierne N, David P, Langlade A, Bonhomme F (2002) Can habitat specialisation maintain a mosaic hybrid zone in marine bivalves?. Mar Ecol Progr Ser 245:157 – 170

Black WC, Baer CF, Antolin MF, DuTeau NM (2001) Population genomics: genome-wide sampling of insect populations. Annu Rev Entomol 46:441 – 469

Boguski MS, Lowe TMJ, Tolstoshev CM (1993) dbEST-database for 'expressed sequence tags'. Nat Genet 4:332 – 333

Bonin A, Taberlet P, Miaud C, Pompanon F (2006) Explorative genome scan to detect candidate loci for adaptation along a gradient of altitude in the common frog (*Rana temporaria*). Mol Biol Evol 23:773 – 783

Boutet I, Ky CL, Bonhomme F (2006) A transcriptomic approach of salinity response in the euryhaline teleost, *Dicentrarchus labrax*. Gene 379:40 – 50

Boutet I, Nebel C, De Lorgeril J, Guinand B (2007) Molecular characterisation and extrapituitary prolactin expression in the European sea bass *Dicentrarchus labrax* under salinity stress. Comp Biochem Physiol D 2:74 – 83

Boutet I, Tanguy A, Le Guen D, Piccino P, Hourdez S, Ravaux J, Shillito B, Legendre P, Jollivet D (2008) Global depression in gene expression as a response to rapid changes of temperature in the hydrothermal vent mussel *Bathymodiolus azoricus*. PLOS Biol 276:3071 – 3079

Bradley B, Theodorakis C (2002) The post-genomic era and ecotoxicology. Ecotoxicology 11: 7 – 9

Brenner S, Johnson M, Bridgham J, Golda G, Lloyd DH, Johnson D, Luo S, McCurdy S, Foy M, Ewan M et al. (2000a) Gene expression analysis by massively parallel signature sequencing (MPSS) on microbead arrays. Nat Biotech 18:630 – 634

Britten RJ, Rowen L, Williams J, Cameron RA (2003) Majority of divergence between closely related DNA samples is due to indels. Proc Natl Acad Sci USA 100:4661 – 4665

Buckley BA (2007) Comparative environmental genomics in non-model species: using heterolo-gous hybridisation to DNA-based arrays. J Exp Biol 210:1602 – 1606

Buckley BA, Owen M-E, Hofmann GE (2001) Adjusting the thermostat: the threshold induction temperature for the heat-shock response in intertidal mussels (genus *Mytilus*) changes as a

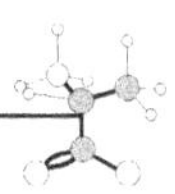

function of thermal history. J Exp Biol 204:3571 – 3579

Buckley BA, Somero GN (2009) cDNA analysis reveals the capacity of the cold adapted Antarctic fish *Trematomus bernacchii* to alter gene expression in response to heat stress. Polar Biol 32: 403 – 415

Caldeira K, Wicket ME (2003) Oceanography: anthropogenic carbon and ocean pH. Nature 425:365

Caldeira K, Wicket ME (2005) Ocean model prediction of chemistry changes from car-bon dioxide emission to the atmosphere and ocean. Geophy Res Lett 110: C09S04. doi: 10.1029/2004JC002671

Company R, Serafim A, Cosson RP, Fiala-Medioni A, Camus L, Colaco A, Serrao-Santos R, Bebianno MJ (2008) Antioxidant biochemical responses to long-term copper exposure in Bathymodiolus azoricus from Menez-Gwen hydrothermal vent. Sci Total Environ 389:407 – 417

Campbell D, Bernatchez L (2004) Generic scan using AFLP markers genes as a means to assess the role of directional selection in the divergence of sympatric whitefish ecotypes. Mol Biol Evol 21:945 – 956

Campbell D, Duchesne P, Bernatchez L (2003) AFLP utility for population assignment studies: analytical investigation and empirical comparison with microsatellites. Mol Ecol 12: 1979 – 1991 Campbell NR, Narum SR (2008) Identification of novel SNPs in Chinook salmon and variation among life history types. Trans Am Fish Soc 137:96 – 106

Canadell JG, Le Quere C, Raupach MR, Field CB, Buitenhuis ET, Ciais P, Conway TJ, Gillett NP, Houghton RA, Marland G (2007) Contributions to accelerating atmospheric CO_2 growth from economic activity, carbon intensity, and efficiency of natural sinks. Proc Natl Acad Sci USA doi: 10.1073/pnas.0702737104

Cao L, Caldeira K, Jain AK (2007) Effects of carbon dioxide and climate change on ocean acidification and carbonate mineral saturation. Geophys Res Lett 34: L05607. doi: 10.1029/2006GL028605

Carninci P, Shibata Y, Hayatsu N, Sugahara Y, Shibata K, Itoh M, Konno H, Okazaki Y, Muramatsu M (2002) Normalization and subtraction of cap-trapper-selected cDNAs to prepare full-length cDNA libraries for rapid discovery of new genes. Genome Res 10:1617 – 1630

Caroll SP (2008) Facing change: forms and foundations of contemporary adaptation to biotic invasions. Mol Ecol 17:361 – 372

Chatziplis D, Batargias C, Tsigenopoulos CS, Magoulas A, Kollias S, Kotoulas G, Volckaert FAM, Haley CS (2007) Mapping quantitative trait loci in European sea bass (*Dicentrarchus labrax*): The BASSMAP pilot study. Aquaculture 272: S172 – S182

Cherry R, Desbruyeres D, Heyraud M, Nolan C (1992) High levels of natural radioactivity in hydrothermal vent polychaetes. C R Acad Sci Paris ser III 315:21 – 26

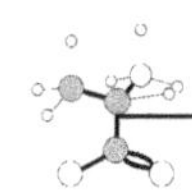

Childress JJ, Fisher CR (1992) The biology of hydrothermal vent animals: physiology, biochem-istry, and autotrophic symbioses. Oceanogr Mar Biol 30:337 – 441

Chini V, Rimoldi S, Terova G, Saroglia M, Rossi F, Bernardini G, Gornati R (2006) EST-based identification of genes expressed in the liver of adult seabass (*Dicentrarchus labrax*, L.). Gene 376:102 – 106

Chistiakhov DA, Hellemans B, Haley CS, Law AS, Tsigenopoulos CS, Kotoulas G, Bertotto D, Libertini A, Volckaert FAM (2005) A microsatellite linkage map of the European sea bass *Dicentrarchus labrax* L. Genetics 170:1821 – 1826

Clark MS, Clarke A, Cockell CS, Convey P, Detrich IIIHW, Fraser KPP, Johnston I, Methe B, Murray AE, Peck LS, Romisch K, Rogers A (2004) Antarctic genomics. Comp Func Genom 5:230 – 238

Clark MS, Fraser KPP, Burns G, Peck LS (2008a) The HSP70 heat shock response in the Antarctic fish *Harpagifer antarcticus*. Polar Biol 31:171 – 180

Clark MS, Fraser KPP, Peck LS (2008b) Antarctic marine molluscs do have an HSP70 heat shock response. Cell Stress Chaperones 13:39 – 49

Clark MS, Fraser KPP, Peck LS (2008c) Lack of an HSP70 heat shock response in two Antarctic marine invertebrates. Polar Biol 31:1059 – 1065

Clark MS, Geissler P, Waller C, Fraser KPP, Barnes DKA, Peck LS (2008d) Low heat shock thresholds inwild Antarctic inter-tidal limpets (*Nacella concinna*). Cell Stress Chaperones 13:51 – 58

Clark MS, Peck LS (2009) Triggers of the HSP70 stress response: environmental responses and laboratory manipulation in an Antarctic marine invertebrate (*Nacella concinna*). Cell Stress Chaperones (in press)

Clarke A, Johnston IA (1996) Evolution and adaptive radiation of Antarctic fishes. Trends Ecol Evol 11:212 – 218

Cohen S (2002) Strong positive selection and habitat-specific amino acid substitution patterns in Mhc from an estuarine fish under intense pollution stress. Mol Biol Evol 19:1870 – 1880

Colosimo PF, Hosemann KE, Balabhadra S, Villarreal G Jr, Dickson M, Grimwood J et al (2005) Widespread parallel evolution in sticklebacks by repeated fixation of ectodysplasin alleles. Science 307:1928 – 1933

Comai L, Young K, Till BJ et al. (2004) Efficient discovery of DNA polymorphisms in natural populations by ecotilling. Plant J 37:778 – 786

Company R, Serafim A, Bebianno MJ, Cosson R, Shillito B, Fiala-Médioni A (2004) Effect of cadmium, copper and mercury on antioxidant enzyme activities and lipid peroxidation in the gills of the hydrothermal vent mussel *Bathymodiolus azoricus*. Mar Environ Res 58:377 – 381

Crawford DL, Oleksiak MJ (2007) The biological importance of measuring individual varia-

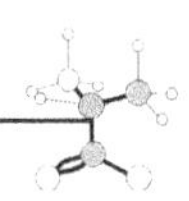

tion. J Exp Biol 210:1613 – 1621

Cresko WA, Amores A, Wilson C, Murphy J, Currey M, Philips P, Bell MA, Kimmel CB, Postlewaith JH (2004) Parallel genetic basis for repeated evolution of armorloss in Alaskan threespine stickleback populations. Proc Natl Acad Sci USA 101:6050 – 6055

Crow JF (1998) 90 years: the beginning of hybrid maize. Genetics 148:923 – 928

Dahlhoff E, O'Brien J, Somero GN, Vetter RD (1991) Temperature effects on mitochondria from hydrothermal vent invertebrates: evidence for adaptation to elevated and variable habitat temperatures. Physiol Zool 64:1490 – 1508

Daib AM, Williams TD, Sabine VS, Chipman JK, George SG (2008) The GENIPOL European flounder *Platichthys flesus* L. toxicogenomics microarray: application for investigation of the response to furunculosis vaccination. Fish Biol 72:2154 – 2169

Darling JA, Blum MJ (2008) DNA-based methods for monitoring invasive species: a review and prospectus. Biol Invasions 9:751 – 765

Dasmahapatra KK, Lacy RC, Amos W (2008) Estimating levels of inbreeding using AFLP markers. Heredity 100:286 – 295

DeVries AL (1970) Freezing resistance in Antarctic fishes. In: Holdgate MW (ed) Antarctic biology. Academic Press, New York

DeVries AL (1982) Biological antifreeze agents in coldwater fishes. Comp Biochem Physiol 73A:627 – 640

DeWoody JA, Avise JC (2000) Microsatellite variation in marine, freshwater and anadromous fishes compared with other animals. J Fish Biol 56:461 – 473

DeWoody YD, DeWoody JA (2005) On the estimation of genome-wide heterozygosity using molecular markers. J Hered 96:85 – 88

DeWoody JA, Nason JD, Hipkins VD (2006) Mitigating scoring errors in microsatellite data from wild populations. Mol Ecol Notes 6:951 – 957

Deere JA, Chown SL (2006) Testing the beneficial acclimation hypothesis and its alternatives for locomotor performance. Am Nat 5:630 – 644

Degnan BM, Degnan SM, Fentenany G, Morse DE (1997) A Mox homeobox gene in the gastropod mollusc Haliotis rufescens is differentially expressed during larval morphogenesis and metamorphosis. FEBS Lett 411:119 – 122

Detrich HWIII, Fitzgerald TJ, Dinsmore JH, Marchese-Ragona SP (1992) Brain and egg tubulins from Antarctic fishes are functionally and structurally distinct. J Biol Chem 267:18766 – 18775

Detrich HWIII, Johnson KA, Marchese-Ragona SP (1989) Polymerization of Antarctic fish tubulins at low temperatures: energetic aspects. Biochem 28:10085 – 10093

di Prisco G, Cocca E, Parker S, Detrich HW III (2002) Tracking the evolutionary loss of hemoglobin expression by the white blooded Antarctic icefishes. Gene 295:185 – 191

Douglas SE, Knickle LC, Kimball J, Reith ME (2007) Comprehensive EST analysis of Atlantic halibut (*Hippoglossus hippoglossus*), a commercially relevant aquaculture species. BMC Genom doi:10.1186/1471-2164-8-144

Dunton K (1992) Arctic biogeography: the paradox of the marine benthic fauna and flora. Trends Ecol Evol 7:183-189

Duperron S, Sibuet M, MacGregor BJ, Kuypers MMM, Fisher CR, Dubilier N (2007) Diversity, relative abundance and metabolic potential of bacterial endosymbionts in three *Bathymodiolus* mussel species from cold seeps in the Gulf of Mexico. Environ Microbiol 9:1423-1438

Dupont S, Havenhand J, Thorndyke W, Peck LS, Thorndyke M (2008) CO_2-driven ocean acidi-fication radically affects larval survival and development in the brittlestar *Ophiothrix fragilis*. Mar Ecol Prog Ser 373:285-294

Ellegren H (2004) Microsatellites: simple sequences with complex evolution. Nat Rev Genet 5:435-445

Ellegren H (2008) Sequencing goes 454 and takes large-scale genomics into the wild. Mol Ecol 17:1629-1631

Ellegren H, Sheldon BC (2008) Genetic basis of fitness differences in natural populations. Nature 452:169-175

Ellis JR, Burke JM (2007) EST-SSRs as a resource for population genetic analyses. Heredity 99:125-132

Enevoldsen LT, Heiner I, DeVries AL, Steffensen JF (2003) Does fish from the Disko Bay area of Greenland possess antifreeze proteins during the summer?. Polar Biol 26:365-370

Engel C, Billard E, Voisin M, Viard F (2008) Conservation and polymorphism of mitochondrial intergenic sequences in brown algae. Eur J Phycol 43:195-205

Erickson DL, Fenster CB, Stenoien HK, Price D (2004) Quantitative trait locus analyses and the study of evolutionary process. Mol Ecol 13:2505-2522

Faure B, Bierne N, Tanguy A, Bonhomme F, Jollivet D (2007) Evidence for a slightly deleterious effect of intron polymorphisms at the EF1a gene in the deep-sea hydrothermal vent bivalve *Bathymodiolus*. Gene 406:99-107

Faure M, David P, Bonhomme F, Bierne N (2008) Genetic hitchhiking in a subdivided population of *Mytilus edulis*. BMC Evol Biol 8:164. doi:10.1186/1471-2148-8-164

Faure B, Jollivet D, Tanguy A, Bonhomme F, Bierne N (2009) Secondary contact zone in the deep sea: a multi-locus analysis of divergence and gene flow between two closely-related species of *Bathymodiolus*. Genetics 4:e6485

Feder ME (2007) Evolvability of physiological and biochemical traits: evolutionary mechanisms including and beyond single-nucleotide mutation. J Exp Biol 310:1653-1660

Feder ME, Mitchell-Olds T (2003) Evolutionary and ecological functional genomics. Nat Rev

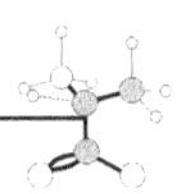

Genet 4:649 – 655

Fernandes JMO, MacKenzie MG, Kinghorn JR, Johnston IA (2007) FoxK1 splice variants show developmental stage-specific plasticity of expression with temperature in the tiger pufferfish. J Exp Biol 210:3461 – 3472

Fernandes JMO, MacKenzie MG, Wright PA, Steele SL, Suzuki Y, Kinghorn JR, Johnston IA (2006) Myogenin in model pufferfish species: comparative genomic analysis and thermal plasticity of expression during early development. Comp Biochem Physiol D 1:35 – 45

Fields PA, Kim Y-S, Carpenter JF, Somero GN (2002) Temperature adaptation in *Gillichthys* (Teleost: Gobiidae) A4-lactate dehydrogenases: identical primary structures produce subtly different conformations. J Exp Biol 205:1293 – 1303

Fields PA, Somero GN (1998) Hot spots in cold adaptation: localised increases in conformational flexibility in lactate dehydrogenase A4 orthologs of Antarctic Notothenioid fishes. Proc Natl Acad Sci USA 95:11476 – 11481

Flannery BG, Wenburg JK, Gharrett AJ (2007) Variation of amplified fragment length polymorphisms in yukon river chum salmon: population structure and application to mixed-stock analysis. Trans Am Fish Soc 136:911 – 925

Ford MJ (2002) Applications of selective neutrality tests to molecular ecology. Mol Ecol 11: 1245 – 1262

Franch R, Louro B, Tsalavouta M, Chatziplis D, Tsigenopoulos CS, Sarropoulou E, Antonello J, Magoulas A, Mylonas CC, Babbucci M, Patarnello T, Power DM, Kotoulas G, Bargelloni L (2006) A genetic linkage map of the hermaphrodite teleost fish *Sparus aurata* L.. Genetics 174:851 – 861

Fuchs Y, Douek J, Rinkevich B, Ben-Shlomo R (2006) Gene diversity and mode of reproduction in the brooded larvae of the coral *Heteroxenia fuscescens*. J Hered 97:493 – 498

Gaill F, Mann K, Wiedemann H, Engel J, Timpl R (1995) Structural comparison of cuticle and interstitial collagens from annelids living in shallow sea-water and at deep-sea hydrothermal vents. J Mol Biol 246:284 – 294

Garoia F, Guarniero I, Grifoni D, Marzola S, Tinti F (2007) Comparative analysis of AFLPs and SSRs efficiency population genetic structure of Mediterranean *Solea vulgaris*. Mol Ecol 16:1377 – 1387

Gibson G (2002) Microarrays in ecology and evolution: a preview. Mol Ecol 11:17 – 24

Gibson G, Riley-Berger R, Harshman L, Kopp A, Vacha S, Nuzhdin S, Wayne M (2004) Extensive sex-specific nonadditivity of gene expression in *Drosophila melanogaster*. Genetics 167: 1791 – 1799

Gort G, Koopman WJM, Stein A (2006) Fragment length distributions and collision probabilities for AFLP markers. Biometrics 62:1107 – 1115

Govoroun M, Le Gac F, Guiguen Y (2006) Generation of a large-scale repertoire of Expressed

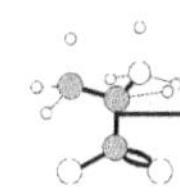

Sequence Tags (ESTs) from normalised rainbow trout cDNA libraries. BMC Genom doi:10. 1186/1471 -2164 -7 -196

Gracey AY, Fraser EJ, Li W, Fang Y, Taylor RR, Rogers J, Brass A, Cossins AR (2004) Coping with cold: An integrative, multitissue analysis of the transcriptome of a poikilothermic vertebrate. Proc Natl Acad Sci USA 101:16970 - 16975

Gruenthal KM, Acheson LK, Burton RS (2007) Genetic structure of natural populations of California red abalone (*Haliotis rufescens*) using multiple genetic markers. Mar Biol 152:1237 - 1248

Gruenthal KM, Burton RS (2008) Genetic structure of natural populations of the California black abalone (*Haliotis cracherodii* Leach, 1814), a candidate for endangered species status. J Exp Mar Biol Ecol 355:47 - 58

Guinand B, Lemaire C, Bonhomme F (2004) How to detect polymorphisms undergoing selection in marine fishes? A review of methods and case studies, including flatfishes. J Sea Res 51:167 - 182

Hadfield MG (1998) Research on settlement and metamorphosis of marine invertebrate larvae: past, present and future. Biofouling 12:9 - 29

Hadfield MG, Carpizo-Ituarte EJ, del Carmen K, Nedved BT (2001) Metamorphic competence, a major adaptive convergence in marine invertebrate larvae. Am Zool 41:1123 - 1131

Halpin PM, Sorte CJ, Hofmann GE, Menge BA (2002) Patterns of variation in levels of Hsp70 in natural rocky shore populations from microscales to mesoscales. Am J Physiol Integ Comp Biol 42:815 - 824

Hamady M, Walker JJ, Harris JK, Gold NJ, Knight R (2008) Error-correcting barcoded primers for pyrosequencing hundreds of samples in multiplex. Nat Meth 5:235 - 237

Hardivillier Y, Denis F, Demattei MV, Bustamante P, Laulier M, Cosson R (2006) Metal influence on metallothionein synthesis in the hydrothermal vent mussel *Bathymodiolus thermophilus*. Comp Biochem Physiol C Toxicol Pharmacol 143:321 - 332

Hayes BJ, Gjuvsland A, Omholt S (2006) Power of QTL mapping experiments in commercial Atlantic salmon populations, exploiting linkage and linkage disequilibrium and effect of limited recombination in males. Heredity 97:19 - 26

Hayes B, Laerdahl JK, Lien S, Moen T, Berg P, Hindar K, Davidson WS, Koop BF, Adzhubei A, Høyheim B (2007) An extensive resource of single nucleotide polymorphism markers associated with Atlantic salmon (*Salmo salar*) expressed sequences. Aquaculture 265:82 - 90

Hedgecock D, Lin JZ, DeCola S, Haudenschild CD, Meyer E, Manahan DT, Bowen B (2007) Transcriptomic analysis of growth heterosis in larval Pacific oysters (*Crassostrea gigas*). Proc Natl Acad Sci USA 104:2313 - 2318

Hedgecock D, McGoldrick DJ, Bayne BL (1995) Hybrid vigor in Pacific oysters: an experi-

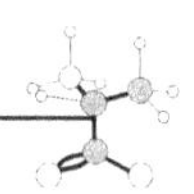

mental approach using crosses among inbred lines. Aquaculture 137:285 – 298

Hoekstra HE, Coyne JA (2007) The locus of evolution: evo-devo and the genetics of adaptation. Evolution 61:995 – 1016

Hoekstra HE, Nachman MW (2003) Different genes underlie adaptive melanism in different populations of rock pocket mice. Mol Ecol 12:1185 – 1194

Hofmann GE (2005) Patterns of gene expression in ectothermic marine organisms on small to large-scale biogeographical patterns. Intergr Comp Biol 45:247 – 255

Hofmann GE, Buckley BA, Airaksinen S, Keen JE, Somero GN (2000) Heat-shock protein expres-sion is absent in the Antarctic fish *Trematomus bernacchii* family Nototheniidae. J Exp Biol 203:2331 – 2339

Hogstrand C, Balesaria S, Glover CN (2002) Application of genomics and proteomics for study of the integrated response to zinc exposure in a non-model fish species, the rainbow trout. Comp Biochem Physiol Mol Biol 133:523 – 535

Holland BS (2000) Genetics of marine bioinvasions. Hydrobiologia 420:63 – 71

Hourdez S, Weber RE (2005) Molecular and functional adaptations in deep-sea hemoglobins. J Inorg Biochem 99:130 – 131

Hubert S, Hedgecock D (2004) Linkage maps of microsatellite DNA markers for the pacific oyster *Crassostrea gigas*. Genetics 168:351 – 362

Hudson ME (2008) Sequencing breakthrough for genomic ecology and and evolutionary biology. Mol Ecol Res 8:3 – 17

IPCC (2007) Climate change 2007: synthesis report. Contribution of work groups I, II and III to the 4th Assessment Report of the Intergovernmental Panel on Climate Change. Core writing team: Pachauri RK and Reisinger A (eds). IPCC, Geneva, Switzerland.

Irwin RD, Boorman GA, Cunningham ML, Heinloth AN, Malarkey DE, Paules RS (2004) Application of toxicogenomics to toxicology: basic concepts in the analysis of microarray data. Toxicol Pathol 32:72 – 83

Jackson DJ, Ellemor N, Degnan BM (2005) Correlating gene expression with larval competence, and the effect of age and parentage on metamorphosis in the tropical abalone *Haliotis asinina*. Mar Biol 147:681 – 697

Jin Y, deVries AL (2006) Antifreeze glycoprotein levels in Antarctic notthenioid fishes inhabiting different thermal environments and the effect of warm acclimation. Comp Biochem Physiol B 144:290 – 300

Johannesson K, Andre C (2006) Life on the margin: genetic isolation and diversity loss in a peripheral marine ecosystem, the Baltic Sea. Mol Ecol 15:2013 – 2029

Jollivet D, Desbruyeres D, Ladrat C, Laubier L (1995) Evidence for differences in the allozyme thermostability of deep-sea hydrothermal vent polychaetes (Alvinellidae): a possible selection

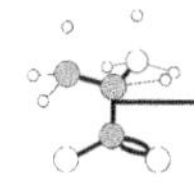

by habitat. Mar Ecol Prog Ser 123:125 – 136

Jorde LB (2000) Linkage disequilibrium and the search for complex disease genes. Gen Res 10:1435 – 1444

Kashi Y, King D, Soller M (1997) Simple sequence repeats as a source of quantitative genetic variation. Trends Genet 13:74 – 78

Kassahn KS (2008) Microarrays for comparative and ecological genomics: beyond single-species applications for array technologies. J Fish Biol 72:2407 – 2434

Kim A, Misra A (2007) SNP genotyping: technologies and biomedical applications. Ann Rev Biomed Eng 9:289 – 320

Kim KS, Ratcliffe ST, French BW, Liu L, Sappington TW (2008) Utility of EST-Derived SSRs as population genetics markers in a beetle. J Hered 99:112 – 124

Klinbunga S, Khetpu K, Khamnamtong B, Menasveta P (2007) Genetic heterogeneity of the blue swimming crab (*Portunus pelagicus*) in Thailand determined by AFLP analysis. Biochem Genet 45:725 – 736

Kumar S, Tsai CJ, Nussinov R (2000) Factors enhancing protein thermostability. Protein Eng 13:179 – 191

Lallias D, Beaumont AR, Haley CS, Boudry P, Heurtebise S, Lapègue S (2007a) A first-generation genetic linkage map of the European flat oyster *Ostrea edulis* (L.) based on AFLP and microsatellite markers. Anim Genet 38:560 – 568

Lallias D, Lapegue S, Hecquet C, Boudry P, Beaumont AR (2007b) AFLP-based genetic linkage maps of the blue mussel (*Mytilus edulis*). Anim Genet 38:340 – 349

Langaee T, Ronaghi M (2005) Genetic variation analyses by Pyrosequencing. Mut Res 573: 96 – 102 Launey S, Hedgecock D (2001) High Genetic Load in the Pacific Oyster *Crassostrea gigas*. Genetics 159:255 – 265

Le Bris N, Zbinden M, Gaill F (2005) Processes controlling the physico-chemical micro-environments associated with Pompeii worms. Deep-Sea Res 52:1071 – 1083

Lee M, Kwon J, Kim SN, Kim JE, Koh WS, Kim EJ, Chung MK, Han SS, Song CW (2003) cDNA microarray gene expression profiling of hydroxyurea, paclitaxel, and p-anisidine, genotoxic compounds with differing tumorigenicity results. Environ Mol Mutagen 42:91 – 97

Lennard Richard ML, Bengten E, Wilson MR, Miller NW, Warr GW, Hikima J (2007) Comparative genomics of transcription factors driving expression of the immunoglobulin heavy chain locus in teleost fish. J Fish Biol 71(Suppl. B):153 – 173

Li L, Guo X (2004) AFLP-based genetic linkage maps of the Pacific oyster C*rassostrea gigas* Thunberg. Mar Biotech 6:26 – 36

Luikart G, Allendorf FW, Cornuet JM, Sherwin WB (1998) Distortion of allele frequency distributions provides a test for recent population bottlenecks. J Hered 89:238 – 247

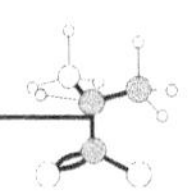

Luikart G, England PR, Tallmon D, Jordan S, Taberlet P (2003) The power and promise of population genomics: from genotyping to genome typing. Nat Rev Genet 4:981 – 994

Luther GW, Rozan TF, Taillefert M, Nuzzio DB, Meo CD, Shank TM, Lutz RA, Cary SC (2001) Chemical speciation drives hydrothermal vent ecology. Nature 410:813 – 816

Lynch M, Walsh B (1998) Genetics and analysis of quantitative traits. Sinauer Associates, Sunderland, MA

Mackay TFC (2001) Quantitative trait loci in *Drosophila*. Nat Rev Genet 2:11 – 20

Mackenzie MG (2006) Characterisation of genes regulating muscle development and growth in two model pufferfish species (*Takifugu rubripes* and *Tetraodon nigroviridis*). Thesis St Andrews University, Scotland, UK

Maldini M, Marzano FN, Fortes GG, Papa R, Gandolfi G (2006) Fish and seafood trace-ability based on AFLP markers: Elaboration of a species database. Aquaculture 261:487 – 494

Marcil A, Dumontier E, Chamberland M, Camper SA, Drouin J (2003) *Pitx1* and *Pitx2* are required for development of hindlimb buds. Development 130:45 – 55

Marden JH (2008) Quantitative and evolutionary biology of alternative splicing: how changing the mix of alternative transcripts affects phenotypic plasticity and reaction norms. Heredity 100:111 – 120

Margulies M et al. (2005) Genome sequencing in microfabricated high-density picolitre reactors. Nature 437:376 – 380

Maynard Smith J, Haigh J (1974) The hitchhiking effect of a favorable gene. Genet Res 23: 23 – 35

Medina M (2009) Functional genomics opens doors to understanding metamorphosis in non-model invertebrate organisms. Mol Ecol 18(5):763 – 764

Meredith MP, King JC (2005) Rapid climate change in the ocean west of the Antarctic Peninsula during the second half of the 20th century. Geophys Lett 32:L19604 – L19609

Meudt HM, Clarke AC (2007) Almost forgotten or latest practice? AFLP applications, analyses and advances. Trends Plant Sci 12:106 – 117

Minic Z, Herve G (2004) Biochemical and enzymological aspects of the symbiosis between the deep-sea tubeworm *Riftia pachyptila* and its bacterial endosymbiont. Eur J Biochem 271: 3093 – 3102

Mitchell-Olds T, Willis JH, Goldstein DB (2007) Which evolutionary processes influence natural genetic variation for phenotypic traits?. Nat Rev Genet 8:845 – 856

Moen T, Hayes B, Nilsen F, Delghandi M, Fjalestad K, Fevolden S-E, Berg PR, Lien S (2008) Identification and characterization of novel SNP markers in Atlantic cod: evidence for directional selection. BMC Genet 9:18

Moore MN (2001) Biocomplexity: the post-genome challenge in ecotoxicology. Aquat Toxicol

59:1 – 15

Morand P, Briand X (1996) Excessive growth of macroalgae: a symptom of environmental disturbance. Botanica Marina 39:491 – 516

Morin PA, Luikart G, Wayne RK SNP Workshop Group (2004) SNPs in ecology, evolution and conservation. Trends Ecol Evol 19:208 – 216

Moy GW, Springer SA, Adams SL, Swanson WJ, Vacquier VD (2008) Extraordinary intraspecific diversity in oyster sperm bindin. Proc Natl Acad Sci USA 105:1993 – 1998

Moylan TJ, Sidell BD (2000) Concentrations of myoglobin and myoglobin mRNA in heart ventricles from Antarctic fishes. J Exp Biol 203:1277 – 1286

Murphy JM, Mitchell JFB (1995) Transient-response of the Hadley-Center coupled ocean-atmosphere model to increasing carbon-dioxide. 2. Spatial and temporal structure of response. J Climate 1:57 – 80

Murray MC, Hare MP (2006) A genomic scan for divergent selection in a secondary contact zone between Atlantic and Gulf of Mexico oysters, *Crassostrea virginica*. Mol Ecol 15:4229 – 4242

Nielsen R (2005) Molecular signatures of natural selection. Ann Rev Genet 39:197 – 218

Nishio Y, Nakamura Y, Kawarabayasi Y, Usuda Y, Kimura E, Sugimoto S, Matsui K, Yamagishi A, Kikuchi H, Ikeo K, Gojobori T (2003) Comparative complete genome sequence analysis of the amino acid replacements responsible for the thermostability of Corynebacterium efficiens. Genome Res 13:1572 – 1579

Nordborg M, Tavare S (2002) Linkage disequilibrium: what history has to tell us. Trends Genet 18:83 – 90

Oetjen K, Reusch TBH (2007a) Genome scans detect consistent divergent selection among subtidal vs. intertidal populations of the marine angiosperm *Zostera marina*. Mol Ecol 16:5156 – 5157

Oetjen K, Reusch TBH (2007b) Identification and characterization of 14 polymorphic EST – derived microsatellites in eelgrass (Zostera marina). Mol Ecol Notes 7:777 – 780

O'Leary DB, Couglan J, McCarthy TV, Cross TF (2006) Application of a rapid method of SNP analysis (glycosylase mediated polymorphism detection) to mtDNA and nuclear DNA of cod *Gadus morhua*. J Fish Biol 69(Suppl. A):145 – 153

Orr JC, Fabry VJ, Aumont O, Bopp L, Doney SC, Feely RA, Gnandesikan A, Gruber N, Ishida A, Joos F, Key RM, Lindsay K, Plattner GK, Rodgers KB, Sabine CL, Sarmiento JL, Schlitzer

R, Slater RD, Totterdell IJ, Weirig MF, Yamanaka Y, Yool A (2005) Anthropogenic ocean acidification over the twenty-first century and its impact on calcifying organisms. Nature 437:681 – 686

Osovitz CJ, Hofmann GE (2005) Thermal history-dependant expression of the hsp70 gene in pur-ple sea urchins: biogeographic patterns and the effect of temperature acclimation. J Exp Mar Bi-

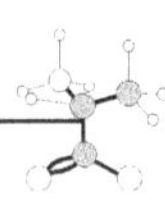

ol Ecol 327:134 – 143

Otsuka M, Arai M, Mori M, Kato M, Kato N, Yokosuka O, Ochiai T, Takiguchi M, Omata M, Seki N (2003) Comparing gene expression profiles in human liver, gastric, and pancreatic tissues using full-length-enriched cDNA libraries. Hepato Res 1:76 – 82

Oudot M-P, Kloareg B, de Goër S (2002) The complete sequence of the mitochondrial genome of *Laminaria digitata* (Laminariales). Eur J Phycol 37:163 – 172

Oudot-Le Secq M-P, Fontaine J-M, Rousvoal S, Kloareg B, Loiseaux-de Goër S (2001) The com-plete sequence of a brown algal mitochondrial genome, the ectocarpale *Pylaiella littoralis* (L.) Kjellm. J Mol Evol 53:80 – 88

Pace DA, Marsh AG, Leong P, Green A, Hedgecock D, Manahan DT (2006) Physiological bases of genetically determined variation in growth of marine invertebrate larvae: a study of growth heterosis in the bivalve *Crassostrea gigas*. J Exp Mar Biol Ecol 353:188 – 209

Palumbi SR (1999) All males are not created equal: fertility differences depend on gamete recognition polymorphisms in sea urchins. Proc Natl Acad Sci 96:12632 – 12637

Pechenik JA (1999) On the advantages and disadvantages of larval stages in benthic marine invertebrate life cycles. Mar Ecol Prog Ser 177:269 – 297

Peck LS, Clark MS, Clarke A, Cockell CS, Convey P, Detrich IIIHW, Fraser KPP, Johnston I, Methe B, Murray AE, Romisch K, Rogers A (2005) Genomics: Applications to Antarctic Ecosystems. Polar Biol 28:351 – 365

Peck LS, Clark MS, Morley SA, Massey A, Rosetti H (2009) Animal temperature limits: effects of size, activity and rates of change. Func Ecol 23:248 – 256

Peck LS, Webb KE, Bailey DM (2004) Extreme sensitivity of biological function to temperature in Antarctic marine species. Func Ecol 18:625 – 630

Pemberton JM (2004) Measuring inbreeding depression in the wild: the old ways are the best. Trends Ecol Evol 19:613 – 615

Pfaffl MW (2001) A new mathematical model for relative quantification in real-time RT-PCR. Nucl Acids Res 29:2002 – 2007

Pfaffl MW, Horgan GW, Dempfle L (2002) Relative expression software tool (REST©) for group – wise comparison and statistical analysis of relative expression results in real-time PCR. Nucl Acids Res 30:1 – 10

Phillips C (2007) Online resources for SNP analysis-a review and route map. Mol Biotechnol 35:65 – 97

Piccino P, Viard F, Sarradin PM, Le Bris N, Le Guen D, Jollivet D (2004) Thermal selection of PGM allozymes in newly founded populations of the thermotolerant vent polychaete *Alvinella pompejana*. Proc Roy Soc Lond B 271:2351 – 2359

Place SP, O'Donnell MJ, Hofmann GE (2008) Gene expression in the intertidal mussel Mytilus

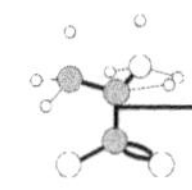

californianus: physiological response to environmental factors on a biogeographic scale. Mar Ecol Prog Ser 356:1 – 14

Podrabsky JE, Somero GN (2004) Changes in gene expression associated with acclimation to constant temperatures and fluctuating daily temperatures in an annual killifish *Austrofundulus limnaeus*. J Exp Biol 207:2237 – 2254

Pompanon F, Bonin A, Bellemain E, Taberlet P (2005) Genotyping errors: causes, consequences and solutions. Nat Rev Genet 6:847 – 859

Price AH (2006) Believe it or not, QTLs are accurate! . Trends Plant Sci 11:213 – 216

Pruski AM, Dixon DR (2003) Toxic vents and DNA damage: first evidence from a naturally contaminated deep-sea environment. Aquat Toxicol 64:1 – 13

Purac′ J, Burns B, Thorne MAS, Grubor-Lajšic′ G, Worland MR, Clark MS (2008) Cold hardening processes in the Antarctic springtail, *Cryptopygus antarcticus*: clues from a microarray. J Insect Physiol 54:1356 – 1362

Qin Y, Liu X, Zhang H, Zhang G, Guo X (2007) Genetic mapping of size-related quantitative trait loci (QTL) in the bay scallop (*Argopecten irradians*) using AFLP and microsatellite markers. Aquaculture 272:281 – 290

Radonic A, Thulke S, Mackay IM, Landt O, Siegert W, Nitsche A (2004) Guideline to reference gene selection for quantitative real-time PCR. Biochem Biophys Res Comm 313:856 – 862

Renn SCP, Aubin-Horth N, Hofmann HA (2004) Biologically meaningful expression profiling across species using hetrerologous hybridisation to a cDNA microarray. BMC Genom doi:10.1186/1471 – 2164 – 5 – 42

Ricciardi A, MacIsaac HJ (2000) Recent mass invasion of the north american great lakes by ponto-caspian species. Trends Ecol Evol 15:62 – 65

Roberts DA, Hofmann GE, Somero GN (1997) Heat shock protein expression in *Mytilus califor-nianus*: Acclimatization (seasonal and tidal height comparisons) and acclimation effects. Biol Bull 192:309 – 320

Roberts S, Romano C, Gerlach G (2005) Characterization of EST derived SSRs from the bay scallop, *Argopecten irradians*. Mol Ecol Notes 5:567 – 568

Robinson-Rechavi M, Alibes A, Godzik A (2006) Contribution of electrostatic interactions, com-pactness and quaternary structure to protein thermostability: Lessons from structural genomics of *Thermotoga maritime*. J Mol Biol 356:547 – 557

Roff DA (2007) Contributions of genomics to life-history theory. Nat Rev Genet 8:116 – 125

Rogers SM, Bernatchez L (2005) Integrating QTL mapping and genome scans towards the characterization of candidate loci under parallel selection in the lake whitefish (*Coregonus clupeaformis*). Mol Ecol 14:351 – 361

Rogers SM, Campbell D, Baird SJE, Danzmann RG, Bernatchez L (2001) Combining the ana-

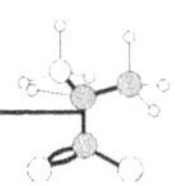

ly-ses of introgressive hybridisation and linkage mapping to investigate the genetic architecture of population divergence in the lake whitefish (*Coregonus clupeaformis*, Mitchill). Genetica 111:25 - 41

Roman J, Darling JA (2007) Paradox lost: genetic diversity and the success of aquatic invasions. Trends Ecol Evol 22:454 - 464

Römisch K, Collie N, Soto N, Logue J, Lindsay M, Scheper W, Cheng C-H C (2003) Protein translocation across the endoplasmic reticulum membrane in cold adapted organisms. J Cell Sci 116:2875 - 2883

Royal Society (2005) Ocean acidification due to increasing atmospheric carbon dioxide. Policy Document 12/05, The Royal Society

Rungis D, Berube Y, Zhang J, Ralph S, Ritland CE, Ellis BE et al. (2004) Robust simple sequence repeat markers for spruce (*Picea* spp.) from expressed sequence tags. Theor Appl Genet 109:1283 - 1294

Ryynänen HJ, Primmer C (2006) Single nucleotide polymorphism (SNP) discovery in duplicated genomes: intron-primed exon-crossing (IPEC) as a strategy for avoiding amplification of duplicated loci in Atlantic salmon (*Salmo salar*) and other salmonid fishes. BMC Genom 7:192

Ryynänen HJ, Tonteri A, Vasemägi A, Primmer CR (2007) A comparison of the efficiency of single nucleotide polymorphisms (SNPs) and microsatellites for the estimates of population and conservation genetic parameters in Atlantic salmon (*Salmo salar*). J Hered 98:692 - 704

Saastamoinen M, Hanski I (2008) Genotypic and environmental effects on flight activity and oviposition in the Glanville fritillary butterfly. Am Nat 171:701 - 712

Saavedra C, Bachere E (2006) Bivalve genomics. Aquaculture 256:1 - 14

Sanchez S, Hourdez S, Lallier F (2007) Identification of proteins involved in the functioning of *Riftia pachyptila* symbiosis by Subtractive Suppression Hybridization. BMC Genomics 8:337

Sax DF, Stachowicz JJ, Brown JH, Bruno JF, Dawson MN, Gaines SD, Grosberg RK, Hastings A, Holt RD, Mayfield MM, O'Connor MI, Rice WR (2007) Ecological and evolutionary insights from species invasions. Trends Ecol Evol 22:465 - 471

Schlotterer C (2003) Hitchhiking mapping-functional genomics from the population genetics perspective. Trends Genet 19:32 - 38

Shank TM, Halanych KM (2007) Toward a mechanistic understanding of larval dispersal: insights from genomic fingerprinting of the deep - sea hydrothermal vent tubeworm *Riftia pachyptila*. Mar Ecol 28:25 - 35

Shapiro MD, Marks ME, Peichel CL, Blackman BK, Nereng BJ, Schluter D et al. (2004) Genetic and developmental basis of evolutionary pelvic reduction in threespine sticklebacks. Nature 428:717 - 723

Sheader DL, Williams TD, Lyons BP, Chipman JK (2006) Oxidative stress response of European flounder (Platichthys flesus) to cadmium determined by a custom cDNA microarray. Mar Environ Res 62:33 – 44

Sicot FX, Mesnage M, Masselot M, Exposito JY, Garrone R, Deutsch J, Gaill F (2000) Molecular adaptation to an extreme environment: origin of the thermal stability of the Pompeii worm collagen. J Mol Biol 302:811 – 820

Slate J, David P, Dodds KG, Veenvliet BA, Glass BC, Broad TE et al. (2004) Understanding the relationship between the inbreeding coefficient and multilocus heterozygosity: theoretical expectations and empirical data. Heredity 93:255 – 265

Smith CT, Antonovich A, Templin WD, Elfstrom CM, Narum SR, Seeb LW (2007) Impacts of marker class bias relative to locus – specific variability on population inferences in chinook salmon: A comparison of single-nucleotide polymorphisms with short tandem repeats and allozymes. Trans Am Fish Soc 136:1674 – 1687

Smith CT, Elfstrom CM, Seeb LW, Seeb JE (2005) Use of sequence data from rainbow trout and Atlantic salmon for SNP detection in Pacific salmon. Mol Ecol 14:4193 – 4203

Somero GN (2002) Thermal physiology and vertical zonation of intertidal animals: optima, limits and costs of living. Am J Physiol Integ Comp Biol 42:780 – 789

Stachowicz JJ, Terwin JR, Whitlatch RB, Osman RW (2002) Linking climate change and biologi – cal invasions: Ocean warming facilitates nonindigenous species invasions. Proc Natl Acad Sci USA 99:15497 – 15500

Stern S, Dror T, Stolovicki E, Brenner N, Braun E (2007) Genome-wide transcriptional plasticity underlies cellular adaptation to novel challenge. Mol Sys Biol Doi:10. 1038/msb4100147

Stewart FJ, Cavanaugh CM (2006) Bacterial endosymbioses in Solemya (Mollusca: Bivalvia) — model systems for studies of symbiont-host adaptation. Antonie Leeuwenhoek 90:343 – 360

Stinchcombe JR, Hoekstra HE (2008) Combining population genomics and quantitative genetics: finding the genes underlying ecologically important traits. Heredity 100:158 – 170

Stolovitzky GA, Kundaje A, Held GA, Duggar KH, Haudenschild CD, Zhou D, Vasicek TJ, Smith KD, Aderem A, Roach JC (2005) Proc Natl Acad Sci USA 102:1402 – 1407

Storz JF (2005) Using genome scans of DNA polymorphism to infer adaptive population divergence. Mol Ecol 14:671 – 688

Streelman JT, Kocher TD (2002) Microsatellite variation associated with prolactin expression and growth of salt-challenged tilapia. Physiol Genom 9:1 – 4

Suzuki Y, Yoshitomo-Nakagawa K, Maruyama K, Suyama A, Sugano S (1997) Construction and characterization of a full length-enriched and a 5-end-enriched cDNA library. Gene 1 – 2:149 – 156

Swanson-Wagner RA, Jia Y, De Cook R, Borsuk LA, Nettleton D, Schnable PS (2006) All

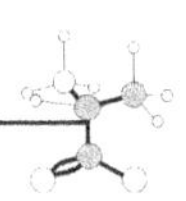

possible modes of gene action are observed in a global comparison of gene expression in a maize F-1 hybrid and its inbred parents. Proc Natl Acad Sci USA 103:6805 – 6810

Tabor HK, Risch NJ, Myers RM (2002) Candidate gene approaches for studying complex genetic traits: practical considerations. Nat Rev Genet 3:391 – 397

Tanguy A, Bierne N, Saavedra C, Pina B, Bachère E, Kube M, Bazin E, Bonhomme F, Boudry P, Boulo V, Boutet I, Cancela L, Dossat C, Favrel P, Huvet A, Jarque S, Jollivet D, Klages S, Lapegue S, Leite R, Moal J, Moraga D, Reinhardt R, Samain J-F, Zouros E, Canario A (2008) Increasing genomic information in bivalves through new EST collections in four species: Development of new genetic markers for environmental studies and genome evolution. Gene 408:27 – 36

Tao WJ, Boulding EG (2003) Associations between single nucleotide polymorphisms in candidate genes and growth rate in Arctic charr (*Salvelinus alpinus* L. Heredity 91:60 – 69

Taris N, Comtet T, Viard F (2009) Inhibitory function of nitric oxide on the onset of metamorphosis in competent larvae of *Crepidula fornicata*: a transcriptional perspective. Mar Genomics 2: 161 – 167

Teranishi KS, Stillman JH (2007) A cDNA microarray analysis of the response to heat stress in hepatopancreas tissue of the porcelain crab *Petrolisthes cinctipes*. Comp Biochem Physiol D 2:53 – 62

Teshima KM, Coop G, Preworski M (2006) How reliable are empirical genomic scans for selective sweeps?. Genome Res 16:702 – 712

Tomanek L (2002) The heat shock response: its variation, regulation and ecological importance in intertidal gastropods (genus *Tegula*). Am J Physiol Integ Comp Biol 42:797 – 807

Tomanek L (2005) Two-dimensional gel analysis of the heat shock response in marine snails (genus *Tegula*): interspecific variation in protein expression and acclimation ability. J Exp Biol 208:3133 – 3143

Tomanek L, Somero GN (2000) Time course and magnitude of synthesis of heat shock proteins in congeneric marine snails (Genus *Tegula*) from differentb tidal heights. Physiol Biochem Zool 73:249 – 256

Toth AL, Varala K, Newman TC, Miguez FE, Hutchison SK, Willoughby DA et al. (2007) Wasp brain gene expression supports an evolutionary link between maternal behavior and eusociality. Science 318:441 – 444

Troester MA, Hoadley KA, Parker JS, Perou CM (2004) Prediction of toxicant-specific gene expression signatures after chemotherapeutic treatment of breast cell lines. Environ Health Perspect 112:1607 – 1613

Ungerer MC, Johnson LC, Herman MA (2008) Ecological genomics: understanding gene and genome function in the natural environment. Heredity 100:178 – 183

Uthicke S, Conand C (2005) Amplified fragment length polymorphism (AFLP) analysis indicates the importance of both asexual and sexual reproduction in the fissiparous holothurian Stichopus chloronotus (Aspidochirotida) in the Indian and Pacific Ocean. Coral Reefs 24:103 – 111

Vasemägi A, Nilsson J, Primmer CR (2005) Expressed Sequence Tag-Linked Microsatellites as a Source of Gene-Associated Polymorphisms for Detecting Signatures of Divergent Selection in Atlantic Salmon (*Salmo salar* L). Mol Biol Evol 22:1067 – 1076

Vasemägi A, Primmer CR (2005) Challenges for identifying functionally important genetic varia-tion: the promise of combining complementary research strategies. Mol Ecol 14:3623 – 3642

Vaughan P (2000) Use of uracil DNA glycosylase in the detection of known DNA mutations and polymorphisms. Glycosylase-mediated polymorphism detection (GMPD-check). Methods Mol Biol (Clifton, NJ) 152:169 – 177

Venter JC, Remington K, Heidelberg JF, Halpern AL, Rusch D, Eisen JA et al. (2004) Environmental genome shotgun sequencing of the Sargasso Sea. Science 304:66 – 74

Vera JC, Wheat CW, Fescemyer HW, Frilander MJ, Crawford DL, Hanski I, Marden JH (2008) Rapid transcriptome characterization for a nonmodel organism using 454 pyrosequencing. Mol Ecol 17:1636 – 1647

Verde C, Lecointre G, di Prisco G (2007) The phylogeny of polar fishes and the structure and molecular evolution of hemoglobin. Polar Biol 30:523 – 539

Voisin M, Engel C, Viard F (2005) Differential shuffling of native genetic diversity across in-tro-duced region in a brown alga: aquaculture vs. maritime traffic effects. Proc Natl Acad Sci USA 102:5432 – 5437

Vos P, Hogers R, Bleeker M, Reijans M, van de Lee T, Hornes M, Frijters A, Pot J, Peleman J, Kuiper M, Zabeau M (1995) AFLP: a new technique for DNA fingerprinting. Nucl Acids Res 23: 4407 – 4414

Ward RD, Woodwark M, Skibinski DOF (1994) A comparison of genetic diversity levels in marine, freshwater, and anadromous fishes. J Fish Biol 44:213 – 232

Wares JP, Blakeslee AMH (2007) Amplified fragment length polymorphism data provide a poor solution to the *Littorina littorea* puzzle. Mar Biol Res 3:168 – 174

Weetman D, Ruggierro E, Mariani S, Shaw PW, Lawler AR, Hauser L (2007) Hierarchical pop-ulation genetic structure in the commercially exploited shrimp *Crangon crangon* identified by AFLP analysis. Mar Biol 151:565 – 575

Weiss E, Bennie M, Hodgins-Davis A, Roberts S, Gerlach G (2007) Characterization of new SSR-EST markers in cod, *Gadus morhua*. Mol Ecol Notes 7:866 – 867

Wenne R, Boudry P, Hemmer-Hansen J, Lubieniecki KP, Was A, Kause A (2007) What role for genomics in fisheries management and aquaculture?. Aquatic Living Res 3:241 – 255

Wilding CS, Butlin RK, Grahame J (2001) Differential gene exchange between parapatric morphs of Littorina saxatilis detected using AFLP markers. J Evol Biol 14:611 – 619

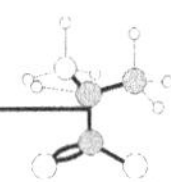

Williams EA, Degnan BM, Gunter H, Jackson DJ, Woodcroft BJ, Degnan SM (2009) Widespread transcriptional changes pre-empt the critical pelagic-benthic transition in the vetigastropod Haliotis asinina. Mol Ecol 18:1006 – 1025

Wilson K, Thorndyke M, Nilsen F, Rogers A, Martinez P (2005) Marine systems: moving into the genomics area. Mar Ecol 26:3 – 16

Wolff WJ, Reise K (2002) Oyster imports as a vector for the introduction of alien species into northern and western European coastal waters. In: Leppäkoski E, Gollasch S, Olenin S (eds) Invasive aquatic species of Europe. Distribution, impacts and management. Kluwer Academic Publishers, Dordrecht/Boston/London, pp 193 – 205

Woodhead M, Russell J, Squirrell J, Hollingsworth PM, Mackenzie K, Gibby M et al. (2005) Comparative analysis of population genetic structure in *Athyrium distentifolium* (Pteridophyta) using AFLPs and SSRs from anonymous and transcribed gene regions. Mol Ecol 14:1681 – 1695

Wray GA (2007) The evolutionary significance of *cis* – regulatory mutation. Nat Rev Genet 8: 206 – 216

Yang Z, Bielawski J (2000) Statistical methods for detecting molecular adaptation. Trends Ecol Evol 15:496 – 503

Yu J et al. (2006) A unified mixed-model method for association mapping that accounts for multiple levels of relatedness. Nat Genet 38:203 – 208

Yu Z, Guo X (2003) Genetic linkage map of the Eastern oyster *Crassostrea virginica* Gmelin. Biol Bull 204:327 – 338

Zachos J, Pagani M, Sloan L, Thomas E, Billups K (2001) Trends, rhythms and aberrations in global climate 65 Ma to present. Science 292:686 – 693

Zane L, Bargelloni L, Patarnello T (2002) Strategies for microsatellite isolation: a review. Mol Ecol 11:1 – 16

Zhao YM, Li Q, Kong LF, Bao ZM, Zhang RC (2007) Genetic diversity and divergence among clam *Cyclina sinensis* populations assessed using amplified fragment length polymorphism. Fish Sci 73:1338 – 1343

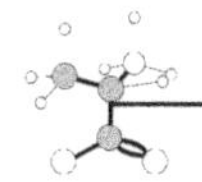

第4章　动物系统发育：基因组有很多话要说

摘要：各方面的证据一直不断地对后生动物的系统进化关系进行更新和完善。以形态和发育为基础的进化学观点主要依赖一些在进化上难以判断的特征，导致其在进化关系的研究上经常受到质疑。分子生物学可以为进化树的构建提供新的独特视角，核糖体RNA亚基和Hox等基因使后生生物进化树的树形有了极大的变化，促进了动物系统发育关系"新观点"的形成。然而，经典的分子生物学方法并不能成功解决动物亲缘关系中一些长期存在的问题。现在，包括海洋生物在内的大量动植物以及微生物基因组测序工作的完成将极大促进系统基因组学技术的发展，因为基于全基因组来研究生物系统进化关系可以有效避免单基因构建进化树的弊端。这种新方法使动物之间的进化关系研究有了进一步的提高，如对脊索动物间亲缘关系的重新评估以及对存疑的较小门类进行更加准确的重新定位。对于动物形态发育进程以及分子系统发育学的前景，也提出了很多新的问题和思考。

4.1　引言

亚里士多德的《动物史》第一次对动物进行了分类，并根据动物的繁殖方式和血液温度将动物分为五大类：哺乳类、卵胎生鲨鱼、鸟类与爬行类、海洋动物类（鱼、头足类和甲壳类）以及昆虫类（Aristotle 1965）。从现代系统发生学的观点来看，只有海洋动物类含有进化树中所有的代表动物（图4.1），表明我们以前对动物的认识局限于陆地，而缺乏对海洋动物的关注，但这也体现出了开展海洋动物多样性研究的重要性。

对后生动物进行分类的主要困难是，在公认的36大类群中存在高度多样的动物形态和身体构造。而这种多样性在海洋生物中是最高的。如最新研究表明，与脊椎动物亲缘关系最近的尾索动物，其成年动物的身体构造与经典的脊索动物差异极大（图4.2；Delsuc et al. 2006）。包括重演理论和聚类方法

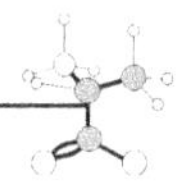

在内的很多研究策略都在不断增加形态学特征分类的层级，但都不能完全解释趋同进化和平行进化。近期进化发育生物学领域的发现也显示，同源基因通路与趋同组织的发育有关，比如节肢动物和脊椎动物的肢体（Shubin et

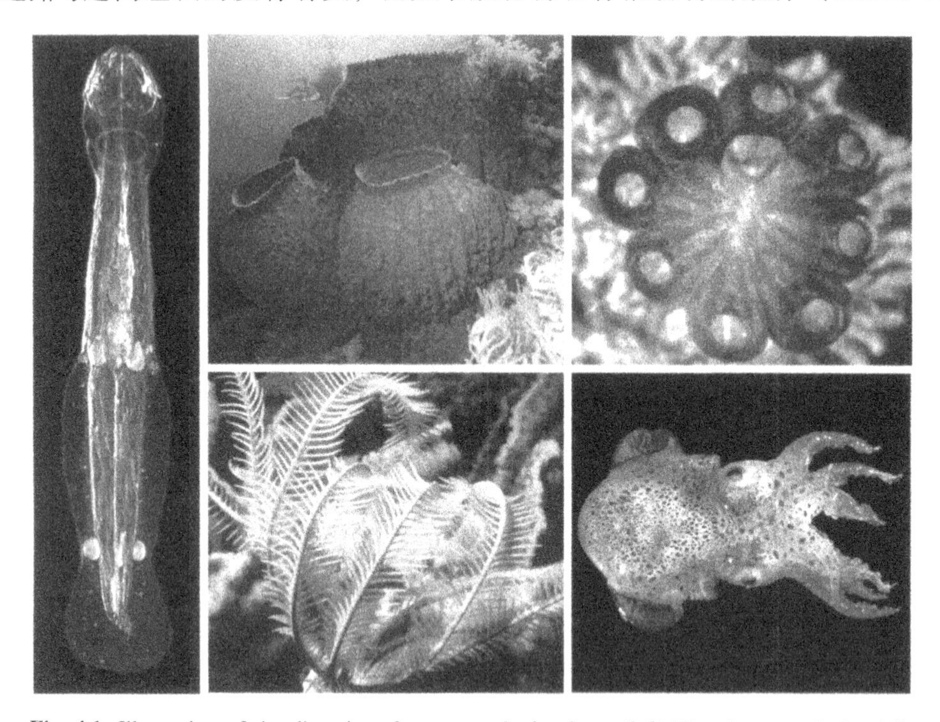

Fig. 4.1 Illustration of the diversity of metazoan body plans. (*left*) The chaetognath *Spadella cephaloptera* is representative of one of the most unique bilaterian phyla. (*middle top*) The massive barrel sponges *Xestospongia testudinaria.* (**c**) Several individuals of the colonial ascidian *Botrylloides leachi,* which belongs to urochordates, the closest relatives of the chordates (*top left*). (**d**) Another member of the deuterostomes, the crinoid *Antedon* (*bottom middle*). (**e**) The bobtail squid *Sepiola atlantica*, a cephalopod that displays numerous innovative features with respect to its body plan (*bottom left*)

图 4.1　后生动物体型多样性说明。(a) 毛颚类动物头翼锄虫是两侧对称动物类群中的特有代表（左）。(b) 巨型圆桶海绵（中上）。(c) 拟菊海鞘个体，属于尾索动物海鞘类，与脊索动物亲缘关系最近（右上）。(d) 另一种后口动物——海羊齿（下中），属于海百合类动物。(e) 短尾乌贼类大西洋耳乌贼，在体型方面有很多新特点的一种头足类动物（右下）

英文注释：Metazoan，后生动物；Chaetognath，毛颚类动物；Spadella，锄虫，锄虫属；Bilaterian，两侧对称动物；Phyla，门；Xestospongia Testudinaria，巨型圆桶海绵；Colonial Ascidian，海鞘，海鞘类；Botrylloides Leachi，拟菊海鞘；Urochordates，尾索动物；Chordates，脊索动物；Deuterostomes，后口动物；Crinoid，海百合纲动物；Antedon，海羊齿；Sepiola Atlantica，大西洋耳乌贼；Cephalopod，头足类动物

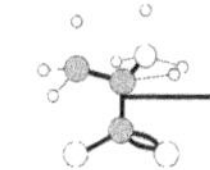

al. 1997）进化。相反，虽然腔肠动物和脊椎动物具有不同的身体结构式样，但他们却共享大量的基因网络，这些基因网络构成了它们身体前后两侧和背腹侧的基本结构模式（Mtindale 2005）。总的来说，这是从分子水平上对困扰许多形态学研究的同源性问题进行了重新描述与分析（Gould 2002，Wagner 2007）。若要准确地解释发育和形态特征，则需要一个独立的参照体系来确定动物在进化过程中所处的状态。对这样参照体系的寻找不仅推动了分子推演法的发展，还促进了人们对一些特定基因特征的研究。作为一个被大量使用的分子标记，18S rRNA 导致了一系列关于动物系统发育新观点（Halanych 2004）的形成，也促进了人们对已有动物分类结果的讨论与重新评价。尽管基于 18S rRNA 的新方法解决了很多动物分类的问题，但并不能完全解决一些动物系统发育中长期存在的争议，比如后生动物进化树基部的聚类问题和一些分类未定的生物之确切分类地位（如毛颚类动物和阿克尔扁形虫，图 4. 1）。这些局限性可能是由以下两种原因造成的：一是随机抽取的标记基因太少造成的随机误差，二是分子进化模型不能对数据进行彻底地解释而造成的系统误差。

越来越多生物基因组数据不断地发表和公开，为动物之间系统关系的研究提供了很多研究内容和线索。基因组数据既可以消除随机误差，也可以帮助鉴定更多的质量分子性状（有时被称为“罕有基因组改变”；Rokas and Holland 2000a）。在这一章里，我们将针对基因组学研究方法的优缺点进行深入探讨，并通过这两方面的内容来展示该方法是如何影响着我们对动物亲缘关系的认识的。随着推理方法的不断改进以及测序技术的出现，使得后生动物进化树发生了深层次的调整，这种调整最终将会改变我们对动物进化的理解。

4.2 动物系统发育的起源

在分子生物学诞生之前，为了区分错综复杂的动物形态，人们一直尝试通过各种系统分类学方法来描述动物的形态特征。

4.2.1 以前的策略是基于体腔进化的假设

1866 年 Haeckel 提出的“重演论”假说认为，通过将动物胚胎发育阶段补充到动物进化史中可以说明动物进化的连续性（Gould 1977）。因此，通过研究生物体在最初胚胎发育阶段的相似性，就能推断出他们之间深层次的进

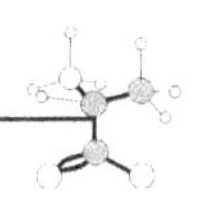

化关系，由此可见胚胎学特征对动物系统发育研究十分重要（Valentine 1997）。于是以体腔的结构和形成为基础，形成了过去大多数关于动物亲缘关系的假说。

在 Libbie Hyman 的研究基础上提出的系统发育学说被认为是“传统教科书”，该学说认为：原口动物和后口动物的体腔都是独立演化而来的（Hyman 1940 – 1967）。由此可以认为，裂肠动物和腔肠动物最初是由一种两侧对称的三胚层无体腔动物通过不同的体腔形成机制进化而来的（参见 Willmer 1990）。在此基础上又形成了一些新的观点，这些观点认为在后生动物的进化过程中，无体腔动物（扁形动物门和纽形动物门）和假体腔动物（如线虫类和轮虫类）形成了早期的进化分支（图 4.2；Barnes 1974）。不过这些观点与 Libbie Hyman 最初的想法并不同，因为后者的“gradist”方法主要用于研究形体的复杂程度，而不是为了准确地解释辐射进化（Jenner 2004）。

基于对体腔进化的不同解释，关于后生动物的亲缘关系可能会形成两种完全对立的观点。比如，拥有三聚体腔的两侧对称的动物祖先是由肠腔动物演化而来这一假想，进一步发展成了德国研究者普遍认可的“archecoelomate”假说。这些多聚体腔应该是来源于一种珊瑚虫样祖先的胃囊（Remane 1963）。因此，一些通过原肠腔的外包缝合方式形成体腔的动物（如半索动物类、脊索动物、触手冠类动物和毛颚类动物等），被认为在两侧对称动物早期就有了分化（Siewing !976；图 4.2）。环节动物和节肢动物的分节躯体特征应该是后来起源形成的，即失去体腔前部，通过体腔后部的分段而成（Tautz 2004）。而无体腔和假体腔动物的身体结构式样则被认为是体腔退化过程多样性的代表（Siewing 1976）。

人们针对这些例子提出了许多种假说，并且对这些假说不断地进行论证和发展（Willmer 1990），但无论哪一种假说都不能完全清楚地解释在动物系统发育研究中遇到的基础性问题（特征描述）。

4.2.2 通过分支系统分析法筛选更多的特征

人们在重建后生动物亲缘关系的过程中，尝试应用了许多动物的发育形态特征，但是这些特征提供的信息仅与进化树中某些节点有关（参照综述 Willmer1990）。有人认为两侧对称动物获得第三胚层（中胚层），其左右对称是基于双胚层，而最基础的后生动物（腔肠动物、海绵动物和栉水母门动物）则是辐射对称。有些动物的特征很难被解释，因为这些特征可能会与他们所属类群的共同特征相互矛盾。譬如，环节动物和扁形动物的形态相当简单，拥

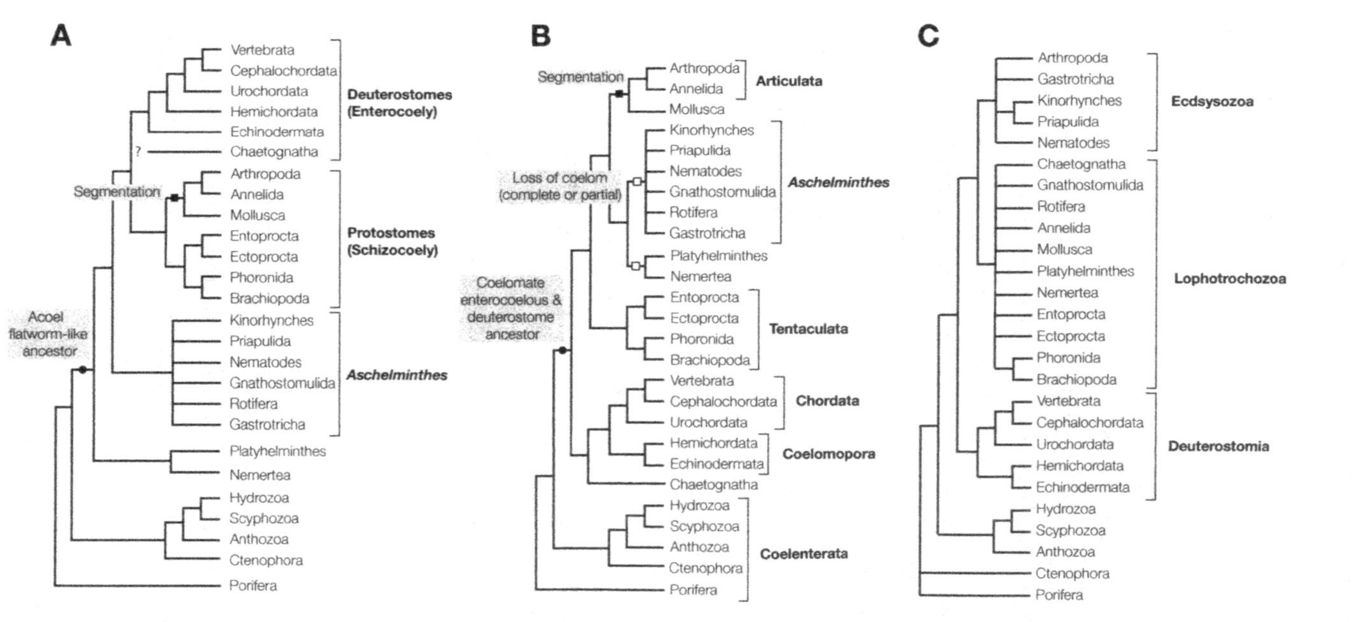

Fig. 4.2 Some historical schemes of metazoan phylogeny. (**a**) An extreme representation of the "traditional" textbook phylogeny assuming the progressive divergence of increasingly complex forms of bilaterian animals from a triploblastic acoelomate ancestor. The main character underlying this view is the establishment of body cavities, from Barnes (1974). (**b**) An archaecoelomate scheme of metazoan phylogeny. This view postulates an early origin of the coelom and enterocoely as the ancestral mode of coelom formation, which results in organisms with such developmental features being at the base of the tree (Marcus 1958). (**c**) The "new view" of metazoan phylogeny that splits all bilaterians into three main clades on the base of evidence from SSU molecular phylogeny and Hox genes, from Adoutte et al. (2000)

图4.2 后生动物系统发育的一些经典模式。(a)传统教科书式系统发育关系的一种极端代表，其展示了由一种三胚层无体腔祖先逐渐演化成日益复杂的两侧对称动物形式的过程。Barnes认为这一演变过程的一个主要特征就是体腔的形成（1974）。(b)后生动物系统发育的古体腔模式。这个观点假设早期的体腔动物和肠腔动物是体腔形成的原始模式，因此导致具有这些发育特征的生物处于进化树的根部(Marcus 1958)。(c) 基于来自SSU和Hox基因的相关证据，后生动物系统发育的“新观点”将所有两侧对称动物分为三个主要进化枝（从Adoutte et al. 2000)

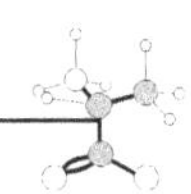

英文注释：Acoel Flatworm-like Ancestor，阿克尔扁形虫样祖先；Segmentation，身体分节；Vertebrata，脊椎动物门；Cephalochordata，头索动物纲；Urochordata，尾索动物门；Hemichordata，半索动物门 ；Echinodermata，棘皮动物门；Chaetognatha，毛颚动物门；Deuterostomes（Enterocoely），后口动物（肠腔动物）；Arthropoda，节肢动物门；Annelida，环节动物门 ；Mollusca，软体动物门；Entoprocta，内肛动物门；Anthozoa，珊瑚虫纲；Ectoprocta，外肛动物门；Phoronida，帚虫动物门；Brachiopoda，腕足动物门；Protostomes（Schizocoely），原口动物（裂腔动物）；Kinorhynches，动吻动物类；Priapulida，三部虫门；Nematodes，线虫属；Gnathostomulida，颚胃动物门；Rotifera，轮虫纲；Gastrotricha，腹毛动物门；*Aschelminthes*，袋形动物门；Platyhelminthes，扁形动物门；Nemertea，纽形动物门；Hydrozoa，水螅纲；Scyphozoa，钵水母纲；Ctenophora，栉水母门；Porifera，海绵动物门；Coelomate Enterocoelous & Deuterostome Ancestor，肠体腔动物和后口动物祖先；Loss of Coelom（Complete or Partial），体腔遗失（完全或部分）；Articulata，有关节类；*Aschelminthes*，袋形动物门；Tentaculata，触手动物纲；Chordata，脊索动物；Coelenterata，腔肠动物门；Lophotrochozoa，冠轮动物超门；Metazoan，后生动物，后生动物的；Phylogeny：系统发生，系统发育；Bilaterian，两侧对称动物；Triploblastic，三胚层的；Acoelomate，无体腔的；Cavities，腔；Coelom，胚腔，体腔；Enterocoely，肠腔动物；Clades，进化枝；Molecular Phylogeny，分子系统发生，分子系统学

有节肢动物没有的一些突出特征，如螺旋卵裂模式和担轮型幼虫阶段（Nielsen 2001）。但根据普通的分段体节分类方法，环节动物和节肢动物却被归类到“Articulés”动物群（Cuvier 1817）。

1950 年 Willi Hennig 建立的分支系统分析法不仅可以更加准确地处理特征进化，而且促进了基于形态学的生物分类（Hennig 1966）。进化枝分析方法并不支持缺乏特定特征的分类，相反，它提出真正的分类（即分支）应该有清晰的共同衍生特征（即共有衍征）。根据这一理论观点，发展出了依据动物形态数据矩阵来构建进化树的最大简约法，极大促进了对动物系统发育中形态特征的研究（Swofford 1990）。例如，前面提到的腕螺类动物与关节类动物对比的例子就是一个里程碑式的研究（Eernisse et al. 1992），它依赖于精确的特征定义并通过特征比对否定了关节动物假说。因此，该研究认为身体分节现象集中在环节动物和节肢动物，并发现了新的进化支（参见下文）。

尽管这项研究获得了很大的成功，但是由于其主要是以形态学特征为基础，因此也受到了颇多质疑。有些研究者简单地认为形态学并不能提供足够多的可信特征，尤其是涉及到与同源性评估有关的问题时（Scotland et al. 2003）。当然，也有人认为，这样的观点太严苛了。方法学的进步使可用于系

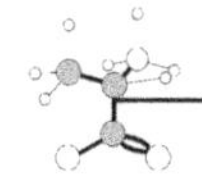

统发育研究的可信形态特征越来越多。最近，利用激光扫描共聚焦显微镜进行的超微结构研究就为一些有争议的课题提供了很多新的思路。例如，尽管星虫动物的肢体并不分节，但研究表明星虫动物和环节动物却有着很近的亲缘关系（Wanninger et al. 2005）。同样4D显微技术的应用也使人们对细胞系和原基分布图的研究变得更加得心应手，也得出了很多出乎意料的新结论，如被囊类尾海鞘纲动物异体住囊虫可能正处于衍生状态（Stach et al. 2008）。

但随着特征矩阵的广泛应用，人们也发现了很多问题，尤其是离散的特征编码与原始进化枝方法所支持的同源性存在很多冲突（Hennig 1966，Jenner 2001）。在parcimony架构下结合分子生物学和形态学对系统关系进行推断时，距离矩阵法特别容易与全部证据研究法产生冲突（Giribet et al. 2000，Jenner 2001）。不管怎样，为了提高基于形态特征推断系统关系的准确性，分子生物学的证据也在争论中越来越被重视。

4.2.3　小核糖体RNA基因和动物系统发生的新观点

分子数据的应用使后生动物进化树产生了深层次的变化。小核糖体RNA（也称为18S rRNA）在动物系统发育领域的最早应用要追溯到20多年前，Fields et al.（1988）首次发表了基于SSU 18s rRNA对22个分类单元动物的系统发育研究。该文破天荒地提出后生动物的多起源性，并认为真核生物中的两侧对称动物和腔肠动物分别有着独立的起源，同时也强调了扁形动物是早期从两侧对称动物中分化出来的（Field et al. 1988）。尽管这是一项开创性研究，但后来对数据的重新解读表明距离法构建系统发育关系可能会引起长枝吸引（Long-branch Attraction）假象，即一种与类群不等速进化有关的常见系统性误差。因此，这一假说也就随之被否定了（Felsenstein 1978）。之后使用其他方法（如进化简约法）对同一数据组的分析确定了后生动物和后口动物的单源性进化，也确定了环节动物和软体动物的近亲关系（Lake 1990）。该评估分子系统发育的新标准的建立，开启了后生动物系统发育研究的新领域，也将不断改变人们对后生动物系统发育的认识。

小核糖体RNA对传统分类观点的第一个挑战是依据帚虫和腕足类动物小核糖体RNA的测序和分析结果将三类触手冠动物归为原口动物（Halanych et al. 1995）。将触手冠动物从后口动物划出并将其归为原口动物这一做法是与它们的某些特征相矛盾的，如体腔的结构与发育特征（Eernisse et al. 1992，Emig 1982）。这使研究人员创造了冠轮动物一词来描述这一新的进化枝，包含触手冠动物和担轮幼虫型动物（如软体动物和环节动物）。但是，触手担轮

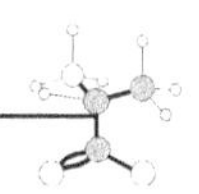

类进化枝中的假体腔动物（此前的袋形动物）和节肢动物的确切分类地位仍未确定，主要是因为它们的进化速度过快而引起了长枝吸引假象（参照 Bergsten 2005）。为了解决这一问题，Aguinaldo et al.（1997）挑选了进化速度非常快的典型原口动物——线虫，并从中选择了保守性最高的 SSU 基因序列。该方法的使用恢复了蜕皮习性动物的进化地位，称为蜕皮动物门，包括节肢动物和部分囊蠕虫，如肢吻动物和线虫。他们的分析结果同时也支持将扁形动物纳入触手担轮类动物，这就意味着原口动物可以分为两大类：蜕皮动物和触手担轮类动物。随后这种拓扑结构被确立为动物系统发育的"新观点"，因为这样的进化树与之前以形态学分类为基础的进化树形成了鲜明的对比，特别是对早期的囊蠕虫门进行了拆分。结果显示，即使是不同身体结构的生物（如线虫和节肢动物或者扁形动物和环节动物）也可能具有较近的进化关系（Adoutte et al. 2000）。

来源于 Hox 基因同源域的分子特征进一步支持了上述关于动物亲缘关系的"新观点"（de Rosa et al. 1999）。Hox 转录因子之所以能组成多基因家族，主要是因为所有的两侧对称动物都存在十多个 Hox 同源基因。这些基因通常都紧密地聚集在基因组上，和动物身体结构发育的顺序一致，按照染色体上基因的排列顺序依次共线性表达（Lemons and McGinnis 2006）。de Rosa et al.（1999）对肢吻类和腕足类动物 Hox 基因的研究表明，后蜕皮动物和触手担轮类动物的 hox 基因可能有着独立的起源，并最终形成了各自具有某些独有特征的 Abd-B 和 Post1/2 两类 hox 基因。而之后的数据也支持 hox 基因可以作为研究两侧对称动物进化的分子标记（Balavoine et al. 2002）。

4.2.4 新观点的局限性

动物系统发育的"新观点"一直影响着我们对动物进化的理解（Adoutte et al. 2000）。但是，很多在进化研究中有争议的节点都不能通过早期的分子生物学数据来解决：①新原口动物进化枝（冠轮动物超门和蜕皮动物超门）的亲缘关系仍未确定；②由于进化速度过快，毛鳄类动物、腹毛类动物、轮虫类、异涡虫类和无肠目动物等类群的分类地位难以确定；③后生动物亲缘关系也没有确定。后来，除了使用分子生物学数据之外，人们还开始尝试使用形态学矩阵数据来解决这些问题（Giribet et al. 2000，Petersonand Eernisse 2001）。虽然使用这些"总证据"解决了一些有疑问的节点，并且形成了一个可信的动物亲缘关系图，但由于不当地使用已发表的数据而重复了之前的错

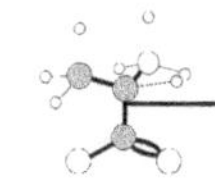

误，也遭到了大量的批评（Jenner 2001）。

解决这个问题的另一个方法就是开展大规模测序来鉴定这些亲缘关系很近的物种，不过这种方法也不能解决所有问题。例如毛颚类动物的 SSU 基因进化速度过快，导致它们最初被认为是后生动物的原始种类（Telford and Holland 1993）。但对大量物种 SSU 基因的深入研究，并不能有效地找出进化更慢的种类（Papillon et al. 2006）。另外，研究结果还显示阿克尔扁形虫是较原始的两侧对称动物，但有时也会因为存在"长枝吸引"假象而被否定（Deutsch 2008，Ruiz-Trillo et al. 1999）。后生动物进化树增加了一些动物类别，包括腔肠动物、多孔动物、栉水母类和神秘的扁盘动物，这些物种形成了像海绵动物侧系这样有趣的假说，但是关于它们各自的亲缘关系还没有统一的结论（Borchiellini et al. 2001，Collins 1998）。

另一个方法是将核糖体 RNA 大亚基（LSU 或 28S）基因作为替代基因，并与其他基因结合分析，如 SSU（Mallatt and Winchell 2002，Winchell et al. 2002）、延伸因子（EFIa；Littlewood et al. 2001）、热休克蛋白（Borchiellini et al. 1998）或者钠钾 ATP 酶 β 亚基（Anderson et al. 2004）等。这为研究后生动物系统发育提供了很多有价值的新思路。但是，使用不同类型基因构建的进化树在拓扑结构之间仍存在明显差异。因此，基于核基因构建新的更可靠的后生动物进化关系，就需要解决这种拓扑结构方面的不一致。

4.3 系统基因组学的优缺点

基因组可以说是所有生物的另一面，其中蕴藏着大量可用于系统发育研究的信息。的确，形态特征往往是在强烈的选择压力之下形成的，反之，基因组的进化更遵循一个中性的过程（参照 Kimura 1983，Lynch 2007）。从基因组数据中能得到两类主要的系统发育证据：第一，从定量来说，可以使用大量来自于基因组数据的核基因序列来推断进化关系。第二，从定性来说，利用调查全基因组数据可以检测到分离的分子标记类型（Philippe et al. 2005a）。

第一种方法是以分析大量基因序列为基础，试图通过比较不同标记基因建立拓扑结构来解决不同进化树之间的差异（同上）。有人提出，通过大量核基因的串联，可能会从源自不同基因的几个拓扑结构中发现真实的物种进化关系（Rokas et al. 2003）。但也有观点认为使用大量位点进行分析会增加系统性偏差，形成具有充分数据支持但却错误的系统进化树（Jeffroy et al. 2006）。偏差的产生主要与进化速度和序列组成的差异有关，并进一步造成类

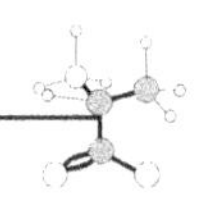

群和节点排列位置的差异（Philippe et al. 2005a）。使用改进的替换模型或增加种群的抽样数目也许可以减少该系统性偏差（Delsuc et al. 2005）。比如，使用最近发展起来的用于说明基因排列位置差异性的 CAT 模型，就可以降低长枝吸引假象出现的概率（Lartillot and Philippe2004）。

对庞大的基因数据集的分析策略有两个：超级树和超级矩阵（Delsuc et al. 2005）。超级树法是根据每个标记基因分别推断一棵进化树，随后通过这些树再计算出一棵能够总结所有进化树的超级树。该方法之所以能降低系统性误差是因为所用的进化模型更适应短的、同质性高且比对效果好的单基因。但是与全数据的系统发育评价相比，该构建方法通常会造成信息丢失（de Queiroz and Gatesy 2007）。超级矩阵方法是将所有的标记基因串联成一个单一的长序列。针对大量位点进行分析，虽然在数学模型的使用和计算方面存在一定困难，但推论法的不断进步极大促进了超级矩阵策略的使用。比如最大似然法（Guindon and Gascuel 2003，Stamatakis 2006）和贝叶斯算法（Huelsenbeck et al. 2001）的出现就极大促进了进化模型的构建。因此超级矩阵策略被视为当前研究动物间亲缘关系的首选方法。

第二种方法，也就是本部分开头提到的系统基因组学方法，主要依赖于基因组水平上的特征鉴定，这些特征可以组成有用的性状分子标签。一些研究者创造出了“稀有基因组变化”这个新术语用来描述这类新性状，并强调他们的目的是应用这种方法来克服基于大规模序列推断系统发育时所遇到的困难，或者将这种方法作为完成此类研究的方法之一（Rokas and Holland 2000a）。有些基因组特征可以为动物系统发育研究提供有价值的信息，下面的例子将具体介绍。多基因家族的结构，尤其是基因丢失或基因获取为解决某些不确定的节点提供了很多重要的线索（Copley et al. 2004，Marlétaz et al. 2006）。随着腔肠动物海葵和环节动物沙蚕中保守内含子的确定，证明了后生动物同源基因内含子区域的保守特性（Putnam et al. 2007，Raible et al. 2005），因此，研究者认为内含子的保守性适用于推断系统发育（Roy and Gilbert 2005）。另外，人们已经对基因序列和基因组重排进行了大量的研究，尤其是在线粒体基因组重组方面（Boore 2006），得到了一些有意思的发现。

随着后生动物可用基因组数据的增加，也许以后可以通过基因组数据建立多种方法来解决后生动物系统发育方面的难题。目前的研究结果对于重建后生动物系统进化关系也做出了很大的贡献。

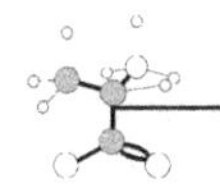

4.4 系统基因组学解析动物亲缘关系

在早期试图使用系统基因组学方法研究后生动物亲缘关系时，经常会得到相反的结果，有时还会对动物系统发育的“新观点”发出挑战。对于解决这些问题，分类采样显得尤为重要。

4.4.1 真体腔动物分类的争议和分类取样的重要性

4.4.1.1 早期的系统基因组学试图挑战“新观点”

蜕皮动物超门假说是动物系统发育“新观点”中最具争议的，因此早期对两侧对称动物的多基因分析一直在试图检验这个假说（Blair et al. 2002，Philip et al. 2005，Wolf et al. 2004）。基于已经完成全基因组测序的物种，研究人员对100到780个核基因进行了合并分析或单独分析（Blair et al. 2002，Philip et al. 2005，Wolf et al. 2004）。虽然这些研究都没有恢复蜕皮动物的进化地位，但是发现了线虫类动物的分化要早于有体腔动物。不过由于线虫的基因组进化速度非常快，该发现被认为受到了“长枝吸引”假象的影响（Aboobaker and Blaxter 2003）。Philippe et al. 2005（a，b）的研究结果支持动物系统发育的“新观点”，对反对“新观点”的研究发起了挑战。这些研究人员最近发布了许多新物种的EST数据，主要包括领鞭虫类、与后生动物最近的亲属类、腔肠动物类、两侧动物的姊妹动物类以及大量线虫与扁形动物这样的类群，其中一些进化速度（例如线虫中旋毛类线虫）比较慢。因此，通过将这些进化速度较慢的特殊类群加入到研究中，可以抑制“长枝吸引”的出现，也支持了“新观点”中的蜕皮动物超门这一分类（图4.3）。在一些研究中，扁形动物和线虫类动物会聚类到一起，但通过去掉一些进化速度快的基因就可以得到与“新观点”一致的结果，即支持扁形动物和环节动物以及软体动物为同一进化枝。如此看来，仅基因取样而非分类学取样所得到的结果可以提高系统发育的准确性（Rokas and Carroll 2005）。

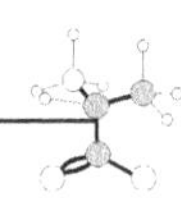

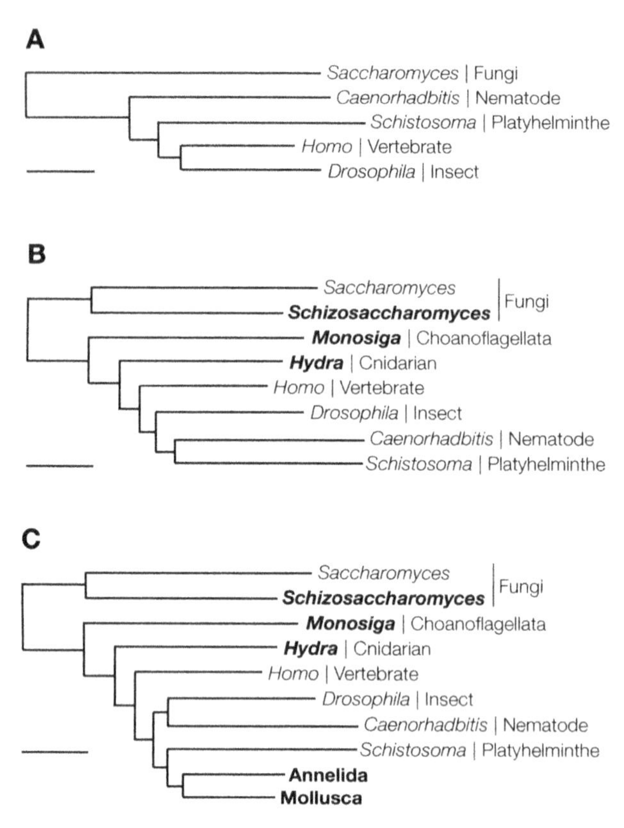

Fig. 4.3 The impact of taxonomic sampling and the demise of the *coelomata* hypothesis. These trees were inferred from 146 nuclear genes using Maximum likelihood inference. One branch length unit represents 0.1 substitutions per site. Progressive inclusion of intermediate taxa shows how long-branch attraction can be overcome. (**a**) The tree with limited taxon sampling supports the *coelomata* hypothesis with early divergence of nematodes and platyhelminthes. (**b**) The inclusion of intermediate bilaterian outgroups – *Hydra* and chaonoflagellates – results in nematodes and platyhelminthes being relocated back within the bilaterians but they still cluster together, in contradiction with the ecdysozoan hypothesis. (**c**) The selection of the least divergent marker genes (70 out of 146) and the addition of annelids and molluscs definitely recovers the "new view" of animal phylogeny (redrawn from Philippe et al. 2005b)

图4.3　分类取样法的重要性和体腔动物假说的终结。图中所示进化树均为使用最大似然法通过对146个核基因的分析推断而来。一个分枝长度单位表示平均每个位点有0.1个核苷酸替换。中间类群的不断加入显示长枝吸引假象可以避免。（a）通过少数类群构建的进化树支持体腔动物假说，线虫和扁形动物在早期发生分化；（b）通过加入外群水螅和领鞭毛虫可使线虫和扁形动物重新归为两侧对称动物，但仍然聚类在一起，与蜕皮动物假说矛盾；（c）通过选择分化最小的标记基因（146个核基因中选出70个）和增加环节动物与软体动物，系统进化树就会很明确地支持动物系统发生的"新观点"（Philippe et al. 2005b）

英文注释：Saccharomyces，酵母属；Fungi，真菌，菌类；Caenorhadbitis，线虫类；Nematode，线虫动物门，线虫类；Schistosoma，裂体吸虫属，血吸虫属；Platyhelminthe，扁形动物门；Homo，人属；Vertebrate，脊椎动物，有脊椎的；Drosophila，果蝇，果蝇属；Insect，昆虫；Schizosaccharomyces，裂殖酵母，裂殖酵母纲；Monosiga，领鞭毛虫；Choanoflagellata，领鞭目；Hydra，水蛭；Cnidarian，刺胞动物；Annelida，环节动物门；Mollusca，软体动物门；Coelomata，体腔动物；Nuclear genes，核基因；Maximum likelihood，最大似然法；Length unit，长度单位；Taxa，分类单位；Taxon，分类学，分类单位；Nematodes，线虫，线虫类；Platyhelminthes，扁形动物门，扁形动物；Bilaterian，两侧对称动物；Outgroups，外群；Cluster，群集，簇；Ecdysozoan，褪皮动物；Marker genes，标记基因；Annelids，环节动物；Molluscs，软体动物；Phylogeny，系统发生，系统发育

4.4.1.2 真体腔动物和稀有基因组变化的解释

最近关于真体腔动物假说的争论突出了精确取样的重要性。该争论源于对两个不同性质的基因组特征的研究："稀有氨基酸替换"和内含子的保守模式。稀有的氨基酸改变被定义为氨基酸替换，只在有限类群中出现，是由几个核苷酸的改变引起的。因此，通过将替换与低概率的趋同性之间建立联系，发现 34 个这样的位点与真体腔动物的进化树一致（Rogozin et al. 2007）。但把一种海葵作为外群加入到这个数据集之后，数据就显示出了许多支持真体腔动物的特征，而这些特征是只有在使用非后生动物（如植物）做外群时才能被发现的。因为这些特征在人、果蝇和海葵中都有发现（Irimia et al. 2007），所以它们可能是后生动物遗传下来的。与此相反，当使用精确的祖先时，13 个稀有氨基酸的改变则支持蜕皮动物超门这一假说。

同样，内含子的保守模式也被当做是系统发育的证据，方法学的进步解释了内含子的丢失存在明显的趋同性：某个物种中丢失的内含子，在其他物种里也有很高的丢失率（Zheng et al. 2007）。最初该结果支持蜕皮动物超门假说（Roy and Gilbert 2005），但后来由于其不能充分解释内含子平行丢失现象而被质疑，而通过相反的分析又支持了真体腔动物（Zheng et al. 2007）。但是有研究表明后来的结果主要倾向于分析稀有氨基酸变化相关的问题，所以之后的研究加入了富含内含子的海葵，又再一次支持了蜕皮动物超门这一分类（Roy and Irimia 2008）。因此，尽管分类取样对以序列为基础的传统发育分子系统学研究的重要性已经被深入讨论，但很明显，类似像稀有基因组变化这样一系列新的系统发育证据也应该重视这一问题（Rokas and Holland2000b）。

4.4.2 系统基因组学是否可以解释动物亲缘关系

对系统基因组学的有效性产生质疑的 Rokas et al.（2003）表示，他们并不

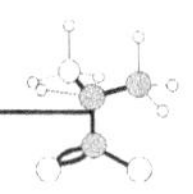

确定系统基因组学是否能完美解析后生动物亲缘关系（Rokas et al. 2005）。他们试图用17个类群和50个基因数据集来恢复几个经典的后生动物进化枝，但均以失败告终。为了解释失败的原因，他们提出后生动物进化树“在时间上被压缩”，也就是说，动物种系发生分化的时间太短而找不到精确的系统演化信号。这个想法虽然新颖大胆，但与其他众多的系统基因组学研究并不相符（Marlétaz et al. 2006，Matus et al. 2006，Philippe et al. 2005b）。针对Rokas et al.（2005）在研究中遇到的难题，后来的研究（Baurain et al. 2007）找到了两个主要的原因：错误的分类取样和错误的进化模型。首先，尽管用剑线虫代替秀丽隐杆线虫更能支持蜕皮动物假说，但是他们在取样过程中忽略了线虫等趋异类群中进化慢的物种（Baurain et al. 2007）。另外，研究也表明使用优化的氨基酸置换模型或者进化树搜索算法可以更准确地修正进化树（Hordijk and Gascuel 2005，Lartillot et al. 2007），如CAT氨基酸模型和SPR搜索法（子树修正和再分支）。最后，低质量的原始序列数据也会影响对节点的推断。通过对保守结构域进行PCR扩增，Rokas et al.（2005）重新得到了大部分序列，去除那些高变区，对这些基因中与系统发育最相关的区域进行分析。而其他关于动物亲缘关系的系统基因组研究主要以EST数据为基础。

系统基因组重建理论与Aguinaldo et al.（1997）基于SSU而提出的系统发育方法密切相关，通过挑选进化缓慢的物种可以极大的提高准确性。但是在使用多基因时，还不清楚单一物种中是否含有分化程度最低的全部标记基因。为解决这个问题，Marlétaz et al.（2006）提出一种新的复合分类策略，即从一个单系群物种中选择目的基因中分化程度最低的拷贝。例如，该方法可以建立一个复合的线虫类群，因为它是一类进化最慢的线虫，其分支比旋毛虫类还要短。

系统基因组学方法与之前基于多种证据而得出的结果基本上能够相互印证，表明该方法有希望成为解决动物亲缘关系的有效手段。不过，系统基因组学方法也容易受到系统误差的影响，所以，为了更准确的分析，有必要对分类取样、推理方法的使用和原始数据质量进行系统的评估。

4.5　后生动物亲缘关系的系统基因组框架图

限制单基因系统发生研究的主要原因是不能依据进化速度和序列组成来处理已经分化的类群。当形态学特征不能很好地解决一些与系统发育密切相关的问题时，这样的类群通常称作“疑问化石”。这些类群中有一类被Libbie

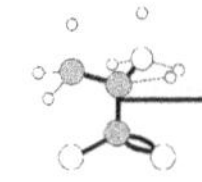

Hyman 称为“小体腔动物类群”的动物，如毛鳄类动物（Hyman 1959）；另一类是极具多样性的袋形动物（从轮虫类至动吻动物类），而清楚地解释这种多样性正是解决动物进化的关键问题之一（Jenner 2000）。系统基因组学提供了更准确定位这些类群在后生动物进化树中位置的机会，所以最近做了很多测序工作（主要是 EST 测序）。EST 方法可以对一个 cDNA 文库中大量的克隆子进行测序，收集、处理后可以得到部分高质量的转录组数据。测序结果显示，得到的 3000 ~ 10000 个 EST 能为一个新类群的系统基因组分析提供一组足够的标记基因（Marlétaz et al. 2008，Philippe and Telford 2006），而在一个给定的类群中缺少一个或多个标记基因并不会对精确重建整个进化树产生影响（Philippe et al. 2004）。

4.5.1 挑战根深蒂固的分类：后口动物

系统基因组学的深入研究引发了人们对一些传统分类假说的重新思考。用尾索动物亚门替代头索动物亚门，并将其归为脊椎动物的姊妹类群，这是对后口动物亲缘关系的首次挑战（Delsuc et al. 2006）。对 146 个核基因和几个关键新类群（如七鳃鳗、八目鳗和尾海鞘纲动物异体住囊虫等）的分析，在统计学上有力地支持了这种分类观点。这个结果与脊椎动物和头索动物的一些近裔共有特征相矛盾，比如分段的肌节（Schubert et al. 2006），但也得到了其他特征的支持，如在被囊类动物中发现了移动的神经嵴样细胞（Jeffery et al. 2004）。该研究支持棘皮动物和头索动物的近亲关系，但还需要进一步的证实。该研究对它们近亲关系的鉴定是通过对包括半索动物在内的动物进行分析得到的，头索类动物被重新纳入脊索类动物，而半索动物和棘皮动物一起组成了一个新类群（称为 Ambulacraria；Bourlat et al. 2006，Marlétaz et al. 2006；图 4.4）。随着异涡虫门作为姊妹群与棘皮动物门、半索动物门聚为一支，一个新后口动物系统发育关系就形成了（Bourlat et al. 2006）。

4.5.2 毛颚动物门归入两侧对称动物进化树

对于那些长期存在问题的类群，系统基因组方法可以更加精准地确定他们在系统发育中的位置。多年来毛颚类的分类地位一直没能确定，因为他们的身体构造具有一些原口动物的特征，但其早期发育过程中却形成肠体腔，这又是典型的后口动物特征（Ball and Miller 2006）。基于 EST 测序的系统基因学组分析表明，毛颚类既不属于蜕皮动物，也不属于触手担轮类动物，而很

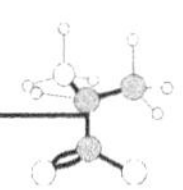

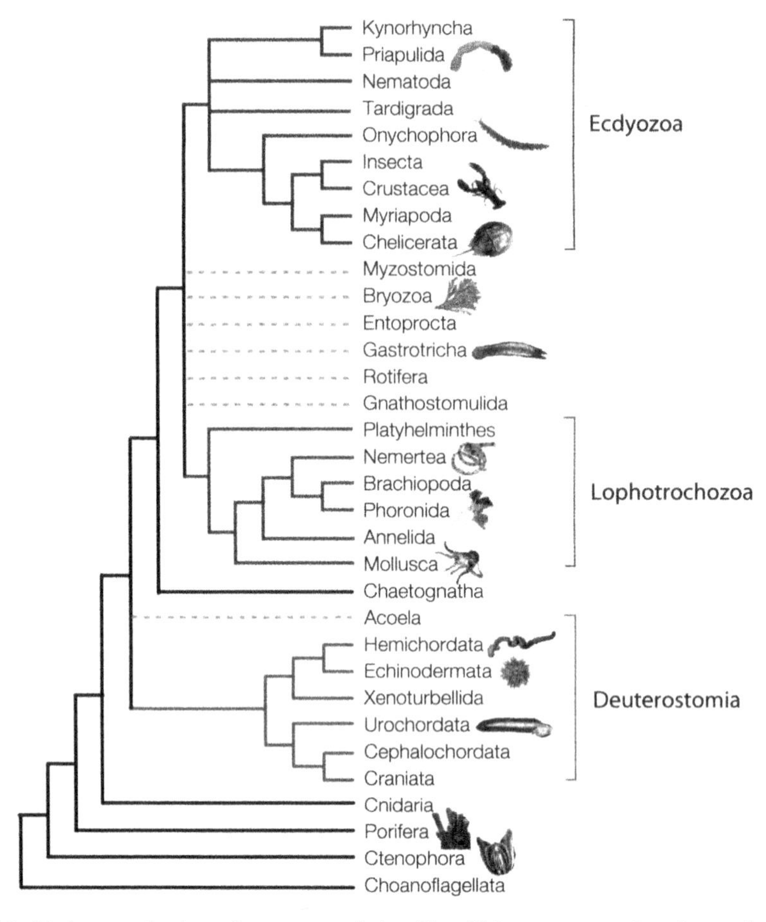

Fig. 4.4 Phylogenomic view of metazoan relationships. This tree summarizes the results of the most recent phylogenomic analyses focused on metazoans. All the nodes presented here are based on firm support values at least with the site-heterogenous CAT model. The overall "New View" of animal phylogeny is recovered with firm support for the deuterostomes, ecdysozoans and lophotrochozoans but a few clades such as chaetognaths, rotifers or acoel flatworms do not fit into this scheme. *Dashed* branches correspond to the most instable taxa whose position remains ambiguous

图4.4　后生动物亲缘关系的系统基因组学观点。该进化树是对最近后生动物系统基因组学分析总结。所有节点都有可靠的模型数据支持。后口动物、蜕皮动物和触手担轮类动物完全支持动物系统发生“新观点”，但如毛颚类、轮虫或阿克尔扁形虫等几个进化枝与“新观点”存在矛盾。虚线表示的进化枝是最不稳定的类群，他们的进化地位仍然不确定

英文注释：Kynorhyncha，动吻动物门（Kinorhyncha）；Priapulida，三部虫门；Nematoda，线虫类；Tardigrada，缓步动物门；Onychophora，有爪动物门；Insecta，昆虫纲；Crustacea，甲壳纲；Myriapoda，多足纲；Chelicerata，螯肢亚门；Myzostomida，吸口虫类；Bryozoa，外肛动物门；Entoprocta，内肛动物门；Gastrotricha，腹毛动物门；Rotifera，轮虫纲；

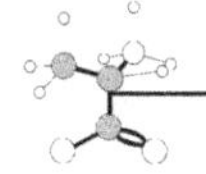

Gnathostomulida，颚胃动物门；Platyhelminthes，扁形动物门；Nemertea，纽形动物门；Brachiopoda，腕足动物门；Phoronida，帚虫动物门；Annelida，环节动物门；Mollusca，软体动物门；Lophotrochozoa，冠轮动物超门；Chaetognatha，毛颚动物门；Acoela，无体腔目；Hemichordata，半索动物门；Echinodermata，棘皮动物门；Xenoturbellida，异涡动物门；Urochordata，尾索动物门；Cephalochordata，头索动物纲；Craniata，脊椎动物门；Deuterostomia，后口动物；Cnidaria，刺胞动物门；Porifera，海绵动物门；Ctenophora，栉水母动物门；Choanoflagellata，领鞭目；Taxa，分类群；Metazoan，后生动物；Phylogenomic，系统发育分析；Nodes，节点；Heterogenous，异源性；Phylogeny，系统发生，系统发育；Deuterostomes，后口动物；Ecdysozoans，蜕皮动物；Lophotrochozoans，触手担轮类；Clades，进化枝；Chaetognaths，毛颚类动物；Rotifers，轮虫类；Flatworms，扁形动物

有可能是其他原口动物的一个姊妹类群（Marlétaz et al. 2008，Marlétaz et al. 2006）。以前对毛颚类线粒体基因组的分析结果支持将他们归为到原口动物（Papillon et al. 2004）。对另一种毛颚类动物的研究也表明，尽管毛颚类与原口动物有着很近的亲缘关系，但更支持毛颚类是一种早期的触手担轮类动物（Matus et al. 2006）或者一种地位未定的原口动物（Dunn et al. 2008）。后来随着毛颚类 EST 中一个特异 DNA 分子标记的发现，进一步支持了毛颚类是原口动物的观点：在原口动物分化之前的所有物种中均发现胍基乙酸 N－甲基转移酶（GMT）基因，相反，在现有的原口动物类群中却未找到该基因，这表明毛颚类从原口动物中分化之后该基因就已经丢失了（Marlétaz et al. 2008，Marlétaz et al. 2006）。虽然也有研究表明毛颚类应该与触手担轮类动物的亲缘关系更近，但多项独立研究都支持将毛颚类归为原口动物的姊妹群（Lartillot and Philippe 2008，Philippe et al. 2007）。这种独特的系统发育地位有两个重要意义：一方面，它表明后口动物样的胚胎发育特别是肠体腔的形成可能是由两侧对称动物祖先遗传而来，这个观点让人联想到 archecoelomate 理论（Remane 1963；图 4.2）；另一方面，这是自毛颚动物门从触手担轮类与蜕皮动物两个分支中划分出来后，对动物系统发育“新观点”第一次清晰地反驳（Adoutte et al. 2000）。

4.5.3 阿克尔扁形虫是最原始的两侧对称动物吗？

基于 SSU 分析，研究人员提出阿克尔扁形虫是最基础的两侧对称动物，从此以后，这就成为了一个极富争议的话题，系统基因组学或许可以帮助它们最终确定系统发育地位（Ruiz-Trillo et al. 1999）。该分类地位表明这些三胚层无体腔扁形虫的身体组织特征可能源自那些两侧对称动物祖先。但因为过

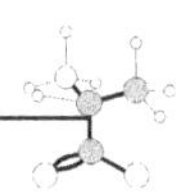

快的进化速度可能会引起“长枝吸引”假象，阿克尔扁形虫被当作是一个中间类群这样的分类地位引起了巨大的争议（Deutsch 2008）。近期，利用系统基因组方法对它们的分类地位进行了重新评估，结论是阿克尔扁形虫不属于扁形动物（Philippe et al. 2007），而其胍基乙酸 *N* – 甲基转移酶（GMT）基因的发现也表明它们不属于原口动物。因为没有找到明显的系统发育特征来确定其分类地位，所以它们在两侧对称动物中的地位仍然模糊不清（图 4.4）。后来通过系统基因组学的研究，终于将阿克尔扁形虫从扁形动物门中划出，并确定了阿克尔扁形虫的特殊分类学地位。

4.5.4 更深入的原口动物亲缘关系研究

原口动物进化枝表现出了极高的类群和身体组织多样性，其身体组织的多样性几乎是所有类群里面最丰富的（Adoutte et al. 2000）。原口动物以前被归为袋形动物门，它们的形态特征一般很难解释。由于进化速度过快，用传统的分子系统发育理论也很难确定它们的分类地位（Passamaneck and Halanych 2006）。为了运用强大的系统基因组学方法去解决这些后生动物的亲缘关系问题，近期在相关物种中开展了大量的测序工作。预期从几个新类群和已知类群中选择进化最慢的几个物种收集 EST 数据，从而帮助解决触手担轮类动物的系统发育问题（Dunn et al. 2008，Hausdorf et al. 2007，Struck and Fisse 2008）。基于对这些新数据的分析，研究人员发现触手担轮类动物间可能存在新的亲缘关系：担轮幼虫动物、软体动物和扩展的环节动物类群（包含螠虫动物门和星虫门）与纽形动物、触手冠类动物间的亲缘关系很近（图 4.4；Dunn et al. 2008）。而纽形动物则被归为触手冠类动物的姊妹群。最后的结果证实了近来的发现：古纽形动物 *Carinomamutabilis* 中帽状的幼虫表现出了与担轮幼虫惊人的相似性（Maslakova et al. 2004）。软体动物、环节动物、触手冠动物和纽形动物都有几丁质刚毛，这些几丁质可能与软体动物壳派生的钙质骨针有着共同起源，反过来又证明了这几类动物群属于同一个进化分枝。纽形动物被定位为触手担轮类动物的姊妹群，但其他与触手担轮类动物相关类群的分类地位依然难以确定（Dunn et al. 2008）。依据早期的假说，有人建议将苔藓动物（也称为外肛动物）和内肛动物归为一类（Hausdorf et al. 2007），但不同的标记基因和分类取样并不支持这种分类。因此这些类群的分类地位依然模糊，与腹毛类动物、轮虫类和吸口虫类一样存在强烈的“分支不稳定性”，即倾向于不同的引导复制，而且在可能性最大的进化树中表现出了替代分支（Dunn et al. 2008）。整个原口动物系统发育关系的成功构建还需

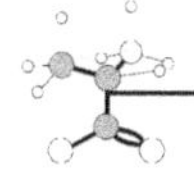

要大量的研究和努力，但近期这些研究进展表明增加分类取样能在很大程度上促进系统发育的研究。

4.6 总结：动物系统发育的前景

本章概述了最近以基因组为基础的系统发育研究，其对当前基于经典形态学分类理论和单基因分子系统发育理论（主要是 SSU rRNA 基因）形成的后生动物亲缘关系的相关观点产生了深远影响。在真体腔动物分类地位存在普遍争论的背景下，建立起了系统发育基因组学这一新领域。该争论主要强调大量分类取样的重要性，包括对定性分子标记和基因组水平特征研究中的分类取样。最终通过对动物系统发育“新观点”进行重新评估才结束了这场争论（Adoutte et al. 2000，Dunn et al. 2008）。随着系统基因组方法的广泛应用，一些长期遗留的动物系统发育问题已经得到了解决，如阿克尔扁形虫、毛颚类动物的分类地位和触手担轮类进化枝内动物间的亲缘关系；当然也带来了一些意料之外的分类结果：被囊动物在进化上被认为是脊椎动物的姊妹群（Delsuc et al. 2005）；无肠目动物和毛颚类动物在进化上不属于蜕皮动物和触手担轮类这一进化枝。这些新类群的发现都对“新观点”发起了挑战。近期 Dunn et al.（2008）发表的文章对小类群动物的大量数据进行了收集与分析，表明还有一些类群的进化地位仍然无法确定。因此后生动物进化树仍然需要继续完善，存在的问题依旧尚待解决。由于很难确定两侧对称动物、多孔动物、腔肠动物和栉水母类的分支顺序，因此，后生动物进化树基部类群间确切的进化关系仍然不清楚，所有的这些类群都可能在后生动物辐射基部开始分化（Dunn et al. 2008）。要对这个问题作出回答就必须先深入了解后生动物的祖先，他们的祖先到底是一种已经拥有中胚层和两侧对称的复杂生物（如栉水母），还是一种复杂性较低的生物（如海绵幼虫：Martindale and Henry 1999，Nielsen 2008）？大量的基因、分类取样与改进的推理模型都可能成为解决这些问题的有效方法。此外，当分子系统发育研究中使用大量可用标记基因对动物进行分类时，许多像基因顺序和共线性关系这样的基因特征可能还需要更深入的研究分析（Philippe et al. 2005a）。

后生动物辐射基部类群间进化关系的例子强调了分子系统发生学对于研究动物主要形态特征演化过程的重要性，尤其是表明了形态学复杂性的形成并没有一般趋势。纽形动物与环节动物的密切关系同脊椎动物与被囊动物一样，而且它们的形态在后生动物进化过程中屡屡发生简化（Delsuc et al.

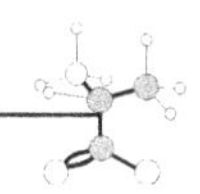

2005，Dunn et al. 2008）。Dunn et al.（2008）提出的栉水母类动物的早期分化这一假设表明所有后生动物的祖先比想象中更复杂，而多孔动物的形态简化是后来形成的。

不过这种形态学转变在基因组上的反映尚不清楚，基因组进化与形态特征进化之间的关系是进化生物学当前尚需解决的问题之一。造成这个问题的主要原因是不同物种在这两个方面经历了完全不同的进化方式，即身体形态的变化更加剧烈而基因组的变化则独立缓慢。发育生物学领域最近发现了宏进化变化的发育遗传学来源，但这些深入的研究仅局限于一些特定的基因通路（Muller 2007）。另一方面，通过对多基因家族和基因组重排的相关研究能够解释一些基因组水平上的发育调控，尤其是使用模式物种的遗传数据时，对发育调控研究的帮助更大。比如，Domazet-Loso et al.（2007）开发的“phylostratigraphic”方法就是使用大量与胚层发育相关的多基因的基因结构和表达模式进行系统发育研究。他们观察到这些基因间的系统发育起源都存在一定差异，例如与外胚层相关的一些基因的起源时间可能更为古老。这一发现使人们开始关注一些此前被忽视的动物类群，这些类群对比较方法的应用和生物特征的定位都极为重要。对一些具有重要系统发育地位的生物（栉水母类和毛颚类等）的基因组测序无疑会使我们能够更加清晰地认识后生动物的进化过程。

参考文献

Aboobaker AA, Blaxter ML (2003) Hox Gene Loss during Dynamic Evolution of the Nematode Cluster. Curr Biol 13:37 - 40

Adoutte A, Balavoine G, Lartillot N, Lespinet O, Prud'homme B, de Rosa R (2000) The new animal phylogeny: reliability and implications. Proc Natl Acad Sci U S A 97:4453 - 4456 Aguinaldo AM, Turbeville JM, Linford LS, Rivera MC, Garey JR, Raff RA, Lake JA (1997)

Evidence for a clade of nematodes, arthropods and other moulting animals. Nature 387:489 - 493

Anderson FE, Cordoba AJ, Thollesson M (2004) Bilaterian phylogeny based on analyses of a region of the sodium-potassium ATPase beta-subunit gene. J Mol Evol 58:252 - 268

Aristotle (1965) De Generatione animalium, tr. Arthur Platt, Clarendon Press, Oxford

Balavoine G, de Rosa R, Adoutte A (2002) Hox clusters and bilaterian phylogeny. Mol Phylogenet Evol 24:366 - 373

Ball EE, Miller DJ (2006) Phylogeny: the continuing classificatory conundrum of chaetog-

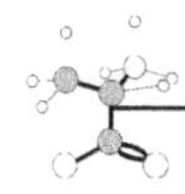

naths. Curr Biol 16:R593 – R596

Barnes RD (1974) Invertebrate zoology. W. B. Saunders Company, Philadelphia.

Baurain D, Brinkmann H, Philippe H (2007) Lack of resolution in the animal phylogeny: closely spaced cladogeneses or undetected systematic errors?. Mol Biol Evol 24:6 – 9

Bergsten J (2005) A reviews of long-branch attraction. Cladistics 21:163 – 193

Blair JE, Ikeo K, Gojobori T, Hedges SB (2002) The evolutionary position of nematodes. BMC Evol Biol 2:7

Boore JL (2006) The use of genome-level characters for phylogenetic reconstruction. Trends Ecol Evol 21:439 – 446

Borchiellini C, Boury-Esnault N, Vacelet J, Le Parco Y (1998) Phylogenetic analysis of the Hsp70 sequences reveals the monophyly of Metazoa and specific phylogenetic relationships between animals and fungi. Mol Biol Evol 15:647 – 655

Borchiellini C, Manuel M, Alivon E, Boury-Esnault N, Vacelet J, Le Parco Y (2001) Sponge paraphyly and the origin of Metazoa. J Evol Biol 14:171 – 179

Bourlat SJ, Juliusdottir T, Lowe CJ, Freeman R, Aronowicz J, Kirschner M, Lander ES, Thorndyke M, Nakano H, Kohn AB, Heyland A, Moroz LL, Copley RR, Telford MJ (2006) Deuterostome phylogeny reveals monophyletic chordates and the new phylum Xenoturbellida. Nature 444: 85 – 88

Collins AG (1998) Evaluating multiple alternative hypotheses for the origin of Bilateria: an analysis of 18S rRNA molecular evidence. Proc Natl Acad Sci U S A 95:15458 – 15463

Conway Morris S, Peel JS (1995) Articulated Halkieriids from the Lower Cambrian of North Greenland and their role in early protostome evolution. Philos Trans Biol Sci 347:305 – 358

Copley RR, Aloy P, Russell RB, Telford MJ (2004) Systematic searches for molecular synapomor-phies in model metazoan genomes give some support for Ecdysozoa after accounting for the idiosyncrasies of Caenorhabditis elegans. Evol Dev 6:164 – 169

Cuvier G (1817) Le règne animal distribué selon son organisation, pour servir de base à l'hisoire naturelle des animaux et d'introduction à l'anatomie comparée. Deterville, Paris.

de Queiroz A, Gatesy J (2007) The supermatrix approach to systematics. Trends Ecol Evol 22:34 – 41

de Rosa R, Grenier JK, Andreeva T, Cook CE, Adoutte A, Akam M, Carroll SB, Balavoine G (1999) Hox genes in brachiopods and priapulids and protostome evolution. Nature 399:772 – 776

Delsuc F, Brinkmann H, Chourrout D, Philippe H (2006) Tunicates and not cephalochordates are the closest living relatives of vertebrates. Nature 439:965 – 968

Delsuc F, Brinkmann H, Philippe H (2005) Phylogenomics and the reconstruction of the tree of life. Nat Rev Genet 6:361 – 375

Deutsch JS (2008) Do acoels climb up the "Scale of Beings"?. Evol Dev 10:135 – 140

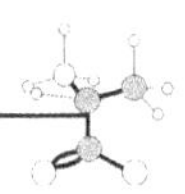

Domazet-Loso T, Brajkovic J, Tautz D (2007) A phylostratigraphy approach to uncover the genomic history of major adaptations in metazoan lineages. Trends Genet 23:533 – 539

Dunn CW, Hejnol A, Matus DQ, Pang K, Browne WE, Smith SA, Seaver E, Rouse GW, Obst M, Edgecombe GD, Sorensen MV, Haddock SH, Schmidt-Rhaesa A, Okusu A, Kristensen RM, Wheeler WC, Martindale MQ, Giribet G (2008) Broad phylogenomic sampling improves resolution of the animal tree of life. Nature 452:745 – 749

Eernisse DJ, Albert JS, Anderson FE (1992) Annelida and arthropoda are not sister taxa: a phylogenetic analysis of spiralian metazoan morphology. Syst Biol 41:305 – 330

Emig CC (1982) The biology of Phoronida. Adv Mar Biol 19:1 – 89

Felsenstein J (1978) Cases in which parsimony or compatibility methods will be positively misleading. Syst Zool 27:401 – 410

Field KG, Olsen GJ, Lane DJ, Giovannoni SJ, Ghiselin MT, Raff EC, Pace NR, Raff RA (1988) Molecular phylogeny of the animal kingdom. Science 239:748 – 753

Giribet G, Distel DL, Polz M, Sterrer W, Wheeler WC (2000) Triploblastic relationships with emphasis on the acoelomates and the position of Gnathostomulida, Cycliophora, Plathelminthes, and Chaetognatha: a combined approach of 18S rDNA sequences and mor-phology. Syst Biol 49:539 – 562

Gould SJ (1977) Ontogeny and phylogeny. Belknap/Harvard, Cambridge, MA.

Gould SJ (2002) The structure of evolutionary theory. Belknap/Harvard, Cambridge, MA.

Guindon S, Gascuel O (2003) A simple, fast, and accurate algorithm to estimate large phylogenies by maximum likelihood. Syst Biol 52:696 – 704

Halanych K (2004) The New View of Animal Phylogeny. Annu Rev Ecol Evol Syst 35:229 – 256

Halanych KM, Bacheller JD, Aguinaldo AM, Liva SM, Hillis DM, Lake JA (1995) Evidence from 18S ribosomal DNA that the lophophorates are protostome animals. Science 267:1641 – 1643

Hausdorf B, Helmkampf M, Meyer A, Witek A, Herlyn H, Bruchhaus I, Hankeln T, Struck TH, Lieb B (2007) Spiralian phylogenomics supports the resurrection of Bryozoa comprising Ectoprocta and Entoprocta. Mol Biol Evol 24:2723 – 2729

Hennig W (1966) Phylogenetic systematics. University of Illinois Press, Urbana.

Hordijk W, Gascuel O (2005) Improving the efficiency of SPR moves in phylogenetic tree search methods based on maximum likelihood. Bioinformatics 21:4338 – 4347

Huelsenbeck JP, Ronquist F, Nielsen R, Bollback JP (2001) Bayesian inference of phylogeny and its impact on evolutionary biology. Science 294:2310 – 2314

Hyman LH (1940 – 1967) The invertebrates. McGraw-Hill, New York.

Hyman LH (1959) The invertebrates, Vol. 5. Smaller Coelomate groups. McGraw-Hill, New York

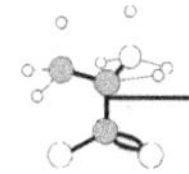

Irimia M, Maeso I, Penny D, Garcia-Fernandez J, Roy SW (2007) Rare coding sequence changes are consistent with Ecdysozoa, not Coelomata. Mol Biol Evol 24:1604 – 1607

Jeffery WR, Strickler AG, Yamamoto Y (2004) Migratory neural crest-like cells form body pigmentation in a urochordate embryo. Nature 431:696 – 699

Jeffroy O, Brinkmann H, Delsuc F, Philippe H (2006) Phylogenomics: the beginning of incongru – ence?. Trends Genet 22:225 – 231

Jenner RA (2000) Evolution of animal body plans: the role of metazoan phylogeny at the interface between pattern and process. Evol Dev 2:208 – 221

Jenner RA (2001) Bilaterian phylogeny and uncritical recycling of morphological data sets. Syst Biol 50:730 – 742

Jenner RA (2004) Libbie Henrietta Hyman (1888 – 1969): from developmental mechanics to the evolution of animal body plans. J Exp Zoolog B Mol Dev Evol 302:413 – 423

Kimura M (1983) The neutral theory of molecular evolution. Cambridge University Press, Cambridge.

Lake JA (1990) Origin of the Metazoa. Proc Natl Acad Sci U S A 87:763 – 766

Lartillot N, Brinkmann H, Philippe H (2007) Suppression of long-branch attraction artefacts in the animal phylogeny using a site-heterogeneous model. BMC Evol Biol 7(Suppl 1): S4

Lartillot N, Philippe H (2004) A Bayesian mixture model for across – site heterogeneities in the amino-acid replacement process. Mol Biol Evol 21:1095 – 1109

Lartillot N, Philippe H (2008) Improvement of molecular phylogenetic inference and the phylogeny of Bilateria. Philos Trans R Soc Lond B Biol Sci 363:1463 – 1472

Lemons D, McGinnis W (2006) Genomic evolution of Hox gene clusters. Science 313:1918 – 1922

Littlewood DT, Olson PD, Telford MJ, Herniou EA, Riutort M (2001) Elongation factor 1 – alpha sequences alone do not assist in resolving the position of the acoela within the metazoa. Mol Biol Evol 18:437 – 442

Lynch M (2007) The origins of genome architecture. Sinauer, Sunderland.

Mallatt J, Winchell CJ (2002) Testing the new animal phylogeny: first use of combined large-subunit and small-subunit rRNA gene sequences to classify the protostomes. Mol Biol Evol 19:289 – 301

Marcus E (1958) On the evolution of the animal phyla. Quart Rev Biol 33:24 – 58

Marlétaz F, Gilles A, Caubit X, Perez Y, Dossat C, Samain S, Gyapay G, Wincker P, Le Parco Y (2008) Chaetognath transcriptome reveals ancestral and unique features among bilaterians. Genome Biol 9: R94

Marlétaz F, Martin E, Perez Y, Papillon D, Caubit X, Lowe CJ, Freeman B, Fasano L, Dossat C, Wincker P, Weissenbach J, Le Parco Y (2006) Chaetognath phylogenomics: a protostome with

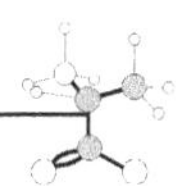

deuterostome-like development. Curr Biol 16:R

Martindale MQ (2005) The evolution of metazoan axial properties. Nat Rev Genet 6:917 - 927 Martindale MQ, Henry JQ (1999) Intracellular fate mapping in a basal metazoan, the ctenophore Mnemiopsis leidyi, reveals the origins of mesoderm and the existence of indeterminate cell lineages. Dev Biol 214:243 - 257

Maslakova SA, Martindale MQ, Norenburg JL (2004) Vestigial prototroch in a basal nemertean, Carinoma tremaphoros (Nemertea; Palaeonemertea). Evol Dev 6:219 - 226

Matus DQ, Copley RR, Dunn CW, Hejnol A, Eccleston H, Halanych KM, Martindale MQ, Telford MJ (2006) Broad taxon and gene sampling indicate that chaetognaths are protostomes. Curr Biol 16:R

Muller GB (2007) Evo-devo: extending the evolutionary synthesis. Nat Rev Genet 8:943 - 949 Nielsen C (2001) Animal Evolution: interelationships of the living phyla. Oxford University Press, New York.

Nielsen C (2008) Six major steps in animal evolution: are we derived sponge larvae?. Evol Dev 10:241 - 257

Papillon D, Perez Y, Caubit X, Le Parco Y (2004) Identification of chaetognaths as protostomes is supported by the analysis of their mitochondrial genome. Mol Biol Evol 21:2122 - 2129

Papillon D, Perez Y, Caubit X, Le Parco Y (2006) Systematics of Chaetognatha under the light of molecular data, using duplicated ribosomal 18S DNA sequences. Mol Phylogenet Evol 38:621 - 634

Passamaneck Y, Halanych KM (2006) Lophotrochozoan phylogeny assessed with LSU and SSU data: evidence of lophophorate polyphyly. Mol Phylogenet Evol 40:20 - 28

Peterson KJ, Eernisse DJ (2001) Animal phylogeny and the ancestry of bilaterians: inferences from morphology and 18S rDNA gene sequences. Evol Dev 3:170 - 205

Philip GK, Creevey CJ, McInerney JO (2005) The Opisthokonta and the Ecdysozoa may not be clades: stronger support for the grouping of plant and animal than for animal and fungi and stronger support for the Coelomata than Ecdysozoa. Mol Biol Evol 22:1175 - 1184

Philippe H, Brinkmann H, Martinez P, Riutort M, Baguna J (2007) Acoel flatworms are not platyhelminthes: evidence from phylogenomics. PLoS ONE 2:e717

Philippe H, Delsuc F, Brinkmann H, Lartillot N (2005a) Phylogenomics. Annu Rev Ecol Evol Syst 36:541 - 562

Philippe H, Lartillot N, Brinkmann H (2005b) Multigene analyses of bilaterian animals cor-rob-orate the monophyly of Ecdysozoa, Lophotrochozoa, and Protostomia. Mol Biol Evol 22:1246 - 1253

Philippe H, Snell EA, Bapteste E, Lopez P, Holland PW, Casane D (2004) Phylogenomics of eukaryotes: impact of missing data on large alignments. Mol Biol Evol 21:1740 - 1752

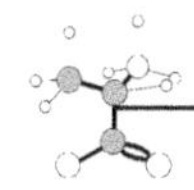

Philippe H, Telford MJ (2006) Large-scale sequencing and the new animal phylogeny. Trends Ecol Evol 21:614 – 620

Putnam NH, Srivastava M, Hellsten U, Dirks B, Chapman J, Salamov A, Terry A, Shapiro H, Lindquist E, Kapitonov VV, Jurka J, Genikhovich G, Grigoriev IV, Lucas SM, Steele RE, Finnerty JR, Technau U, Martindale MQ, Rokhsar DS (2007) Sea anemone genome reveals ancestral eumetazoan gene repertoire and genomic organization. Science 317:86 – 94

Raible F, Tessmar-Raible K, Osoegawa K, Wincker P, Jubin C, Balavoine G, Ferrier D, Benes V, de Jong P, Weissenbach J, Bork P, Arendt D (2005) Vertebrate-type intron-rich genes in the marine annelid *Platynereis dumerilii*. Science 310:1325 – 1326

Remane A (1963) The enterocelic origin of the coelom. In: Dougherty EC (ed) The lower metazoa. University of California Press, Berkeley, CA, pp 78 – 90

Rogozin IB, Wolf YI, Carmel L, Koonin EV (2007) Ecdysozoan clade rejected by genome-wide analysis of rare amino acid replacements. Mol Biol Evol 24:1080 – 1090

Rokas A, Carroll SB (2005) More genes or more taxa? The relative contribution of gene number and taxon number to phylogenetic accuracy. Mol Biol Evol 22:1337 – 1344

Rokas A, Holland PW (2000a) Rare genomic changes as a tool for phylogenetics. Trends Ecol Evol 15:454 – 459

Rokas A, Holland PW (2000b) Rare genomic changes as a tool for phylogenetics. Trends Ecol Evol 15:454 – 459

Rokas A, Kruger D, Carroll SB (2005) Animal evolution and the molecular signature of radiations compressed in time. Science 310:1933 – 1938

Rokas A, Williams BL, King N, Carroll SB (2003) Genome-scale approaches to resolving incongruence in molecular phylogenies. Nature 425:798 – 804

Roy SW, Gilbert W (2005) Resolution of a deep animal divergence by the pattern of intron conservation. Proc Natl Acad Sci U S A 102:4403 – 4408

Roy SW, Irimia M (2008) Rare genomic characters do not support Coelomata: intron loss/gain. Mol Biol Evol 25:620 – 623

Ruiz-Trillo I, Riutort M, Littlewood DT, Herniou EA, Baguna J (1999) Acoel flatworms: earliest extant bilaterian Metazoans, not members of Platyhelminthes. Science 283:1919 – 1923

Schubert M, Escriva H, Xavier-Neto J, Laudet V (2006) Amphioxus and tunicates as evolutionary model systems. Trends Ecol Evol 21:269 – 277

Scotland RW, Olmstead RG, Bennett JR (2003) Phylogeny reconstruction: the role of morphology. Syst Biol 52:539 – 548

Shubin N, Tabin C, Carroll S (1997) Fossils, genes and the evolution of animal limbs. Nature 388:639 – 648

Siewing R (1976) Probleme und neuere Erkenntnisse in der Groβsystematik der Wirbellosen.

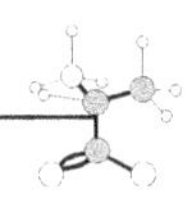

Verh Dtsch Zool Ges 70:59 - 83

Stach T, Winter J, Bouquet JM, Chourrout D, Schnabel R (2008) Embryology of a planktonic tunicate reveals traces of sessility. Proc Natl Acad Sci U S A 105:7229 - 7234

Stamatakis A (2006) RAxML - VI - HPC: maximum likelihood-based phylogenetic analyses with thousands of taxa and mixed models. Bioinformatics 22:2688 - 2690

Struck TH, Fisse F (2008) Phylogenetic position of Nemertea derived from phylogenomic data. Mol Biol Evol 23:2058 - 2071

Swofford DL (1990) PAUP: Phylogenetic analysis using parsimony, Version 3.0. Illinois Natural History Survey, Champaign

Tautz D (2004) Segmentation. Dev Cell 7:301 - 312

Telford MJ, Holland PW (1993) The phylogenetic affinities of the chaetognaths: a molecular analysis. Mol Biol Evol 10:660 - 676

Valentine JW (1997) Cleavage patterns and the topology of the metazoan tree of life. Proc Natl Acad Sci U S A 94:8001 - 8005

Wagner GP (2007) The developmental genetics of homology. Nat Rev Genet 8:473 - 479

Wanninger A, Koop D, Bromham L, Noonan E, Degnan BM (2005) Nervous and muscle system development in Phascolion strombus (Sipuncula). Dev Genes Evol 215:509 - 518

Willmer P (1990) Invertebrates relationships: patterns in animal evolution. Cambrige University Press, Cambridge

Winchell CJ, Sullivan J, Cameron CB, Swalla BJ, Mallatt J (2002) Evaluating hypotheses of deuterostome phylogeny and chordate evolution with new LSU and SSU ribosomal DNA data. Mol Biol Evol 19:762 - 776

Wolf YI, Rogozin IB, Koonin EV (2004) Coelomata and not Ecdysozoa: evidence from genome-wide phylogenetic analysis. Genome Res 14:29 - 36

Zheng J, Rogozin IB, Koonin EV, Przytycka TM (2007) Support for the Coelomata clade of animals from a rigorous analysis of the pattern of intron conservation. Mol Biol Evol 24:2583 - 2592

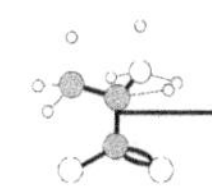

第5章 后生动物的复杂性

摘要：进化通常被视为从简单到复杂的过程，这种观点对进化论影响颇深。然而，新海洋模式系统的研究及分析水平的提高对这一观点提出质疑，因为新的研究结果显示复杂性水平可能与个体组织复杂性背道而驰。本章分析了不同动物类群的分子遗传学研究进展及其决定主要进化过渡（如多细胞的过渡、胚层的起源等）相关的分子改变机制。

5.1 复杂性的途径

海克尔（Haeckel）绘制的“人的谱系”（Pedigree of Man）（图5.1）是一张关于共同进化的精美图谱，能有效阐明动物系统发育问题。从这幅图可以看出，动物从位于树底部的简单生命形态（如蠕形动物门的蠕虫）稳步向更复杂的形态直至位于树顶端的哺乳动物进化。不同动物类群分支也有类似的进化趋势。例如，“鲶鱼”（mud fish）靠近主干，而硬骨鱼类处于该分支的顶端，这与硬骨鱼比鲶鱼进化程度高的直观概念相吻合。

进化论中“复杂性”稳步增加这一概念一直很有影响力，能从许多不同层面解释和评估包括基因组特征和组织类型的数量与特征在内的生物学数据。然而，随着时间的推移，越来越多的特例使复杂性稳步增加（并且灵长类代表复杂动物）这一推测看起来既奇怪又有悖常理。例如，在非编码DNA发现之前，“C值矛盾”常用于描述基因组大小和动物的复杂性之间并没有明显的相互关系（Gregory 2005）。肺鱼（Pedersen 1971）的基因组大小约是高度进化的硬骨鱼河豚（Jaillon et al. 2004）基因组的400倍，是人类的40倍。

当非编码DNA的数量差异解决了C值矛盾问题之后，科学家们推测，随着复杂性的增加，复杂动物的编码DNA数量（或基因数量）要多于简单动物编码DNA的数量。脊索动物在进化过程中，Hox基因簇发生复制，有人认为Hox基因簇的复制是两次全基因组复制（ZR-hyprthesis）的结果，这一观点与现象均支持另一基本概念，即基因加倍是发展进化的主要来源（参阅综述Taylor and Raes

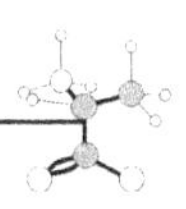

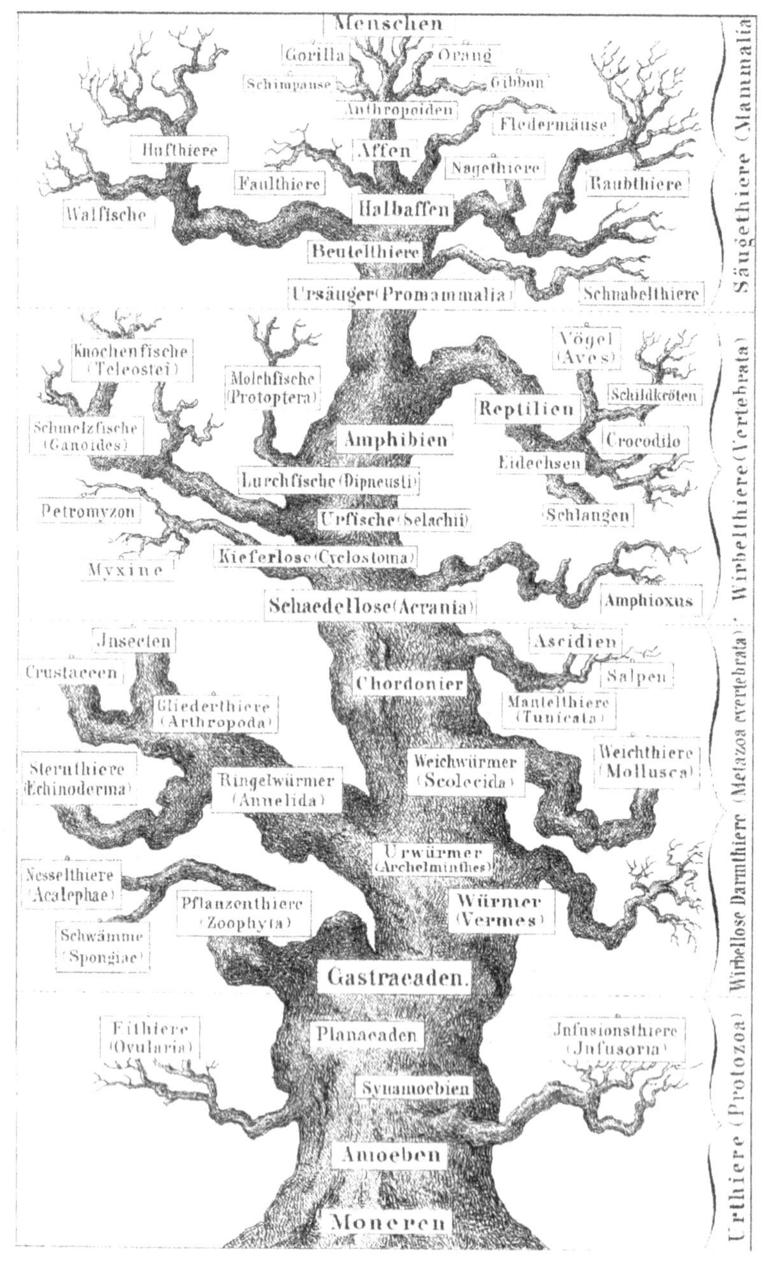

Fig. 5.1 Increasing complexity as a topological principle in Haeckel's "Stammbaum des Menschen" ("Pedigree of Man"). Animals with presumed simple organization are placed close to the bottom and stem of the tree, whereas more "elaborate" forms are found in the tips of the branches. Reproduced from (Haeckel 1903)

图5.1 Haeckel根据拓扑原理构建的解释动物复杂性的生命之树——"Stammbaum des Menschen"。结构相对简单的动物位于树底或靠近树干，而相对"更高级"的物种位于树枝的顶端

英文注释：Complexity，复杂性；Elaborate，精心制作的、详尽的、更高级的

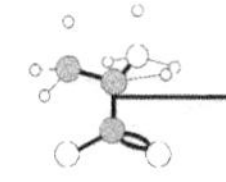

2004）。然而出人意料的是，目前已完成基因组测序的脊椎动物的编码基因总数仅仅比经典无脊椎模式动物（线虫和果蝇）略微多一点点（Claverie 2001），不同的脊椎动物所含的基因数都大致相同。此外，发育模式的某些重要方面（例如转录因子 pax6 调控感光结构的形成、Hox 基因在前后轴线分化中的重要作用）在进化早期就已形成，为古老核心概念“发育遗传工具箱”在动物发育中的作用提供证据（Cañestro et al. 2007）。鉴于该工具箱不同组件之间的联系对动物的发育极其重要，动物复杂性的研究不能像基因或基因功能的研究那样简单量化，而需更深入的分析。

从形态学水平上讲，动物复杂性也可通过统计成体细胞类型数量来评估（Bell 1997，Sempere et al. 2006，Valentine et al. 1994），有时也通过是否存在其他形态或胚胎特征进行补充（Aburomia et al. 2003，Heimberg et al. 2008）。但这些方法也存在一些缺陷：首先，有些动物比其他动物拥有更多的形态学数据，但往往没有对应的高复杂性（Bonner 1988）。其次，形态学特征不足以描述复杂性的其他要素，如发育复杂性和行为复杂性中涉及的生活史策略（直接或变态发育）和细胞类型的个体发育（Bonner 1988，Valentine 2000，Valentine et al. 1994）。最后，形态学特征很难进行量化和比较，复杂性的测定需要更简单、可量化的新方法。

在过去的数十年，分子生物学为复杂性的评估提供了许多新方法。除了基因谱系（基因或基因家族数量）的总体比较之外，保守基因家族的多样性差异也成为不同物种之间分子特征比较的切入点，对越来越多的物种实施这些分析手段的同时还加入表达动力学和个体基因位点活性研究等分子技术作为补充，以利于研究者从分子水平来描述个体的细胞群体和组织特征。同样，基因敲除技术也应用于越来越多的物种，为控制动物系统的调控网络提供越来越详细的分析数据。

然而，发展最快的还属测序技术，全基因组测序或是转录组测序计划为研究动物多样化之间的分子进化提供了更好的视野。我们将在本章回顾不同动物类群在分子层面的最新进展，探讨它们能够揭示动物进化中复杂特征起源的哪些问题，并尤其关注多细胞动物（后生动物）从单细胞真核生物起源和多样性演化过程中所发生的进化改变，以及两侧对称动物类群的复杂性差异。

5.2　领鞭毛虫类：后生动物的多细胞进化

领鞭毛虫类单独组成一个门，是有鞭毛单细胞动物的总称。它们顶端被

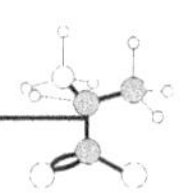

鞭毛，周身环绕着用于捕食细菌碎屑的放线状微绒毛，细胞外形成硅质骨针，其形态及功能很容易让人联想到海绵的领细胞（Clark 1866，1868）。分子系统发生研究证实，领鞭毛虫类是整个后生动物的单系姊妹群（Carr et al. 2008，Clark 1868，Haeckel 1874，King et al. 2008，Shalchian-Tabrizi et al. 2008）。在现存的一些领鞭毛虫中，有的形成群体，有的分化成变形虫样单个细胞或生殖囊（Bütschli 1883-1887，Leadbeater 1983，Siewing 1985）。因此，领鞭毛虫类和后生动物的单细胞类共同祖先很可能具有细胞分化或群体形成的能力，因而能够表现出多细胞的原始特征（King et al. 2008，Lang et al. 2002）。真若如此，则现存群体形式的领鞭毛虫类可能与原始的多细胞生命形式类似，这对于研究后生动物进化线上所有动物的进化是有益的。又或者，在领鞭毛虫与后生动物分开之后，群体形成和细胞分化上各自独立进化。为了区别这两种情况，科学家们围绕多细胞发生（如细胞粘附、细胞间通讯、分化细胞的分工）相关的基因开展了一系列的实验，研究对象为严格单细胞领鞭毛虫（用 EST 和基因组数据）和群体形式的类原绵虫（用 EST 数据；Abedin and King 2008，King et al. 2003，2008）。

在最新的领鞭毛虫基因组注释中，总共发现 78 种后生动物特有而植物、真菌和粘菌中没有的蛋白质结构域（虽然有些与细菌蛋白质结构域类似），说明领鞭毛虫类与后生动物有密切联系（King et al. 2008）。这些结构域存在于后生动物细胞的粘附蛋白（例如细胞外钙粘蛋白、糖结合 C 型凝集素、免疫球蛋白和整合蛋白 α 结构域）和细胞外基质（ECM）组件（例如由几种胶原蛋白和层粘连蛋白组成的基底膜元件）的纤连蛋白中，说明这些独立的功能结构域在领鞭毛虫与后生动物从进化树上分离前就已经形成，虽然很多结构域仅存在于领鞭毛虫的组件中，例如含有细胞外钙粘蛋白结构域（ECD）的蛋白。但是，一些 ECD 在脂肪型钙粘蛋白（存在于海绵、刺胞动物、两侧对称动物）和 Hedgehog 相关蛋白（仅存在于海绵和刺细胞动物）中相当保守（图 5.2；Abedin and King 2008，King et al. 2008），尤其是典型后生动物型钙粘蛋白的胞浆结构域，这一结构域并不存在于领鞭毛虫中（Abedin and King 2008），并且整合蛋白 β、后生动物特有细胞粘附受体域和 ECM 组件层黏连蛋白 B－IV 也未在领鞭毛虫细胞内发现（King et al. 2008）。

虽然领鞭毛虫拥有一套相当丰富的后生动物细胞粘附和 ECM 结构域系统，但缺失大部分与后生动物胞间通讯相关的胞内信号级联系统，即使有也与后生动物的差别很大。后生动物拥有庞大的 Wnt、TGF-β 配体或核激素受体同源基因，但领鞭毛虫的基因组中并不存在这些基因，并且另外一些信号通

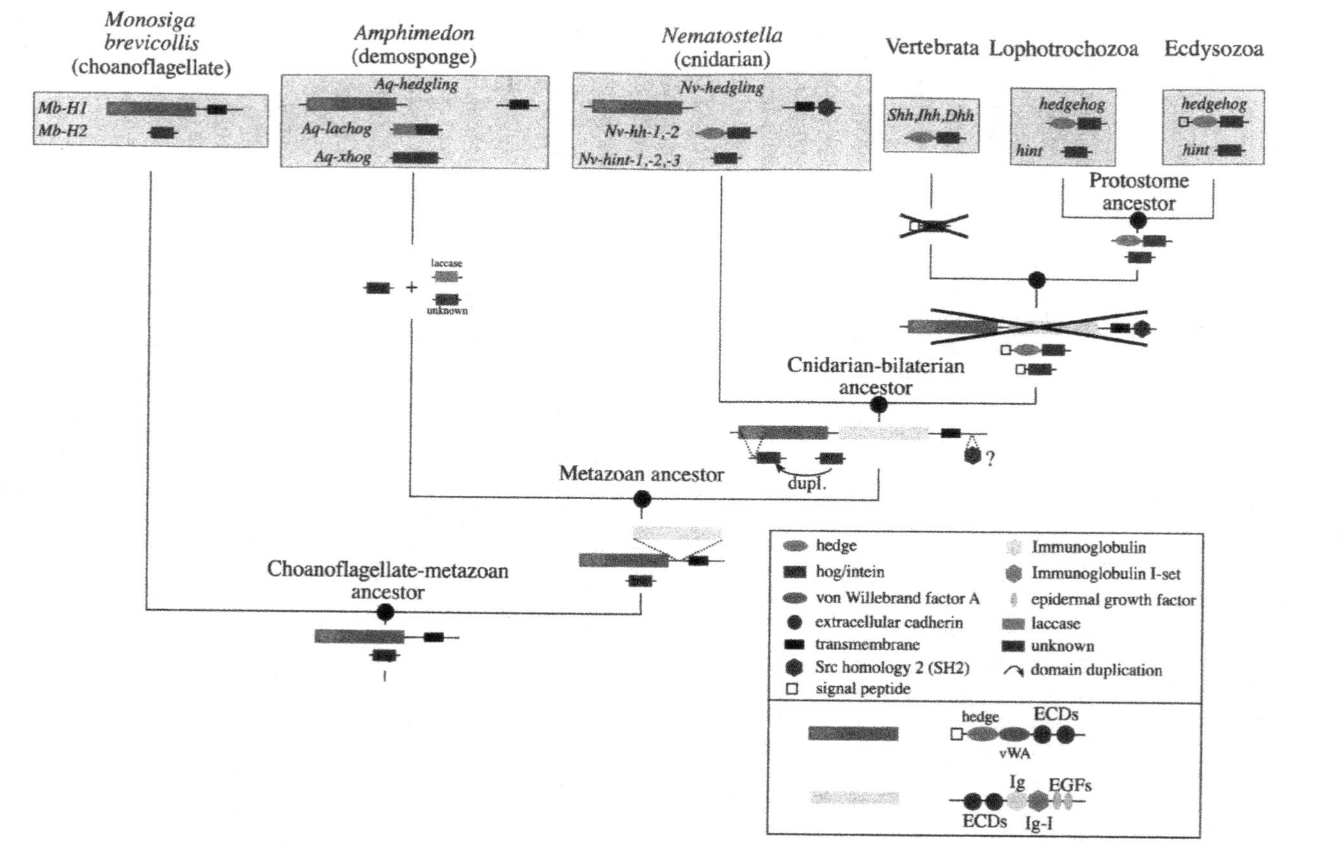

Fig. 5.2 Domain evolution. Scenario of Hedgehog domain evolution by domain shuffling, duplication and loss during the evolution of metazoans (based on Adamska et al. 2007b, King et al. 2008, Matus et al. 2008, Snell et al. 2006). Schematics at roots depict protein and domain arrangements present in the common ancestor. Schematics within terminal branches depict putative rearrangements not present in the ancestor (derived). Grey boxes highlight protein orthologues present in extant organisms

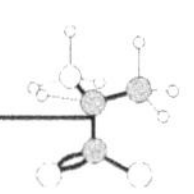

图5.2 结构域的进化。后生动物进化过程中Hedgehog结构域进化方案：结构域重排、复制、缺失（Adamska et al. 2007，King et al. 2008，Snell et al. 2006）。图表底部显示共同祖先的蛋白和结构域编排方式，图表的分支显示祖先里不存在的可能重排方式，灰色底纹标注显示现存物种中的蛋白或同源基因

英文注释：Shuffling，重排；Duplication，重复、复制；Metazoans，后生动物、多细胞动物；Orthologues，同源基因

路（JAK/STAT，Hedgehog，Delta/Notch）也是不完整的。例如，后生动物的一些多结构域信号通路组件（如Notch或Hedgehog蛋白）在领鞭毛虫中编码的是单个结构域，而领鞭毛虫的细胞粘附蛋白大多不存在于后生动物里。再如，领鞭毛虫体内的酪氨酸激酶信号转导通路拥有数量最多的酪氨酸激酶和受体、调控磷酸酶和磷酸化酪氨酸结合SH2域蛋白（信号传感器），但这些酪氨酸激酶通路与后生动物的通路相去甚远（King et al. 2003，2008，Manning et al. 2008）。综上，领鞭毛虫缺乏具有代表性的后生动物同源基因，在调节酪氨酸激酶信号上又与后生动物存在差异，并且在后生动物蛋白中也没有发现如领鞭毛虫般庞大的酪氨酸激酶结构域组件（King et al. 2008，Manning et al. 2008，Pincus et al. 2008，Segawa et al. 2006）。但也有一个例外：细胞内酪氨酸激酶结构域和细胞外钙粘蛋白带有胞浆SH2结构域和EGF域相结合，说明酪氨酸激酶信号通路可以传递胞外信号（Abedin and King 2008，King et al. 2003），检测胞外食物供应的功能实验证明酪氨酸激酶信号与细胞周期调控相关（King et al. 2003）。

Hedgehog（Hh）配体的进化过程很好地阐明了结构域的重排在多细胞动物蛋白进化过程中的重要性（图5.2）。两侧对称动物的Hh蛋白是由两个功能结构域组成：N端的分散受体结构域和C端的自动催化结构域（Bijlsma et al. 2004），虽然两个结构域均存在于领鞭毛虫中，但分别由两个独立的基因编码（King et al. 2008，Snell et al. 2006）。在海绵动物和刺胞动物，N端Hedge结构域是多域蛋白（该蛋白为Hedgling）的一部分，而领鞭毛虫的C端Hog/Intein结构域在一个单域蛋白上（Adamska et al. 2007b，King et al. 2008，Matus et al. 2008），另一种领鞭毛虫（*M. ovata*）中的C端结构域以及潜在的纤维素结合结构域同时存在于同一个蛋白上（Carr et al. 2008，Snell et al. 2006）。如果独立的Hh结构域是原始祖先特征，那么两侧对称动物的Hh蛋白则是特殊结构域复制的结果，并且这种复制是在海绵动物/真后生动物和刺胞动物/两侧对称动物从进化树上分离之前就已发生。因此，在刺胞动物/两侧对称动物祖先的两个含Hedge结构域的蛋白中，相对古老的那个蛋白

(Hedgling) 在早期的两侧对称动物中发生丢失，仅保留了 *bona fide* Hh 蛋白。

细胞外结构域和信号传导结构域的存在暗示领鞭毛虫之间存在强大的相互作用，转录因子谱系则反映了调控系统和细胞类型分化的复杂性。领鞭毛虫基因组包含了所有主要的且普遍存在的转录因子基序（如锌指结构、同源或螺旋蛋白；King et al. 2008），少数后生动物特有的转录因子（如 p53、Myc）在领鞭毛虫中也存在，它们在细胞周期或转录中有普遍的调控作用（Nedelcu and Tan 2007）。仅有的两个同源框基因特异地聚类于 Meis/Prep/TGIF 基因（TALE 超类），但不能比对上其他的 TALE 基因（如 Iropois；King et al. 2008），表明早在领鞭毛虫出现之前，至少 Iropois 相关基因和非 TALE 同源框因子就已丢失（Derelle et al. 2007，Mukherjee and Bürglin 2007），并且领鞭毛虫缺乏后生动物特有的 Ets、Hox，POU 和 T-box 家族（King et al. 2008）。

总之，后生动物特有的蛋白结构域（如与细胞粘合及 ECM 互作蛋白的结构域）的高保守性预示着领鞭毛虫在相互作用及其与环境互作方面存在极大潜能，并主要反映在某些领鞭毛虫的群落形成和附着能力（Leadbeater 1983，Siewing 1985）。然而，后生动物信号级联蛋白的缺失或片段化暗示了其在细胞通讯方面是相当受限的。拿酪氨酸激酶蛋白来说，现有的信号结构域是通过调节胞内加工来改变环境条件，而不是通过细胞间相互作用。尽管大多数保守的细胞粘合和信号结构域都是存在的，但它们在多区域蛋白中的独特组合使得我们很难推断出它们在领鞭毛虫 - 后生动物祖先中的原始功能。蛋白结构域的广泛重排导致了后生动物和领鞭毛虫在进化路线上产生新功能。领鞭毛虫的某些蛋白或类原棉虫（群体生活的领鞭毛虫）和后生动物的某些蛋白（如脂型钙粘蛋白）能建立明显的同源性关系，进一步的功能分析则利于区分“原始后生动物”型（能形成群落且有细胞识别和/或粘附功能）和“单细胞”型（早于多细胞生物的进化，如细菌的捕食和吞噬作用、环境应激反应）。前一种情况是，群体形成减少之后，这些结构域在领鞭毛虫中进化出了特异的功能。后一种情况是，领鞭毛虫某些蛋白结构域（如整合蛋白 β）发生缺失，有些结构域发生重排形成新蛋白（如“经典”钙粘蛋白的形成），这些都是单细胞向多细胞进化的重要步骤。

就细胞类型多样性而言，领鞭毛虫缺失了后生动物特有的主要转录因子家族，从而导致其较低的细胞分化程度。然而，hox 基因二次缺失则表明，在领鞭毛虫进化过程中，转录因子复杂性和细胞类型多样性也可能经历二次缺失。将分子分析扩展至更大数量的领鞭毛虫物种，同样的分析也扩展至形成

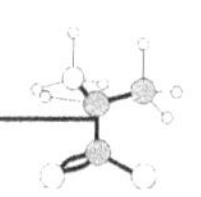

领鞭毛虫加后生动物集群的外类群 filasterean 和 ichtyosporean choanozoans（例如 *Ministeria*，*Capsaspora* or *Sphaeroforma*），将会阐明领鞭毛虫和后生动物进化过程中基因的得失模式。

5.3　海绵动物的进化：体轴、细胞类型和皮层

海绵动物大多栖息于海洋中，属于固着、滤食性动物，被视为最“简单”的动物门。它的形体横剖面主要由两层上皮细胞组成：单层或双层扁平细胞形成保护外层，有纤毛的领细胞形成领细胞层，水从孔道流入体内，从出水口排出。扁平细胞和领细胞中间是基质，变形细胞与细胞外基质形成结缔组织。领细胞几乎占满整个内胚层。海绵的简单性还表现在细胞类型上：寻常海绵纲仅由 12 种组织学上可分辨的成熟细胞组成，其中包括了形成骨骼的造骨细胞和可收缩的肌细胞（Siewing 1985）。另一些细胞类型可能是幼体特有的（例如原生感光细胞）或者以形成特殊小泡的形式进一步分化。海绵动物缺少真正的肌肉细胞和神经细胞。

一般来说，海绵动物门（多孔动物门）可细分为六放海绵纲、寻常海绵纲和钙质海绵纲，它们都是由同一祖先进化而来（Philippe et al. 2009）。有些学者认为这三个亚纲的海绵是一类与其他后生动物亲缘更近的并系群，甚至比其他海绵的亲缘关系还要近（Borchiellini et al. 2004，2001，Haen et al. 2007，Sperling and Peterson 2007；想了解更多，请参看第 4 章）。无论如何，海绵动物被认为是现存最古老的多细胞群体（参看第 4 章以及后文中丝盘虫相关内容），也是最早出现胚胎和幼体发育的动物。通过对比不同海绵群体间细胞粘着蛋白和细胞外基质相互作用蛋白，有助于确定多细胞进化过程中的关键事件。此外，对发育相关的转录因子和信号分子的比较研究，有助于深入了解包括生殖层分化、体轴特征和胚胎分化在内的早期胚胎演化与幼体发育过程。

正如引言所述，动物的复杂性往往与细胞的类型多样性有关。细胞类型由转录因子特征性组合（分子指纹，molecular fingerprint）来界定（Arendt 2005），转录因子家族可以表明细胞类型的多样性（Vogel and Chothia 2006）。这个假设正通过在现有的海绵动物中转录因子谱系的确定进行验证，这有助于重建他们在动物进化中的发生顺序。目前已经证明在寻常海绵纲参考物种中存在主要的后生动物转录因子。系统发生分析可以更加详细地解释海绵的 bHLH、Fox、Sox、T-box、Paired-like、Antennapedia、NK、TALE、Six、POU 和 LIM 同源性蛋白与其他动物的关系（Adell et al. 2003，Adell and Müller

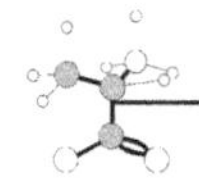

2004，2005，Jager et al. 2005，Larroux et al. 2007，2006，2008，Manuel and Le Parco 2000，Manuel et al. 2004，Simionato et al. 2007）。这种方法已经被证明可有效应用于基于后生动物进化的基因家族范围的确定。某些海绵基因和一些两侧对称动物基因家族（如 ARNT/bamal 或 NK2/4；图 5.3a）相对应（Larroux et al. 2007，Peterson and Sperling 2007，Simionato et al. 2007），这就意味着海绵动物的基因代表一种祖先基因，在真后生动物进化过程中发生复制或形成新的家族。海绵的其他基因可以和真后生动物的单基因家族归为一组（图 5.3b；Larroux et al. 2008，Simionato et al. 2007），表明在海绵真后生祖先里不同基因家族已经分化。某些情况下，有的家族缺少明显的海绵同源基因，但却存在单源相关性，可以推测海绵中这些基因出现缺失，同样通过对 NK hox 基因的研究表明，有些 NK 基因家族在寻常海绵纲里发生缺失（Peterson and Sperling 2007）。通过与其他海绵物种的相应基因进行比对，能直接证明这些基因发生了缺失（图 5.3b；Larroux et al. 2008）。由于目前其他海绵动物的序列信息相当少，海绵动物祖先真实存在的转录因子的数量可能比目前预测的要多。

虽然海绵的成熟细胞类型不多，海绵个体的形态又很简单，但令人惊奇的是，所有主要的真后生动物转录因子家庭的代表在海绵中都有发现。这一矛盾也许可以这样解释：因为大部分海绵动物生命周期包括有纤毛的幼虫期和之后的变形期（Leys and Ereskovsky 2006），早期胚胎发育的高度变异导致了目前有关海绵中胚层及原肠胚形成及定义的相对立观点。（Ereskovsky and Dondua 2006，Leys 2004）。然而相比于成熟期，许多幼体有明显的双胚层和单一体轴（Ereskovsky and Dondua 2006，Leys and Ereskovsky 2006）。表达分析显示，海绵幼体沿体轴发展的分子模式是被 Wnt 和 TGF-β 成员调节（Adamska et al. 2007a）。体轴直接与游动的方向有关，具有趋光性（Leys and Degnan 2001）。虽然缺乏神经元细胞和纤毛或感状束感觉器，一些感光幼体在长长的纤毛上具有色素细胞，这些细胞可以作为感受器，因此也可以在分化成光感受细胞之前代替光感受器细胞（Arendt 2008，Leys and Degnan 2001）。海绵基因组存在一套几乎完整的两侧对称动物的后突触蛋白，说明虽然缺乏神经系统、突触和密集的后突触，但具有一个简单的感觉系统也能正常工作（Sakaraya 2007）。更多的功能、生理和表达分析将能阐明海绵动物细胞类型是否比目前预测的更多。实际上，通过形态学标准鉴定的某个单一类型细胞可能包含几个功能不同的细胞，而这些细胞的区别只能通过不同的分子组成来区分。

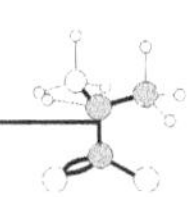

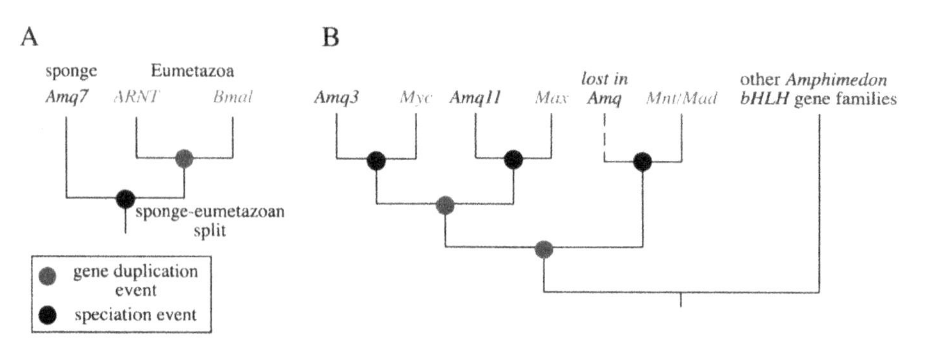

Fig. 5.3 Evolution of gene families. Evolution of gene families by gene duplication and speciation events (based on Simionato et al. 2007). (**a**) Evolution of the eumetazoan bHLH genes *ARNT* and *bmal* by gene duplication from a single ancestor present in the last sponge-eumetazoan ancestor. (**b**) Evolution of the bHLH genes *myc*, *max* and *mnt/mad* by gene duplication prior to the eumetazoan/sponge split implies that the absence of *mnt/mad* in *Amphimedon* is due to secondary loss. This is confirmed by the presence of a *mnt/mad* ortholog in the homoscleromorph sponge *Oscarella*. See text for further details. *Blue*: eumetazoan orthologs; *red*: *Amphimedon* orthologs

图 5. 3　基因家族的进化。基因复制导致基因家族的进化和物种形成（基于 Simionato et al. 2007）。（a）真后生动物的 bHLH 基因 *ARNT* 和 *bmal* 通过基因复制从同一祖先（海绵动物—真后生动物祖先）进化而来，（b）bHLH 基因 *myc*、*max* 和 *mnt/mad* 的复制先于真后生动物与海绵动物的分化，表明在海绵中 *mnt/mad* 的缺失是基因 secondary loss 的结果，多孔海绵中 *mnt/mad* 直系同源基因的存在也证实以上观点。详细信息参见文章相关内容。蓝色代表真后生动物直系同源基因；红色代表海绵直系同源基因

英文注释：Eumetazoan，真后生动物的；Secondary loss，次级损耗

5.4　丝盘虫：生而简单还是高度简化

直到最近，扁盘动物门仍仅包含两种丝盘虫（*Trichoplax adhaerens* 和只被描述过一次的 *Treptoplax reptans*），但分子分析却发现了其他的神秘物种（形态学上难以区分；Grell 1971b，Voigt et al. 2004）。目前，只观察到分裂生殖或出芽生殖这两种无性繁殖方式的存在（Siewing 1985），但能够分裂至 64 细胞期（Grell 1971a，1972，Grell and Ruthman 1991）的假定卵母细胞的发现、重组和性别的分子特征都标志着有性生殖的存在，因此，丝盘虫生活史的描述很可能是不完全的，在未知寄主身上的中间寄生阶段和幼虫阶段也必须考虑在内（Miller and Ball 2005）。丝盘虫展示了一种独特的食藻模式（“跨上皮细胞吞噬”），它通过将藻输送到上部的缺口，经单纤毛上皮细胞再到内纤维细胞的吞噬（Wenderoth 1986）来完成食藻过程。此外，腹部的非纤毛上皮也许能分泌消化酶，但不能进行吞噬作用。

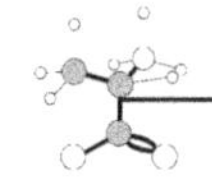

因其只含四种类型的体细胞（包括三个细胞层和一个上下型的单轴），缺少基底膜、细胞外基质、口、内脏和神经系统，丝盘虫的身体构造通常被认为是所有后生动物中结构最简单的（Syed and Schierwater 2002）。细胞层由纤维细胞分隔成上下两部分的上皮细胞构成，而这个有收缩性的纤维细胞极有可能控制类似变形运动（Syed and Schierwater 2002）。在丝盘虫这种简单的轴向生物基因组中，发现一整套功能性的 Wnt/TGF－β 信号组件（在两侧对称动物体轴模式中是保守的）是相当不可思议的（Srivastava et al. 2008）。

基于线粒体序列得到的系统发生结果显示，扁盘动物类群和海绵及刺胞动物归类在一起，与两侧对称动物形成单源发生的姐妹门（Dellaporta et al. 2006，Haen et al. 2007，Signorovitch et al. 2007），而基于对丝盘虫和其他后生动物基因组测序（Srivastava et al. 2008）的大数据的进化分析，支持扁盘动物门的出现是晚于多孔动物门（Borchiellini et al. 2001，Collins 1998，da Silva et al. 2007）。目前，还没有任何分子系统发生能够说明丝盘虫是其他后生动物的姊妹群，因此丝盘虫表型的形态学简化很有可能是通过二次简化而逐步形成的。因细胞外基质和基底膜存在于同硬形海绵亚纲的海绵中，说明这些特征在丝盘虫中出现二次缺失。

所有 ANTP 同源框基因谱系也表明丝盘虫中存在二次缺失现象。基因组测序（Schierwater et al. 2008，Srivastava et al. 2008）鉴定出一些新的 ANTP 同源异型盒基因，是除去目前已知的 Hox/ParaHox（*Trox-2*；Jakob et al. 2004）、NK-like（*Dlx*、*Hm*、*Not*；Martinelli and Spring 2004，Monteiro et al. 2006）和延伸的 Hox（*Mnx*）（Monteiro et al. 2006）家族成员之外的新成员。然而即便如此，ANTP 基因数仍保持相当低的数量。ANTP 基因丢失的证据包括有：①一些 NK 基因（*msx*、*bar*H）的缺失，而这些基因出现在了海绵中（Schierwater et al. 2008）；② ANTP 大亚类中的某些单个成员被归类到不同两侧对称的刺胞动物基因家族中（比如 *trox*2 作为一个 *cnox/gsx* 的直系同源包括在 Hox/ParaHox 亚类中；Monteiro et al. 2006，Schierwater et al. 2008）。无论这个独特的家族是所有亚类的鼻祖（在 *cnox/gsx* 的情况下不大可能），还是 Hox/ParaHox 的延伸（之前在海绵动物章节中有相关介绍），基因缺失现象在丝盘虫的进化过程中确有发生（Jakob et al. 2004，Monteiro et al. 2006，Peterson and Sperling 2007）。

许多丝盘虫转录因子，如 T-box 基因 *Tbx*2/3 和 *brachyury*、成对盒基因 *TriPaxB*、感觉细胞标志基因的假定前体基因 PaxA/B/C（刺胞动物）、Pax2/5/8 和 Pax4/6（两侧对称动物）都在外边界表达，这与 RF-amide（在刺胞动

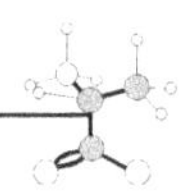

物神经肽中含量丰富）的表达相似（Martinelli and Spring 2003，Schuchert 1993）。该区域包含多种形态不常见的细胞，这些细胞代表了原始的神经元或多能干细胞，它们在刺胞动物和两侧对称动物的进化过程中分化为不同的细胞类型（Jakob et al. 2004，Martinelli and Spring 2003）。丝盘虫基因组中存在许多基本合成组件，包括神经递质合成、释放和摄取以及突触形成、感光和刺激的电传导等，这些组件的存在进一步支持了假定神经前体细胞的存在。

虽然丝盘虫在很多方面都简化了，但小丝盘虫基因组（98 Mb）结构仍具有祖先原始特征，大丝盘虫和脊椎动物基因组区域之间的高度保守连锁性（同线性）、古老内含子的保留、内含子－外显子边际的高保守性抑制了基因组的二次缩减（线虫、果蝇和住囊虫均发生二次基因缩减）。对丝盘虫生活史和细胞类型的进一步研究可以帮助阐明模糊的发育阶段和未发现的细胞类型多样性是否可以从基因含量和结构方面对基因组复杂性做出解释。

5.5　刺胞动物：身体简单，基因复杂

刺胞动物构成一个物种丰富的海洋动物类群，由珊瑚虫类（例如海葵和珊瑚）、箱形水母类（如箱水母）、真水母类（如海黄蜂）和水螅类（如水螅）组成（Siewing 1985）。很多珊瑚类和水螅类具有复杂的生命周期，包括被纤毛的浮浪幼体期（Nielsen 2001）。另外，珊瑚类除外的其他所有类群都有一个自由游动的水母期。虽然水母阶段被认为是二次创新，但它可能是在珊瑚类之后出现，或者可能在珊瑚类起源后发生二次丢失（Collins 2002）。和他们复杂的生命周期相比，刺胞动物的身体构造却非常简单，仅含外胚层和内胚层两个胚层，中间由包含细胞外基质的非细胞结构中胶层隔开。刺胞动物由胚孔发育成单一身体开口，兼具嘴和肛门的功能。在浮浪幼体另一端是纤毛顶毛丛，推测其可能行使感觉功能。虽然简单的辐射对称广泛存在于刺胞动物类，但许多珊瑚类拥有沿着口到对口线的第二体轴（“定向轴”）形成双侧对称（Stephenson 1928）。其明显由内胚层褶皱（肠系膜）肌肉形成并在类狭缝咽两端具更长的纤毛（Siewing 1985，Stephenson 1935）。刺胞动物和两侧动物轴的同源性将在后面章节讨论。

总的来说，刺胞动物的神经系统简单而且弥散（Bullock and Horridge 1965，Siewing 1985），而许多水母有富含轴突的神经环，真水母和箱形水母有类似眼晶状体和平衡器的感觉器官，被称为感棍（Nilsson et al. 2005，Piatigorsky and Kozmik 2004，Skogh et al. 2006）。前期刺细胞动物的神经发生分

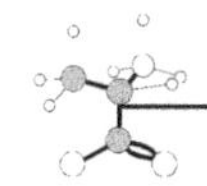

子机制以及神经细胞多样性和特异性的研究主要局限在淡水水螅中，近期才通过海葵的相关研究逐渐揭示珊瑚幼体神经系统的发育机制（Marlow et al. 2009），而更多的对神经结构发育的特定分析研究主要集中于感光系统（Kozmik et al. 2003，Stierwald et al. 2004，Suga et al. 2008）及顶器模式（Matus et al. 2007，Pang et al. 2004，Rentzsch et al. 2008）。我们认为，对这些细胞类神经系统发育和神经细胞多样性的深入分子研究，将有助于理解大部分两侧对称动物复杂神经系统的进化。

淡水水螅是研究对称轴和干细胞的模式生物，它的成功运用促进了人们利用刺胞动物来研究非两侧对称动物的发育和基因组研究的兴趣。另外，近10年来，珊瑚类和水母类等用于比较基因组学和发育的研究在逐渐兴起。

5.5.1　海葵基因组

海葵基因组的测序验证了之前基于海葵和鹿角珊瑚 EST 分析的假设，即珊瑚虫类基因组比两侧对称动物基因组要复杂得多（Kortschak et al. 2003，Miller and Ball 2008，Miller et al. 2005，Putnam et al. 2007，Technau et al. 2005）。例如海葵的基因比昆虫和线虫的多，基因组也更大（Miller and Ball 2008）；另外，它的基因组具两侧对称动物所有的信号通路和几乎所有转录因子基因家族的典型编码序列（Larroux et al. 2008，Putnam et al. 2007，Ryan et al. 2006，Technau et al. 2005）。一些特定基因家族的研究表明，相对于昆虫和线虫，海葵与脊椎动物类的同源基因更多（Kusserow et al. 2005，Matus et al. 2006a，Rentzsch et al. 2006）。另外，在海葵中多个 Fox 基因发生2次缺失（Larroux et al. 2008）。同时，刺胞动物基因组似乎也比很多两侧对称动物复杂：就保守基因而言，80%的人类基因的内含子保守地存在于海葵中，甚至比线虫或果蝇还多（Putnam et al. 2007）。海葵和两侧对称动物的高度复杂性，不仅体现在基因含量上，也体现在基因构成上，支持了刺胞动物和两侧对称动物祖先形态结构复杂的假说。但是，刺胞动物和两侧对称动物的巨大形态差异，让目前的比较研究困难重重。在以下章节，我们将重点列举一些例子，通过比较保守模式系统来阐明两侧动物独特的特征，如两侧动物的体轴和中胚层的起源，以及其从刺胞—两侧对称动物祖先进化而来的发展过程。

5.5.2　刺胞动物 BMP 模式和两侧对称的背—腹轴的进化

分泌型骨形态形成蛋白（BMP）和它们的拮抗因子腱蛋白、头蛋白、Gremlin 蛋白和滤泡稳定蛋白，在两侧对称动物背腹轴形成过程中发挥重要而

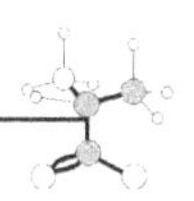

保守的作用（Grunz 2004）。胚胎在神经一侧（原肢类在腹侧、脊椎类在背侧）表达 BMP 拮抗因子，相对的一侧表达 BMP 配体。与两侧对称动物不同的是，珊瑚类并不存在沿轴线表达的 BMP 拮抗因子和配体模式（Arendt and Nübler-Jung 1997，De Robertis and Sasai 1996）。珊瑚类 BMP2/4 和 BMP5/6/7/8 早期会不对称地在定向轴（在胚孔和咽一侧）和口—反口轴（仅在口侧）表达，但后期却只在定向轴一侧表达（Hayward et al. 2002，Matus et al. 2006b，Rentzsch et al. 2006）。虽然海葵的腱蛋白和 Gremlin 蛋白能在异源的斑马鱼中拮抗海葵 BMP2/4，但它们的表达却与两侧对称动物大不相同：*gremlin* 是唯一被报道的在定向轴抑制 *bmp2/4* 表达的拮抗因子，而 *chordin* 的表达则类似于 *noggin*，它们和 *bmp2/4* 有大部分重叠。拮抗因子 *follistatin* 的径向表达进一步反驳了空间上明显的激动剂—拮抗剂模式（Matus et al. 2006a，b. Rentzch et al. 2006）。由于很多拮抗因子和 BMPs 一样在原肠胚口侧特定表达，因而也没有发现明显的口—反口轴的拮抗。总之，海葵胚轴的非一致性表达说明 BMP 信号通路是复杂的，同时，结合两侧对称动物类似的 BMP 拮抗系统，对珊瑚类的单胚轴模式提出质疑（Rentzsch et al. 2006）。BMP 和它们拮抗因子形成背腹轴似乎是两侧—刺胞动物分化之后起源的，因此这些结果不能用于两侧对称动物背腹轴和任何刺胞动物体轴的直接同源性分析。

5.5.3　刺胞动物 Hox 基因和前后轴的演化

比较 Hox 基因的表达不仅能用于研究信号通路，而且在阐明刺胞类和两侧对称动物体轴的关系上有积极意义。在两侧对称类物种中，Hox 基因簇是保守的前后轴模式系统的典型代表，基因簇内部基因的位置反映果蝇和脊椎动物体轴的前后轴位点的相对表达（McGinnis and Krumlauf 1992）。两侧对称动物 Hox 基因通常依据它们的相关性而分成不同的组：“前端组”（Hox 1－2）、Hox 3 组、中间组（Hox 4－8）和“后端组”（Hox 9－13；Garcia-Fernàndez 2005）。虽然刺胞类可能存在 Hox 同源基因，但它们的基因组结构和表达与两侧对称类存在着显著差异：

（1）海葵有“前端组”Hox 的同源性基因，但是没有 Hox 3 同源基因。早期研究将一些海葵 Hox 基因归类为“后端组”（Hox 9－13；Finnerty and Martindale 1999，Finnerty et al. 2004），但是最近的研究表明这些和“中间组”（Hox 4－8）及“后端组”Hox 基因高度相似的基因并不具有明确的同源关系（Chourrout et al.，Ryan et al. 2006）。

（2）海葵类 Hox 和非 Hox 基因（*Evx*、*Mnx* 和 *rough*）的染色体连锁在两

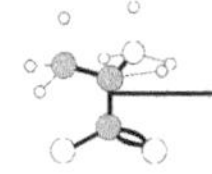

侧对称动物中或多或少有些保守，但海葵的所有其他 Hox 基因簇的出现则可能是物种特异性的基因加倍结果（Chourrout et al. 2006，Kamm et al. 2006），因此，和两侧对称动物对应的 Hox 基因簇可能在海葵类中缺失。虽然偶尔有些不符合该规则，但海葵基因组和两侧对称动物连锁群具有高度保守性。因此，刺胞动物的原始 Hox 基因簇的种内特异性二次缺失不太可能造成 Hox 基因簇的缺失。

（3）绝大多数刺胞类 Hox 基因在内胚层表达（Finnerty et al. 2004，Ryan et al. 2007），相反大部分两侧对称动物 Hox 基因在外胚层表达。

（4）大部分海葵 Hox 基因沿着定向轴交错表达，但是仅少部分在口—反口轴不同部位表达（Finnerty et al. 2004，Ryan et al. 2007）。另外，在珊瑚类浮浪幼体、海葵和水螅虫中，同源基因的表达并不存在进化上的保守性，同源基因也可以在口—反口轴相对立的两端表达（Finnerty et al. 2004，Kamm et al. 2006，Masuda-Nakagawa et al. 2000，Yanze et al. 2001）。这些研究推翻了刺胞类 Hox 基因沿着口—反口轴表达的进化保守模式。

5.5.4　刺胞类与两侧对称动物体轴的同源性比较

Hox 基因在刺胞类和两侧对称类的这些基本差异，并不能否认 Hox 基因在刺胞类胚轴形成中的重要作用，但也说明不了两侧对称类前—后轴与刺胞类口—反口轴或定向轴的直系同源关系。同样，其他的两侧对称胚轴标记基因（例如 BMP）和它们的拮抗基因（见前）、背腹轴标记基因 *goosecoid* 皆表达于海葵前内胚层定向轴两端（Matus et al. 2006a），前端组的同源标记基因 *otx* 则弥散地分布在整个前后轴，这些研究也能得出相同结论（de Jong et al. 2006，Mazza et al. 2007）。刺胞动物和两侧对称动物的胚轴没有直系同源性，可能是因为两侧对称类祖先的躯干发生进化。刺胞类的胚轴起始于胚孔，再经过对侧和顶孔，直接组成前—后轴（Keller et al. 2000，Nielsen 2001），但两侧对称类却大不相同，在进化早期，它们只有一个胚轴即前—后胚孔（AB）轴，随后 AB 轴通过原肠运动（例如会聚延伸运动）转换为 AP 和 DV 轴（Fioroni 1992，Keller et al. 2000，Steinmetz et al. 2007）。例如，在鱼类或青蛙原肠胚孔边缘早期的“组织者”区域神经与中胚衔接处发育为前脑和其后的脊髓（Hirose et al. 2004，Keller 1975，Woo and Fraser 1995）。另外在原肢类的旋毛虫，2d 分裂球在早期原肠胚孔边缘的一端出现，形成整个外胚层体节（Ackermann 2002，Nielsen 2004，Shankland and Seaver 2000）。原肠胚孔虽然在两侧对称动物和刺胞类动物中保守，但是在后口动物中则形成前部的

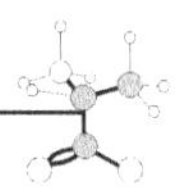

口，而在原口类动物中形成后部的肛门，或者在端盘吸虫（例如一些线虫和环节动物）中形成口和肛门（Arendt and Nübler-Jung 1997，Holland 2000，Nielsen 2001）。因此，就像刺胞类动物 AB 轴一样，两侧对称动物的 AB 轴也无法与 AP 和 DV 轴相提并论（见图 5.4）。

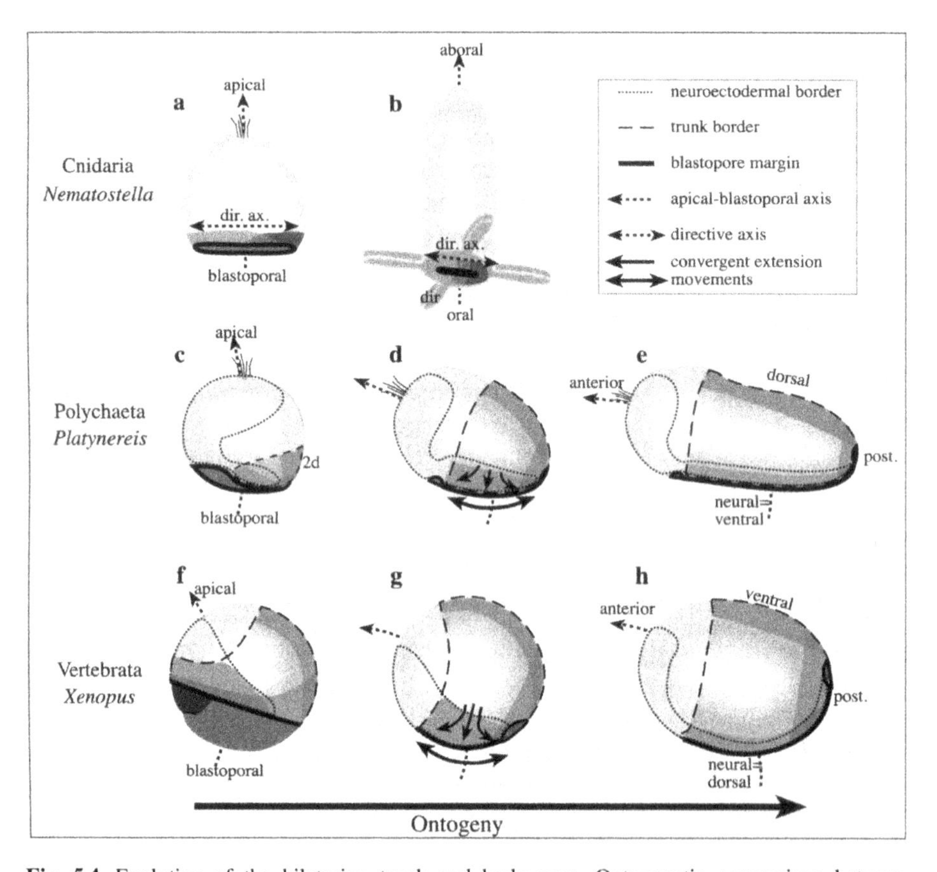

Fig. 5.4 Evolution of the bilaterian trunk and body axes. Ontogenetic comparison between *Nematostella* (**a**, **b**), *Platynereis* (**c–e**) and *Xenopus* (**f–h**) (Keller 1975) as representatives of Cnidaria, Protostomia and Deuterostomia to explain the evolution of the trunk and main body axes in Bilateria (based on Arendt 2004, Arendt and Nübler-Jung 1997, Denes et al. 2007, Shankland and Seaver 2000, Steinmetz et al. 2007). Selected species are considered relatively ancestral and prototypic but do not represent stem species of phyla. Note that in early bilaterian embryos (**c**, **f**), the blastopore margin (*thick black lines*) unifies ventral (*green*), posterior (*blue*) and dorsal (*orange*) fates that get separated by convergent extension movements (**d, e, g, h**). Also note the bending of the initial apical-blastoporal axis (*dotted arrow*) by the passive anterior tilting of the head (*yellow*) due to the proliferation and convergent extension of the 2d descendents that form and elongate the trunk. Orange and blue colouring of the blastopore rim in cnidarians represents the region on one end of the directive axis homologous to the trunk-forming region of Bilateria. *Purple*: endoderm and mesoderm. *Brown*: organizer region

图 5.4　两侧对称动物体节和体轴的演化。通过对海葵（刺胞动物）（a，b）、沙蚕（原口动物）（c-e）和非洲爪蟾（后口动物）（f-h）个体发育的比较（Keller 1975）阐明两

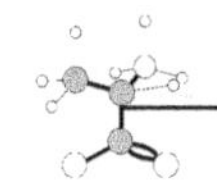

侧对称动物体节和主要体轴的演化过程（基于 Arendt 2004，Arendt and Nübler-Jung 1997，Denes et al. 2007，Shankland and Seaver 2000，Steinmetz et al. 2007）。选择的物种代表祖先类群和原始群，但不代表祖系物种。两侧对称类胚胎早期（c，f）的胚孔边缘（粗黑线）、整合腹侧（绿色部分）、后部（蓝色部分）和背侧（桔色部分）会随着集中延伸运动被分开（d，e，g，h）。AB 轴（虚线箭头）因头前部被动前倾（黄色）而发生弯曲。橙色和蓝色标记的是刺胞类的胚孔边缘，说明其定向轴的一端与两侧对称类的体节形成区域同源。紫色代表内胚层和中胚层；棕色代表组织区

英文注释：Bilaterian，两侧对称动物；Trunk，体节；body Axes，体轴；Ancestral，祖先的；Blastopore，胚孔；Convergent Extension movement，集中延伸运动；Cnidarians，刺胞动物；Endoderm，内胚层；Mesoder，中胚层

相反，刺胞动物的 AB 轴或许和两侧对称类的 AB 轴存在同源性，因为 *Brachyury*（Technau 2001）、*forkhead*（Fritzenwanker et al. 2004）和许多 Wnt（Kusserow et al. 2005）等保守基因都在胚孔处表达，细胞核定位的 β－联蛋白在内胚层内陷点处表达，并且第二 AB 轴的诱导发生在原肠胚孔边缘，以上种种现象均证实这一同源关系的存在（Kraus et al. 2007）。前核的分别重定位可以轻易解答刺胞动物的原肠胚孔和两侧对称动物的植物极孔位置矛盾问题（Lee et al. 2007，Martindale 2005）。体轴演化假说可进一步通过比较早期的刺胞动物口对口轴模式和两侧对称动物前原肠期 AB 轴形成进行验证。

5.5.5　刺胞类中胚层的演化

通过比较刺胞类和两侧对称类的发育过程可以发现，中胚层的起源是动物获得新特性的又一例证，它是两侧对称动物的第三个胚层，起源于双胚层刺胞两侧对称类祖先。虽然内胚层在刺胞动物中完全缺失，但它们具有肌肉细胞，这些细胞主要由两侧对称类的中胚层发育而来（Siewing 1985）。可区分的几种常见刺胞类肌肉细胞有：肌上皮类是刺胞类独有的，并具收缩和感觉、分泌或消化多种功能；水螅水母和水螅珊瑚虫有更独特的肌肉细胞类型，常常呈条纹状，在细胞内有重复的特定收缩单元（Amerongen and Peteya 1980，Schuchert et al. 1993，Siewing 1985）。刺胞类和两侧对称类肌肉结构基因的组成将揭示两侧对称类平滑肌和横纹肌的进化关系，并有助于阐明刺胞类肌肉的类型。

刺胞类和两侧对称类不仅都具有类似于“中胚层”的肌肉组织，很多双侧中胚层和肌肉形成的特定功能转录因子都共同存在于这两个类群中。水螅类的中胚层特异基因 *twist*、*mef2*（Spring et al. 2002）和 *msx*（Galle et al.

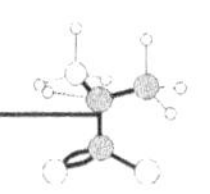

2005）在水螅珊瑚成体形成阶段的“内钟”（entocodon，一种在外胚和内胚之间增值的内胚层细胞团）里表达，“内钟”发育为平滑肌和横纹肌，并且被认为是两侧对称类中胚层的同源物（Seipel and Schmid 2005）。然而，“内钟”是成体组织，形成于无性生殖阶段，因此很难归类为第三胚层。但是钵水母和立方水母肌肉发育阶段不存在“内钟”（Burton 2007）。

海葵的多个“中胚层”转录因子（*mef2*、*twist*、*gata*、*muscle-LIM*）在浮浪幼虫内胚层有不同程度的表达（Martindale et al. 2004）。珊瑚类的内胚层大部分由肌皮细胞组成，而更特化的肌肉（例如咽牵缩肌）连同生殖细胞存在于“中肠组织”，形成内胚层褶皱伸入体腔（Siewing 1985）。“中肠”转录因子功能分析和与两侧对称类同源性比较，对于分析和指导双侧和刺胞类肌肉和中胚层发育之间的基因调控网络的保守性非常必要。有个假说就认为，中胚层是在内胚层和外胚层中间形成的内胚层折叠后的继续延伸，两侧对称动物的肠腔中胚层就是如此形成的（例如文昌鱼、海胆、半索类、环节动物类；Arendt 2004，Remane 1950，Sedgwick 1884，Tautz 2004），因而比较海葵和两侧对称类中肠形态发生的模式组成将会非常有趣。另外，中胚层可能由内胚层细胞形成的间充质细胞进化而来，这类细胞广泛存在于螺旋卵裂动物中（Technau and Scholz 2003）。

5.5.6 刺胞动物“神秘的”复杂性

刺胞动物乍看起来形态很简单，因而当发现海葵基因组比昆虫和线虫的更复杂时，多少让人感到惊讶。然而，仔细观察则表明，一些刺细胞动物（如珊瑚类）的形态可能并不是那么简单，发育的调节、浮浪（幼体）幼虫阶段的形成、两个非对称的体轴的存在均表明刺细胞动物发育的复杂性。另外，很多的细胞类群目前无法通过形态来区分，可能需要分子技术的进一步验证，例如不同神经递质配体和受体的表达定位实验将会显著增加神经细胞的种类。再者，刺胞动物自身扩张了一些基因家族，例如感光的视蛋白（Suga et al. 2008）或者肌肉收缩调节肌球蛋白的轻链，这些或许会引发特定刺胞类群细胞种类的增加。

5.6 蜕皮动物：现有系统之外的动物

蜕皮动物是指所有会产生蜕皮现象动物的总称，其中还包括线虫、节肢动物和蠕虫（Aguinaldo et al. 1997），两个被研究得最为透彻的的分子生物学

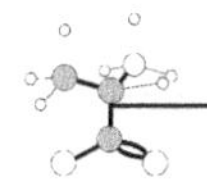

模式生物线虫和果蝇都属于蜕皮动物，在强大的遗传学研究技术支持下，这两种模式生物为生物学研究做出了巨大贡献，其中涵盖了从细胞基本原理到系统学分析等一系列主题，而且，这些研究对我们理解生物学的重要性毋庸置疑。虽然上面罗列了这些研究的意义，但越来越多的证据正在挑战以下这些不言而喻的假说：这些研究结果能适用于大部分的无脊椎动物，或可以通过这些研究结果来推断较简单的祖先动物的代谢过程。相反，这些数据表明，现存的模式蜕皮动物的一些简单特征是祖先复杂结构发生二次简化的结果。

尽管一些蜕皮动物类群很少保留祖先的复杂性，但它们仍进化出迷人的复杂特征。例如，果蝇唐氏综合征细胞粘附分子（*Dscam*）基因的内含子含量很高，在所有已鉴定的生物中剪接变异体数目最多（Schmucker et al. 2000）。黑腹果蝇的 Dscam 能将特定的物质传给迁移神经元，并协调同种抗原排斥，这是建立神经回路的重要前提。Dscam 基因的转录产物在一些群体（包括甲壳动物水蚤）中表现出较高的多样性，表明 Dscam 基因可变剪接导致细胞多样性的形成，这可能是节肢动物拥有的一个基本功能（Brites et al. 2008）。与此相反，脊椎动物中 Dscam 同源基因似乎并没有出现广泛的可变剪接，虽然它们似乎也存在有同种抗原相互作用现象（见综述 Hattori et al. 2008）。

这个例子说明，在衡量动物的复杂性时，即使是基于定量分子分析，仍需要区分它们是保留了祖先的复杂特征还是二次得到的复杂特征。目前还不清楚这两种变化方式的相互关系，因此判定某个物种或类群的复杂性时需要估计它们的进化起源时间。同样值得注意的是，昆虫和线虫都具有非常高的多样性，而且昆虫是动物中物种最丰富的类群（Brusca and Brusca 2003），尽管昆虫丢失了一些复杂的特征，它们却进化出了非常高的形态复杂性和多样性。但是，在海洋环境中昆虫的多样性并不算很高，其他蜕皮动物类群（包括甲壳类动物和自由生活的线虫）在海洋中丰度更高，然而适用于它们的海洋模型和分子学数据仍然非常稀缺。

最近在节肢动物中发现一个新的海洋模式生物——端足目动物，它是研究发育过程的模式动物，已完成基因组测序。与传统的节肢动物模型不同，明钩虾属的早期分裂是完整、不均等和不变的（Gerberding et al. 2002），因此，它的早期分裂类型与果蝇完全不同，后者的早期核分裂会产生合胞体，使得决定分裂类型的分子在其中进行扩散。未来有必要展开深入的遗传学分析，以探讨节肢动物的遗传机制的适用范围。通过与后期亲缘关系较远的节肢动物类群进行比较，可以清楚发现早期节肢动物类群的进化机制相当复杂。同样，明钩虾属的研究可以揭示泛甲壳类原始的四肢发育模式，从而为比较

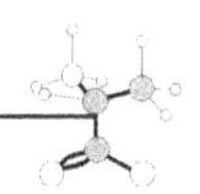

生物学甚至进化学提供基因调控方面的数据（Prpic and Telford 2008），并且大部分的研究均可套用已有的候选基因定位方法，但想全面得到端足目动物的基因组成信息，还将依赖于基因组测序技术。然而明钩虾属基因组比较大（约3.6 Gbp），另一端足目动物 *Jassa slatteryi* 的基因组则相对较小（约690 Mbp），故后者已被提议测序，这将为端足目动物的基因组学研究提供依据。

5.7 冠轮动物：引出新观点的进化分支

如第4章所述，冠轮动物是一个超门，其中还包括软体动物和环节动物（Halanych 2004，Halanych et al. 1995），和蜕皮动物一起代表了大部分的原口动物。虽然冠轮动物代表了两侧对称动物的一个重要分支，但很长时间里一直缺少它们的分子生物学信息，这主要是因为线虫和果蝇这两种蜕皮动物经常被用作研究分子遗传的模式生物。最近几年，表达序列标签（EST）、细菌人工染色体（BAC）和全基因组测序等工具和技术开始探索冠轮动物的基因组信息，从而填补了这个空白。虽然还没有任何一个项目的结果发表，但这些新兴的数据已经提供了冠轮动物生物学一些有趣的信息，也有助于我们了解动物的复杂性。

5.8 海兔：从神经回路到神经转录组学

软体动物是动物界中一个非常大的门类，仅次于节肢动物（Brusca and Brusca 2003），许多软体动物生活在海洋中，此外，软体动物还包括了头足类，这是最高等的无脊椎动物。海兔已成为神经生物学研究的经典模式生物，主要是因为它具有简单的神经回路，且已经对其组成、调节及一些行为的分子介质进行了开拓性的分析（参见综述 Kandel 2001）。同时，海兔具有较大且容易识别的神经元（有些直径能达到1毫米），极大地促进了对其神经系统的研究。最近有研究如何提取这种细胞 RNA，以确定海兔神经系统中特定的转录组，并根据海兔的单种神经元细胞（如大脑细胞 MCC，甚至它的神经元突起）建立了特定文库（Moroz et al. 2006）。作为正在进行的海兔基因组测序的补充办法，EST 取样发现了许多有趣的成分。这种方法能指出一般软体动物的基因谱系，还找到了蜕皮动物模型中漏掉的基因，如锌指转录因子 *churchill*（在 Kortschak et al. 2003 也有论述）、与疼痛和突触可塑性相关的 P2X 受体基因、与 RNA 结合能力相关的硒蛋白 N 同系物或穹窿体主蛋白。此外，还

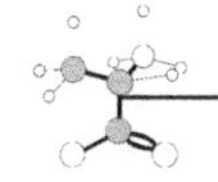

发现了DNA甲基转移酶相关的基因序列、DNA甲基转移酶结合蛋白和转录抑制因子甲基化CpG结合蛋白2。综合这些因子来看，相比于秀丽隐杆线虫或昆虫，海兔拥有一个CpG甲基化通路，是调节复杂基因的一个特殊途径。

这些EST数据集现在可以用来确定细胞类型特异性的转录组，以进一步研究两种神经元之间的差异，或者神经元在刺激前后的差异。这些数据不仅可以解决特定的神经生物学问题，对复杂性比较也有着广泛影响。目前，尚不清楚新的细胞类型是如何在进化中出现的，也不清楚需要多少不同的分子特征才能产生具有独特功能的细胞类型。此外，转录组学也被广泛应用于细胞类型的研究，与相对主观的传统方法相比，转录组学能更好地对细胞中的物质进行定性和定量分析，比传统方法客观得多。反过来，用分子生物学方法区分个体细胞类型时的经验，也可以为评估细胞类型的进化提供新的手段。

5.9 沙蚕：祖先细胞的复杂性和基因组特征

在过去的十年里，海洋环节动物沙蚕逐步成为研究分子进化的合适模型。当前的分子资源包括多于7万条高品质的EST序列和挑选出来的BAC序列，预计2010年底完成全部EST数据集和全基因组测列。沙蚕在很早之前便已成为实验模型系统，目前已经建立起它的基因传递和基因干扰工具，为研究基因表达提供了技术支持（Jekely and Arendt 2007，Tessmar-Raible et al. 2005）。

最近，基于已有的EST和BAC序列，沙蚕的部分转录组和基因组第一次被用于系统性评估其基因结构和蛋白质组演变。这项研究发现沙蚕和人类基因都有一个明显简化的过程，都与它们的外显子/内含子的组成和编码蛋白的进化速度有关（Raible et al. 2005）。这些数据第一次证明了人类大量内含子的原始性，这个观点可通过刺胞动物基因组学分析（*Nematostella*）得到证实（Putnam et al. 2007），也可以通过分析Pax6基因来说明这种现象及对调节复杂性进化的影响，因为Pax6的一个重要功能是参与原口动物和后口动物的感光系统细胞的特化过程（kozmik 2005）。

此外，小鼠缺乏这种基因，这表明了pax 6基因在胰腺α细胞（分泌胰高血糖素）的分化中至关重要（St-Onge et al. 1997）。两侧对称动物的基因结构具有较高的保守性，例如沙蚕的Pax 6基因在配对域含有一个内含子，脊椎动物在这个位置也有一个内含子（见Raible et al. 2005 和图5.5）。这个内含子帮助脊椎动物生成两个功能不同的剪接变异体，但在海鞘和沙蚕基因组中却没有发现，因此该内含子一直被认为是脊椎动物所特有的（Callaerts et al.

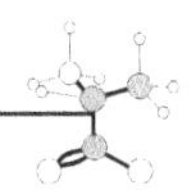

1997)。沙蚕的相关数据表明，原始两侧对称动物的 Pax 6 基因已经有内含子，以便在该位点产生可变剪接变异体，但还没有任何实验证明在沙蚕中存在明确的剪接变异体。然而在果蝇中观察到两组同源基因的 N－末端部分非常独特，且与果蝇基因组中的原始剪接变异体相一致（图 5.5）。此外，脊椎动物 Pax 6（*5a*）基因可以替代果蝇的 *eyg* 基因（Dominguez et al. 2004），说明这两个基因经过平行进化后行使相同功能。值得注意的是，最近有研究者分离到了文昌鱼可变剪接变异体，这与配对域的原始剪接发生在脊椎动物出现之前的观点相吻合（Short and Holland 2008）。

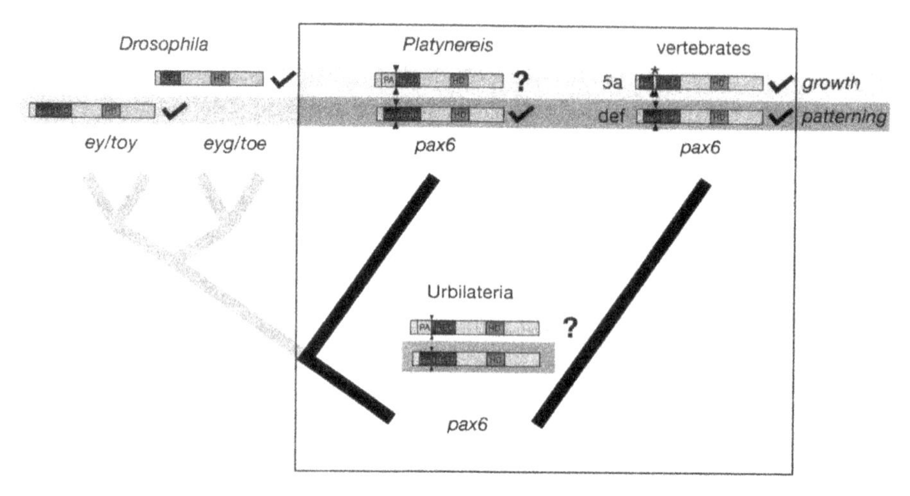

Fig. 5.5 Complex ancestral gene structures and the splicing potential of regulatory genes. Distinct *pax6* variants (*yellow* and *purple backround*, respectively) differ in the integrity of the N-terminal DNA binding domain and are preferentially associated with growth and patterning, respectively. Vertebrate variants are generated as splice isoforms, making use of the separation of the N-terminus into two exons, one of which encodes most of the N-terminal PAI domain that is structurally changed upon insertion of the additional 5a exon (*asterisk*). In contrast, *Drosophila* possesses distinct sets of orthologues lacking the respective intron site. Comparative analysis of the *Platynereis pax6* locus (*middle*) indicates that the intron (*black arrowheads*) within the N-terminal region (PAI) of the PAIRED domain is ancestral for Bilateria and was secondarily lost in the four *Drosophila pax6* orthologs (*left*). Therefore, one alternative scenario is that Urbilateria already generated a functional equivalent of Pax6(5a) (marked by "?") that became fixed as independent variants in the fly genome, but is subject to alternative splicing in both annelids and vertebrates

图 5.5　祖先基因的复杂结构和调控基因的拼接潜能。不同的 Pax6 变体（分别为黄色和紫色背景）的 N 端 DNA 结合域的完整性不同，并分别与生长和成型紧密相关。脊椎动物的变体以剪接亚型的形式产生，其 N 端包含两个外显子，其中一个编码大部分的 N 端 PAI 域，会随着额外 5a 外显子（星号）的插入而发生结构改变。相比之下，果蝇有类似的同源基因，但缺少相应的内含子位点。通过比较分析沙蚕 Pax6 基因的位置（中间）可以看出，配对区域中 N 端（PAI）内含子（黑色箭头所指）遗传到了两侧对称动物中，但在果蝇 Pax6 基因的四个同源序列中发生二次缺失（左边）。另一种情况是，原始两侧对称

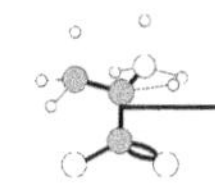

动物已经产生了一种与 Pax6 - 5a（“?”所标志）功能相同的基因，在果蝇的基因组中成为独立的同源基因稳定表达，但在环节动物和脊椎动物中被选择性剪接

英文注释：N-terminal，氨基末端；Splicing，拼接，剪接；Integrity，完整性；Functional Equivalent，功能对等；Annelids，环节动物

5.10 可变剪接：调节基因组复杂性的基本层面

除了上面提到的具体的例子，与动物进化相关的复杂基因结构不断被发现，从而引出了一个更一般性的问题，即转录组复杂性的进化及其与动物复杂性的关系。从早期基因组的计算到第一个基因组计划的完成，越来越多的数据显示无论是基因组的规模还是蛋白质编码基因的数量都和动物表型的复杂性没有很大关系：复杂的后生动物基因组比原生动物的基因组要小（C 值矛盾），复杂的脊椎动物的总基因数目不一定比无脊椎动物多（n 或 g 值矛盾）（Claverie 2001，Hahn and Wray 2002）。可变剪接在不改变基因数量的基础上直接调节蛋白质组复杂性（见综述 Maniatis and Tasic 2002），成为调节细胞类型复杂性的有效手段。虽然难以获得动物全部的可变转录产物，但第一个全球性的分析至少表明了即便关系很近的物种（如人类和黑猩猩）在 6%～8% 的同源基因外显子的剪接中也存在着显著的差异（Calarco et al. 2007），相比于所记录的两个物种间获得或丢失基因的案例，这个数据已经非常庞大（黑猩猩的测序和分析，2005），这表明可变剪接的变化可能有助于加大物种间复杂性的变化，甚至有助于物种本身的形成。因为已有的模型系统所涉及的剪接变异体数目极其有限，并且受到其他因素的影响（如前文提到泛甲壳类动物 Dscam 的亚型多样性是不同品系的剪接变异体独立发展的结果），所以从较长的进化时间尺度上来比较可变剪接变异体受到很大限制（Boue et al. 2003，Brett et al. 2002）。随着高通量测序技术的进步，剪接变异体数目较少的问题不那么重要，而判断不同组织水平上物种的可变剪接的范围是否存在系统性的差异变得更加有趣。

5.11 海胆：后口动物基部意想不到的功能类群

海胆是经常被海洋生物学家使用的最古老的实验模型之一。20 世纪下半叶，生化和分子水平上出现了很多用于阐明转录调控网络等开创性工作的实验方法，紫色海胆因此成为分析发育中基因调控网络的一个主要的模型系统。

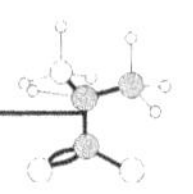

近期对大连紫海胆的基因组测序（海胆基因组测序等，2006）进一步推动海胆的研究，这也是第一次深入探讨棘皮动物的基因组。本章主要关注三个方面的问题：第一，海胆基因组草图显示海胆的基因数量极多，包括222个Toll样受体，数目至少是人类基因组的20倍，根据结构域的组成来判断它们可能与先天免疫相关（Hibino et al. 2006，Rast et al. 2006），虽然还没有对这些受体进行直接的功能检测，但这些数字已经表明在海胆种系进化过程中Toll样受体家族经历了剧烈的二次扩张。第二，海胆基因组包含数百个快速进化的G蛋白偶联受体（GPCRs），它们的组织模式和表达类型说明它们可能是化学感应受体（Raible et al. 2006）。通过对比免疫相关基因发现，G蛋白偶联受体家族的剧烈扩张与之前认为海胆只拥有一套基本的感觉受体的观念相悖。

海胆基因组的第三个基本问题是，它同时具有两边对称的幼体时期和放射对称的胚胎后期，胚胎后期的细胞与幼体时期存在很大不同，相当于一套相同基因组编码出两种完全不同的表型。虽然一些研究试图分析胚胎后期发育中基因的活动，但仍然知之甚少。然而转录组分析表明，与信号传导相关的一系列基因相类似，海胆基因组编码的近80%的转录因子早在胚胎发生期就已相当活跃（Howard-Ashby et al. 2006）。这些数据支持相同的基因组可以通过不同的使用方式来编码完全不同外形的观点，说明不能将基因组成与复杂的基因调控过程简单地联系起来。

5.12 文昌鱼和脊索动物原型

文昌鱼是头索动物的一个小类群，是经典的模式生物之一。因为文昌鱼在系统位置上处于无脊索的后口动物和脊椎动物中间，它们在研究比较进化上包含有大量的信息，并长期用来追溯许多脊椎动物特征的起源（见综述Garcia-fernàndez，Bentio-gutiérrez. 2009）。文昌鱼与脊椎动物有许多共同的特征，如背神经索以及脊索（为游泳幼虫提供稳定性），但他们缺乏脊椎动物中保护背神经索的脊柱。此外，文昌鱼有穿孔咽（咽缝）和双边分段的肌肉，称为肌节，这可与脊椎动物的肌节相媲美。但文昌鱼缺少四肢、神经嵴细胞、成对的感受器官和复杂的前脑，与这些特征出现的时间较晚相吻合。

最近发布的文昌鱼基因组（Holland et al. 2008，Putnam et al. 2008）是对一系列有趣研究的延伸，进一步揭示了沿后口动物—脊索动物—脊椎动物的分子进化谱系：文昌鱼只有一个Hox基因簇（garcia-fernàndez and Holland 1994），而脊椎动物有四个Hox基因簇，这表明头索动物基因组在进行两轮全

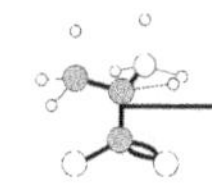

基因组复制（“2R 假说”）之前仍是“原始的”基因组。同样，文昌鱼有一个完整的 Para-Hox 基因簇（Brooke et al. 199），其他种系中该基因簇发生了退化，证明文昌鱼的基因组保留了原始的特征，但其他种系则发生了二次改变。对文昌鱼全基因组测序也验证了这一观点。文昌鱼的基因位点与脊椎动物的相应基因位点有显著的共线性，比较文昌鱼和人类的基因组可以检测出它们的染色体之间具有宏观共线性，并且脊索动物祖先至少有 17 个连锁群（Putnam et al. 2008），每个都包含四个脊椎动物物种，再次验证了“2R 假说”，并强调文昌鱼可以很好地代表原始的脊索动物。

在文昌鱼基因组中另一个重大发现是，脊椎动物基因组中至少存在 50 个高度保守的非编码元件（Putnam et al. 2008），这个数字已经排除保守的基因 UTR 序列，但包括超级保守的增强子。在比较脊椎动物的基因组时鉴定了至少 3000 个类似的元件，它们可能跟发育相关（Pennacchio et al. 2006，Woolfe et al. 2005）。通过评估这些元件在其他模式生物（如小鼠）中的调控能力，可以检验它们能否代表所有脊索动物的关键调控关系。

5.13 海鞘：发育过程中的改变和不变

玻璃海鞘常被用于动物复杂性的比较。系统发育分析表明，被囊动物与脊椎动物的亲缘关系最近（Delsuc et al. 2006，2008），这意味着它们是检测脊椎动物演化特征的合适模型。值得注意的是，虽然海鞘幼体像其他的脊索动物（如青蛙）一样呈蝌蚪型，但只有近2600个细胞，要比其他脊椎动物少至少一个数量级。另外，海鞘基因组大小只有小鼠基因组的 5%，因此与脊椎动物相比，海鞘似乎具有较少的非编码基因和细胞类型，却拥有组织复杂性相似的幼体。比较海鞘和脊椎动物模式生物的调控过程有助于我们理解调控过程的复杂性和发育过程的可塑性。

海鞘的早期基因表达模式可以通过细胞解析方法来绘制（Tassy et al. 2006），外加已有的功能分析工具，不仅可以分析和比较海鞘中发育标记基因的表达情况，还可以系统地评估海鞘基因的功能。高解析度的基因表达数据和功能分析技术的联合使用已成为分析海洋脊索动物发育过程中基因调控网络的强大工具，获得的信息还可以与其他进化支相比较（Imai et al. 2006，Satou and Satoh 2006，Shoguchi et al. 2008）。更重要的是，这种比较可用于发现脊索动物中基因调控网络不同程度的保守性。一些因子似乎在海鞘和脊椎动物胚胎发育过程中发挥了相同的作用，例如在两个系统中，成纤维细胞生

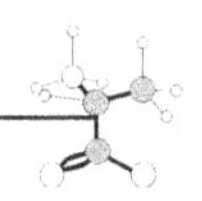

长因子可以促进 T-box 转录因子 Brachyury 的表达和脊索的形成，表明脊索动物特征性发育过程由一套保守的调控系统调控（Imai et al. 2002）。同样，编码脊索结构成分（如 II 型胶原或蛋白聚糖）的操纵基因在海鞘和脊椎动物中十分保守（Hotta et al. 2008）。相反，不同生物间该过程的上游调节因子及它们之间的关系差异似乎非常大（Imai et al. 2006，Lemaire 2006）。基因调控网络保守性和差异性两方面的对比引发了一个问题，即早期发育过程是如何“衔接”的，为什么发育过程中某些水平的变化比其他水平多？虽然发育过程的最终“结果”都是游泳的蝌蚪（这里指幼体性状极像蝌蚪），但这些问题的答案很可能为探究形态和细胞复杂性的分子联系提供见解。

5.14 展望

研究新的海洋生物模型为我们理解动物复杂性提供了新的观点，作为对传统复杂性特征（如分化细胞类型的数目）的补充，基因组学为生物复杂性提供了不同的研究手段。从基础水平上讲，即对生物体遗传网络下所有蛋白质结构域、基因、内含子和转录本的数目研究，这些特征易于比较，并有助于对不同动物类群中存在的结构域、基因和内含子进行更深入的追踪溯源，这些研究结果具有重要的理论意义。

丢失原始复杂性的进化原理：新模型系统的研究反复验证了存在于原始基因组的某些特征在某些进化谱系中逐渐丢失。这一发现改变了以往有关动物进化的观点。显然，丢失原始特征是进化过程中很正常的现象，正如基因复制和修饰是新基因的来源，包括调控基因（如转录因子）以及在动物发育中具有重要影响的细胞外信号和因子（功能分析方面）。因此，原始复杂性的丢失对进化过程的影响是一个非常有趣的问题。

不同的进化速度：我们注意到，基因丢失是导致物种进化的原因，同时不同的动物类群在丢失程度上也有明显的差别。例如，环节动物和头索动物的进化速度比较慢，海鞘的进化速度非常快。这些海洋生物的例子说明进化的快慢与生物是否生活在大海没有直接关系，但把重点放在海洋物种上有助于系统地填补动物在系统发育中的差距，这更有助于我们理解哪些是原始特征，哪些是后来获得的特征。

调控网络：现代基因组学方法更多的关注简单特性（如基因和内含子的存在与否），但正如引言中所说，这种层面的研究只是衡量复杂性的第一步。单一个体如何聚集在一起形成调控网络？调控网络各成分的复杂性怎样影响

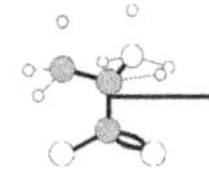

调控网络本身的复杂性？解决这类问题需要一系列可以结合表达分析和基因干扰的方法（如基因芯片或高通量测序），前文也提到这些方法已成功应用于一些海洋生物模型（如玻璃海鞘、海葵、海胆、文昌鱼或沙蚕）并帮助阐明不同门类生物的调节网络的基本构成及复杂性。

复杂性的经典特征间的分子联系：另一个有待解决的问题是分析复杂性的分子学新方法、衡量形态复杂性的传统方法和细胞类型多样性之间的联系。例如，使多细胞发生演变的分子相关因素是什么？是复杂的发育过程调控网络相关的细胞分化类型的数目，还是动物体内有独立的"调控能力"（如依靠个体的发育模式）？到目前为止，我们对细胞分化类型的了解仅限于少数细胞（如淋巴细胞和胰腺β细胞），仍需要做很多基础工作以便在分子水平上确定不同细胞类型间的联系。反过来，这些分析也可以确定外形相似的细胞是否在分子水平上代表了不同的细胞类型。

不同调控层次上的进化：越来越多的证据表明，非编码基因（以前称为"垃圾基因"）有着重要的调控功能，如提供转录因子的结合位点、作为后期基因修饰的底物、产生具有调控功能的非编码转录产物等等。这些方面都有助于正确地理解基因组和形态的复杂性（Hahn and Wray 2002，Levine and Tjian 2003，Mattick 2007）。例如，在双翅目昆虫的翼中，对顺式调控元件的修饰与其复杂的色素类型相关（见综述 Prud homme et al. 2007）。此外，非编码的 miRNA 已被证实具有转录后调控的功能。鉴于许多 miRNA 在不同组织或细胞类型中有特定的表达方式，它们可能是细胞复杂性的一个指标，甚至是决定因素。一些研究试图将 miRNA 的存在与否与动物复杂程度关联起来，用以重建复杂脊椎动物的 miRNA 清单，代替在基部有较多分支的不完整清单（Grimson et al. 2008，Heimberg et al. 2008）。由于缺乏可比较的不同类群 miRNA 数据库，很难评估实验覆盖范围多大程度上能与这种分析结果相一致。

评估同源性和形态进化的新方法：测序技术的革命性进步以及可以使用新序列数据的分子模型系统的建立，为再次使用分子生物学技术处理之前的诸多问题铺平了道路。一个基本的影响是更精确地描述了同源性，获得的分子标记有助于在分子水平上区分和比较组织和细胞，例如追踪刺胞动物中胚层的特点，并使用新数据区分其他类群中不同情况下中胚层与各组织的关系，在这些基础上用分子生物学技术分析组织的发育调控网络、鉴别生物形态发生、改变以及表型相关的分子学因素。最终，这些研究可将基因组与其编码的能用于物种鉴定的特征联系起来。

了解动物多样性：本章所列举的新海洋生物模型大大增加了我们对进化

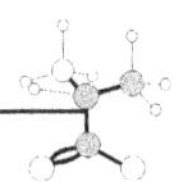

的认识，但它们仍然只代表了丰富的海洋物种中的很小一部分。使用新的基因组学方法来解决动物多样性问题具有非常大的潜力。一方面，模式动物的序列信息可以帮助我们理解表型差异（如形态或颜色）是如何通过基因编码而形成的，另一方面，即使密切相关、形态相似的物种，在遗传网络方面也似乎存在着不同。在这两个方面，基因组学技术已经提供了一个非常有帮助的手段，在未来的研究里将大有用武之地。这些研究在一定程度上将有助于我们把海克尔的“生命之树”描绘得更加精确，并根据不同动物的进化路径来更好地反映出它们的起源。

参考文献

Abedin M, King N (2008) The premetazoan ancestry of cadherins. Science 319:946 – 948

Aburomia R et al (2003) Functional evolution in the ancestral lineage of vertebrates or when genomic complexity was wagging its morphological tail. J Struct Funct Genomics 3:45 – 52

Ackermann C (2002) Markierung der Zellinien im Embryo von *Platynereis*. In: Fachbereich biologie, ed. Mainz: Johannes Gutenberg-Universität

Adamska M et al (2007a) Wnt and TGF – β expression in the sponge *Amphimedon queenslandica* and the origin of metazoan embryonic patterning. PLOS One 2:e1031

Adamska M et al (2007b) The evolutionary origin of hedgehog proteins. Curr Biol 17: R836 – R837 Adell T et al (2003) Isolation and characterization of two T-box genes from sponges, the phylogenetically oldest metazoan taxon. Dev Genes Evol 213:421 – 434

Adell T, Müller WEG. (2004) Isolation and characterization of five Fox (Forkhead) genes from the sponge *Suberites domuncula*. Gene 334:35 – 46

Adell T, Müller WEG. (2005) Expression pattern of the Brachyury and Tbx2 homologues from the sponge *Suberites domuncula*. Biol Cell 97:641 – 650

Aguinaldo AM et al (1997) Evidence for a clade of nematodes, arthropods and other moulting animals. Nature 387:489 – 493

Amerongen HM, Peteya DJ (1980) Ultrastructural study of two kinds of muscle in sea anemones: the existence of fast and slow muscles. J Morphol 166:145 – 154

Arendt D (2004) Comparative aspects of gastrulation. In: Stern C (ed) Gastrulation, edn. Cold Spring Harbor Laboratory Press, Cold SPring Harbor, New York

Arendt D (2005) Genes and homology in nervous system evolution: comparing gene functions, expression patterns, and cell type molecular fingerprints. Theory Biosci 124:185 – 197

Arendt D (2008) The evolution of cell types in animals: emerging principles from molecular studies. Nat Rev Genet 9:868 – 882

Arendt D, Nübler-Jung K (1997) Dorsal or ventral: similarities in fate maps and gastrulation patterns in annelids, arthropods and chordates. Mech Dev 61:7 – 21

Bell G (1997) Size and complexity among multicellular organisms. Biol J Linnean Soc 60: 345 – 363

Bijlsma MF et al (2004) Hedgehog: an unusual signal transducer. Bioessays 26:387 – 394

Bonner JT. (1988) The evolution of complexity. Princeton University Press, Princeton, NJ

Borchiellini C et al (2004) Molecular phylogeny of demospongiae: implications for classification and scenarios of character evolution. Mol Phylogenet Evol 32:823 – 837

Borchiellini C et al (2001) Sponge paraphyly and the origin of Metazoa. J Evol Biol 14:171 – 179

Boue S et al (2003) Alternative splicing and evolution. Bioessays 25:1031 – 1034

Brett D et al (2002) Alternative splicing and genome complexity. Nat Genet 30:29 – 30

Brites D et al (2008) The Dscam homologue of the crustacean *Daphnia* is diversified by alternative splicing like in insects. Mol Biol Evol 25:1429 – 1439

Brooke NM et al (1998) The ParaHox gene cluster is an evolutionary sister of the Hox gene cluster. Nature 392:920 – 922

Brusca RC, Brusca GJ. (2003) Invertebrates. Sinauer Associates. Sunderland, Massachusetts. http://www. sinauer. com/detail. php? id = 0973

Bullock TH, Horridge GA (1965) Structure and function in the nervous system of invertebrates. San Francisco: Freeman

Burton PM (2007) Inisghts from diploblasts; the evolution of mesoderm and muscle. J Exp Zool (Mol Dev Evol) 308B:1 – 10

Bütschli O. (1883 – 1887) Klassen und Ordnungen des Thier-Reichs. Winter, C. F., Leipzig

Calarco JA et al (2007) Global analysis of alternative splicing differences between humans and chimpanzees. Genes Dev 21:2963 – 2975

Callaerts P et al (1997) PAX – 6 in development and evolution. Ann Rev Neurosci 20:483 – 532

Cañestro C et al (2007) Evolutionary developmental biology and genomics. Nat Rev Genet 8: 932 – 942

Carr M et al (2008) Molecular phylogeny of choanoflagellates, the sister group to Metazoa. PNAS 105:16641 – 16646

Chimpanzee Sequencing and Analysis C (2005) Initial sequence of the chimpanzee genome and comparison with the human genome. Nature 437:69 – 87

Chourrout D et al (2006) Minimal ProtoHox cluster inferred from bilaterian and cnidarian Hox complements. Nature 442:684 – 687

Clark H (1866) Note on the infusoria flagellate and the spongiae ciliatae. Am J Sci 1:113 –

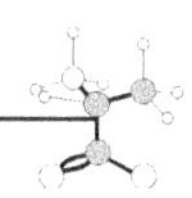

114

Clark H (1868) On the Spongiae ciliatae as *Infusoria flagellata*, or observations on the structure, animality and relationship of *Leucosolenia botryoides* Bowerbank. Ann Mag Nat Hist 4:133 - 142, 188 - 215, 250 - 264

Claverie JM (2001) Gene number. What if there are only 30000 human genes? Science 291: 1255 - 1257

Collins AG (1998) Evaluating multiple alternative hypotheses for the origin of Bilateria: an analysis of 18S rRNA molecular evidence. Proc Natl Acad Sci USA 95:15458 - 15463

Collins AG (2002) Phylogeny of Medusozoa and the evolution of cnidarian life cycles. J Evol Biol 15:418 - 432

da Silva FB et al (2007) Phylogenetic position of Placozoa based on large subunit (LSU) and small subunit (SSU) rRNA genes. Genet Mol Biol 30:127 - 132

de Jong DM et al (2006) Components of both major axial patterning systems of the Bilateria are differentially expressed along the primary axis of a 'radiate' animal, the anthozoan cnidarian *Acropora millepora*. Dev Biol 298:632 - 643

De Robertis EM, Sasai Y (1996) A common groundplan for dorsoventral patterning in Bilateria. Nature 380:37 - 40

Dellaporta SL et al (2006) Mitochondrial genome of *Trichoplax adhaerens* supports placozoa as the basal lower metazoan phylum. Proc Natl Acad Sci USA 103:8751 - 8756

Delsuc F et al (2006) Tunicates and not cephalochordates are the closest living relatives of vertebrates. Nature 439:965 - 968

Delsuc F et al (2008) Additional molecular support for the new chordate phylogeny. Genesis 46:592 - 604

Denes AS et al (2007) Molecular architecture of annelid nerve cord supports common origin of nervous system centralization in Bilateria. Cell 129:277 - 288

Derelle R et al (2007) Homeodomain proteins belong to the ancestral molecular toolkit of eukaryotes. Evol Dev 9:212 - 219

Dominguez M et al (2004) Growth and specification of the eye are controlled independently by Eyegone and Eyeless in *Drosophila melanogaster*. Nat Genet 36:31 - 39

Ereskovsky AV, Dondua AK (2006) The problem of germ layers in sponges (Porifera) and some issues concerning early metazoan evolution. Zoologischer Anzeiger 245:65 - 76

Finnerty JR, Martindale MQ (1999) Ancient origins of axial patterning genes: Hox genes and ParaHox genes in the Cnidaria. Evol Dev 1:16 - 23

Finnerty JR et al (2004) Origins of bilateral symmetry: Hox and dpp expression in a sea anemone. Science 304:1335 - 1337

Fioroni P (1992) Allgemeine und vergleichende Embryologie. Springer, Berlin, Heidelberg,

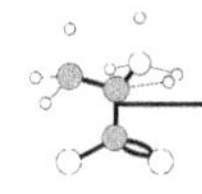

New York

Fritzenwanker JH et al (2004) Analysis of *forkhead* and *snail* expression reveals epithelial-mesenchymal transitions during embryonic and larval development of *Nematostella vectensis*. Dev Biol 275:389 – 402

Galle S et al (2005) The homeobox gene Msx in development and transdifferentiation of jellyfish striated muscle. Int J Dev Biol 49:961 – 967

Garcia-Fernàndez J (2005) The genesis and evolution of homeobox gene clusters. Nat Rev Genet 6:881 – 892

Garcia-Fernàndez J, Bentio-Gutiérrez E (2009) It's a long way from amphioxus: descendants of the earliest chordate. Bioessays 31:665 – 675

Garcia-Fernàndez J, Holland PWH (1994) Archetypal organization of the amphioxus hox gene-cluster. Nature 370:563 – 566

Gerberding M et al (2002) Cell lineage analysis of the amphipod crustacean *Parhyale hawaiensis* reveals an early restriction of cell fates. Development 129:5789 – 5801

Gregory TR (2005) Genome size evolution in animals. In: Gregory TR (ed) The evolution of the genome, 1st edn. Elsevier, San Diego

Grell KG (1971a) Embryonalentwicklung bei *Trichoplax adherens* F. E. Schulze. Naturwiss 58:507

Grell KG (1971b) *Trichoplax adherens*: F. E. Schulze und die Entstehung der Metazoen. Naturwiss Rundschau 24:160 – 161

Grell KG (1972) Eibildung und Furchung von *Trichoplax adherens* F. E. Schulze (Placozoa). Z Morph Tiere 73:297 – 314

Grell KG, Ruthman A (1991) Placozoa, Porifera, Cnidaria and Ctenophora. In Harrisson FW, Westfall JA (eds) Microscopic anatomy of invertebrates. Wiley-Liss, New York

Grimson A et al (2008) Early origins and evolution of microRNAs and Piwi – interacting RNAs in animals. Nature 455:1193 – 1197

Grunz H (2004) The vertebrate organizer. Springer, Berlin Heidelberg

Haeckel E (1874) Die Gastraea-Theorie, die phylogenetische Classification des Thierreiches und die Homologie der Keimblätter. Jena Z. Naturwiss 8:1 – 55

Haeckel E (1903) Anthropogenie oder Entwickelungsgeschichte des Menschen. Keimes – und Stammes – Geschichte. Wilhelm Engelmann, Leipzig

Haen KM et al (2007) Glass sponges and bilaterian animals share derived mitochondrial genomic features: a common ancestry or parallel evolution? Mol Biol Evol 24:1518 – 1527

Hahn MW, Wray GA (2002) The g-value paradox. Evol Dev 4:73 – 75

Halanych KM (2004) The new view of animal phylogeny. Ann Rev Ecol Evol Sys 35:229 – 256

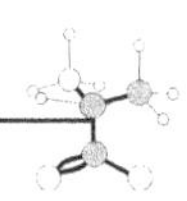

Halanych KM et al (1995) Evidence from 18S ribosomal DNA that the lophophorates are protostome animals. Science 267:1641 - 1643

Halder G et al (1995) Induction of ectopic eyes by targeted expression of the *eyeless* gene in *Drosophila*. Science 267:1788 - 1792

Hattori D et al (2008) Dscam-mediated cell recognition regulates neural circuit formation. Annu Rev Cell Dev Biol 24:597 - 620

Hayward DC et al (2002) Localized expression of a dpp/BMP2/4 ortholog in a coral embryo. Proc Natl Acad Sci USA 99:8106 - 8111

Heimberg AM et al (2008) MicroRNAs and the advent of vertebrate morphological complexity. Proc Natl Acad Sci USA 105:2946 - 2950

Hibino T et al (2006) The immune gene repertoire encoded in the purple sea urchin genome. Dev Biol 300:349 - 365

Hirose Y et al (2004) Single cell lineage and regionalization of cell populations during Medaka neurulation. Development 131:2553 - 2563

Holland LZ (2000) Body-plan evolution in the Bilateria: early antero-posterior patterning and the deuterostome-protostome dichotomy. Curr Opin Genet Dev 10:434 - 442

Holland LZ et al (2008) The amphioxus genome illuminates vertebrate origins and cephalochordate biology. Genome Res 18:1100 - 1111

Hotta K et al (2008) Brachyury-downstream gene sets in a chordate, *Ciona intestinalis*: integrating notochord specification, morphogenesis and chordate evolution. Evol Dev 10:37 - 51

Howard-Ashby M et al (2006) High regulatory gene use in sea urchin embryogenesis: Implications for bilaterian development and evolution. Dev Biol 300:27 - 34

Imai KS et al (2006) Regulatory blueprint for a chordate embryo. Science 312:1183 - 1187

Imai KS et al (2002) Early embryonic expression of FGF4/6/9 gene and its role in the induction of mesenchyme and notochord in *Ciona savignyi* embryos. Development 129:1729 - 1738

Jager M et al (2005) Expansion of the SOX gene family predated the emergence of the Bilateria. Mol Phylogenet Evol 39:468 - 477

Jaillon O et al (2004) Genome duplication in the teleost fish *Tetraodon nigroviridis* reveals the early vertebrate proto-karyotype. Nature 431:946 - 957

Jakob W et al (2004) The Trox-2 Hox/ParaHox gene of *Trichoplax* (Placozoa) marks an epithelial boundary. Dev Genes Evol 214:170 - 175

Jekely G, Arendt D (2007) Cellular resolution expression profiling using confocal detection of NBT/BCIP precipitate by reflection microscopy. Biotechniques 42:751 - 755

Kamm K et al (2006) Axial patterning and diversification in the cnidaria predate the Hox system. Curr Biol 16:920 - 926

Kandel ER (2001) The molecular biology of memory storage: a dialogue between genes and

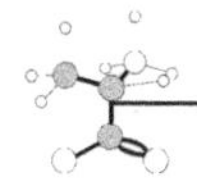

synapses. Science 294:1030 – 1038

Keller R et al (2000) Mechanisms of convergence and extension by cell intercalation. Philos Trans R Soc Lond B Biol Sci 355:897 – 922

Keller RE (1975) Vital dye mapping of the gastrula and neurula of *Xenopus laevis*. I. Prospective areas and morphogenetic movements of the superficial layer. Dev Biol 42:222 – 241

King N et al (2003) Evolution of key cell signaling and adhesion protein families predates animal origins. Science 301:361 – 363

King N et al (2008) The genome of the choanoflagellate *Monosiga brevicollis* and the origin of metazoans. Nature 451:783 – 788

Kortschak RD et al (2003) EST analysis of the cnidarian *Acropora millepora* reveals extensive gene loss and rapid sequence divergence in the model invertebrates. Curr Biol 13:2190 – 2195

Kozmik Z (2005) Pax genes in eye development and evolution. Curr Opin Genet Dev 15:430 – 438

Kozmik Z et al (2003) Role of Pax genes in eye evolution: a cnidarian *PaxB* gene uniting Pax2 and Pax6 functions. Dev Cell 5:773 – 785

Kraus Y et al (2007) The blastoporal organiser of a sea anemone. Curr Biol 17:R874 – R876

Kusserow A et al (2005) Unexpected complexity of the Wnt gene family in a sea anemone. Nature 433:156 – 160

Lang BF et al (2002) The closest unicellular relatives of animals. Curr Biol 12:1773 – 1778

Larroux C et al (2007) The NK homeobox gene cluster predates the origin of Hox genes. Curr Biol 17:706 – 710

Larroux C et al (2006) Developmental expression of transcription factor genes in a demosponge: insights into the origin of metazoan multicellularity. Evol Dev 8:150 – 173

Larroux C et al (2008) Genesis and expansions of metazoan transcription factor classes. Mol Biol Evol 25:980 – 996

Leadbeater BSC (1983) Life-history and ultrastructure of a new marine species of *Proterospongia* (Choanoflagellida). J Mar Biol Assoc UK 63:135 – 160

Lee PN et al (2007) Asymmetric developmental potential along the animal-vegetal axis in the anthozoan cnidarian, *Nematostella vectensis*, is mediated by Dishevelled. Dev Biol 310:169 – 186

Lemaire P (2006) Developmental biology. How many ways to make a chordate? Science 312: 1145 – 1156

Levine M, Tjian R (2003) Transcription regulation and animal diversity. Nature 424: 147 – 151

Leys SP (2004) Gastrulation in sponges. In Stern CD (ed) Gastrulation. Cold Spring Harbor Laboratory Press, Cold Sping Harbor, New York

Leys SP, Degnan BM (2001) Cytological basis of photoresponsive behavior in a sponge larva.

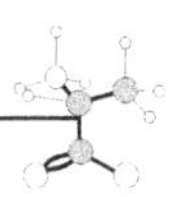

Biol Bull 201:323 – 338

Leys SP, Ereskovsky AV (2006) Embryogenesis and larval differentiation in sponges. Can J Zool 84:262 – 287

Maniatis T, Tasic B (2002) Alternative pre-mRNA splicing and proteome expansion in metazoans. Nature 418:236 – 243

Manning G et al (2008) The protist, *Monosiga brevicollis*, has a tyrosine kinase signaling network more elaborate and diverse than found in any known metazoan. PNAS 105:9674 – 9679

Manuel M, Le Parco Y (2000) Homeobox gene diversification in the calcareous sponge, *Sycon raphanus*. Mol Phylogenet Evol 17:97 – 107

Manuel M et al (2004) Comparative analysis of Brachyury T – domains, with the characterization of two new sponge sequences, from a hexactinellid and a calcisponge. Gene 340:291 – 301

Marlow HQ et al (2009) Anatomy and development of the nervous system of *Nematostella vectensis*, an anthozoan cnidarian. Dev Neurobiol 69:235 – 254

Martindale MQ (2005) The evolution of metazoan axial properties. Nat Rev Genet 6:917 – 927 Martindale MQ et al (2004) Investigating the origins of triploblasty: ‘mesodermal’ gene expression in a diploblastic animal, the sea anemone *Nematostella vectensis* (phylum, Cnidaria; class, Anthozoa). Development 131:2463 – 2474

Martinelli C, Spring J (2003) Distinct expression patterns of the two T-box homologues Brachyury and Tbx2/3 in the placozoan *Trichoplax adhaerens*. Dev Genes Evol 213:492 – 499

Martinelli C, Spring J (2004) Expression pattern of the homeobox gene Not in the basal metazoan *Trichoplax adhaerens*. Gene Expr Patterns 4:443 – 447

Masuda-Nakagawa LM et al (2000) The HOX – like gene Cnox2 – Pc is expressed at the anterior region in all life cycle stages of the jellyfish *Podocoryne carnea*. Dev Genes Evol 210:151 – 156

Mattick JS (2007) A new paradigm for developmental biology. J Exp Biol 210:1526 – 1547

Matus DQ et al (2008) The Hedgehog gene family of the cnidarian, *Nematostella vectensis*, andimplications for understanding metazoan Hedgehog pathway evolution. Dev Biol 313:501 – 518

Matus DQ et al (2006a) Molecular evidence for deep evolutionary roots of bilaterality in animal development. Proc Natl Acad Sci USA 103:11195 – 11200

Matus DQ et al (2006b) Dorso/ventral genes are asymmetrically expressed and involved in germ-layer demarcation during cnidarian gastrulation. Curr Biol 16:499 – 505

Matus DQ et al (2007) FGF signaling in gastrulation and neural development in *Nematostella vectensis*, an anthozoan cnidarian. Dev Genes Evol 217:137 – 148

Mazza ME et al (2007) Genomic organization, gene structure, and developmental expression of three clustered otx genes in the sea anemone *Nematostella vectensis*. J Exp Zoolog B Mol Dev Evol 308:494 – 506

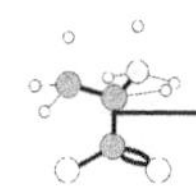

McGinnis W et al (1984) A homologous protein-coding sequence in *Drosophila* homeotic genes and its conservation in other metazoans. Cell 37:403 –408

McGinnis W and Krumlauf R. (1992) Homeobox genes and axial patterning. Cell 68:283 –302 Miller DJ, Ball EE (2005) Animal evolution: the enigmatic phylum placozoa revisited. Curr Biol 15:R26 –R28

Miller DJ, and Ball EE (2008) Cryptic complexity captured: the *Nematostella* genome reveals its secrets. Trends Genet 24:1 –4

Miller DJ et al (2005) Cnidarians and ancestral genetic complexity in the animal kingdom. Trends Genet 21:536 –539

Monteiro AS et al (2006) A low diversity of ANTP class homeobox genes in Placozoa. Evol Dev 8:174 –182

Moroz LL et al (2006) Neuronal transcriptome of aplysia: neuronal compartments and circuitry. Cell 127:1453 –1467

Mukherjee K, Bürglin TR (2007) Comprehensive analysis of animal TALE homeobox genes: new conserved motifs and cases of accelerated evolution. J Mol Evol 65:137 –153

Nedelcu AM, Tan C (2007) Early diversification and complex evolutionary history of the p53 tumor suppressor gene family. Dev Genes Evol 217:801 –806

Nielsen C (2001) Animal Evolution. Interrelationships of the Living Phyla. Oxford University press, Oxford

Nielsen C (2004) Trochophora Larvae: Cell-Lineages, Ciliary Bands, and Body Regions. 1. Annelida and Mollusca. J Exp Zool (Mol Dev Evol) 302B:35 –68

Nilsson DE et al (2005) Advanced optics in a jellyfish eye. Nature 435:201 –205

Pang K et al (2004) The ancestral role of COE genes may have been in chemoreception: evidence from the development of the sea anemone, *Nematostella vectensis* (Phylum Cnidaria; Class Anthozoa). Dev Genes Evol 214:134 –138

Pavlopoulos A, Averof M (2005) Establishing genetic transformation for comparative devel-opmental studies in the crustacean *Parhyale hawaiensis*. Proc Natl Acad Sci USA 102:7888 –7893

Pedersen RA (1971) DNA content, ribosomal gene multiplicity, and cell size in fish. J. Exp. Zool. 177:65 –78

Pennacchio LA et al (2006) In vivo enhancer analysis of human conserved non-coding sequences. Nature 444:499 –502

Peterson KJ, Sperling EA (2007) Poriferan ANTP genes: primitively simple or secondarily reduced? Evol Dev 9:405 –408

Philippe H et al (2009) Phylogenomics revives traditional views on deep animal relationships. Curr Biol 19:706 –712

Piatigorsky J, Kozmik Z (2004) Cubozoan jellyfish: an Evo/Devo model for eyes and other

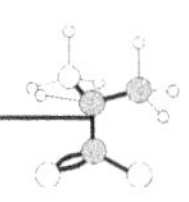

sensory systems. Int J Dev Biol 48:719 - 729

Pincus D et al (2008) Evolution of the phospho-tyrosine signaling machinery in premetazoan lineages. PNAS 105:9680 - 9684

Prpic NM, Telford MJ (2008) Expression of homothorax and extradenticle mRNA in the legs of the crustacean *Parhyale hawaiensis*: evidence for a reversal of gene expression regulation in the pancrustacean lineage. Dev Genes Evol 218:333 - 339

Prud'homme B et al (2007) Emerging principles of regulatory evolution. Proc Natl Acad Sci USA 104:8605 - 8612

Putnam NH et al (2008) The amphioxus genome and the evolution of the chordate karyotype. Nature 453:1064 - 1071

Putnam NH et al (2007) Sea anemone genome reveals ancestral eumetazoan gene repertoire and genomic organization. Science 317:86 - 94

Raible F et al (2006) Opsins and clusters of sensory G-protein-coupled receptors in the sea urchin genome. Dev Biol 300:461 - 475

Raible F et al (2005) Vertebrate-type intron-rich genes in the marine annelid *Platynereis dumerilii*. Science 310:1325 - 1326

Rast JP et al (2006) Genomic insights into the immune system of the sea urchin. Science 314:952 - 956

Remane A (1950) Die Entstehung der Metamerie der Wirbellosen. Vh Dt Zool Ges Mainz:16 - 23

Rentzsch F et al (2006) Asymmetric expression of the BMP antagonists *chordin* and *gremlin* in the sea anemone *Nematostella vectensis*: Implications for the evolution of axial patterning. Dev Biol 296:375 - 387

Rentzsch F et al (2008) FGF signalling controls formation of the apical sensory organ in the cnidarian *Nematostella vectensis*. Development 315:1761 - 1769

Ryan JF et al (2006) The cnidarian-bilaterian ancestor possessed at least 56 homeoboxes: evidence from the starlet sea anemone, *Nematostella vectensis*. Genome Biol 7:R64

Ryan JF et al (2007) Pre-bilaterian origins of the Hox cluster and the Hox code: evidence from the sea anemone, *Nematostella vectensis*. PLOS One 2:e153

Sakaraya O et al (2007) A post-synaptic scaffold at the origin of the animal kingdom. PLoS One 2(6):e506

Samanta MP et al (2006) The transcriptome of the sea urchin embryo. Science 314:960 - 962 Satou Y, Satoh N (2006) Gene regulatory networks for the development and evolution of the chordate heart. Genes Dev 20:2634 - 2638

Schierwater B et al (2008) The early ANTP gene repertoire: Insights from the placozoan genome. PLOS One 3:e2457

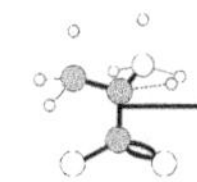

Schmucker D et al (2000) Drosophila Dscam is an axon guidance receptor exhibiting extraordinary molecular diversity. Cell 101:671 – 684

Schuchert P (1993) *Trichoplax adhaerens* (Phylum Placozoa) has cells that react with antibodies against the neuropetide RFamide. Acta Zoologica (Stockholm). 74:115 – 117

Schuchert P et al (1993) Life stage specific expression of a myosin heavy chain in the hydrozoan *Podocoryne carnea*. Differentiation 54:11 – 18

Sea Urchin Genome Sequencing C et al (2006) The genome of the sea urchin *Strongylocentrotus purpuratus*. Science 314:941 – 952

Sedgwick A (1884) On the origin of metameric segmentation and some other morphological questions. Q J Microsc Sci 24:43 – 82

Segawa Y et al (2006) Functional development of Src tyrosine kinases during evolution from a unicellular ancestor to multicellular animals. Proc Natl Acad Sci USA 103:12021 – 12026

Seipel K, Schmid V (2005) Evolution of striated muscle: Jellyfish and the origin of triploblasty. Dev Biol 282:14 – 26

Sempere LF et al (2006) The phylogenetic distribution of metazoan microRNAs: insights into evolutionary complexity and constraint. J Exp Zoolog B Mol Dev Evol 306:575 – 588

Shalchian-Tabrizi K et al (2008) Multigene phylogeny of choanozoa and the origin of animals. PLOS One 3:e2098

Shankland M, Seaver EC (2000) Evolution of the bilaterian body plan: what have we learned from annelids? Proc Natl Acad Sci USA 97:4434 – 4437

Shoguchi E et al (2008) Genome-wide network of regulatory genes for construction of a chordate embryo. Dev Biol 316:498 – 509

Short S, Holland LZ (2008) The evolution of alternative splicing in the Pax family: the view from the Basal chordate amphioxus. J Mol Evol 66:605 – 620

Siewing R (1985) Lehrbuch der Zoologie. Systematik. Gustav Fischer Verlag, Stuttgart, New York Signorovitch AY et al (2007) Comparative genomics of large mitochondria in placozoans. PLoS Genet 3:e13

Signorovitch AY et al (2005) Molecular signatures for sex in the Placozoa. Proc Natl Acad Sci USA 102:15518 – 15522

Simionato E et al (2007) Origin and diversification of the basic helix-loop – helix gene family in metazoans: insights from comparative genomics. BMC Evol Biol 7:33

Skogh C et al (2006) Bilaterally symmetrical rhopalial nervous system of the box jellyfish *Tripedalia cystophora*. J Morphol 267:1391 – 1405

Snell EA et al (2006) An unusual choanoflagellate protein released by Hedgehog autocatalytic processing. Proc R Soc B 273:401 – 407

Sperling EA, Peterson KJ. (2007) Poriferan paraphyly and its implication for precambrian pa-

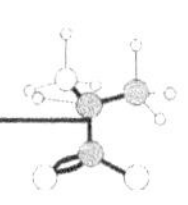

leobi - ology. In Vickers - Rich P, Komarower P (eds) The rise and fall of the ediacaran biota. Geological Society, London

Spring J et al (2002) Conservation of Brachyury, Mef2, and Snail in the myogenic lineage of jellyfish: a connection to the mesoderm of bilateria. Dev Biol 244:372 - 384

Srivastava M et al (2008) The *Trichoplax* genome and the nature of placozoans. Nature 454: 955 - 960

St-Onge L et al (1997) Pax6 is required for differentiation of glucagon-producing alpha-cells in mouse pancreas. Nature 387:406 - 409

Steinmetz PR et al (2007) Polychaete trunk neuroectoderm converges and extends by mediolateral cell intercalation. Proc Natl Acad Sci USA 104:2727 - 2732

Stephenson TA (1928) The British Sea Anemones. Dulau & Co, London

Stephenson TA (1935) The British Sea Anemones. Dulau & Co, London

Stierwald M et al (2004) The Sine oculis/Six class family of homeobox genes in jellyfish with and without eyes: development and eye regeneration. Dev Biol 274:70 - 81

Suga H et al (2008) Evolution and functional diversity of jellyfish opsins. Curr Biol 18:51 - 55 Syed T, Schierwater B (2002) *Trichoplax adherens*: discovered as a missing link, forgotten as a hydrozoan, re-discovered as a key to metazoan evolution. Vie Milieu 52:177 - 187

Tassy O et al (2006) A quantitative approach to the study of cell shapes and interactions during early chordate embryogenesis. Curr Biol 16:345 - 358

Tautz D (2004) Segmentation. Dev Cell 7:301 - 312

Taylor JS, Raes J (2004) Duplication and divergence: The evolution of new genes and old ideas. Ann Rev Genet 38:615 - 643

Technau U (2001) *Brachyury*, the blastopore and the evolution of the mesoderm. BioEssays 23:788 - 794

Technau U et al (2005) Maintenance of ancestral complexity and non-metazoan genes in two basal cnidarians. Trends Genet 21:633 - 639

Technau U, Scholz CB (2003) Origin and evolution of endoderm and mesoderm. Int J Dev Biol 47:47

Tessmar-Raible K et al (2005) Fluorescent two color whole-mount in situ hybridization in *Platynereis dumerilii* (Polychaeta, Annelida), an emerging marine molecular model for evo-lution and development. BioTechniques 39:460 - 464

Valentine JW (2000) Two genomic paths to the evolution of complexity in bodyplans. Paleobiology 26:513 - 519

Valentine JW et al (1994) Morphological complexity increase in metazoans. Paleobiology 20: 131 - 142

Vogel C, Chothia C (2006) Protein family expansions and biological complexity. PLOS Com-

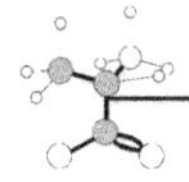

put Biol 2:e48

Voigt O et al (2004) Placozoa-no longer a phylum of one. Curr Biol 14:R944 – R945

Wenderoth H (1986) Transepithelial cytophagy by *Trichoplax adherens* F. E. Schulze (Placozoa) feeding on yeast. Zeitschrift für Naturforschung. Section C, Biosciences 41:343 – 347

Woo K, Fraser S (1995) Order and coherence in the fate map of the zebrafish nervous system. Development 121:2595 – 2609

Woolfe A et al (2005) Highly conserved non-coding sequences are associated with vertebrate development. PLoS Biol 3:e7

Yanze N et al (2001) Conservation of Hox/ParaHox-related genes in the early development of a cnidarian. Dev Biol 236:89 – 98

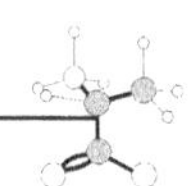

第6章　海洋藻类基因组

摘要：藻类是一群极度多样的生物，表现在以下几个方面，包括系统发育、基础生物学、生物化学及其所展现的不同程度复杂性和对各种栖息地的适应性等方面。事实上，藻类的研究涉及各个方面，包括从作为海洋生态系统中的关键物种到新兴生物分子和生物过程的资源。在过去的几年时间里，有两个关键进展加快了这方面的研究，一个是高通量测序技术在全基因组测序和海洋环境宏基因组中的开发应用，另一个是，随着在几个研究较少的藻类品系中模式生物的发现，基因组学在藻类研究中有效地发挥了作用。本章将讲述如何从基因组学的角度去探索藻类生物学的各个方面，并指出未来的研究趋势。

缩写词：

DD：	(2E，4E/Z) -decadienal	(2E，4E/Z) 癸二烯醛
Kbp	kilobase pairs	千碱基对
Mbp	megabase pairs	兆碱基对
NO	nitric oxide	一氧化氮
PCR	polymerase chain reaction	聚合酶链式反应
pg	picogramme	皮克
rDNA	ribosomal DNA	核糖体 DNA

6.1　什么是藻类？

藻类并不是一个明确的分类学词汇，仅指不同群体的光合生物。在本章如何定义这个术语就显得非常重要，这里我们定义藻类为除有胚植物之外的所有能进行光合作用的（Keeling 2004)）真核生物。虽然这个定义并不包括光合原核生物，不过仍然涵盖了很大部分的类群，因为它囊括真核的具有叶绿体的所有生物。注意，尽管有胚植物亦被定义为陆生植物，但也包含了一些回归到海洋环境中的物种，虽然在本章会对它们作简要讨论，但我们不将

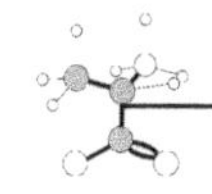

这些生物体定义为藻类。

如何将这些多样的生物归类，依据其质体的起源。真核藻类的质体均被认为直接或间接起源于原始单一的内共生事件，衍生于绿藻、红藻和灰胞藻类的共同祖先（见下文）。

6.2 为什么说藻类是有趣的？

科学家在生物学、生态学和演化历史等各个方面的探索促进了藻类的研究。许多单细胞藻类是浮游藻类的重要成员，它们对全球化学循环作出了重大贡献。同时，浮游藻类也是海洋食物链上的关键初级生产者，有积极作用。藻类也有其危害性，例如藻华会产生毒素。在沿海生态系统中，大型红、绿、褐藻通常是最主要的生物群，有时可形成浓密的群落覆盖大片海床，令人印象深刻。这些生物代表了海岸生态系统的重要组成成分，被认为是生态的拓荒者（Coleman and Williams 2002）。举例来说，大型褐藻群不仅作为初级生产者，而且在海洋食物链的次级生产中发挥着显著作用，同时为众多的动物和藻类提供栖息地（*Duggins* et al. 1989）。同样，藻类的系统发育多样性也是研究的热点，许多在生命进化树中占据关键发育地位的藻类能为真核进化中的古老事件提供很多重要的依据。系统发育多样性也是藻类普通生物学中非常有趣的一个方面。因此，由于它们有非常多样性的演化历史，不同组群的藻类往往在新陈代谢、细胞生物学、形态学和生活史等方面各有特点。这些多样化的特征既可以用来进行基础性的研究，也适用于各种新型生物技术的应用。

近几年，基因组学方法的应用使很多藻类研究领域取得了重大进展。很多情况下，基因组学方法的应用给藻类的生物学研究带来令人惊喜和激动人心的新见解。接下来的几节将会详细介绍藻类研究的几个领域，同时讨论基因组学的影响。

6.3 内共生学说和藻类的起源

乍看起来，藻类在真核生命之树中的广泛分布令人惊讶。广布于6个真核生物界中的4个（图6.1），即植物界（plantae）、囊泡藻界（chromalveolates）、古虫界（excavates）和有孔虫界（rhizaria）。

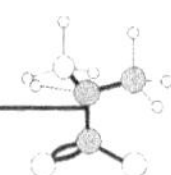

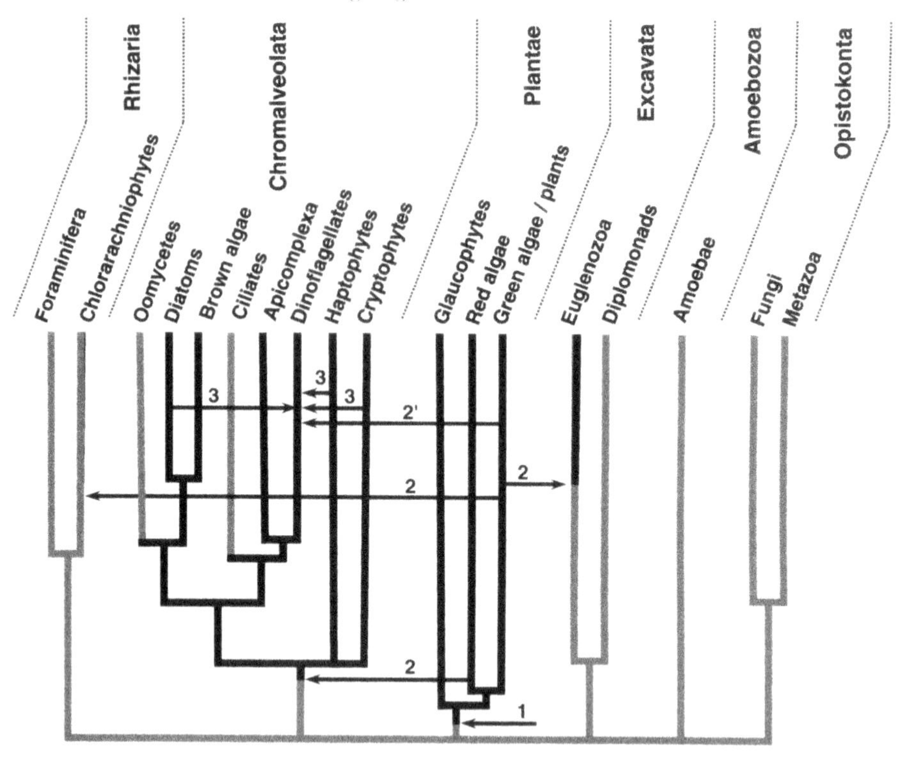

Fig. 6.1 Acquisition of plastids through endosymbiosis during the evolution of the eukaryotes. *Thick bars* indicate the evolutionary relationships between the groups shown; these are in *grey* for lineages without plastids and in *black* for lineages with plastids. *Horizontal arrows* indicate the endosymbiotic events that led to plastid acquisition. 1. Primary endosymbiosis involving the capture of a cyanobacterium by a heterotrophic ancestor of the Plantae group, 2. secondary endosymbiosis involving the enslavement of either a red or a green alga, 3. tertiary endosymbiosis involving the replacement of a plastid from a secondary endosymbiosis with another plastid also derived from a secondary endosymbiosis, 2′. serial secondary endosymbiosis in which the process of secondary endosymbiosis has occurred twice in the same lineage, the second event leading to the elimination of the plastid from the first event. Note that the relationships between the major groups of eukaryotes are poorly supported in many cases. There is accumulating evidence, for example, that the Rhizaria fall within the Chromalveolate group (Yoon et al. 2008, Baldauf 2008)

图6.1　真核生物进化过程中内共生性吞噬获得质体。粗线显示图中组群的进化关系；灰色的为无质体；黑色的为有质体。水平箭头表示获得质体的内共生事件：1. 原始内共生，对蓝藻的捕获；2. 第二级内共生，异养型植物组群祖先对红藻或绿藻的吞噬；3. 第三次内共生，来自第二次内功生的质体被第二次内共生质体替代，2′. 系列的第二次内共生，在同一个系谱中第二次内共生事件发生两次，从而导致来自于首次内共生事件质体的剔除。注意，其实真核生物中主要组群的关系在很多情况下尚缺乏有力的证据支持。比如说目前有越来越多的证据表明，有孔虫界属于囊泡藻界。

英文注释：Rhizaria，有孔虫界；Chromalveolata；囊泡藻界；Plantae，植物界；Excavata，古虫界；Opisthokonta，后鞭毛界；Amoebozoa，变形虫界；Foraminifera，有孔虫门；Oomycetes，卵菌纲；Diatoms，硅藻类；Brown algea，褐藻；Ciliates，纤毛虫类；Apicomplexa，顶复门；Dinoflagellates，鞭毛藻类；Haptophytes，定鞭藻类；Cryptophytes，隐藻类；Glaucophytes，灰胞藻类；Red algae，红藻；Green algae，绿藻；Euglenozoa，眼虫门；Diplomonads，双滴虫类；Amoebae，变形虫；Fungi，真菌；Metazoa，后生动物

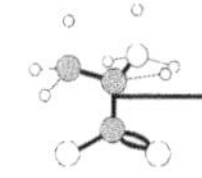

最近几年，在认识光合生物如何产生的问题上有长足的进展，其中基因组分析起了关键作用（Reyes-Prieto et al. 2007）。人们普遍认为所有物种的质体都是起源于16亿年前一个异养生物捕获球状蓝细菌的原始内共生事件（Yoon et al. 2006），然而最近有证据证明存在另外一个时间更近的、发生在丝状有壳变形虫亚目阿米巴原虫的原始内共生事件（Nowack et al. 2008）。蓝细菌在被吞噬成为质体后的进化历程中，其基因组逐渐减少，同时基因向宿主细胞核转移。这个过程需要蛋白靶向系统的进化，以利于蛋白质能够通过双层膜进入质体。

古老的原始内共生事件导致了植物界内出现三个不同的类群，即灰胞藻门（*glaucophyte*）、红藻和绿藻。其他大类群，如囊泡藻类（*chromalveolates*）、古虫类（*excavates*）和有孔虫类（*rhizarians*），都是在第二次或第三次内共生中衍生，真核细胞捕获红藻或绿藻类光合真核生物，然后以类似于原始内共生过程进化成为质体（Keeling 2004，Yoon et al. 2006）。

囊泡藻类假说设定鞭藻类（*haptophytes*）、隐藻类（*cryptophytes*）、不等鞭毛类（*heterokonts or stramenopiles*）和囊泡虫类（*alveolates*）属于同一个超类群。它们的共同祖先捕获了一个红藻，随后分别进化成现今如硅藻、褐藻、一些甲藻和类球菌藻等物种中的质体。顶复虫类（*apicomplexans*）的顶复体（apicoplast）也是起源于该事件。如果第二个质体在很早时候就被共同祖先捕获，那么，在纤毛虫和卵菌等没有质体的类群中应该是又发生了质体的丢失。现有证据支持这个观点，例如在纤毛虫和卵菌中可能来源于第二次内共生的基因已被鉴定（Tyler et al. 2006，Reyes－Prieto et al. 2008）。然而囊泡藻假说仍然存在着较大争议，该假说不能排除另一个假说，即可能是该大类群中某些物种独自获得了质体。此外，囊泡藻界的分类缺乏基于核基因系统发育的支持证据，因此我们必须牢记一个可能性，就是这个大类群是人为划分的结果，其最有利的支持实际上可能只是基于相似但独立的第二次内共生事件。

在隐藻类植物和 *chlorarachniophytes*（一种生有鞭毛的阿米巴状细胞）中可以找到令人信服的第二次内共生事件证据，它们保留了内共生体的核残片，称为类核体（Archibald 2007）。其质体由四层膜包围，两层外膜大概对应到原始细胞的囊泡膜和其内共生体的细胞膜，类核体在两层内膜和两层外膜之间。四层膜是次级质体的共同特征，但眼虫藻和大多数甲藻似乎丢失了其中的一层膜。多层膜的存在使蛋白转运到质体的过程也更复杂，因此这类生物具有复杂的转运系统，它们通过双靶标多肽进行蛋白识别（Lang et al. 1998）。

如果认为第二次内共生的故事还不足够复杂的话，那么一些甲藻类看起

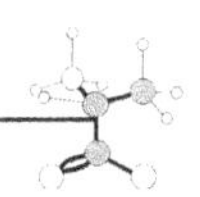

来是经历了更复杂的事件，不论是其连续的第二次内共生导致绿藻起源的质体取代了红藻起源的质体，还是它们参与第三次内共生，捕获了已经发生第二次内共生的定鞭毛藻类、隐藻类或不等鞭毛藻类（Keeling 2004）。*Chlorarachniophytes* 和眼虫藻则在独立的第二次内共生事件中，通过捕获绿藻获得其质体（图 6.1）。

水平或横向基因转移过程中，生物个体的基因整合到第二个生物个体的基因组中，随后与受体细胞的遗传物质一起遗传，两个个体可以是远缘的物种（Keeling and Palmer 2008）。水平基因转移的一个很好的例子就是内共生事件导致的基因转移，大量的基因从内共生体中转移到宿主细胞核内。但是，这也不是发生在真核生物中唯一的水平基因转移。近几年来，对真核生物特别是原生生物中水平基因转移的认识变得越来越清晰，原生生物可从与之联系紧密的生物体中获得基因（Nosenko and Bhattacharya 2007，Bowler et al. 2008）。

近年来，基因组数据在真核生物的进化以及该过程中的内共生事件的研究中，起了重要作用。大量来自真核生物关键物种的基因组数据，使我们可以挑选最近缘的基因序列来做进化分析，并且可用多个基因的组合来分析，从而显著提高分析的分辨率。同时，基因组数据能为内共生过程提供详细信息，关于特定内共生事件中涉及的宿主和内共生体，以及内共生吞噬过程，特别是内共生体基因向宿主细胞核的转移。这些领域中的一些关键基因组计划罗列在表 6.1 中，下文将会有所介绍。

6.4 藻类和海洋生态系统

以下各节将会讨论基因组学方法在藻类生物学研究中的广泛应用，特别是藻类在海洋生态系统中的重要作用。

6.4.1 地球演化中浮游生物的多样性

浮游生物是指生活在海水和淡水中可自由漂浮的生物。在这些生态系统中，能进行光合作用的生物统被称为浮游植物。虽然它们只占总初级生产者生物量很少的一部分（0.2%），却对地球的初级生产作用有着重大的贡献，估计约占地球总初级生产总量的 45%（Field et al. 1998）。许多真核藻类属于浮游植物，但浮游植物中数量最多的是蓝细菌。从进化的角度来看，蓝细菌是浮游植物中最古老的成员。该生物体解释了光合产氧的诞生，导致地球从古

表6.1 藻类基因组测序计划

Table 6.1 Algal genome sequencing projects

Species	Strain	Phylogenetic group	Marine	Genome size (Mbp)	Status or Reference	URL
Bathycoccus prasinos	Bban7	Plantae, Prasinophyceae (green alga)	Yes		Pending	http://www.cns.fr/externe/English/corps_anglais.html
Ostreococcus tauri	OTH95	Plantae, Prasinophyceae (green alga)	Yes	12.6	Derelle et al. (2006)	http://bioinformatics.psb.ugent.be/genomes/
Ostreococcus "lucimarinus"	CC9901	Plantae, Prasinophyceae (green alga)	Yes	13.2	Palenik et al. (2007)	http://www.jgi.doe.gov/genome-projects/
Ostreococcus sp.	RCC809	Plantae, Prasinophyceae (green alga)	Yes		Available	http://www.jgi.doe.gov/genome-projects/
Micromonas pusilla	RCC827	Plantae, Prasinophyceae (green alga)	Yes	15	Worden et al. (2009)	http://www.jgi.doe.gov/genome-projects/
Micromonas pusilla	CCMP1545	Plantae, Prasinophyceae (green alga)	Yes	15	Worden et al. (2009)	http://www.jgi.doe.gov/genome-projects/
Dunaliella salina	CCAP 19/18	Plantae, Chlorophyceae (green alga)	No	130	Sequencing	http://www.jgi.doe.gov/genome-projects/
Chlorella vulgaris	C-169	Plantae, Trebouxiophyceae (green alga)	No	40	Annotation	http://www.jgi.doe.gov/genome-projects/
Chlorella sp.	NC64A	Plantae, Chlorellaceae (green alga)	No	46.2	Available	http://genome.jgi-psf.org/ChlNC64A_1/ChlNC64A_1.home.html

Table 6.1 (continued)

Species	Strain	Phylogenetic group	Marine	Genome size (Mbp)	Status or Reference	URL
Chlamydomonas reinhardtii	CC-503 cw92 mt+	Plantae, Chlorophyceae (green alga)	No	120	Merchant et al. (2007)	http://genome.jgi-psf.org/Chlre3/Chlre3.home.html
Volvox carteri		Plantae, Chlorophyceae (green alga)	No	140	Available	http://genome.jgi-psf.org/Volca1/Volca1.home.html
Zostera marina		Plantae, Zosteraceae (sea-grass)	Yes		Pending	http://www.jgi.doe.gov/genome-projects/
Cyanophora paradoxa		Plantae, Glaucophyta	No		Assembly	http://www.biology.uiowa.edu/cyanophora/cyanophora_home.htm
Cyanidioschyzon merolae	10D	Rhodophyta, Cyanidiaceae (red alga)	No	16.5	Matsuzaki et al. (2004)	http://merolae.biol.s.u-tokyo.ac.jp/
Galdieria sulphuraria		Rhodophyta, Cyanidiaceae (red alga)	No		Ongoing	http://genomics.msu.edu/galdieria
Chondrus crispus		Rhodophyta, Florideophyceae (red alga)	Yes	150	Ongoing	http://www.genoscope.cns.fr/spip/spip.php?lang=en
Porphyra umbilicalis		Rhodophyta, Bangiophyceae (red alga)	Yes	300–400?	Ongoing	http://www.jgi.doe.gov/genome-projects/
Thalassiosira pseudonana	CCMP1335	Heterokonta, Bacillariophyceae (diatom)	Yes	34.5	Armbrust et al. (2004)	http://www.jgi.doe.gov/genome-projects/

Table 6.1 (continued)

Species	Strain	Phylogenetic group	Marine	Genome size (Mbp)	Status or Reference	URL
Phaeodactylum tricornutum	CCAP1055/1	Heterokonta, Bacillariophyceae (diatom)	Yes	20	Bowler et al. (2008)	http://www.jgi.doe.gov/genome-projects/
Pseudo-nitzschia multiseries	CLN-47	Heterokonta, Bacillariophyceae (diatom)	Yes	250	Sequencing	http://www.jgi.doe.gov/genome-projects/
Fragilariopsis cylindrus	CCMP 1102	Heterokonta, Bacillariophyceae (diatom)	Yes	35	Sequencing	http://www.jgi.doe.gov/genome-projects/
Aureococcus anophagefferens	CCMP1984	Heterokonta, Pelagophyceae	Yes	32	Available	http://genome.jgi-psf.org/Auran1/Auran1.home.html/
Ochromonas	CCMP1393	Heterokonta, Chrysophyceae	No		Ongoing	http://www.jgi.doe.gov/sequencing/why/nanoflagellates.html
Ectocarpus siliculosus	Ec 32	Heterokonta, Phaeophyceae (brown alga)	Yes	200	Completed	http://www.genoscope.cns.fr/spip/Ectocarpus-siliculosus.html
Phaeocystis globosa		Haptophyta, Prymnesiophyceae	Yes		Pending	http://www.jgi.doe.gov/genome-projects/
Phaeocystis antarctica		Haptophyta, Prymnesiophyceae	Yes		Pending	http://www.jgi.doe.gov/genome-projects
Emiliania huxleyi	CCMP1516	Haptophyta, Prymnesiophyceae	Yes	220	Available	http://www.jgi.doe.gov/genome-projects/
Guillardia theta	CCMP2712	Cryptophyta, Cryptophyceae	Yes		Assembly	http://www.jgi.doe.gov/sequencing/why/50026.html/
Bigelowiella natans	CCMP2755	Rhizaria, Chlorarachniophyceae	Yes		Assembly	http://www.jgi.doe.gov/sequencing/why/50026.html

英文注释：Species，物种；Strain，品系；Phylogenetic group，系统发育组群；Prasinophyceae，绿枝藻纲；Chlorophyceae，绿藻纲；Trebouxiophyceae，共球藻纲；Chlorellaceae，小球藻科；Zosteraceae，大叶藻科；Rhodophyta，红藻门；Florideophyceae，真红藻纲；Bangiophyceae，红毛菜纲；Bacillariophyceae，硅藻纲；Pelagophyceae，浮生藻纲；Chrysophyceae，金藻纲；Prymnesiophyceae，定鞭金藻纲；Cryptophyceae，隐藻纲

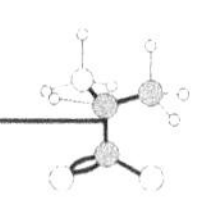

老的无氧世界转变成现代的有氧环境。如上所述，蓝细菌被吞噬而成为内共生体导致光合真核生物的出现，从而增加了浮游植物的多样性。化石记录表明，红藻最晚出现于12亿年前；还有证据表明，真核浮游植物可能在16亿～18亿年前就已经出现（Falkowski et al. 2004）。中古生代时期，绿藻可能是海洋浮游植物的优势类群，但到了三叠纪，这个种群的丰度和多样性均下降，并很大程度上被甲藻、球菌藻和硅藻三个现在浮游植物生态系统中的优势藻类种群所替代（Falkowski et al. 2004）。这三个种群衍生自第二次内共生事件，大多数情况下是捕获红藻，说明拥有这样的质体是有利于宿主的，可能与摄取微量元素有关（Falkowski et al. 2004）。在较近的地质年代，甲藻、球菌藻和硅藻的相对丰度都有所波动，这可能与生态位偏好性相关：甲藻和球菌藻更适应于稳定的环境，而硅藻则倾向于养分供给频繁变化的环境。

6.4.2 藻类：浮游植物中的重要成分

众所周知，从系统发育范围上看现存生物中浮游植物是最复杂的群体。这与认为藻类生物在同一生态位的有限资源中相互竞争这一理论是互相矛盾的。一些假说被提出来解释这种情况，包括 Hutchinson 提出的观点，但实际上由于对生态系统本身特性了解还很有限，从而限制了对悖论的解释。

过去的十年间，从分子技术到后来基因组技术的发展，都显著影响了对浮游生物群体组成和单个浮游植物物种的研究。这些技术在浮游生物研究中提供了更多可供选择的方法，特别是对至今仍未被分离培养的物种。直接对环境样本 rDNA 测序为研究真核浮游生物物种的系统发育多样性提供了强大的工具。“环境克隆”（environmental cloning）方法最先创立于原核浮游生物的研究（Giovannoni et al. 1990）。比起真核生物，这种方法在原核浮游生物中的应用更广泛。然而，该方法在浮游真核生物上的应用，揭示了浮游真核生物有极高的多样性，且导致了几类新真核生物的发现，特别是来自不等鞭毛类（*heterokonts*）和囊泡虫类（*alveolates*）（Moreira and López-García 2002）的物种。另外，这些研究只是从生态系统中抽取一部分样品，未来，预计深度测序会发现更多的新序列。目前面临的一个挑战是如何应用这些从环境中获得的序列来鉴定物种，从而了解新发现物种的生态地位。这对改良培养技术的发展和基于如荧光定位技术（FISH）等活体检测方法的提高有重大意义。还有一个重要的问题是，上述新发现的不等鞭毛类和囊泡虫类是自养还是异养生物尚待明确。

环境样品的 PCR 扩增和 rDNA 序列克隆显著扩大了对生物的研究范围。

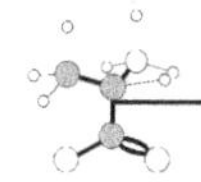

其他类似的基因组学研究方法同样在发展，例如利用寡核苷酸芯片检测生物样品和评估环境样品的生物多样性（Medlin et al. 2006，Metfies and Medlin 2008）。但是，这些方法都有其自身的局限性，尤其他们都是基于 PCR 的技术，因此 DNA 序列间的相似性和简并引物都会对序列的检测有很大影响。近几年对克隆的 DNA 片度的高通量随机测序方法的开发，为探索不同生态系统的遗传复杂性提供了新方法。该方法在海洋系统研究方面的应用在第一章里已作了概述。这里，我们将集中于具体例子，来说明该方法如何应用于浮游生态系统特性及浮游藻类组成成分的研究。

6.4.3　基于高通量测序技术对浮游生态系统的探索

Craig Venter 和他的团队（Venter et al. 2004，Rusch et al. 2007）利用基于 Sanger 的高通量测序技术分析用 0.8μm 滤膜过滤海水后所获得的微生物，这些样品是从马尾藻海（Sargasso Sea）的几个站点采集来的，后续研究中将样品采集地扩展到北大西洋的东部沿岸，从巴拿马运河一直到西太平洋，这是科考船 Sorcerer II 号的其中一次全球航程。用这些样本的 DNA 构建了插入片段为 2～6 kbp 的文库，产生了总共超过 930 万条测序序列，超过 63 亿个碱基对。该方法提供了特定环境中生物多样性的大量信息。仅仅是马尾藻海的样品中，估计就有 1800 个物种，其中包括 148 个未知的细菌种类。然而，高通量的测序成本并没有限制对物种多样性的研究。测序数据同时提供了基因水平的信息和群落全部基因的信息概况，从而有利于新基因的发现。收集样本中所有基因信息亦称为宏基因组，通过深入研究样本，比如在特定环境中具有代表性且相对重要的不同功能途径达到。

Craig Venter 等主要研究生态系统中的细菌组成。为此，他们用滤器过滤更小的生命体（比如病毒）和大的生物体（如大的真核细胞）（Venter et al. 2004，Rusch et al. 2007）。然而，样品中还是保留有少量同细菌一样大小的超微型真核生物，在 Sorcerer II 的那次航程所得的结果中，真核生物的序列约为总量的 2.8%。这些序列代表的物种广泛分布于真核生命之树（Piganeau et al. 2008）。因此，这些研究给我们带来可观的序列数据量和浮游藻类的信息。更重要的是，这两项分析无疑展现了高通量测序技术作为探索海洋藻类及其生态系统多样性的潜力。此外，该调查中还有其他过滤孔径收集的样品，为以后该生态系统中藻类成分的进一步系统研究提供了可能。

同所有的新技术一样，高通量环境测序也会带来新的问题。尤其是对高生物多样性的生态系统，比如浮游生物的生态系统，测序数据组装是个严峻

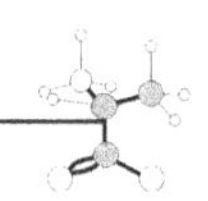

的挑战。因为尽管测序产生大量的序列数据，但测序的深度远远小于单个基因组的测序深度。解决这个问题的最直接的方法是通过研究生物个体来弥补信息量的不足。例如，在马尾藻海调查项目开始的时候，绿枝藻的 *Ostreococcus* 属中就已经有了两个完整的基因组，有关的基因组序列已被应用于马尾藻海数据库关联序列的恢复，从而发现样品中至少有 *Ostreococcus* 属中的 2 个物种（Piganeau and Moreau 2007）。两个层面的数据具有很强的互补性，基因组序列有利于环境数据库中特异性个体数据的恢复，环境数据库可同时提供已测序生物个体的生态学及其基于序列比较的基因组分子进化两方面的信息。

6.4.4 浮游生态系统的多样性和动态性

到目前为止，高通量测序只应用于有限的样本，因此在鉴定新物种和研究新生态系统方面仍有巨大的潜力。高通量测序证实一个浮游生态系统就具有显著的多样性，而海洋环境不同区域间的浮游生态系统均存在着显著性差异。

远洋区的寡营养生态环境明显不同于沿海地区的中等营养条件（Guillou et al. 1999，Massana et al. 2004，Romari and Vaulot 2004），北极和暖水海域发现的生物群体也有显著的差异（Lovejoy et al. 2006）。有一个值得研究的环境是海冰（Mock and Thomas 2005）。生物体在极端条件下生存，这些环境异常寒冷，同时盐度，pH 和光照的变化范围很大，但是可以作为很好的样本运用生物技术来研究地球生命起源，因此科学家们对此很有兴趣。未来几年，高通量测序必定在不同海洋生态系统中有更广阔的应用。

另外一个重要的因素是浮游生态系统的动态变化。受到物理和化学因素（比如光照、温度、有效营养成分）和生物之间相互作用的影响，生物种类和密度在其生态系统中会产生剧烈变化。被其他浮游生物如被鞭毛虫类原生动物等捕食，由病毒感染引起的细胞溶解，是影响浮游植物群体的两大主要作用。捕食作用可导致营养（比如碳）以颗粒沉降，或在更大生物体内富集固定；反之，病毒裂解倾向于释放更多的营养元素到自然环境中（Suttle 2005）。总体而言，地球上的海洋水域被认为包含有约 4×10^{30} 个病毒颗粒，其中可能包括了绝大部分海洋生物的病原体。由此可见，这些捕食者很可能在海洋碳循环中起着关键作用，但是对它们的基本生物学信息却知之甚少。海洋微藻易受多种不同的病毒感染（Nagasaki 2008），基因组学方法已开始应用于研究这类问题。比如颗石藻病毒（Coccolithovirus）EhV－86 能感染赫氏圆石藻（Wilson et al. 2005），OtV5 病毒可感染绿枝藻（Derelle et al. 2008）。此外，美国的联合基因组研究所（JGI）最近发起了对异养微型鞭毛藻基因组的测序。这种藻捕食微藻和一些无法培养的病毒，不久的将来有望得到更多关于这种藻类的重要信息。

藻华是显示浮游生态系统动态变化的一个特别例子，其涉及一个或很少一部分浮游藻类的快速增殖，对人类的活动有着重大影响，尤其是当其中的藻类产生毒素的时候。很多有毒的藻华都是由于夏天的环境条件，比如高温、高营养和水流停滞，有利于甲藻类的快速增长。赤潮爆发时，每公升海水中可包含高达数百万的细胞（Guiry and Guiry 2008），其毒素可以通过食物链的生物富集作用进而影响到鱼类、鸟类和哺乳动物，因此对环境产生巨大的影响。人类摄入富集在鱼类和滤食性贝类中的高浓度毒素后，会造成肠胃功能紊乱、永久性神经损伤甚至死亡（Faust and Gulledge 2002）。塔玛亚历山大藻和链状亚历山大藻产生的贝类毒素可导致麻痹性中毒，而短裸甲藻产生的短裸甲藻毒素可导致神经毒性中毒（Faust and Gulledge 2002）。甲藻类的 *Heterocapsa triquetra* 的大量繁殖，会导致鱼类死亡，并非是因为毒素，而是由于藻类细胞分解时消耗大量的氧（Guiry and Guiry 2008）。一种前沟藻（*Amphidinium gibbosum*）产生的细胞毒素具有潜在抗癌活性。其他能产生毒素的种群还包括不等鞭毛藻类和定鞭毛藻类。硅藻中的拟菱形藻属（*Pseudo-nitzschia*）可产软骨藻酸，这是一种神经毒素，当人类误食被污染的贝类时可引起失忆性中毒。海金藻类中的褐潮藻被认为能分泌毒素至其胞外多糖鞘。定鞭毛藻类中的小定鞭金藻分泌的普林藻素毒素（prymnesin toxins）对鱼有很大的毒性（La Claire 2006）。基因组学方法（比如 EST 测序），已应用于引起有毒藻华的生物研究中，主要目的是解析产生毒素的机理（http://genome.imb-jena.de/ESTTAL/cgi-bin/Index.pl）。

6.4.5 基于生物个体途径探索浮游藻类生物学

以下讨论的方法会提供更多关于浮游生物中生物体的细节信息。基于高通量测序的宏基因组序列，提供了该生态系统中的代谢过程和细胞过程的信息。不过，要真正了解生物群体，其生态系统中单个物种详细的生物学信息必不可少。传统的生物学方法，特别是浮游植物单个谱系的分离培养及研究依然对浮游植物的认识有极其重要的贡献。但是，最近几年，以全基因组测序技术为手段的强大基因组学方法及模式生物的建立对这些传统技术的不足进行了补充和完善。

最初，基因组测序仅限于少数的藻类。虽然还要考虑到其他因素，比如生态关联性、系统发育地位、无菌培养的可行性及其他生物学特性，但基因

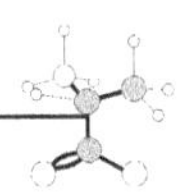

组测序的首要选择条件是小基因组的物种。然而，随着测序技术的发展，关于藻类基因组计划的数量快速增长，而且很大部分的项目已在最近完成或接近完成（表6.1，图6.2）。至今，已公布了8个藻类基因组序列，它们全部来自单细胞藻类，当中包括6个海洋物种：2种硅藻，海链藻（*Thalassiosira pseudonana*，Armbrust et al. 2004）和三角褐指藻（*Phaeodactylum tricornutum*，Bowler et al. 2008）；2种绿枝藻类，*Ostreococcus tauri*，Derelle et al. 2006和*O. lucimarinus nomen nudum*，Palenik et al. 2007；2种微胞藻（*Micromonas*）。另外2种是淡水藻类，即红藻（*Cyanidioschyzon merolae*，Matsuzaki et al. 2004）及一种绿藻-莱茵衣藻（*Chlamydomonas reinhardtii*，Merchant et al. 2007）。除了莱茵衣藻有较大的基因组（120 Mbp）外，这些藻类的基因组都很小（从*Ostreococcus*的12 Mbp到*Thalassiosira*的34 Mbp）。下面将介绍基因组技术在特定海洋藻类方面的应用细节，特别着重于硅藻中的三角褐指藻（*P. tricornutum*）和绿枝藻中的*O. tauri*，这2个物种已发展为模式生物。

6.4.5.1 硅藻基因组

中心硅藻纲中的海链藻（*T. Pseudonana*）是第一个被测序的海洋藻类（Armbrust et al. 2004）。该浮游植物具有重要的生态学意义，遍布全球海洋。硅藻属于不等鞭毛藻类，所以跟动物和绿色植物的亲缘关系疏远，结果却发现其基因组有后两者的特征，对光合生物来说这是出人意料的。至少从基因组水平和特定代谢途径两者来说这是真实存在的，比如在海链藻（*Thalassiosira*）中发现具有完整的尿素循环，这是动物才有的典型生化特征。

海链藻的基因组序列现在被用于研究硅藻的一些更为独特的生物学特性。很多硅藻能形成具有精致图案的二氧化硅细胞壁，这种细胞壁的制造也是一个极有意思的生物学过程，也是潜在的纳米技术应用。最近，结合芯片技术的全基因组研究已用于鉴定在构建细胞壁中起作用的候选基因。该方法包括寡核苷酸微阵列的构建，寡聚核苷酸序列覆盖了所有染色体区域，从而保证基因组的任何转录部分能通过与带标记的cDNA杂交而被检测。海链藻的芯片中的寡核苷酸长度为36个碱基，对应到染色体上，每两个相邻的区域中间有10bp的缺口（Mock et al. 2008）。该实验鉴定了75个在二氧化硅为限制因素时被诱导表达的基因，另外一些基因是在二氧化硅和铁离子浓度同时受限时被诱导表达，说明两种代谢可能存在联系，或者说铁离子是二氧化硅代谢所需的辅助因子。该实验同时还鉴定了3000个未被注释的新基因，其中包括非编码RNA和反义RNA。

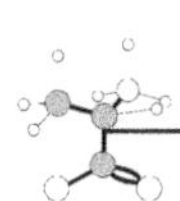

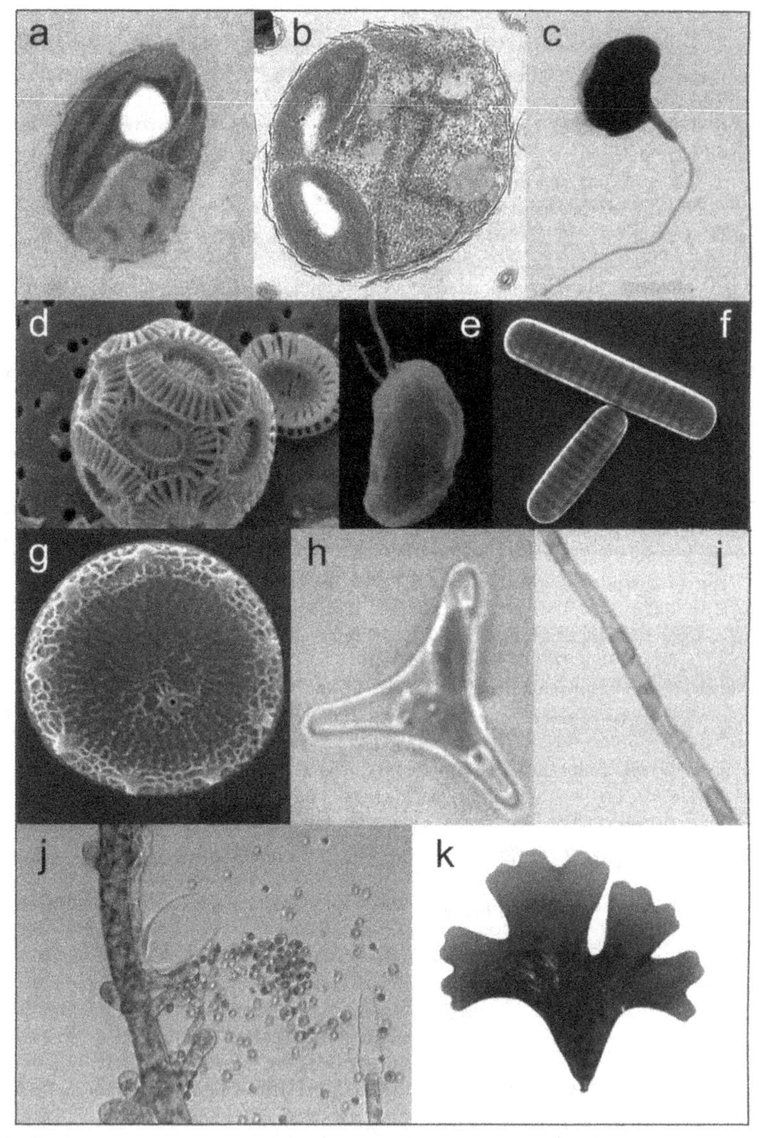

Fig. 6.2 Some examples of marine algae for which genome sequencing projects have been carried out or are currently in process. (**a**) *Ostreococcus* sp., (**b**) *Batycoccus* sp. (photograph courtesy of Marie-Josèphe Dinet), (**c**) *Micromonas*sp. (from Guillou et al. 2004), (**d**) *Emiliania huxleyi* (photograph courtesy of Jeremy R. Young, The Natural History Museum, London, UK), (**e**) *Guillardia theta* (photograph courtesy of Geoff McFadden, University of Melbourne, Australia), (**f**) *Fragilariopsis cylindrus* (photograph courtesy of Gerhard Dieckmann Alfred Wegener Institute for Polar and Marine Research, Bremerhaven, Germany), (**g**) *Thalassiosira pseudonana* (photograph courtesy of Virginia Armbrust, University of Washington, USA), (**h**) *Phaeodactylum tricornutum* (photograph courtesy of Alessandra De Martino and Chris Bowler, Ecole National Supérieur, Paris, France), (**i**) *Pseudo-nitzschia multiseries* (photograph courtesy of the Joint Genome Institute, USA), (**j**) detail of *Ectocarpus siliculosus* thallus showing release of meiospores from a unilocular sporangium. (**k**) *Chondrus crispus* plantlet (Photograph courtesy of Jonas Collén, Station Biologique de Roscoff)

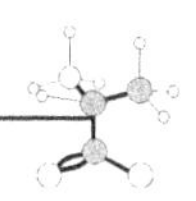

图6.2　海洋藻类基因组测序项目中已测序或正在测序的物种。(a) 绿枝藻 *Ostreococcus* sp.；(b) *Batycoccus* sp.（照片由 Marie-Josèphe Dinet 提供）；(c) 微胞藻 *Micromonas* sp.（图片来自 Guillou et al. 2004）；(d) 赫氏颗石藻 *Emiliania huxleyi*（图片由 Jeremy R. Young 提供）；(e) 蓝隐藻 *Guillardia theta*（照片由 Geoff McFadden 提供）；(f) 圆柱拟脆杆藻 *Fragilariopsis cylindrus*（图片由 Gerhard Dieckmann 提供）；(g) 海链藻 *Thalassiosira pseudonana*（照片由 Virginia Armbrust 提供）；(h) 三角褐指藻 *Phaeodactylum tricornutum*（图片由 Alessandra De Martino and Chris Bowler 提供）；(i) 多列拟菱形藻 *Pseudo - nitzschia multisseries*（图片来源 the Joint Genome Institute）；(j) 长囊水云 *Ectocarpus siliculosus* 菌体的单室孢子囊释放减数孢子细节；(k) 皱波角叉菜 *Chondrus crispus* 小植株（由 Jonas Collé 提供）。

硅藻可被分成两类主要的形态学组群，即中心硅藻纲和羽纹硅藻纲。海链藻属于中心硅藻纲，最近还完成了第二个硅藻—羽纹硅藻纲的三角褐指藻（*P. tricornutum*）的基因组测序（Bowler et al. 2008）。对该硅藻基因组分析的最突出的成果是鉴定了数百个可能是通过水平基因转移获得的细菌基因。这些基因至少占到了基因组的5%，从而表明水平基因转移在硅藻中的程度比在以往研究的浮游真核生物高出一个数量级。

通过两个硅藻的基因组数据能够对硅藻新陈代谢各方面（包括碳、氮代谢和类胡萝卜素的生物合成）进行比较分析（Allen et al. 2006，Coesel et al. 2008，Kroth et al. 2008），结果表明，海洋环境中硅藻的生长不受 CO_2 浓度的限制。这两种硅藻都能编码 C_4 光合作用必需的所有酶（Kroth et al. 2008），从而产生了一个非常有趣的可能性，即尽管该途径耗能高，在 CO_2 受限环境中，比如浮游植物大量繁殖时，C_4 光合作用就具有明显的优势。此外，最近的实验表明大部分硅藻可能拥有基于 C_4 途径的 CO_2 浓缩机制（McGinn and Morel 2008）。

由于褐指藻属（*Phaeodactylum*）在实验中的可控性，其基因组测序存在强烈的争议。它可被转化（Zaslavskaia et al. 2000），因此很多分子工具已被开发用于研究其基因表达和基因功能（Siaut et al. 2007）。最新的基于乙醛的细胞间信号传导技术被用于研究硅藻在海洋生态中的关键作用。硅藻能产生活性醛类物质用于防御（Ianora et al. 2004），在褐指藻的实验室培养过程中额外加入醛基（2E，4E/Z）- 葵二烯醛（DD）会诱导细胞内钙离子浓度瞬变（calcium transients）和一氧化氮的产生，从而使细胞死亡。但是，低浓度乙醛处理过的藻细胞，会产生抗性，以抵御高浓度乙醛的毒性作用。该结果表明这类藻细胞中存在一个特殊系统，能通过分泌物质进行细胞间的交流，这可能在藻华的种群动态变化中发挥重要作用。

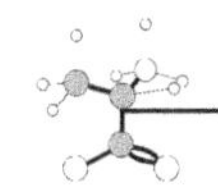

褐指藻也是研究如何运输蛋白至硅藻质体的实验系统。如上所述，由于质体有四层同心状的膜，蛋白运输是个复杂的过程。核编码的蛋白对质体的识别过程中有双靶标序列的参与（Lang et al. 1998）。大规模的测序数据使我们可以对大量识别质体的蛋白进行比较，从而鉴定双靶标序列的保守模块（motif）（Kilian and Kroth 2005）。这些保守区域已通过褐指藻野生型和突变型细胞融合蛋白的表达实验得到研究。

生物技术在褐指藻的研究中得到了很好的应用。一个重要的突破是，该藻通过转化获得了葡萄糖转运基因后，即使在黑暗中也能在碳源培养基上生长（Zaslavskaia et al. 2001）。相对于多数藻类依赖于光照才能生长，这个能力有利于在更简单的条件下进行藻类的培养，特别是大规模培养。

在未来的几年，基因组数据结合分子工具应用于后基因组时代的基因功能分析，将进一步推动硅藻的生物学研究。同样，其他正在进行中的硅藻基因组测序计划，也将有助于我们进一步了解这些硅藻在特定环境中的重要作用。这些测序计划中，包括有毒藻华硅藻拟菱形藻（*Pseudo-nitzschia multieries*）和海上冰块硅藻圆柱拟脆杆藻（*Fragilariopsis cylindrus*）（表6.1）。此外，值得一提的是其他不等鞭毛藻类的基因组也已经获得，包括非光合作用卵菌纲（Tyler et al. 2006）的其中一个种和褐藻类的长囊水云（*Ectocarpus siliculosus*；见下文）。另一个不等鞭毛微藻中的海金藻类的褐潮藻（*Aureococcus anophagefferens*），最近也已完成测序（表6.1）。

6.4.5.2 绿枝藻类基因组

绿枝藻是植物界中的古老支系，是海洋浮游植物中的一个大类。*Ostreococcus. tauri* 是第一个进行基因组测序的绿枝藻类物种（Derelle et al. 2006），紧跟着是 *O. ucimarinus*（Lenik et al. 2007）。这个属有几个特点：*Ostreococcus* 属是已知的最小浮游真核生物，直径小于0.8 μm；每个细胞仅有一个线粒体和叶绿体；它们的基因组极小（*O. tauri* 为2.56 Mbp）而且高度紧凑（例如 *O. tauri*的基因间序列平均仅为196 bp）。*Ostreococcus* 藻类发生基因组精简的条件似乎与寄生虫（如小孢子虫）基因组产生的机制明显不同：前者保留了一整套的基因，本质上是通过消除重复序列和减少非编码区（基因间和内含子）DNA来精简其基因组（Keeling 2007）；而寄生生物体中基因组的缩减则趋向于去除非必需基因，由其宿主执行该基因的功能。这两个藻种还有一个共同的神秘特征；它们各自存在两条特殊的染色体，其成分与组成基因组的其他染色体明显不同，同时还包含大量的转座元件。它们的起源和功能至今仍知之甚少。有趣的是，在两个 *Ostreococcus* 藻种的基因组比较中，发现这两个物

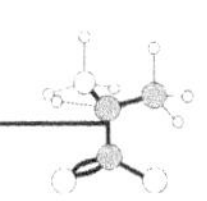

种的基因组染色体表现出较高的共线性，而这两条染色体却只有很低的共线性（Palenik et al. 2007）。基于该结果和基因结构的对比差异，发现其中的一条即2号染色体可能与进化有关（Palenik et al. 2007）。

O. tauri 和 *O. lucimarinus* 是隐秘种的特殊例子。尽管无法通过电子显微镜进行形态学区分，且18S rDNA序列同源性高达99.8%，但两物种间直系同源基因的氨基酸序列同源性只有70%（Palenik et al. 2007）。所以，在基因组水平上它们是明显分开的物种，甚至可划分为单独的属。最近，第三个 *Ostreococcus* 藻种已完成基因组分析，结果也表现出相同的情况。因此，浮游藻类如果只从单细胞形态学上判断，可能严重低估了其多样性。

O. tauri 和 *O. lucimarinus* 的基因组序列给绿枝藻类的生物学研究提供了多方面的信息。*Ostreococcus* 的基因组中存在编码 C_4 光合作用所需酶的全部基因，表明其可能跟硅藻一样，也是用这个途径进行 CO_2 浓缩。这可能是浮游微藻类在多起源进化中的适应性策略。这两个 *Ostreococcus* 的基因组中包含大量的硒蛋白酶基因（Palenik et al. 2007），这似乎是海洋微藻类（包括硅藻）的普遍特征（Lobanov et al. 2007）。与不含硒的蛋白相比，等量的硒蛋白具有更高的酶催化活性，这可能更有利于这类生物在海洋环境中生存。相比于陆生生境来说，水生环境中的生物更偏向于利用硒蛋白（Lobanov et al. 2007）。

另外两株微胞藻的基因组近来也完成了测序（Worden et al. 2009）。它们形态学上一致，被认为是细小微胞藻，但基因组测序表明两者仅有90%的基因是相同的。微胞藻属物种的基因组（20.9和21.9 Mbp）比 *Ostreococcus* 属要大，并且其基因家族数量更多。其中一个基因组上有个非常有趣的特征，就是含有丰富的内含子重复序列，几乎延伸至内含子的供体和受体位点。

绿枝藻类的 *O. tauri* 也用作模式生物，以便通过实验对绿枝藻类进行生物学研究，这类似于通过三角褐指藻（*P. tricornutum*）来研究硅藻。作为模式生物，*O. tauri* 有很多优点，其中最重要的一点是其基因组比较简单，尤其大多数基因是唯一的，而且不是基因家族中的冗余基因，不同于经典的陆生模式植物比如拟南芥，因此是一个研究绿色植物细胞内各种过程的理想系统；另外，其基因间隔区短，有利于启动子区的研究；该物种是单倍体，在遗传学研究方法上具有潜在的优势，即使目前仍无法在实验室条件下完成其增殖周期。目前，已经建立了研究 *O. tauri* 的工具，包括在平板上分离单克隆、携带报道基因进行遗传转化、特定生长条件下的基因表达检测和基因组规模的芯片分析。这些方法已用于研究藻类细胞有丝分裂周期与昼夜节律调控之间的关系（Moulager et al. 2007）。

如同硅藻一样，大量的基因组序列信息可用于绿枝藻类植物的研究，合适的模式生物可用来做后续的基因功能分析。其他即将完成的基因组数据亦将应用于研究不同环境中分离得到的各种绿枝藻类物种（表6.1；Rodríguez et al. 2005，Slapeta et al. 2006）。这些物种的比较基因组信息，将为未来野外和实验室研究提供新的假说。

6.4.5.3 其他微藻类基因组计划

颗石藻（*Coccolithophores*）属于单细胞的定鞭藻类，分布于海洋寡营养区域。该藻可形成大片的藻华，细胞外的碳酸钙颗石保护层在光线的反射下，看起来就像大块的绿松石。这些数量巨大的生物体在生物地球化学循环中起着重要作用，随着个体的死亡，碳经由细胞碎片沉降到海底沉积物中得到固定。一种颗石藻赫氏圆石藻（*Emiliania huxleyi*）已完成基因组测序（表6.1），目前的数据分析有利于对其生物学特点的深入认识。定鞭毛藻类棕囊藻属（*Phaeocystis*）的两个物种也正在计划测序（表6.1），而定鞭金藻（*Prymnesium parvum*）已有EST序列（La Claire 2006）。

最近启动的基因组计划的研究目标是原始或第二次内共生事件中的几种关键生物（图6.1），很快就会得到更多关于内共生事件的信息。灰胞藻不属于海洋藻类，但是其中的蓝载藻（*Cyanophora paradoxa*）（http://www.biology.uiowa.edu/cyanophora/cyanophora_home.htm）也是测序目标之一且占有重要地位，因为发生在该类群中的单一的原始内共生事件导致真核生物中所有藻类获得了质体。隐藻类（cryptophytes or cryptomonads）属于单细胞藻类，在淡水和海洋环境中均有发现。这个类群非常有趣，其质体来源于第二次内共生事件，仍然保留内共生体的核残余（类核体，nucleomorph）。其中一个分布于沿海地区的物种蓝隐藻（*Guillardia theta*），将进行基因组测序（表6.1）。*Chlorarachniophytes* 是海洋中的变形鞭毛生物，属于丝足虫类（Cercozoa）超群，类似于隐藻类，也保留有源自第二次内共生事件的带有类核体的叶绿体。该类群中的 *Bigelowiella natans*（表6.1）的基因组测序计划已经启动。*Bigelowiella natans* 已经有大量的EST序列数据，有趣的是，对这些数据的分析发现了大量基因水平转移的证据，这些基因来自细菌和其他真核植物、不等鞭毛藻类和红藻（Archibald et al. 2003）。据我们所知，目前还没有计划对光合生物眼虫门（*Euglenozoa*）进行基因组测序，尽管其中的淡水藻 *Euglena gracilis* 已经有EST序列（Durnford and Gray 2006）。

6.4.5.4 甲藻

尽管只有约一半的甲藻（Dinoflagellates）具有叶绿体，能够进行光合作

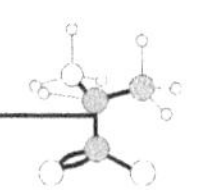

用，但总的来说，这个类群是海洋中的重要初级生产者。大多数甲藻的叶绿体被认为是起源于红藻参与的第二次内共生事件。甲藻的叶绿体有几个与众不同的特点，比如有辅助色素多甲藻黄素（peridinin）和叶绿体的三层膜结构（Nassoury et al. 2003，Patron et al. 2005）。另外，叶绿体的基因组高度简化，其独立的小环状染色体只保留小部分基因（Zhang et al. 1999）。其他叶绿体功能所需基因，包括核编码的变形杆菌型 II RuBisCO 基因，已转移至细胞核里。正如前面所述，某些甲藻丢失了起源于红藻的叶绿体后，代之可能是通过第三次内共生吞噬来自光合真核生物（如定鞭藻或硅藻）的质体（Inagaki et al. 2000，Bhattacharya et al. 2004）。同样，甲藻的细胞核也表现出独特的特征，其 DNA 含量极大（3～250 pg，或相当于 3000～215000 Mbp/细胞），组成上百条的染色体，例如塔玛亚历山大藻（*Alexandrium tamarense*）有 143 条染色体（Hackett et al. 2005）。由于细胞核中 DNA 浓度高，超过了 10 倍折叠所需的基本 DNA 结合蛋白浓度，导致 DNA 浓缩成液晶态并附着在核膜上，形成了独特的"甲藻核"式细胞核形态（Spector 1984，Gautier et al. 1986，LaJeunesse et al. 2005，Hackett et al. 2005）。甲藻的大基因组严重阻碍了基因组方法的应用，以致于目前还没有开展该组群基因组的测序。虽然如此，还是可以通过对 cDNA 克隆和基因组中分离的单个基因进行测序获得可观的信息。目前，几个甲藻物种已有 EST 数据，特别是那些能形成有毒代谢产物的物种（Tanikawa et al. 2004，Hackett et al. 2005，Bachvaroff and Place 2008）。

6.4.6 巨藻基因组

不同于浮游藻类群体以单细胞藻类形式为主，沿海生态系统中的红、褐、绿等大型藻类属于优势种群，在沿岸生态系统中发挥重要作用。它们不仅作为初级生产者，同时也给种类繁多的海洋生物提供栖息地。从进化角度上看，大型海藻也非常重要。褐、红和绿大型藻类代表了真核系统发育树中的不同分支，每个类群独立进化成复杂的多细胞生物体。如真核中其他三个支系（后生动物、陆生植物和真菌），这三个海藻群的复杂多细胞系统的独立进化在真核生物的进化地位中亦尤其重要。因此，海洋大型藻是认识复杂发育系统进化的重要资源。

海藻还有很多其他特性。不同于被子植物的受精是发生在几层母体组织内（Brownlee et al. 2001），褐藻是体外受精，配子释放到周围海水中进行融合，这是研究早期发育事件的一大优势。大型藻的生活周期也具有很高的多样性，且大多数都非常复杂，这使它们为研究在不同生活周期如何进化及在

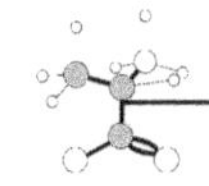

进化时如何保持稳定状态提供了理想的模型（Coelho et al. 2007）。最后，海藻也是食品和一些工业化合物（如海藻酸盐、琼脂、卡拉胶）的重要来源。海藻是重要的资源，应用于食物、化妆品和化肥工业等，估计每年全球总产值达45亿欧元（McHugh 2003）。同时，作为活性生物分子的新来源，它们越来越受到重视。例如IODUS 40来源于掌状海带（*Laminaria digitata*）的细胞壁提取物，能激发农作物的先天性免疫应激（Klarzynski et al. 2000）。

下面将会讲述基因组学方法如何应用于大型藻3个系统发育组群的生物学研究。

6.4.6.1 褐藻

褐藻纲中拥有进化最高级、形态最复杂的物种，属于海带目（*Laminariales*）或墨角藻目（*Fucales*），包括生活在潮间带的墨角藻类和在潮下带形成密林的巨大海带。以往对褐藻的研究很大部分集中于这两个类群，这不仅是因为其生态学上的重要性，也由于在工业应用上的重要性（主要是海带；McHugh 2003，Bartsch et al. 2008），或因其已经发展成为基础研究的模型（墨角藻目；见下文）。就基因组分析方法在这两个目的应用而言，从掌状海带（*laminaria digitata*）中产生的EST序列总共有3000个，包括来自于生活史的孢子和配子世代中表达的基因和由孢子体衍生的原生质体的表达基因。这些序列信息用于研究不同生化过程，包括二氧化碳浓缩机制、细胞壁的生物合成、卤元素代谢和应激反应等。与之相比，尽管墨藻类海藻早已广泛用于早期胚胎发生及生态学的研究（Serrao et al. 1996，Coyer et al. 2007，Muhlin et al. 2008），但该类群的EST序列信息直到近期才获得。

虽然EST方法可以深入了解海带目和墨角藻目物种的生物学特点，但目前来说，这些生物体并不适合使用其他的基因组学工具，比如基因组测序或功能基因分析。主要有两方面原因：第一，这两类生物体的基因组大（例如掌状海带有650 Mbp，锯齿形墨角藻有1095 Mbp；Le Gall et al. 1993，Peters et al. 2004）。第二，它们都是生活周期长的大生物个体，限制了在实验室研究中的应用。为了突破局限性，近来开展了褐藻类模式生物的研究，从而能更好地进行基因组学研究。该研究选择了丝状长囊水云（*Ectocarpus siliculosus*）作为褐藻的模式生物（Peters et al. 2004）。

长囊水云属于水云目（*Ectocarpales*），是褐藻门中进化最高等物种中的一员，与海带关系密切。水云属（*Ectocarpus*）的实验室研究始于19世纪（参阅Charrier et al. 2008）。早期研究包括对该物种的生物学描述，随后是生殖生物学和生活史的研究。还包括超微结构、光合作用、碳固定作用、信息素的

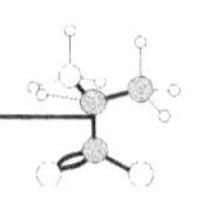

产生、配子识别和与病原体的相互作用，特别是与病毒 EsV－1 的相互作用（该病毒感染长囊水云后可整合到其基因组上）。

水云属具有多个特点，使其更适合应用基因组学和遗传学工具，因此作为研究褐藻的模式生物。其中尤为重要的是其基因组仅有 200 Mbp，明显小于海带和墨角藻类褐藻的基因组。还有其他一些适用于实验室工作的特性，包括其个体小、可在实验室条件下于培养皿中完成生活周期、高繁殖力、快速生长（生活周期只需 2 个月）和易于进行遗传物质交换（Peters et al. 2004）。基于该生物建立一套体系，不仅可获得简单的基因信息，更可以应用遗传工具进行深入的基因功能研究。

对水云属基因组的测序已在 2007 年完成，现在也有大量其他可利用的基因组工具，包括基因组嵌合芯片数据和不同时期生活史中的 91000 个 cDNA 序列。传统的实验室遗传操作，比如诱变、突变系的筛选、遗传交换和互补分析，都可开展。一些细胞生物学技术工具，包括体内体外成像技术、显微注射和原生质体再生技术等，也可有效应用。另外，目前正在开发其他的技术，譬如遗传图谱、遗传转化程序、基于 RNA 干扰的基因敲除技术和突变位点的定点克隆等。

基于以上可使用的技术工具，水云属被广泛用于藻类生物学研究。最近关于生活周期调控的研究工作就是个很好的例子（Coelho et al. 2007，Peters et al. 2008）。水云属有单倍—二倍体生活周期现象，涉及到孢子体和配子体的世代交替（Müller 1967）。通过干扰该生活过程筛选突变体以寻找两个世代交替的控制开关。在一个突变体 *immediate upright*（*imm*）中，孢子体部分转变为配子体，表现出类似于早期发育配子体的模式。但产生的还是孢子而非配子。通过芯片杂交和定量 PCR 分析突变体表达基因，发现大量正常情况下在配子体世代中表达的基因在该突变体的孢子体中也表达。未来的工作包括对其他突变株系的分析、在基因组层面上分析突变体基因调控变化、以及通过定点克隆来鉴定突变基因定等。在未来的几年，有望在控制生活周期的调控机制方面获得更深刻的认识。

类似方法已开始应用于水云属其他生物学特点的研究中，包括性别决定、形态发生、细胞壁生物合成以及对生物与非生物胁迫的应激性方面。很多获得的信息都与褐藻这个类群相关，具有普遍性，在鉴定新的生物分子和未来基于基因组信息进行海藻育种方面的应用具有重大潜力。

6.4.6.2 红藻

大型红藻广泛分布于海岸线区域的栖息地，从海岸的上限一直到透光层

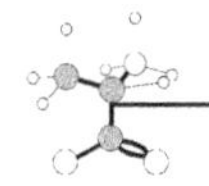

的下限。其中几个物种可开发成食品，或具有工业用途。红藻的产量每年约208万吨，价值约20亿美金（FAO 2003）。紫菜属（*Porphyra*）的主要作为海苔食品，而麒麟菜属（*Eucheuma*）和卡帕藻属（*Kappaphycus*）则用来制造卡拉胶。红藻是古老的真核生物类群，估计起源于15亿年前（Yoon et al. 2004）。表征红藻的多细胞发育和性别的化石最早可追溯到大概12亿年前，说明该类群可能是最早进化成复杂多细胞体的生物（Butterfield 2000）。同褐藻一样，红藻的演化历史有很多独特的性质，例如复杂的生活周期和异常的代谢途径，主要是指氧化脂类、细胞多糖和卤代化合物的产生（Siegel and Siegel 1973，Manley 2002，Bouarab et al. 1999，Coelho et al. 2007）。如上所述，红藻在光合作用的演化历史中也起着关键作用，尤其是通过内共生作用导致次级叶绿体的进化。

红藻的这些特征也引起了一些争议，即是否使用基因组学工具进行红藻的基因组研究。现在有一些红藻已经获得了EST序列，包括条斑紫菜（*Porphyra yezoensis*）（20000条EST；Nikaido et al. 2000，Asamizu et al. 2003）、皱波角叉菜（*Chondrus crispus*）（4056条EST；Collén et al. 2006b）、细基江蓠（*Gracilaria tenuistipitata*）（3000 EST；Pi Nyvall 个人通讯）和 高氏红藻（*Griffithsia okiensis*）（1104 EST；Lee et al. 2007）。cDNA芯片已用于检测皱波角叉菜（*Chondrus crispus*）（Collén et al. 2006a）和条斑紫菜（*Porphyra yezoensis*）（Kitade et al. 2008）的基因表达。但是，目前唯一已完成基因组测序的红藻物种 *C. merolae*，是来源于酸性热泉、具有高度简化和罕见的基因组的一个单细胞生物体。第二个红藻基因组测序计划，*Galdieria sulphuraria* 的测序正在进行（http://genomics.msu.edu/galdieria/about.html），但它是非海洋的极端环境中的单细胞红藻。以上两个物种都有着很小而高度分化的基因组，因此限制了人们对红藻普遍特征的认识。因此，对更“典型”红藻的基因组测序，无论是在红藻的生物学研究或作为参照追溯内共生事件中获得基因的起源，都极其有用。在这些研究需求的驱动下，最近启动了两个红藻的基因组计划，即真红藻类物种皱波角叉菜（*Chondrus crispus*）和红毛菜类物种条斑紫菜（*Porphyra yezoensis*）。

这两个物种的基因组计划很可能有高度的兼容性。这两个海藻都在特定栖息地中发挥着重要的生态学作用。在实验室方面，这两个物种的操作相对简易，而且都有着广泛的实验室研究基础。紫菜属（*Porphyra*）物种可能最适合于实验室工作的开展，以往的研究曾报道突变株的分离（Ohme and Miura 1988，Mitman and van der Meer 1994，Yan et al. 2000）、遗传标记的鉴定（Park

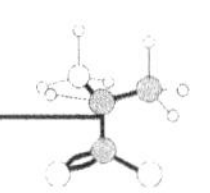

et al. 2007)、原生质体的制备和再生（Waaland et al. 1990），整体原位杂交技术（Shimizu et al. 2004）和遗传转化体系构建（Cheney et al. 2001，He et al. 2001，Lin et al. 2001）的进展。实际上，相关的物种条斑紫菜（*Porphyra yezoensis*），已被推荐为大型藻模式生物的候选者（Kitade et al. 2004，Waaland et al. 2004）。紫菜属的物种基因组约400 Mbp（Kapraun2005），相对脐形紫菜（*P. umbilicalis*）来说，真红藻皱波角叉菜（*Chondrus crispus*）更具优势，因为其基因组更小，只有150 Mbp（ Peters et al. 2004)，且更能代表"典型"红藻（大多数的红藻属于真红藻类）。同时，皱波角叉菜（*Chondrus crispus*）和脐形紫菜（*P. umbilicalis*）基因组在应用性研究方面非常有意思，皱波角叉菜（*Chondrus crispus*）有助于认识卡拉胶的生物合成，而脐形紫菜（*P. umbilicalis*）则因为紫菜品种（海苔）数十亿美元产业而意义重大。另外，很重要的一点是，皱波角叉菜（*Chondrus crispus*）和脐形紫菜（*P. umbilicalis*）的共同祖先可追溯到约14亿年前，因此这两个物种代表了进化上不同的组群（Yoon et al. 2004)。

6.4.6.3 绿藻

在海洋绿色大型藻方面，基因组方法的应用还没达到像红藻和褐藻的程度。可能很大部分原因是其很少能应用在工业上。不过，绿色大型藻中比如石莼属（*Ulva*）能导致生物污染，这是因为在高营养条件下（比如化肥的污染加上夏季温带地区的高温及高光通量），石纯会大量生长而形成危害。在绿色海洋大型藻中，石纯属的研究最广泛，并建立了一些 EST 序列（Bryhni 1974，Fjeld and Løvle 1976，Reddy et al. 1992)。不过，目前还没有计划要对该类群的物种进行全基因组测序。

海草属于被子植物，所以不属于藻类的范畴。这里需要注意的是，虽然这里提及海草，主要是其跟大型藻一样在很多沿海地区是具有重要生态作用的光合生物。海草能在沙地或泥泞的海岸线上形成广阔的"草地"，为更多的其他生物提供栖息地。但是，这个重要的生态系统正在世界各地受到人类活动的威胁。目前已启动对一种海草（鳗草 *Zostera marina*）的基因组测序测序（表6.1)。

6.5 展望

当前，基因组学方法应用于藻类生物学的广泛研究中，包括真核生物起源的基本问题、各种远洋和沿岸生态系统中藻类的重要作用、深入了解藻类

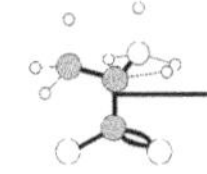

生物学特性。未来的研究手段工作将会大大丰富对这些领域的认识。虽然序列数据仍然零碎，很多主要的真核生物在分子水平上的认识依然匮乏。未来的工作会填补这些方面的空缺，这将给真核生物的系统发育提供强有力的支持。不过，主要的真核群体可能还是有待确定（Not et al. 2007），到时需要对目前真核生物进化树中的超群重新进行调整。新的信息同样有利于研究进化史上起着重要作用的各个内共生事件。

海洋藻类生态系统研究中，基因组学方法的应用才刚起步。例如，浮游性微藻已经获得了少量基因组序列信息，虽然这些信息对这些生物如何作用于其生态系统提供重要的信息，但是这些信息也仅触及这些复杂生态系统的表层。未来的研究将会结合环境样本测序数据和单个物种的基因组信息，以更好地诠释系统的生物多样性和其中组成物种的特征。这些方法同时也会结合基因表达分析的方法，比如 cDNA 测序或芯片分析，不仅可获得生态系统中基因的总量，也可以知道在特定条件下活跃基因的信息。类似的研究计划已在沿海环境的大型藻类中开展。就目前而言，具有大基因组的大型藻，限制了基因组测序计划的开展，但是在未来几年，这些技术的快速发展将有望改善这种情况。总的来说，这些未来的研究手段将会为藻类在海洋生态系统中的地位提供更多更详细的描述，特别是其在生物地球化学循环中的关键性影响、对气候变化和其他人为影响的应激反应。

藻类在系统中分布广泛，横跨几个非常古老的真核生物组群，从中可得到一个重要的结论，就是藻类在发现新的生物进程（包括代谢途径、信号通路、细胞过程、发育调控等方面）具有巨大的潜力。藻类基因组计划中有一个令人意外的结果是，总有高比例的基因数据不能匹配到公共数据库。例如，在硅藻类的海链藻（*T. pseudonana*）基因组中（Armbrust et al. 2004），大概有一半的基因信息未能比对上数据库。因此，未来的一个主要挑战是研究这些未知基因的功能，而通过建立大量微藻和大型藻类的模式生物则是迈向这个目标的重要一步。

参考文献

Allen AE, Vardi A, Bowler C (2006) An ecological and evolutionary context for integrated nitro-gen metabolism and related signaling pathways in marine diatoms. Curr Opin Plant Biol 9:264 –273

Archibald JM (2007) Nucleomorph genomes: structure, function, origin and evolution. Bioessays 29:392 –402

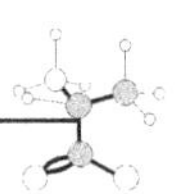

Archibald JM, Rogers MB, Toop M, Ishida K, Keeling PJ (2003) Lateral gene transfer and the evo-lution of plastid-targeted proteins in the secondary plastid-containing alga *Bigelowiella natans*. Proc Natl Acad Sci U S A 100:7678 - 7683

Armbrust EV, Berges JA, Bowler C, Green BR, Martinez D, Putnam NH, Zhou S, Allen AE, Apt KE, Bechner M, Brzezinski MA, Chaal BK, Chiovitti A, Davis AK, Demarest MS, Detter JC, Glavina T, Goodstein D, Hadi MZ, Hellsten U, Hildebrand M, Jenkins BD, Jurka J, Kapitonov VV, Kröger N, Lau WW, Lane TW, Larimer FW, Lippmeier JC, Lucas S, Medina M, Montsant A, Obornik M, Parker MS, Palenik B, Pazour GJ, Richardson PM, Rynearson TA, Saito MA, Schwartz DC, Thamatrakoln K, Valentin K, Vardi A, Wilkerson FP, Rokhsar DS (2004) The genome of the diatom *Thalassiosira pseudonana*: Ecology, evolution, and metabolism. Science 306:79 - 86

Asamizu E, Nakajima M, Kitade Y, Saga N, Nakamura Y, Tabata S (2003) Comparison of RNA expression profiles between two generations of *Porphyra yezoensis* (Rhodophyta), based on expressed sequence tag frequency analysis. J Phycol 39:923 - 330

Bachvaroff TR, Concepcion GT, Rogers CR, Herman EM, Delwiche CF (2004) Dinoflagellate expressed sequence tags data indicate massive transfer of chloroplast genes to the nuclear genome. Protist 155:65 - 78

Bachvaroff TR, Place AR (2008) From stop to start: tandem gene arrangement, copy number and trans - splicing sites in the dinoflagellate *Amphidinium carterae*. PLoS ONE 3:e2929

Baldauf SL (2008) An overview of the phylogeny and diversity of eukaryotes. J Syst Evol 46: 263 - 273

Bartsch I, Wiencke C, Bischof K, Buchholz CM, Buck BH, Eggert A, Feuerpfeil P, Hanelt D, Jacobsen S, Karez R, Karsten U, Molis M, Roleda M, Schubert H, Schumann R, Valentin K, Weinberger F, Wiese J (2008) The genus *Laminaria* sensu lato: recent insights and developments. Eur J Phycol 43:1 - 86

Bauer I, Maranda L, Young KA, Shimizu Y, Fairchild C, Cornell L, MacBeth J, Huang S (1995) Isolation and structure of caribenolide I, a highly potent antitumor macrolide from a cul-ture free-swimming Caribbean dinoflagellate, *Amphidinium* sp. S1 - 36 - 5. J Org Chem 60: 1084 - 1086

Berger F, Taylor A, Brownlee C (1994) Cell fate determination by the cell wall in early *Fucus* development. Science 263:1421 - 1423

Bhattacharya D, Yoon HS, Hackett JD (2004) Photosynthetic eukaryotes unite: endosymbiosis connects the dots. Bioessays 26:50 - 60

Bouarab K, Potin P, Correa J, Kloareg B (1999) Sulfated oligosaccharides mediate the interaction between a marine red alga and its green algal pathogenic endophyte. Plant Cell 11(9):1635 - 1650

Bouget FY, Berger F, Brownlee C (1998) Position dependent control of cell fate in the *Fucus* embryo:role of intercellular communication. Development 125:1999 – 2008

Bowler C, Allen AE, Badger JH, Grimwood J, Jabbari K, Kuo A, Maheswari U, Martens C, Maumus F, Otillar RP, Rayko E, Salamov A, Vandepoele K, Beszteri B, Gruber A, Heijde M, Katinka M, Mock T, Valentin K, Verret F, Berges JA, Brownlee C, Cadoret JP, Chiovitti A, Choi CJ, Coesel S, De Martino A, Detter JC, Durkin C, Falciatore A, Fournet J, Haruta M, Huysman MJ, Jenkins BD, Jiroutova K, Jorgensen RE, Joubert Y, Kaplan A, Kröger N, Kroth PG, La Roche J, Lindquist E, Lommer M, Martin-Jézéquel V, Lopez PJ, Lucas S, Mangogna M, McGinnis K, Medlin LK, Montsant A, Secq MP, Napoli C, Obornik M, Parker MS, Petit JL, Porcel BM, Poulsen N, Robison M, Rychlewski L, Rynearson TA, Schmutz J, Shapiro H, Siaut M, Stanley M, Sussman MR, Taylor AR, Vardi A, von Dassow P, Vyverman W, Willis A, Wyrwicz LS, Rokhsar DS, Weissenbach J, Armbrust EV, Green BR, Van de Peer Y, Grigoriev IV (2008) The Phaeodactylum genome reveals the evolutionary history of diatom genomes. Nature 456:239 – 244

Brownlee C, Bouget FY, Corellou F (2001) Choosing sides:establishment of polarity in zygotes of fucoid algae. Semin Cell Dev Biol 12:345 – 351

Bryhni E (1974) Control of morphogenesis in the multicellular alga *Ulva mutabilis.* Defect in cell wall production. Dev Biol 37:273 – 277

Butterfield NJ (2000) *Bangiomorpha pubescens* n. gen. , n. sp. ; implications for the evolution of sex, multicellularity, and the Mesoproterozoic/Neoproterozoic radiation of eukaryotes. Paleobiol 26:386 – 404

Charrier B, Coelho SM, Le Bail A, Tonon T, Michel G, Potin P, Kloareg B, Boyen C, Peters AF, Cock JM (2008) Development and physiology of the brown alga *Ectocarpus siliculosus*:two centuries of research. New Phytol 177:319 – 332

Cheney D, Metz B, Stiller J (2001) *Agrobacterium* – mediated genetic transformation in the macroscopic marine red alga *Porphyra yezoensis.* J Phycol 37:11

Coelho S, Peters AF, Charrier B, Roze D, Destombe C, Valero M, Cock JM (2007) Complex life cycles of multicellular eukaryotes:new approaches based on the use of model organisms. Gene 406:152 – 170

Coelho SM, Taylor AR, Ryan KP, Sousa-Pinto I, Brown MT, Brownlee C (2002) Spatiotemporal patterning of reactive oxygen production and Ca(2 +) wave propagation in *Fucus* rhizoid cells. Plant Cell 14:2369 – 2381

Coesel S, Oborník M, Varela J, Falciatore A, Bowler C (2008) Evolutionary origins and functions of the carotenoid biosynthetic pathway in marine diatoms. PLoS ONE 3:e2896

Coleman FC, Williams SL (2002) Overexploiting marine ecosystem engineers:potential consequences for biodiversity. Trends Ecol Evol 17:40 – 44

Collén J, Hervé C, Guisle-Marsollier I, Leger J, Boyen C (2006a) Expression profiling of *Chondrus crispus* (Rhodophyceae) after exposure to methyl jasmonate. J Exp Bot 57:3869 – 3881

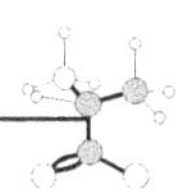

Collén J, Roeder V, Rousvoal S, Collin O, Kloareg B, Boyen C (2006b) An expressed sequence tag analysis of thallus and regenerating protoplasts of *Chondrus crispus* (Gigartinales, Rhodophyceae). J Phycol 42:104 – 112

Corellou F, Bisgrove SR, Kropf DL, Meijer L, Kloareg B, Bouget FY (2000) A S/M DNA replication checkpoint prevents nuclear and cytoplasmic events of cell division including cen-trosomal axis alignment and inhibits activation of cyclin-dependent kinase-like proteins in fucoid zygotes. Development 127:1651 – 1660

Corellou F, Brownlee C, Kloareg B, Bouget FY (2001) Cell cycle-dependent control of polarised development by a cyclin-dependent kinase-like protein in the *Fucus* zygote. Development 128: 4383 – 4392

Coyer JA, Hoarau G, Stam WT, Olsen JL (2007) Hybridization and introgression in a mixed pop – ulation of the intertidal seaweeds *Fucus evanescens* and *F. serratus*. J Evol Biol 20:2322 – 2333

Crépineau F, Roscoe T, Kaas R, Kloareg B, Boyen C (2000) Characterisation of complementary DNAs from the expressed sequence tag analysis of life cycle stages of *Laminaria digitata* (Phaeophyceae). Plant Mol Biol 43:503 – 513

Derelle E, Ferraz C, Escande ML, Eychenié S, Cooke R, Piganeau G, Desdevises Y, Bellec L, Moreau H, Grimsley N (2008) Life-cycle and genome of OtV5, a large DNA virus of the pelagic marine unicellular green alga *Ostreococcus tauri*. PLoS ONE 3:e2250

Derelle E, Ferraz C, Rombauts S, Rouze P, Worden AZ, Robbens S, Partensky F, Degroeve S, Echeynie S, Cooke R, Saeys Y, Wuyts J, Jabbari K, Bowler C, Panaud O, Piegu B, Ball SG, Ral JP, Bouget FY, Piganeau G, De Baets B, Picard A, Delseny M, Demaille J, Van de Peer Y, Moreau H (2006) Genome analysis of the smallest free-living eukaryote *Ostreococcus tauri* unveils many unique features. Proc Natl Acad Sci U S A 103:11647 – 11652

Duggins DO, Simenstad CA, Estes JA (1989) Magnification of secondary production by kelp detritus in coastal marine ecosystems. Science 245:170 – 173

Durnford DG, Gray MW (2006) Analysis of Euglena gracilis plastid-targeted proteins reveals different classes of transit sequences. Eukaryot Cell 5:2079 – 2091

FAO (2003) A guide to the seaweed industry. FAO Fisheries technical paper 441

Falkowski PG, Katz ME, Knoll AH, Quigg A, Raven JA, Schofield O, Taylor FJ (2004) The evolution of modern eukaryotic phytoplankton. Science 305:354 – 360

Faust MA, Gulledge RA (2002) Identifying harmful marine dinoflagellates. Contributions from the United States National Herbarium 42:1 – 144

Field CB, Behrenfeld MJ, Randerson JT, Falkowski P (1998) Primary production of the biosphere: integrating terrestrial and oceanic components. Science 281:237 – 240

Fjeld A, Løvle A (1976) Genetics of multicellular algae. In: Lewin, RA (ed) The genetics of

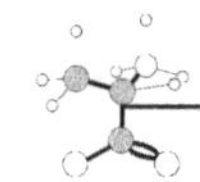

algae, Blackwell Scientific Publications, Oxford, pp 219 – 235

Gautier A, Michel-Salamin L, Tosi – Couture E, McDowall AW, Dubochet J (1986) Electron microscopy of the chromosomes of dinoflagellates in situ: confirmation of Bouligand's liquid crystal hypothesis. J Ultrastruc Mol Struct Res 97: 10 – 30

Giovannoni SJ, Britschgi TB, Moyer CL, Field KG (1990) Genetic diversity in Sargasso Sea bacterioplankton. Nature 345: 60 – 63

Goddard H, Manison NF, Tomos D, Brownlee C (2000) Elemental propagation of calcium signals in response-specific patterns determined by environmental stimulus strength. Proc Natl Acad Sci USA 97: 1932 – 1937

Guillou L, Eikrem W, Chrétiennot-Dinet MJ, Le Gall F, Massana R, Romari K, Pedrós – Alió C, Vaulot D (2004) Diversity of picoplanktonic prasinophytes assessed by direct nuclear SSU rDNA sequencing of environmental samples and novel isolates retrieved from oceanic and coastal marine ecosystems. Protist 155: 193 – 214

Guillou L, Moon-Van Der Staay SY, Claustre H, Partensky F, Vaulot D (1999) Diversity and abundance of Bolidophyceae (Heterokonta) in two oceanic regions. Appl Environ Microbiol 65: 4528 – 4536

Guiry MD, Guiry GM (2008) AlgaeBase. Worldwide electronic publication, National University of Ireland, Galway. http://www. algaebase. org.

Hackett JD, Scheetz TE, Yoon HS, Soares MB, Bonaldo MF, Casavant TL, Bhattacharya D (2005) Insights into a dinoflagellate genome through expressed sequence tag analysis. BMC Genomics 6: 80

Hackett JD, Yoon HS, Soares MB, Bonaldo MF, Casavant TL, Scheetz TE, Nosenko T, Bhattacharya D (2004) Migration of the plastid genome to the nucleus in a peridinin dinoflagellate. Curr Biol 14: 213 – 218

He P, Yao Q, Chen Q, Guo M, Xiong A, Wu W, Ma J (2001) Transferring and expression of glucose oxidase gene-gluc in *Porphyra yezoensis*. J Phycol 37(suppl): 22

Hutchinson GE (1961) The paradox of the plankton. Amer Nat 95: 137 – 145

Ianora A, Miralto A, Poulet SA, Carotenuto Y, Buttino I, Romano G, Casotti R, Pohnert G, Wichard T, Colucci – D'Amato L, Terrazzano G, Smetacek V (2004) Aldehyde suppression of copepod recruitment in blooms of a ubiquitous planktonic diatom. Nature 429: 403 – 407

Inagaki Y, Dacks JB, Doolittle WF, Watanabe KI, Ohama T (2000) Evolutionary relation-ship between dinoflagellates bearing obligate diatom endosymbionts: insight into tertiary endosymbiosis. Int J Syst Evol Microbiol 50: 2075 – 2081

Kapraun DF (2005) Nuclear DNA content estimates in multicellular green, red and brown algae: phylogenetic considerations. Ann Bot 95: 7 – 44

Keeling PJ (2004) Diversity and evolutionary history of plastids and their hosts. Amer J Bot

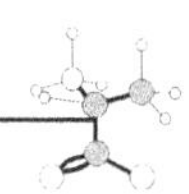

91:1481 - 1493

Keeling PJ (2007) *Ostreococcus tauri*: seeing through the genes to the genome. Trends Genet 23:151 - 154

Keeling PJ, Palmer JD (2008) Horizontal gene transfer in eukaryotic evolution. Nat Rev Genet 9:605 - 628

Kilian O, Kroth PG (2005) Identification and characterization of a new conserved motif within the presequence of proteins targeted into complex diatom plastids. Plant J 41:175 - 183

Kitade Y, Asamizu E, Fukuda S, Nakajima M, Ootsuka S, Endo H, Tabata S, Saga N (2008) Identification of genes preferentially expressed during asexual sporulation in Porphyra yezoen-sis gametophytes (bangiales, rhodophyta). J Phycol 44:113 - 123

Kitade Y, Iitsuka O, Fukuda S, Saga N (2004) *Porphyra yezoensis* as a model plant for genome sciences. Jpn J Phycol 52:129 - 131

Klarzynski O, Plesse B, Joubert J - M, Yvin J - C, Kopp M, Kloareg B, Fritig B (2000) Linear b-1,3 glucans are elicitors of defense responses in tobacco. Plant Physiol 124:1027 - 1037

Kroth PG, Chiovitti A, Gruber A, Martin-Jezequel V, Mock T, Parker MS, Stanley MS, Kaplan A, Caron L, Weber T, Maheswari U, Armbrust EV, Bowler C (2008) A model for carbohy-drate metabolism in the diatom *Phaeodactylum tricornutum* deduced from comparative whole genome analysis. PLoS ONE 3:e1426

La Claire JW 2nd. (2006) Analysis of expressed sequence tags from the harmful alga, *Prymnesium parvum* (Prymnesiophyceae, Haptophyta). Mar Biotechnol 8:534 - 546

LaJeunesse TC, Lambert G, Andersen RA, Coffroth MA, Galbraith DW (2005) *Symbiodinium* (Pyrrhophyta) genome sizes (DNA content) are smallest among dinoflagellates. J Phycol 41:880 - 886

Lang M, Apt KE, Kroth PG (1998) Protein transport into "complex" diatom plastids utilizes two different targeting signals. J Biol Chem 273:30973 - 30978

Le Gall Y, Brown S, Marie D, Mejjad M, Kloareg B (1993) Quantification of nuclear DNA and G - C content in marine macroalgae by flow cytometry of isolated nuclei. Protoplasma 173: 123 - 132

Lee H, Lee HK, An G, Lee YK (2007) Analysis of expressed sequence tags from the red alga *Griffithsia okiensis*. J Microbiol 45:541 - 546

Lin CM, Larsen J, Yarish C, Chen T (2001) A novel gene transfer in *Porphyra*. J Phycol 37: 31

Lobanov AV, Fomenko DE, Zhang Y, Sengupta A, Hatfield DL, Gladyshev VN (2007)

Evolutionary dynamics of eukaryotic selenoproteomes: large selenoproteomes may associate with aquatic life and small with terrestrial life. Genome Biol 8(9): R198

Lovejoy C, Massana R, Pedrós - Alió C (2006) Diversity and distribution of marine microbial

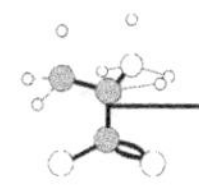

eukaryotes in the Arctic Ocean and adjacent seas. Appl Environ Microbiol 72:3085 – 3095

Manley SL (2002) Phytogenesis of halomethanes: A product of selection or a metabolic accident? Biogeochem 60:163 – 180

Massana R, Balague V, Guillou L, Pedros – Alio C (2004) Picoeukaryotic diversity in an oligotrophic coastal site studied by molecular and culturing approaches. FEMS Microbiol Ecol 50:231 – 243

Massana R, Guillou L, Díez B, Pedrós – Alió C (2002) Unveiling the organisms behind novel eukaryotic ribosomal DNA sequences from the ocean. Appl Environ Microbiol 68:4554 – 4558

Matsuzaki M, Misumi O, Shin-I T, Maruyama S, Takahara M, Miyagishima SY, Mori T, Nishida K, Yagisawa F, Nishida K, Yoshida Y, Nishimura Y, Nakao S, Kobayashi T, Momoyama Y, Higashiyama T, Minoda A, Sano M, Nomoto H, Oishi K, Hayashi H, Ohta F, Nishizaka S, Haga S, Miura S, Morishita T, Kabeya Y, Terasawa K, Suzuki Y, Ishii Y, Asakawa S, Takano H, Ohta N, Kuroiwa H, Tanaka K, Shimizu N, Sugano S, Sato N, Nozaki H, Ogasawara N, Kohara Y, Kuroiwa T (2004) Genome sequence of the ultrasmall unicellular red alga *Cyanidioschyzon merolae* 10D. Nature 428:653 – 657

McGinn PJ, Morel FM (2008) Expression and inhibition of the carboxylating and decarboxylating enzymes in the photosynthetic C4 pathway of marine diatoms. Plant Physiol 146:300 – 309

McHugh DJ (2003) A guide to the seaweed industry. FAO Fisheries Technical Paper No. 441. FAO, Rome, 105 pp.

Medlin LK, Metfies K, Mehl H, Wiltshire K, Valentin K (2006) Picoeukaryotic plankton diversity at the Helgoland time series site as assessed by three molecular methods. Microb Ecol 52:53 – 71

Merchant SS, Prochnik SE, Vallon O, Harris EH, Karpowicz SJ, Witman GB, Terry A, Salamov A, Fritz-Laylin LK, Maréchal-Drouard L, Marshall WF, Qu LH, Nelson DR, Sanderfoot AA, Spalding MH, Kapitonov VV, Ren Q, Ferris P, Lindquist E, Shapiro H, Lucas SM, Grimwood J, Schmutz J, Cardol P, Cerutti H, Chanfreau G, Chen CL, Cognat V, Croft MT, Dent R, Dutcher S, Fernández E, Fukuzawa H, González-Ballester D, González-Halphen D, Hallmann A, Hanikenne M, Hippler M, Inwood W, Jabbari K, Kalanon M, Kuras R, Lefebvre PA, Lemaire SD, Lobanov AV, Lohr M, Manuell A, Meier I, Mets L, Mittag M, Mittelmeier T, Moroney JV, Moseley J, Napoli C, Nedelcu AM, Niyogi K, Novoselov SV, Paulsen IT, Pazour G, Purton S, Ral JP, Riaño-Pachón DM, Riekhof W, Rymarquis L, Schroda M, Stern D, Umen J, Willows R, Wilson N, Zimmer SL, Allmer J, Balk J, Bisova K, Chen CJ, Elias M, Gendler K, Hauser C, Lamb MR, Ledford H, Long JC, Minagawa J, Page MD, Pan J, Pootakham W, Roje S, Rose A, Stahlberg E, Terauchi AM, Yang P, Ball S, Bowler C, Dieckmann CL, Gladyshev VN, Green P, Jorgensen R, Mayfield S, Mueller-Roeber B, Rajamani S, Sayre RT, Brokstein P, Dubchak I, Goodstein D, Hornick L, Huang YW, Jhaveri J, Luo Y, Martínez D, Ngau WC, Otillar B, Poliakov A, Porter A, Szajkowski L, Werner G, Zhou K, Grigoriev IV, Rokhsar

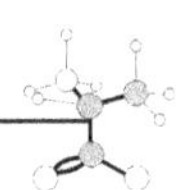

DS, Grossman AR (2007) The Chlamydomonas genome reveals the evolution of key animal and plant functions. Science 318(5848):245 – 250

Metfies K, Medlin LK (2008) Feasibility of transferring fluorescent in situ hybridization probes to an 18S rRNA gene phylochip and mapping of signal intensities. Appl Environ Microbiol 74:2814 – 2821

Mitman GG, van der Meer JP (1994) Meiosis, blade development, and sex determination in Porphyra purp*urea* (Rhodophyta). J Phycol 30:147 – 159

Mock T, Samanta MP, Iverson V, Berthiaume C, Robison M, Holtermann K, Durkin C, Bondurant SS, Richmond K, Rodesch M, Kallas T, Huttlin EL, Cerrina F, Sussman MR, Armbrust EV (2008) Whole-genome expression profiling of the marine diatom Thalassiosira pseudonana identifies genes involved in silicon bioprocesses. Proc Natl Acad Sci U S A 105(5):1579 – 1584

Mock T, Thomas DN (2005) Recent advances in sea-ice microbiology. Environ Microbiol 7: 605 – 619

Moreira D, López-García P (2002) The molecular ecology of microbial eukaryotes unveils a hidden world. Trends Microbiol 10:31 – 38

Morse D, Salois P, Markovic P, Hastings WJ (1995) A Nuclear-Encoded Form II RuBisCO in Dinoflagellates. Science 268:1622 – 1624

Moulager M, Monnier A, Jesson B, Bouvet R, Mosser J, Schwartz C, Garnier L, Corellou F, Bouget FY (2007) Light-dependent regulation of cell division in *Ostreococcus*: evidence for a major transcriptional input. Plant Physiol 144:1360 – 1369

Muhlin JF, Engel CR, Stessel R, Weatherbee RA, Brawley SH (2008) The influence of coastal topography, circulation patterns, and rafting in structuring populations of an intertidal alga. Mol Ecol 17:1198 – 1210

Müller DG (1967) Generationswechsel, Kernphasenwechsel und Sexualität der Braunalge *Ectocarpus siliculosus* im Kulturversuch. Planta 141:39 – 54

Nagasaki K (2008) Dinoflagellates, diatoms, and their viruses. J Microbiol 46:235 – 243

Nassoury N, Cappadocia M, Morse D (2003) Plastid ultrastructure defines the protein import pathway in dinoflagellates. J Cell Sci 116:2867 – 2874

Nikaido I, Asamizu E, Nakajima M, Nakamura Y, Saga N, Tabata S (2000) Generation of 10154 expressed sequence tags from a leafy gametophyte of a marine red alga, *Porphyra yezoensis*. DNA Res 7:223 – 227

Nosenko T, Bhattacharya D (2007) Horizontal gene transfer in chromalveolates. BMC Evol Biol 7:173

Not F, Valentin K, Romari K, Lovejoy C, Massana R, Töbe K, Vaulot D, Medlin LK (2007) Picobiliphytes: a marine picoplanktonic algal group with unknown affinities to other eukary-otes. Science 315:253 – 255

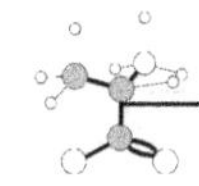

Nowack EC, Melkonian M, Glöckner G (2008) Chromatophore genome sequence of *Paulinella* sheds light on acquisition of photosynthesis by eukaryotes. Curr Biol 18:410 – 418

Ohme M, Miura A (1988) Tetrad analysis in conchospore germlings of *Porphyra yezoensis* (Rhodophyta, Bangiales). Plant Sci 57:135 – 140

Palenik B, Grimwood J, Aerts A, Rouzé P, Salamov A, Putnam N, Dupont C, Jorgensen R, Derelle E, Rombauts S, Zhou K, Otillar R, Merchant SS, Podell S, Gaasterland T, Napoli C, Gendler K, Manuell A, Tai V, Vallon O, Piganeau G, Jancek S, Heijde M, Jabbari K, Bowler C, Lohr M, Robbens S, Werner G, Dubchak I, Pazour GJ, Ren Q, Paulsen I, Delwiche C, Schmutz J, Rokhsar D, Van de Peer Y, Moreau H, Grigoriev IV (2007) The tiny eukaryote *Ostreococcus* provides genomic insights into the paradox of plankton speciation. Proc Natl Acad Sci USA 104:7705 – 7710

Parfrey LW, Barbero E, Lasser E, Dunthorn M, Bhattacharya D, Patterson DJ, Katz LA (2006) Evaluating support for the current classification of eukaryotic diversity. PLoS Genet 2:e220

Park E-J, Fukuda S, Endo H, Kitade Y, Saga N (2007) Genetic polymorphism within *Porphyra yezoe*nsis (Bangiales, Rhodophyta) and related species from Japan and Korea detected by cleaved amplified polymorphic sequence analysis. Eur J Phycol 42:29 – 40

Patron NJ, Waller RF, Archibald JM, Keeling PJ (2005) Complex protein targeting to dinoflagellate plastids. J Mol Biol 348(4):1015 – 1024

Peters AF, Marie D, Scornet D, Kloareg B, Cock JM (2004) Proposal of *Ectocarpus siliculosus* as a model organism for brown algal genetics and genomics. J Phycol 40:1079 – 1088

Peters AF, Scornet D, Ratin M, Charrier B, Monnier A, Merrien Y, Corre E, Coelho SM, Cock JM (2008) Life-cycle-generation-specific developmental processes are modified in the immediate upright mutant of the brown alga *Ectocarpus siliculosus*. Development 135:1503 – 1512

Piganeau G, Desdevises Y, Derelle E, Moreau H (2008) Picoeukaryotic sequences in the Sargasso Sea metagenome. Genome Biol 9:R5

Piganeau G, Moreau H (2007) Screening the Sargasso Sea metagenome for data to investigate genome evolution in *Ostreococcus* (Prasinophyceae, Chlorophyta). Gene 406:184 – 190

Reddy CRK, Iima M, Fujita Y (1992) Induction of fastgrowing and morphologically dif-ferent strains through intergeneric protoplast fusions of *Ulva* and *Enteromorpha* (Ulvales, Chlorophyta). J Appl Phycol 4:57 – 65

Reyes – Prieto A, Moustafa A, Bhattacharya D (2008) Multiple genes of apparent algal origin suggest ciliates may once have been photosynthetic. Curr Biol 18:956 – 962

Reyes – Prieto A, Weber AP, Bhattacharya D (2007) The origin and establishment of the plastid in algae and plants. Annu Rev Genet 41:147 – 168

Rodriguez F, Derelle E, Guillou L, Le Gall F, Vaulot D, Moreau H (2005) Ecotype diversity in the marine picoeukaryote *Ostreococcus* (Chlorophyta, Prasinophyceae). Environ Microbiol 7:853 – 859

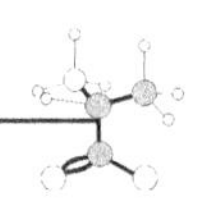

Roeder V, Collen J, Rousvoal S, Corre E, Leblanc C, Boyen C (2005) Identification of stress gene transcripts in *Laminaria digitata* (Phaeophyceae) protoplast cultures by expressed sequence tag analysis. J Phycol 41:1227 – 1235

Romari K, Vaulot D (2004) Composition and temporal variability of picoeukaryote communities at a coastal site of the English Channel from 18S rDNA sequences. Limnol Oceanogr 49:784 – 798

Rusch DB, Halpern AL, Sutton G, Heidelberg KB, Williamson S, Yooseph S, Wu D, Eisen JA, Hoffman JM, Remington K, Beeson K, Tran B, Smith H, Baden-Tillson H, Stewart C, Thorpe J, Freeman J, Andrews – Pfannkoch C, Venter JE, Li K, Kravitz S, Heidelberg JF, Utterback T, Rogers YH, Falcón LI, Souza V, Bonilla-Rosso G, Eguiarte LE, Karl DM, Sathyendranath S, Platt T, Bermingham E, Gallardo V, Tamayo-Castillo G, Ferrari MR, Strausberg RL, Nealson K, Friedman R, Frazier M, Venter JC (2007) The Sorcerer II global ocean sampling expedition: northwest Atlantic through eastern tropical Pacific. PLoS Biol 5:398 – 430

Serrao EA, Pearson G, Kautsky L, Brawley SH (1996) Successful external fertilization in turbulent environments. Proc Natl Acad Sci U S A 93:5286 – 5290

Shimizu Y, Kitade Y, Saga N (2004) A nonradioactive whole-mount in situ hybridization protocol for *Porphyra* (Rhodophyta) gametophytic germlings. J Appl Phycol 16:329 – 333

Siaut M, Heijde M, Mangogna M, Montsant A, Coesel S, Allen A, Manfredonia A, Falciatore A, Bowler C (2007) Molecular toolbox for studying diatom biology in *Phaeodactylum tricornutum*. Gene 406:23 – 35

Siegel BZ, Siegel SM (1973) The chemical composition of algal cell walls. CRC Crit Rev Microbiol 3:1 – 26

Slapeta J, López-García P, Moreira D (2006) Global dispersal and ancient cryptic species in the smallest marine eukaryotes. Mol Biol Evol 23:23 – 29

Spector DL (1984) Dinoflagellate nuclei. In: Spector DL (ed) Dinoflagellates, Academic Press Inc, New York, pp 107 – 147

Suttle CA (2005) Viruses in the sea. Nature 437:356 – 361

Tanikawa N, Akimoto H, Ogoh K, Chun W, Ohmiya Y (2004) Expressed sequence tag analysis of the dinoflagellate *Lingulodinium polyedrum* during dark phase. Photochem Photobiol 80:31 – 35

Tyler BM, Tripathy S, Zhang X, Dehal P, Jiang RH, Aerts A, Arredondo FD, Baxter L, Bensasson D, Beynon JL, Chapman J, Damasceno CM, Dorrance AE, Dou D, Dickerman AW, Dubchak IL, Garbelotto M, Gijzen M, Gordon SG, Govers F, Grunwald NJ, Huang W, Ivors KL, Jones RW, Kamoun S, Krampis K, Lamour KH, Lee MK, McDonald WH, Medina M, Meijer HJ, Nordberg EK, Maclean DJ, Ospina-Giraldo MD, Morris PF, Phuntumart V, Putnam NH, Rash S, Rose JK, Sakihama Y, Salamov AA, Savidor A, Scheuring CF, Smith BM, Sobral BW, Terry A, Torto-Alalibo TA, Win J, Xu Z, Zhang H, Grigoriev IV, Rokhsar DS, Boore JL (2006) *Phytophthora* genome sequences un-

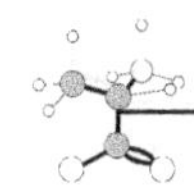

cover evolutionary origins and mechanisms of pathogenesis. Science 313:1261 – 1266

Vaulot D, Eikrem W, Viprey M, Moreau H (2008) The diversity of small eukaryotic phytoplankton (≤3 mum) in marine ecosystems. FEMS Microbiol Rev 32:795 – 820

Venter JC, Remington K, Heidelberg JF, Halpern AL, Rusch D, Eisen JA, Wu D, Paulsen I, Nelson KE, Nelson W, Fouts DE, Levy S, Knap AH, Lomas MW, Nealson K, White O, Peterson J, Hoffman J, Parsons R, Baden-Tillson H, Pfannkoch C, Rogers YH, Smith HO (2004) Environmental genome shotgun sequencing of the Sargasso Sea. Science 304:66 – 74

Waaland JR, Dickson LG, Watson BA (1990) Protoplast isolation and regeneration in the marine alga Porpyra nereocystis. Plantation 181:522 – 528

Waaland JR, Stiller JW, Cheney DP (2004) Macroalgal candidates for genomics. J Phycol 40: 26 – 33

Wilson WH, Schroeder DC, Allen MJ, Holden MTG, Parkhill J, Barrell BG, Churcher C, Hamlin N, Mungall K, Norbertczak H, Quail MA, Price C, Rabbinowitsch E, Walker D, Craigon M, Roy D, Ghazal P (2005) Complete Genome Sequence and Lytic Phase Transcription Profile of a Coccolithovirus. Science 309:1090 – 1092

Worden AZ, Lee J – H, Mock T, Rouzé P, Simmons MP, Aerts AL, Allen AE, Cuvelier ML, Derelle E, Everett MV, Foulon E, Grimwood J, Gundlach H, Henrissat B, Napoli C, McDonald SM, Parker MS, Rombauts S, Salamov A, Von Dassow P, Badger JH, Coutinho PM, Demir E, Dubchak I, Gentemann C, Eikrem W, Gready JE, John U, Lanier W, Lindquist EA, Lucas S, Mayer KFX, Moreau H, Not F, Otillar R, Panaud O, Pangilinan J, Paulsen I, Piegu B, Poliakov A, Robbens S, Schmutz J, Toulza E, Wyss T, Zelensky A, Zhou K, Armbrust EV, Bhattacharya D, Goodenough EW, Van de Peer Y, Grigoriev IV (2009) Green Evolution and Dynamic Adaptations Revealed by Genomes of the Marine Picoeukaryotes. Science 324:268 – 272

Yan X, Fujita Y, Aruga Y (2000) Induction and characterization of pigmentation mutants in *Porphyra yezoensis* (Bangiales, Rhodophyta). J Appl Phycol 12:69 – 81

Yoon HS, Grant J, Tekle YI, Wu M, Chaon BC, Cole JC, Logsdon JM Jr, Patterson DJ, Bhattacharya D, Katz LA (2008) Broadly sampled multigene trees of eukaryotes. BMC Evol Biol 8:14

Yoon HS, Hackett JD, Bhattacharya D (2006) A genomic and phylogenetic perspective on endosymbiosis and algal origin. J Appl Phycol 18:475 – 481

Yoon HS, Hackett JD, Ciniglia C, Pinto D, Bhattacharya D (2004) A molecular timeline for the origin of photosynthetic eukaryotes. Mol Biol Evol 21:809 – 818

Zaslavskaia LA, Lippmeier JC, Kroth PG, Grossman AR, Apt KE (2000) Transformation of the diatom *Phaeodactylum tricornutum* (Bacillariophyceae) with a variety of selectable marker and reporter genes. J Phycol 36:379 – 386

Zaslavskaia LA, Lippmeier JC, Shih C, Ehrhardt D, Grossman AR, Apt KE (2001) Trophic conversion of an obligate photoautotrophic organism through metabolic engineering. Science 292:

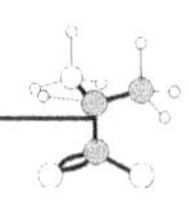

2073 – 2075

Zhang Z, Green BR, Cavalier-Smith T (1999) Single gene circles in dinoflagellate chloroplast genomes. Nature 400:155 – 159

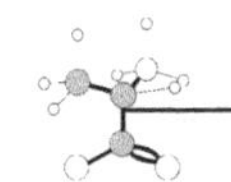

第7章　基因组学技术在水产养殖与渔业上的应用

摘要：近十年来，尽管世界范围内对鱼类和贝类养殖的发展投入巨大，基因组技术在水产养殖和渔业中的应用仍然进展缓慢。促进前沿的基因组学技术在各种水产养殖系统及产业化过程中的应用，仍然是这个领域的主要挑战，而且需要科学的研究策略。本章首先综述现有的基因组学技术和资源，然后探讨基因组学技术在鱼类和贝类养殖上的应用（如育种、繁殖、生长、营养和产品质量等），包括种质多样性的评估和在选育程序上的使用。另外，本章还探讨了基因组学技术在野生鱼类和贝类种群研究与检测中的应用，并且进一步阐明种群与生态系统之间的相互作用。

7.1　前言

虽然基因组学早期更多地应用于人类和生物医学模型的哺乳动物，但目前，它已经扩展到其他种类，包括各种鱼类，特别是在水生系统占据多样化环境的硬骨鱼类。鱼类经历过多种进化途径，因而成为有价值的实验模型生物，这不仅是因为它们具有重要的经济价值，而且比较基因组学能开拓我们了解分子和物种进化的新视野。此外，目前研究的物种经历了硬骨鱼类特异的全基因组复制，因此研究人员正在探讨如此大量的基因拷贝是如何进化来的，特别是：①组成多结构域蛋白的新序列元件的产生；②导致蛋白亚型和变体冗余的复杂RNA加工机制的多样性和增殖；③导致调节机制多样化和基因表达时空变化的新调控因子如何产生；④非编码区的功能。

最近，一些鱼类（包括斑马鱼、青鳉、日本河鲀、黑斑鲀、大西洋鲑鱼、三棘鱼）和少数贝类（牡蛎和贻贝）的基因组数据已经公布（贝类只有一部分核基因组数据），并且另外一些种类正在测序。同时，近期欧洲海洋基因组学研究联盟（Marine Genomics Europe network of excellence）已完成的许多鱼类（如金头鲷、欧洲海鲈）和贝类（如太平洋牡蛎、紫贻贝）大量的EST

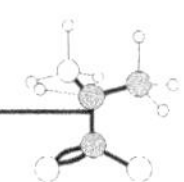

（表达序列标签）序列。在基因组草图和互补 EST 的基础上，就可以对这些基因组序列进行比对，从而推断出基因结构和相应的蛋白序列。然而，一旦识别并注释出某个物种的基因组序列，人们就会关心特定基因/蛋白/DNA 原件的功能是否跟预测的一致，或者某些基因/调控因子受到正选择的原因。相对于近期发展较快的鱼类与贝类基因组测序和比较分析，功能基因组学研究手段还是比较落后，针对鱼类与贝类体内和体外的功能分析方法尚待大力发展。

尽管在过去的十年里，世界范围内的水产养殖取得巨大发展，基因组学技术在水产养殖上的应用还是明显滞后。促进前沿基因组学技术在各种水产养殖系统及在产业化过程中的应用，仍然是这个领域开展新项目的主要挑战。本章首先综述现有的基因组学研究方法和资源，然后探讨基因组学技术在鱼类和贝类养殖中的作用，包括种质多样性的评估和在选育程序上的使用。另外，本章还探讨了基因组学技术在野生鱼类和贝类种群研究与检测中的应用，并进一步阐明种群与生态系统之间的相互作用（图 7.1）。

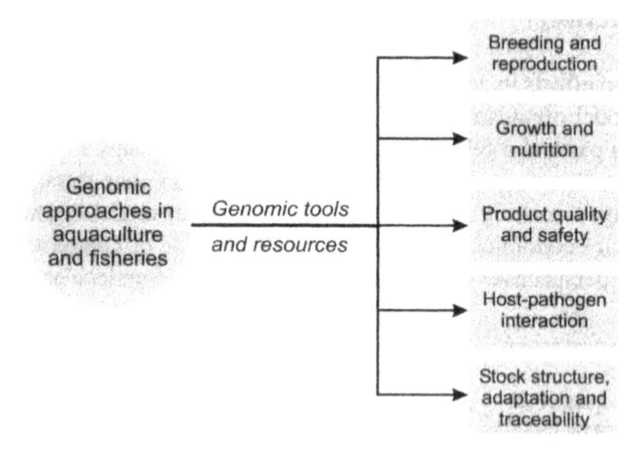

Fig. 7.1 Overview of genomic approaches in aquaculture and fisheries (Figure credit: V. Laizé)

图 7.1　基因组学技术在水产养殖和渔业上的应用。

英文注释：Breeding and Reproduction，育种与繁殖；Growth and Nutrition，生长与营养；Product Quality and Safety，产品质量与安全；Host-pathogen Interaction，宿主—病原互作；Stock Structure，Adaptation and Traceability，群体结构，适应与溯源

7.2　基因组学技术和资源

在这一节中，主要描述海水鱼类和贝类的结构和功能基因组学研究中现

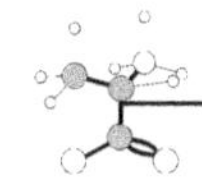

有的或即将可用的技术方法。大部分方法已经应用于经济种类的养殖或捕捞。由于基因组学技术的发展需要相应的经济投入，因此目前只限于在具有重要经济地位和模式物种中有所应用。尽管如此，新技术的引入，如新一代高通量测序技术（Margulies et al. 2005）已经显著降低EST、BAC文库和全基因组测序的成本。因此，越来越多的海水鱼类和贝类研究将会进入基因组学时代。

7.2.1 遗传连锁图谱

在历史上，首次对动物基因组的描述是以遗传连锁图谱的方式。任何多态位点（如至少可观察到2个等位基因的位点），均可用于遗传图谱定位（图7.2）。在过去的30年里，不同类型的标记（同工酶、微卫星、RAPD、AFLP、SSR和SNP）被广泛使用（Schlötterer 2004；见词汇表和第三章有关于这些标记的详细叙述）。由于同工酶标记在大多数物种中（特别是在双壳类中）具有丰富的多态性（Solé-Cava and Thorpe 1991），因此被广泛采用。同工酶标记适合大群体的遗传学研究并且仍被采用，特别是与其他类型的标记结合使用（Nikula et al. 2008）。采用同工酶标记时，经常在野生或养殖群体中观察到杂合子缺陷，以及杂合率与相关性状的不平衡关系（Raymond et al. 1997；Bierne et al. 2000）。因此，非中性的同工酶标记在双壳类中的应用受到强烈的质疑（McDonald et al. 1996）。

大多数标记至今都是基于DNA技术。如今最受欢迎的是微卫星标记（即SSR位点）。SSR位点由短序列（2～6核苷酸）重复组成。SSR一般是均匀地相对高效率分布在基因组上（1.5～6 kbp存在1个位点；Zane et al. 2002，Chistiakov et al. 2006）。SSR核心序列的重复性质使一个或几个重复单元的增加或缺失，产生了不同长度的等位基因（Ellegren 2004）。但是，非完美微卫星大量存在，特别是在甲壳类中，每个位点的等位基因数目非常多（>50；Huvet et al. 2004）。因此，大多数SSR位点具有丰富的多态性，非常适用于遗传连锁图谱的构建（Schlötterer 2004）。海洋鱼类和贝类遗传连锁图的研究进展可参考Wenne et al（2007）的综述。大部分图谱采用SSR位点，有一些（如海鲈、斑点叉尾鮰、贻贝、欧洲牡蛎）也采用AFLP。由于SSR位点首先是在非编码区出现，因此有必要识别并将SSR与EST序列对应起来（EST-SSR）。近期产生的多个物种的EST序列有助于增加可用的EST-SSR个数（Yu and Li 2008）。另外一种可以用来定位遗传连锁图谱上基因的标记是SNP位点。当SSR位点被作为遗传标记广泛使用时，SNP可作为大多数动物基因组学研究的辅助工具。迄今为止，海洋生物遗传连锁图谱使用SNP标记的还

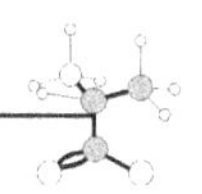

比较少，不过随着高通量SNP基因分型的发展，在不久的将来用SNP作图会显著增加。相对于SSR位点，SNP在基因组中出现的频率更高，不论是编码区还是非编码区，并且是双等位基因型，这样更易在不同实验室间共享且适合于高通量的分析和自动化。SNP在高效种群规模（每一代都有一个有效的育种数量）的海洋蚌类里出现的频率特别高（如编码区平均60 bp就有1个SNP，非编码区平均40 bp就有1个SNP；Sauvage et al. 2007）。SNP的多态性不如SSR，但对于大多数物种，其出现的频率更高，并且采用高通量技术可以识别成千上万个SNP。同样，目前对一百万个SNP位点进行基因分型已成为可能（如Illumina Infinium HD Human 1 M）。至于小一点的规模，SNP分离和基因分型也有广泛应用，特别是对于大西洋鲑鱼，同时其他海洋生物的研究正在迅速推进（如大西洋鳕鱼，欧洲鲈鱼和太平洋牡蛎）。

无论是用什么类型的标记进行作图，都需要足够数量的独立的减数分裂事件来检测位点间的重组率。一般来说，这需要好的作图群体，如来源于F1的试验群体、回交或者是遗传差异较大的亲本产生的F2群体。对于海洋鱼类和贝类，这些类型的作图群体比较难以获得，因此通常是采用来源于野生亲本交配产生的F1群体。虽然存在不足，但是大多数海洋生物（有效群体规模较大）具有较高的杂合率，从而增加可用于作图的多态性标记的数量。

连锁图其实就是可用于基因组学研究标记的框架图，这些标记可用于全基因组范围内具有经济或科研价值的表型所对应基因型的筛查。除了用于识别某些养殖种类的数量性状位点（QTL），三棘鱼同域群体之间形态差异与基因位点间关系的辨别，就是这个方法得到运用的典型案例。Peichel et al.（2001）基于中等密度的遗传连锁图，分析了来自海洋和不列颠哥伦比亚Priest湖三棘鱼骨骼盔甲和摄食形态进化的遗传学机制。

现有的遗传连锁图谱最主要的不足，是位点上编码基因的信息较少，并且不同物种之间连锁图谱的可比性较差。近十年来，某些具有重要水产养殖价值的鱼类和贝类已构建出低—中等密度的遗传连锁图（Wenne et al. 2007）。现有的图谱大部分是采用杂交的F1群体和AFLP标记。其中有一些是属于“草图”，并且通过增加共线性标记来进行完善（如太平洋鲍鱼，Liu et al. 2006）。微卫星标记连锁图（如太平洋牡蛎，88个标记；Hubert and Hedegock 2004；太平洋鲍鱼，176个标记；Sekino and Hara 2007）更需要进一步开发，并且其标记密度往往比AFLP还低。SNP标记连锁图目前正在一些物种中展开。

比较基因组学通过分析不同物种间的保守区域，可以比对和比较物种间

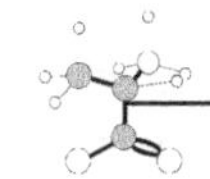

的基因组信息差异。如果有部分 SSR 位点是保守的，即可进行不同基因组的定位（Stemshorn et al. 2005，Franch et al. 2006）。但是典型的微卫星标记具有物种特异性，并且旁系同源区域在其他基因组上往往不发生重组。这种现象在具有丰富核苷酸多态性而导致不同物种难以重复检测的贝类中普遍存在。

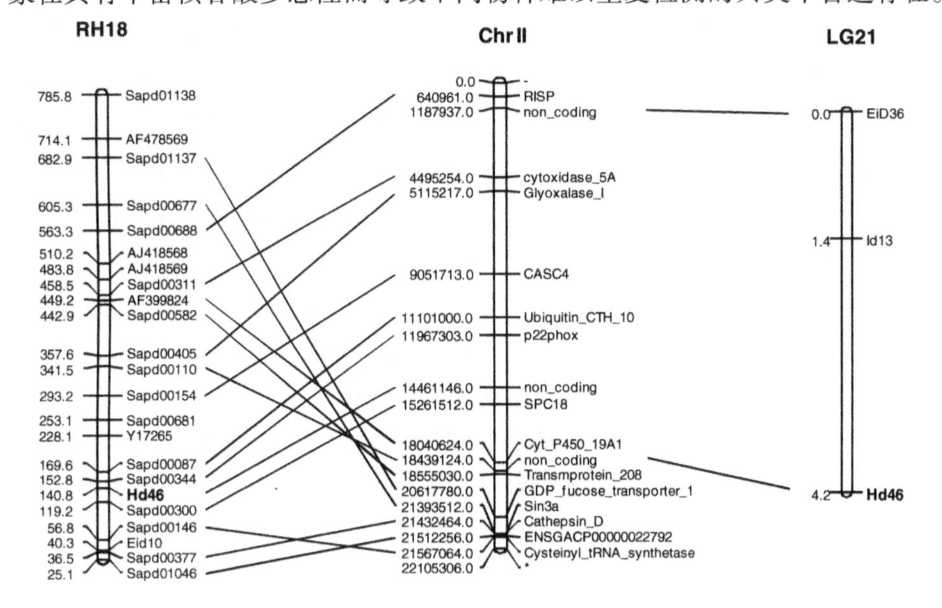

Fig. 7.2 Comparative mapping. RH18 (radiation hybrid group 18) from seabream radiation hybrid map (Sarropoulou et al. 2007), LG21 (linkage group 21) from seabream genetic linkage map (L. Bargelloni, unpublished data) and Chr II (chromosome II) from stickleback genome (www.ensembl.org). Marker names on RH map correspond to unique transcripts from SAPD database (identified with prefix Sapd), publicly available genes or SSR markers (identified with GenBank accession numbers) and unpublished SSR markers (Eid10 and Hd46). Any PCR-amplifiable sequence can be mapped onto the RH map (see text), allowing for a higher marker density. On the other hand, only polymorphic SSR markers (Eid36, Id13 and Hd46) can be located on the linkage map. In *bold*, the single SSR locus present in both maps. Correspondences between individual seabream markers and putative homologues of seabream markers in stickleback genome are shown. Several changes/errors in gene order between the two species are evident. (Figure credit: L. Bargelloni)

图 7.2　比较作图。金头鲷辐射杂交图谱中的 RH18（Sarropoulou et al. 2007）、金头鲷遗传连锁图谱的连锁群 21（L. Bargelloni，待发表）和三棘鱼染色体 II（www. ensembl. org）。RH 图的标记来自 SAPD 数据库（以前缀 Sapd 表示）的单一转录本、已公开的基因或 SSR 标记（以 GenBank 登录号表示）和待发表的 SSR 标记（Eid10 和 Hd46）。为了提高标记密度，任意可 PCR 扩增的序列均比对到 RH 图谱（见正文）。另一方面，只有多态 SSR 标记（Edi36、Id13 和 Hd46）能定位到遗传图谱。加粗表示的一个 SSR 标记（Hd46）同时存在于 2 个图谱。金头鲷的标记和已知的三棘鱼基因组同源的金头鲷标记的相关性如图所示。这 2 个物种的基因顺序还是存在着一些明显的差异或错误

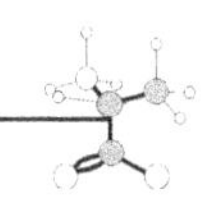

7.2.2 辐射性杂交（RH）作图

为了克服遗传连锁图谱的缺陷，辐射性杂交作图应运而生（Walter et al. 1994）。关于RH作图的细节超出了本书的内容，读者可参考相关文献（如Kwok et al. 1998）。简单地说，先建立杂交细胞系，其中包含供体（所研究的物种）基因组随机片段与受体细胞（一般是仓鼠细胞系）的全基因组。一组独立的细胞系基因组覆盖了数倍的目标物种基因组。采用PCR方法对每个细胞系的标记和基因进行分析。然后对每条染色体上标记的线性排列顺序和每个位点的置信值进行统计分析。来自供体基因组且能被PCR选择性扩增出来的区域（如编码区、SSR、EST）均能被定位在RH图上（图7.2）。这样，就实现了不具有多态性位点的作图。而且，蛋白编码区甚至在亲缘关系较远的物种中也具有很好的保守性，使得RH图上的大多数标记能在其他物种的基因组上找到同源区域，方便开展比较基因组学分析。

迄今，技术问题（宿主和供体细胞之间的兼容性、供体细胞对辐射的敏感性）限制了RH作图在非哺乳类脊椎动物中（特别是鱼类）的应用。已完成RH作图的只有2种硬骨鱼类，即斑马鱼（Geisler et al. 1999，Hukriede et al. 1999）和金头鲷（Senger et al. 2006，Sarropoulou et al. 2007）。至于第3个物种欧洲海鲈，基于改进过的方法（不采用细胞培养）制作RH嵌板，其RH作图正在开展。这些技术的改进有望使RH嵌板的制作更加迅速，并更容易在任意脊椎动物中使用。同时，采用高通量的方法对RH嵌板进行信息采集（McKay et al. 2007，Park et al. 2008），能显著简化RH作图过程中最昂贵和费时的步骤（比如杂交细胞系的分子标记分型）。快速且相对便宜的RH作图，将有利于更大范围的基因组比较以及完整或者高质量基因组图谱向信息的转化（Sarropoulou et al. 2008），从而加快对生物学重要表型相关位点的识别。同时，中—高密度RH图谱结合低覆盖（1－2 X）基因组图谱将能提供与高覆盖（7 X以上）基因组图谱相当的基因定位信息（Hitte et al. 2008）。太平洋牡蛎RH作图的可行性目前正在论证（F. Galibert，私人通讯）。

7.2.3 基于BAC的物理图谱

对于两个位点的距离，遗传图谱提供了间接的估计，基因组范围的物理图谱则能以碱基长度的形式做出真实的估计。由于RH图谱标记间的关联频率由辐射决定而不是遗传重组，RH图谱可以认为既是一种物理图谱也可认为是类似于连锁图谱。物理图谱一般包含诸如细菌人工染色体（BAC）等长片

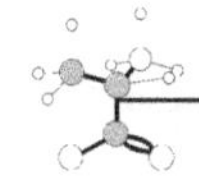

段插入克隆的排列集合。物理图谱与遗传信息是独立的，但如果与遗传连锁图结合，甚至与基因组数据结合，作用会非常大。现有的物理图谱是基于检测 BAC 之间的重叠区域，最常见的就是基于酶切指纹图谱的 BAC 图谱（Meyers et al. 2004）。一些海水鱼类的 BAC 文库已经构建，但高密度的 BAC 物理图谱还没完成。低密度的 BAC 图谱，连接 84 个 BAC 文库至连锁图谱（Wang et al. 2007b），在澳洲肺鱼（*Neoceratodus forsteri*）或亚洲鲈鱼（*Lates calcarifer*）已有报道（Wang et al. 2008）。BAC DNA 制备技术的改进结合高通量测序技术，使得狼鲈（*Dicentrarchus labrax*）BAC 文库（Whitaker et al. 2006）和金头鲷（*Sparus aurata*）BAC 文库中超过 5 万个克隆的双末端测序得以完成。在这种情况下，通过与三棘鱼高质量基因组序列信息进行比较，BAC 末端序列即可得到排序。与亲缘关系较近物种的基因组进行比较来定位克隆，能对其他方法提供互补的框架结构（Gregory et al. 2002），并且是 BAC 克隆初步组装的好方法。

BAC 文库在软体贝类中的应用还较少。Cunningham et al.（2006）首次报道了太平洋牡蛎和美洲牡蛎的 BAC 文库。最近，Zhang et al.（2008）构建了栉孔扇贝（*Chlamys farreri*）的 2 个 BAC 文库，并将其应用于天然免疫基因的识别。一个覆盖太平洋牡蛎基因组 10X 的 BAC 文库的制备和指纹采集也于近期完成，该文库作为由 P. Gaffney（University of Delaware，USA）协调倡导的项目的一部分。

7.2.4 高质量基因组图谱

迄今，有 5 种硬骨鱼的基因组已得到深度测序，包括斑马鱼、河鲀、黑斑鲀、青鳉和三棘鱼。这些基因组数据均公布在 www. ensembl. org。其中，河鲀和三棘鱼都属于海水鱼类，但只有河鲀是完全海水鱼，三棘鱼具有淡水和海水两种亚型，不过属于海水亚型的鱼也要返回淡水进行繁殖。这两种鱼基因组测序覆盖度轻微不同，河鲀 8.7X，而三棘鱼覆盖 11X。对这两个物种进行比较可以阐述不同基因组序列的异同。河鲀基因组大小为 393 Mb，由 7213 条 scaffold 组成，最大的 scaffold 为 7 Mb。另一方面，三棘鱼的基因组大一些（大约为 460 Mb），可组装成 21 条染色体（连锁群）和 1822 条未能定位的 supercontigs。很显然，三棘鱼基因组为其他鱼类基因组的比较提供更好更有意义的方法。实际上，其他鱼类的连锁图谱和 RH 图谱均会与三棘鱼基因组做比较（Franch et al. 2006，Sarropoulou et al. 2007，2008，Bouza et al. 2007；图 7.2），结果显示了较好的共线性，且说明基因组重组主要是发生在染色体

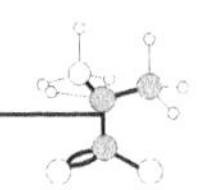

内。正如之前所说，新一代 DNA 测序技术使得非模式物种的全基因组测序成为可能，但相应的工具（如连锁图谱和物理图谱）对高质量的全基因组组装还是非常必要的。由中国（中国科学院海洋研究所和华大基因）和国际牡蛎基因组联盟联合开展的太平洋牡蛎基因组项目，正是采用这种策略。在帽贝（*Lottia gigantea*）测序（见 JGI 网站）之后，牡蛎将会是首个完成基因组测序的冠轮动物。

7.2.5 功能基因组学技术

在基因组时代，核心就是基于大规模测序获取基因组信息，从而阐明生物学特征和表型变异与基因组结构之间的关系。以 EST 序列信息为例，EST 序列数据集结合公共数据库的可用资源，将 EST 序列组装成 contigs，然后利用基因芯片或 SAGE（Serial Analysis of Gene Expression 基因表达系列分析）或 MPSS（Massively Parallel Signature Sequencing 大规模平行测序技术）来获得转录，它可用于研究海洋动物生理学。功能基因组主要研究基因组的动态变化，如基因的转录、翻译和蛋白质之间的相互作用。在这里，我们只是从狭义上把功能基因组学定位为转录组学。引入 DNA 芯片技术后（Schena et al. 1995），便可以从单个实验来研究整个基因组的转录。真核细胞的转录组比预期的还要复杂，因此很难从单个实验就能得到完整展示（Birney et al. 2007），而 DNA 芯片确实为基因表达研究提供新的切入点。基因表达 DNA 芯片存在一些技术可选参数，如探针既可以采用点阵的 cDNA 和寡核苷酸，也可以采用不同合成技术原位合成的不同长度寡核苷酸（Holloway et al. 2002）。目前，适用于海洋鱼类和贝类的芯片平台为 cDNA 点阵芯片。另外也可采用寡核苷酸探针，通过合成排布于玻片上或直接在玻片上原位合成。

一个国际合作小组已经对北美牡蛎（*C. virginica*）和长牡蛎（*C. gigas*）构建了 cDNA 芯片，其中4460条序列来源于北美牡蛎、2320条序列来源于长牡蛎（Jenny et al. 2007）。这个芯片已被用于研究长牡蛎家系的热应激反应（Lang et al. 2008）。在欧洲，欧洲海洋基因组学研究联盟和欧洲水产项目组（Aquafirst European project）联合制作了太平洋牡蛎转录组的 10X 芯片。该芯片用于研究牡蛎对夏季死亡率表现出的抗性或易感的生物学机制（Boudry et al. 2008; E Fleury，私人通讯）。常用于海洋环境污染指示生物的紫贻贝，其首张 cDNA 芯片包含1714个探针，可识别贻贝组织中约 50 种重要污染物相关剂量的信号（Venier et al. 2006）。一张包含 24 个贻贝基因的低密度寡聚核苷酸芯片已经完成，基因的筛选是基于它们与污染物和异生物质应答机制有关

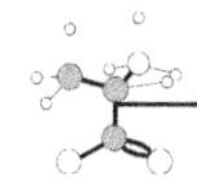

系（Dondero et al. 2006）。Place et al.（2008）采用芯片分析了从潮间带不同分布地点采集的加利福尼亚贻贝，强调在研究海洋潮间带双壳类生态学方面基因芯片这种转录组学工具的有效性。

公用数据库上多种海洋生物具有越来越丰富的 EST 数据资源，如大西洋鲑鱼有433 337条，三棘鱼有 276 992条，鳕鱼有 181 734 条，狼鲈有 32 755 条和长牡蛎有 56 327 条，从而推动了寡核苷酸—DNA 芯片更广泛的应用。欧洲海洋基因组学研究联盟已经完成一个构建两种海水鱼类（金头鲷和狼鲈）芯片平台的初步计划。采用基于喷墨技术的原位合成，每个独立的转录本设计 2 个非重叠的寡核苷酸，从而实现包含金头鲷 19 715 个基因转录本的芯片平台（http：// enne. cribi. unipd. it：5555/biomart/martview）。这个芯片平台已经被批准验证过，并可免费使用（Ferraresso et al. 2008）。另外，包含狼鲈 19 048 个独立转录本的寡核苷酸芯片也已完成（L. Bargelloni，私人通讯）。（P. Martinez，私人通讯）和塞内加尔鳎（Cerda et al. 2008）成功构建了包含转录本数较少的芯片平台。大规模 DNA 测序成本的显著下降，将增加更多种海洋鱼类的 EST 数据，使得功能基因组学工具将被更多的研究人员所采用。最终，直接测序也会成为转录组研究的方法（Sultan et al. 2008）。

功能强大的 MPSS 方法（Brenner et al. 2000），已被用于研究双壳类中广泛存在的杂种优势。MPSS 的主要原理在于从复杂的 RNA 样品中产生短序列标签，从而分析基因表达模式。MPSS 对于缺少基因组数据资源的物种具有优势：①无需参考序列信息，可识别新转录本，如果有大量 EST 数据效果会更理想；② MPSS 是一种无偏好、综合的定量方法，其基于对标签序列的计数进而实现转录本丰度的定量；③ MPSS 能识别低丰度表达信息（丰度为 $3/10^6$ 的转录本）。直至目前为止，4. 5 Mb 长的牡蛎序列标签包含 23 274 个特异基因，覆盖了 350 个与生长杂种优势有关的候选基因（Hedgecock et al. 2007）。SAGE 是另外一种研究转录表达谱的方法（Velculescu et al. 1995），与 MPSS 具有相似的前提假设和优点。至今，还没有海洋贝类 SAGE 方面的报道，但 SAGE 即将被应用于牡蛎血淋巴细胞和免疫组织细胞的基因组范围的表达谱研究（E. Bachère，私人通讯）。最后，关于这 3 种方法的比较研究，Nygaard et al.（2008）总体指出寡核苷酸芯片能提供定量转录本的关键点，这些点能与 MPSS 和 SAGE 数据能结合起来使用，但是在绝对定量方面，这 3 种方法存在差异。

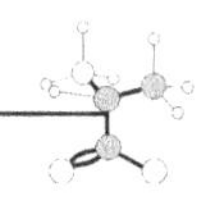

7.3 基因组学技术在水产生物育种和繁殖研究中的应用

鱼类和贝类选育的最终目标是提高水产养殖的可持续性和经济效益，同时保持养殖群体的遗传多样性以及减少对野生资源和环境的影响。为了使有效群体规模能最大化，并且充分利用具有亲缘关系的个体从而提高预测育种值的精度，我们需要制定有效的育种策略，这需要有谱系信息，但往往我们就是缺乏这些信息。分子标记能达到这个目的，它在基因组水平检测与目标性状相关的区域（如 QTL）。本节将主要介绍分子标记信息用于鱼类和贝类育种的可行性以及初步的应用实例，包括系谱追溯和遗传多样性的维持，QTL 定位和标记辅助选育（MAS）。

实际上，鱼类和贝类选育的两个主要障碍是：①大多数物种在幼体时期由于个体太小很难被同时逐个标记；②空间和技术上限制了同时培育大量符合系谱要求的个体。为了解决这个问题，对家系来源不同、规格一致的个体混合起来培育，从而消除家系特异的环境效应。在表型鉴定之后，利用分子标记对混合的个体进行系谱鉴定。由于具有丰富的遗传变异和很高的个体间特异性，SSR 是当前亲缘分析和系谱追溯最常用的标记类型（Herbinger et al. 1995，Fishback et al. 2002，Vandeputte et al. 2004）。不同的育种群体规模会有差异，一般来讲，要对一个全同胞家系 95% 以上的个体进行亲权分析时需要 10～20 个标记（Vandeputte et al. 2006）。但是，在贝类（Hedgecock et al. 2004）和鱼类（Castro et al. 2004，2006）一般会产生很多无效等位基因，这导致分析上的困难和偏差。一些物种在实验水平上已经采用微卫星标记进行亲权鉴定的研究，比如对有限亲本数目的贝类（Boudry et al. 2002，Taris et al. 2006，Li and Kijima 2006，McAvoy et al. 2008）、鲷（Castro et al. 2007）、大西洋鳕鱼（Herlin et al. 2007，2008）、大西洋鲑鱼（Norris et al. 2000）、虹鳟（Mcdonald et al. 2004）、大菱鲆（Borrell et al. 2004）和舌鳎（Porta et al. 2006）等的亲权鉴定。然而，即使有大量的分子标记可用，这种方法在混合家系选育过程的可行性还没有得到完全证实（Li et al. 2003a）。正如 SNP、AFLP（Gerber et al. 2000）同样被用于亲权鉴定（Anderson and Garza 2006），但是高通量 SNP 基因分型的成本较高（Hayes et al. 2005）。由于牡蛎 SNP 平均密度很高（Curole and Hedgecock 2005，Sauvage et al. 2007），并且其他海洋双壳类也具有类似的多态性水平，因此潜在的可用的 SNP 标记数目极大。

对于养殖场逃逸物种的鉴定，水产品市场疾病或有毒物质来源的物种的识别，可追溯性已成为重要的原则（可参见本章 7.5 节及本书第 1 章）。

分子标记其他方面潜在的应用主要有溯源选择（Li et al. 2003b，Sonesson 2005）和亲权辅助选择等方法（Lynch and Walch 1998）。某些贝类由于较低的自交水平导致遗传参数估计值偏差增大。这种现象在扇贝中和欧洲平牡蛎确实存在（Martinez and di Giovanni 2007），（Lallias et al. 2008）。因此，分子标记可用于区分自交和杂交家系以及同一家系内的个体。

育种程序创始阶段的决策对育种能否最终成功至关重要，这主要体现在育种基础群体的构建。缺乏足够的基础群体是某些鱼类（Gjedrem 2000）和贝类（Naciri-Graven et al. 2000）在育种过程中选择力较低的主要原因。在育种过程中，增加有效群体规模能降低随机遗传漂变和提高选择反应的概率。分子标记能鉴别作为候选群体中个体之间的亲缘关系（Hayes et al. 2006），从而避免近亲交配和自交（Gallardo et al. 2004，Camara et al. 2008）。因此，分子标记是检测牡蛎（Hedgecock and Davis 2007）、鲷（Blanco et al. 2007）、鳟（Was and Wenne 2002，Gross et al. 2007）、鲑鱼（Norris et al. 1999，Koljonen et al. 2002，Rengmark et al. 2006）、鳕鱼（Pampoulie et al. 2006）和牙鲆（Liu et al. 2005b）等繁殖群体遗传多样性水平的常规检测方法。另外，在欧洲平牡蛎（Launey et al. 2001）和日本牙鲆（Sekino et al. 2002）中，采用微卫星证实了在大规模群体选育过程中遗传多样性的丢失。

目前，MAS 在农业育种的遗传改良项目中还未发挥显著的作用，特别是在水产养殖领域。一般来讲，鱼类和贝类选育的目标性状要能易于在个体上进行表型鉴定并且能通过大规模选育得到改进（体重和体长等）。尽管如此，在分子标记和高通量基因分型技术发展的推动下，难以进行鉴定的表型，成本较高或鉴定较耗时的表型，或在个体发育后期产生的表型（抗病能力、肉质、摄食率和性成熟等）也可以成为 MAS 的候选目标性状。进一步讲，相对于遗传改造（如转基因），MAS 不受限制，就如同野外试验、商业交易一样，不会为公众舆论所诟病。

开展 MAS 的先决条件就是要知道影响目标性状（表型）的基因型与分子标记的关联性，寻找这种关联性就是定位 QTL。目前，在贝类中只有 2 篇关于识别 QTL 的文献报道，而在鱼类则有 20 多篇。例如，在鱼类中，影响一些罗非鱼的耐寒和体重性状（Cnaani et al. 2003，Moen et al. 2004）、鳟鱼的体

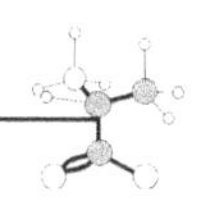

重性状（Reid et al. 2005）、鲑鱼的抗病能力（Ozaki et al. 2001，Moen et al. 2004，Rodriguez et al. 2004，Khoo et al. 2004，Cnaani et al. 2004）和耐热能力（Somorjai et al. 2003 and references therein）方面的QTLs均有报道。在贝类中，Yu and Guo（2006）识别了北美牡蛎抵抗鞭孢簇虫（*Perkinsus marinus*）性状的QTLs，Liu et al.（2007）定位了皱纹盘鲍（*Haliotis discus hannai*）壳重、肌肉重量、性腺重量、消化道重量和鳃重量的QTL。另外，欧洲平牡蛎抗寄生虫（*Bonamia ostreaea*；Lallias et al. 2008），以及太平洋牡蛎抗夏季死亡（C. Sauvage；私人通讯）和生长杂交优势（D. Hedgecock；私人通讯）的QTL定位研究也有所开展。

候选基因法是研究基因型与表型相关性的另外一种方法，通过分析已知位点上基因的变异来阐述基因型与目标性状表型之间的关系。基因多态性与生长表型差异的直接相关性在牡蛎中已有报道（Prudence et al. 2006）。通过研究淀粉酶等位基因与淀粉酶活性的相关性，再分析淀粉酶活性与摄食相关性状的联系，即可将基因型与表型联系在一起（Huvet et al. 2008）。类似地，David et al.（2007）报道了谷氨酰胺合成酶（氨基酸代谢）和Δ9去饱和酶（脂类代谢）基因与牡蛎抗夏季死亡的相关性。在北极红点鲑（*Salvelinus alpinus*）中，利用与生长激素轴相关的10个保守基因序列研究生长性状相关的候选基因（Tao and Boulding 2003）并发现1个与生长速度相关联的SNP。在大西洋鲑鱼的主要组织相容复合体（MHC）中，特异等位基因或杂合子对传染性组织坏死病毒的抗性和易感性也有报道（Langefors et al. 2001，Lohm et al. 2002，Arkush et al. 2002，Grimholt et al. 2003，Bernatchez and Landry 2003）

除以上这些方法外，差异表达基因研究也能提供新的候选基因。通过对夏季死亡具有抗性或易感性（Huvet et al. 2004），以及是否暴露于污染物（Boutet et al. 2004，Tanguy et al. 2005）的牡蛎进行差异表达基因研究，发现大量的候选EST。直至目前为止，抑制性消减杂交（SSH）是研究个体之间差异表达基因最常用的方法。尽管如此，新一代高通量转录组分析方法，比如基因芯片（Jenny et al. 2007）、MPSS（Hedgecock et al. 2007）、SAGE在不久的将来将会极大地促进基因差异表达的研究。基因芯片已被用于研究鲑鱼是否感染寄生虫的基因表达差异（Rise et al. 2004）。最后，通过QTL和候选基因的共同定位，将可提高项目研究的效率。

尽管分子标记已被用于鱼类和贝类的育种项目，比如用于追踪养殖个体、

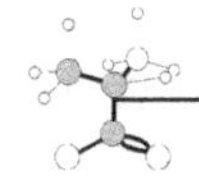

估计逃离率和使繁殖种群所需的数量最优化，但相对而言，水产领域 QTL 作图和 MAS 的进展明显落后于陆生的动植物品种。尽管如此，结合遗传学和基因组学的方法来研究影响鱼类和贝类复杂性状的变异，将会提高 MAS 的可行性。

7.4 基因组学技术在水产生物生长和营养研究中的应用

7.4.1 前言

鱼类养殖主要的目标之一就是使鱼类获得最佳生长速度。野生条件下，鱼类群体的整体适合度（特别是繁殖性能），取决于鱼类能否获得特定的生长速度。从生理学的角度而言，鱼类的生长是一个复杂的过程，与其他的生理过程（如发育、营养和代谢）相互关联。鱼类的生长一般被认为是鱼类体重的增加，特别是肌肉重量与其对应的骨骼和器官的增加。占鱼类体重 50% 以上的骨骼肌是生长的最重要部位，并且也是人类的消费品。大部分鱼类在整个生活周期均会持续地生长（Mommsen 2001）。生长取决于肌纤维的产生与发育以及代谢水平，而代谢主要依靠食物的摄入、运输以及营养物质的利用。因此，与营养相关的器官（如肠道、肝脏和脂肪组织）在骨骼肌的生长过程中起着重要作用。肝脏由于具有储存碳水化合物（如糖原），并通过糖异生生成葡萄糖以及合成和存储脂肪，因此肝脏对于骨骼肌代谢应是最重要的器官。既然生长是由多种组织参与的生理过程，必须全局地用多种方法联合的策略来研究肌肉生长的各种影响因素。目前，这种整合的生理学研究策略（如系统生物学）适用于综合处理来自转录组、蛋白质组和代谢组的信息，并重构出肌肉生长过程的功能途径和网络。

7.4.2 与肌肉生长相关的骨骼肌转录变化

最近几年，采用高通量技术研究鱼类肌肉生长已成为一种趋势。已有一些关于肌肉生长的转录组研究，但大多数是关于鲑鳟鱼类的研究。研究鱼类肌肉生长的转录组，首要的途径是分析转生长激素（GH）基因的鲑鱼的基因表达。GH 会导致肌肉增生，通过抑制性消减杂交可检测到转录因子、肌纤维生成和肌肉结构相关的基因表达上调（Hill et al. 2000）。近期有研究表明，转 GH 的鲑鱼能在体外直接被 GH 诱导出更多的肌肉干细胞数量和更高的增值

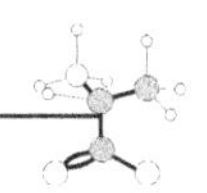

率，这两个过程（体内和体外）均与不同肌源性因子的 mRNA 表达变化有关（Levesque et al. 2008）。尽管如此，至今还没有采用高通量的方法对转 GH 鲑鱼的肌肉做全景式的转录组研究。采用 GRASP 16 K cDNA 芯片研究外源牛的 GH 对虹鳟肌肉转录组的影响已有报道（Gahr et al. 2008）。在该项研究中，对经选育的生长快和生长慢的两个虹鳟家系注射 GH，识别出白肌中调节细胞进程（如细胞周期、免疫反应、代谢或蛋白质降解）中相关上调或下调的基因。基于另外一种生长模型（饥饿和再投喂），采用 INRA-GADIE 资源中心制作的包含 9023 条基因序列的 cDNA 芯片来研究肌肉转录组也有报道（Rescan et al. 2007）。在肌肉增长复苏的早期，涉及 RNA 加工、转录和细胞增殖以及后期涉及高尔基体，内质网变化和肌肉重构的基因诱导过程也有报道。

目前，在已公布全基因组序列的模式生物(如斑马鱼和河鲀)中，采用基因芯片研究肌肉生长的基因表达模式还未见报道。已有研究报道利用抑制性消减杂交和荧光定量 PCR 研究河鲀幼鱼和成鱼骨骼肌转录差异（Fernandes et al. 2005）。成年河鲀骨骼肌停止肌纤维生长，抑制了肌纤维的再生，而幼鱼骨骼肌具有纤维再生活性。Fernandes et al.（2005）在成年河鲀快肌中识别出 4 个高表达的新基因（Fernandes et al. 2005）。这些新基因可能属于生长抑制因子，但与已知的基因没有显著的同源性。该研究为进一步采用高通量策略研究河鲀的生长提供了案例，因为河鲀已经有不同 EST 文库可用来合成 cDNA 或寡聚核苷酸芯片。奇怪的是，斑马鱼有许多商业性的寡聚核苷酸芯片（如安捷伦）并且其基因组测序已经完成并完成部分注释，而基于基因芯片的斑马鱼肌肉生长（快速生长）的研究却还未见报道。

7.4.3 外界因素影响下的骨骼肌转录组变化

外界因素（比如温度或者病原生物）会影响不同的生理过程，从而影响肌肉的生长。采用高通量技术，研究环境温度（Cossins et al. 2006，Gracey et al. 2004，Malek et al. 2004）卵黄生产作用诱导的骨骼肌萎缩（Salem et al. 2006）和接种疫苗（Purcell et al. 2006）的变化对骨骼肌转录组的影响均有报道。首先在斑马鱼中报道温度对骨骼肌转录组的影响（Malek et al. 2004）。10℃的温差变化会引起某些基因表达上调，这些基因涉及线粒体代谢、氧化应激以及热激蛋白 70 和 90。但是，这些基因的表达变化与表现为肥大的肌肉生长的变化没有关联。采用包含13 349条序列的常规 cDNA 芯片在鲤鱼（*Cyprinus carpio*）中开展了一项有趣的关于温度适应性研究（Gracey et al. 2004）。研究人员研究了 7 种组织，发现温度下降会引起所有组织代谢相关的反应，以及额外更特异地与各组织机能相关的反应。尤其是骨骼肌涉及收缩系统的基因表达下调，而涉及蛋白质降解的基因表达上调。关于这个主题可参考Gracey及其同事撰写的综述（Cossins et al. 2006，Gracey 2007）。关于温度适应对转录的影响有一个普遍接受的假说，即肌肉的活性降低并且伴随着导致萎缩的蛋白质降解。基于虹鳟肌肉萎缩的生理模型，采用GRASP 16K芯片平台可检测基因的表达变化。Salem et al.（2006）识别了占芯片基因总数1%的差异表达基因。卵苗生成期间，鳟鱼骨骼肌产生了明显的萎缩，通过不同细胞进程的转录变化反映出来，其中蛋白质降解是相关性最高的现象之一。这个模型在识别基因或基因表达模式时大有用处，而基因或基因表达模式对评价骨骼肌功能具有重要作用。

由于操作简单并且能产生大量抗体，骨骼肌是注射DNA疫苗的最佳组织。对于生长，肌肉中抗体的过量表达会激发免疫系统，从而导致生长缺陷。Purcell et al.（2006）采用GRASP 16K芯片研究虹鳟接种传染性造血组织坏死病病毒（IHNV）DNA疫苗之后的基因表达变化。疫苗会引发显著的免疫反应，这可以从参与骨骼肌抗原表达和病毒应答的基因表达上调方面得到说明。另外，白细胞内标记基因表达上调表明肌肉组织内白细胞发生炎症反应。因此，接种疫苗骨骼肌的芯片分析结果，阐明了组织中转录的激活会改变其代谢和功能状态。

7.4.4 基因组学技术在肝功能研究中的应用

在硬骨鱼中，肝脏作为合成脂类、碳水化物和蛋白质等能量储备合成的重要器官，在营养和代谢方面起关键的作用。另外，它是应对饥饿的必不可少的器官。肝脏通过代谢分解能量储备物质，并经血液把营养物质输送至各个器官，这对保持体内平衡具有重要作用。作为维持体内平衡的关键角色，肝脏是代谢类激素（如胰岛素和胰高血糖素）的主要靶器官之一。肝脏通过合成和分泌类胰岛素生长因子I（IGF-1）（一种重要的GH生长促进效应传递因子）来应答GH，因此肝脏在生长过程中同样发挥重要作用。由于肝脏具有关键的生理功能并且其功能和调节机制非常复杂，基因组学技术就非常适用于研究肝脏的功能及其与生理因素或环境因素的相互作用。目前，硬骨鱼类生长和营养有关的肝脏转录组和蛋白质研究已有一些报道。

7.4.4.1 与生长和营养有关的肝脏转录组变化

最近有研究采用GRASP 16K芯片平台已经检测GH对鲑鱼生长促进效应的肝脏转录组应答。对于转GH鲑鱼，转录组最主要的变化之一就是线粒

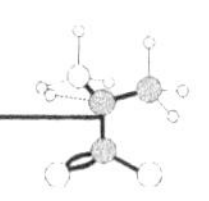

体活性相关的基因表达上调（Rise et al. 2006），这应该与转 GH 鲑鱼较高的生长速度需要更高的代谢速率有关。另外，在转 GH 鲑鱼的肝脏中，血红蛋白基因表达上调，这可能也是跟较高代谢速率需要合成较多血红蛋白有关。出人意料的是，这张芯片检测的基因只有一小部分是肝脏酶类的编码基因，大部分却是脂肪酸去饱和酶和前列腺素 D 合成酶。采用同样的基因芯片，近期一项关于虹鳟肝脏在 GH 短期（3 天）处理条件下的转录组变化，也是检测到一小部分与肝脏酶类相关基因的表达发生变化（Gahr et al. 2008）。该项研究中，应答 GH 的是代谢和免疫这两大功能类群的相关基因。应答 GH 的肝脏转录组信息非常少，可能的原因是大部分调节是在反应或反应之后发生的。因此，有必要采用蛋白质学技术来研究肝脏的生理反应。

从营养学的角度来看，适应植物油替代的肝脏转录变化已有研究。利用来源于大西洋鲑 EST 的基因芯片，可研究大西洋鲑对含有 75% 植物油饲料的肝脏转录组变化（Jordal et al. 2005）。结果表明，与肝脏代谢相关的基因表达量明显发生变化。一方面，Δ5 和 Δ9 脂肪酸去饱和酶基因表达上调。另一方面，为了适应植物油丰富的饲料，鱼肝中长链乙酰辅酶 a 合成酶基因表达下调。另外，对于植物油含量较高的饲料，线粒体外膜的一些线粒体基因和蛋白的表达下调。这些结果表明，在营养条件改变的情况下，这些鱼肝脏中 β 氧化的活性降低。从开始投喂就以 100% 植物油代替鱼油饲养虹鳟 62 周的结果表明，与脂类代谢相关的 2 个基因（脂肪酸合成酶和长链脂肪酸延长酶）表达下调（Panserat et al. 2008b）。除了脂肪合成的相关基因发生变化外，植物油全替代还会影响肝脏中与类固醇合成、解毒、蛋白质代谢和转录调节相关的基因表达。有趣的是，鱼的体重、摄食效率和摄食量均不发生变化（Panserat et al. 2008a, b ）。相比较而言，采用植物蛋白替代动物蛋白（鱼糜），虹鳟的生长速度会下降，而摄食量会增加，与蛋白质和氨基酸代谢相关的基因表达会发生变化（Panserat et al. 2008b）。可想而知，饥饿 3 周之后虹鳟肝脏也会出现类似的情况：与蛋白质生物合成相关的基因表达下调，而蛋白质降解的反应增强（Salem et al. 2007）。

7.4.4.2 与生长和营养有关的肝脏蛋白质组变化

硬骨鱼类肝脏对生长和营养刺激的转录应答有限，因此有必要对转录后的蛋白质组进行研究。目前，已有报道研究虹鳟在饥饿以及喂食植物来源蛋白替代动物蛋白后的肝脏蛋白质组变化。饥饿实验中，在虹鳟的肝脏中发现有 24 个表达差异显著的蛋白质，这其中包括与较高能量需求相关的烯醇化酶

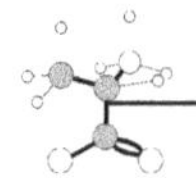

和细胞色素C氧化酶，以及与蛋白质降解相关的组织蛋白酶D（Martin et al. 2001）。采用同样的蛋白质组学研究方法，发现鳟鱼在植物蛋白替代鱼糜的情况下，生长性能和肝脏蛋白质含量发生变化（Martin et al. 2003，Vilhelmsson et al. 2004）。在其中一项研究中，采用大豆蛋白部分替代（30%）鱼糜饲喂鳟鱼12周，发现鱼的生长速度不变而蛋白质代谢增强，蛋白质周转效率和蛋白质代谢率升高，有33个蛋白质的表达丰度发生变化（Martin et al. 2003）。尽管没有鉴别所有的蛋白质，该研究报道了结构蛋白（如角蛋白和微管蛋白）、脂质结合蛋白（如载脂蛋白A）和最为明显的热激蛋白的丰度发生改变，热激蛋白是大豆的抗营养因子可能引发的应激反应的指示物（Martin et al. 2003）。在随后的一项研究中，用混合植物蛋白替代鱼糜的饲料饲喂鳟鱼，鳟鱼生长速度下降，尽管摄食量不变，但摄食效率降低，这可能是蛋白质利用效率降低导致的（Vilhelmsson et al. 2004）。这些营养学参数的变化伴随着蛋白质产量的变化，特别是与初级能量代谢相关的蛋白质（如NADPH和ATP的产量），以及表征蛋白质降解率升高的2个蛋白酶亚基。总而言之，含植物蛋白饲料饲喂的鱼，其肝脏中蛋白质丰度的变化表明，这些鱼比完全用鱼糜饲喂的鱼具有更高的能量需求。

7.4.5 结论与展望

目前，转录组与蛋白质组技术被用于研究鱼类的生长和营养，尽管是刚刚开始，但在不久的将来，这些技术将应用于水产养殖与渔业中的以下几方面：①识别相关的基因并用于鱼类生长和营养学特征选育的分子标记辅助选育项目；②用于诊断鱼类生长和营养状态；③辅助开发全新的并且环境友好型鱼类饲料。尽管总的趋势是积极的，但已发表的芯片分析数据主要集中在一小部分基因，这与现有芯片平台的通量特点有关，与相应的生物信息分析能力有限也有关系。因此，大多数研究无法开展高通量的基因表达分析。为了充分利用转录组技术的力量，未来水产养殖和渔业的基因组学研究必须对功能基因类群进行深度的信息分析，更重要的是，对不同条件或处理的样本可采用大量的芯片实验来开展宏分析。正是使用宏分析手段才能阐明特定生物学过程的分子机制以及正确描述鱼类的生理状态。最后必须强调的是，要把转录组学和蛋白质组学技术应用到非鲑鳟类的其他鱼类研究中。

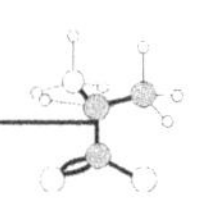

7.5 基因组学技术在海产品质量和安全研究中的应用

近几十年来，健康饮食习惯的发展提高了人们对海产品的需求，因为鱼类和贝类含有诸多的重要营养成分，包括蛋白质（营养价值高，易吸收）、维生素（A、D 和 B12）、微量元素（硒和碘）和脂肪酸［长链 n-3 多不饱和脂肪酸 n-3 PUFA 或者 ω-3 脂肪酸，以十二碳五烯酸 EPA（C20：5 n-3）和二十二碳六烯酸 DHA（C22：6 n-3）为主］。这些营养成分（特别是 n-3 PUFA）的摄入，能有效地预防心血管疾病、2 型糖尿病和神经退行性疾病（Bourre 2005，Calder 2008）

人口的增长以及经济增长的刺激，特别是人均消费水平还在中等以下的国家，鱼类和贝类的消费水平将会持续地增长。至 2020 年，发展中国家人均海产品的消费水平会持续增加，而发达国家将无显著变化（Delgado et al. 2003）。尽管海产品的捕获量会下降且部分是不可持续的，但人工养殖的海产品将会满足需求。各个季节里，安全健康海产品的有效供应将会激活消费者对海产品的消费潜力。

7.5.1 影响海产品质量的各种因素

作为消费者的期望及需求和食品内外在质量属性的结合点，品质观念尤为重要。需考虑的质量属性日渐增加，全球化进程加快以及不同国家消费习惯的差异，使得质量观念的贯彻日益困难。鱼类的品质是一个广泛且复杂的概念，包含多个因素，且这些因素在生产者、加工者、批发商、经销商、餐饮业者、消费者、管理机构和立法机构具有不同的重要性。研究表明，安全认证、质量标识和养生价值是消费者对海产品最关注的地方（Pieniak et al. 2006，Werbeke et al. 2007）。鱼类的肉质依品种而异，其受到多种内在因素的影响，譬如肌肉化学成分（脂肪含量、脂肪酸组成、糖原组分、抗氧化性和颜色）和肌肉细胞特性，同时肉质也受到多种外在因素的影响，例如饲料、屠宰前后的处理、加工和储存程序。由于肉质的遗传力较低，对其进行表型鉴定较困难且成本较高，一般要在屠宰后才能进行，所以肉质性状难以通过选育得以提高。另外，可以明确地讲，肉质受到多种因素的影响而且受未知的 QTL 控制。

在下面这一节中，我们将总结基因组学和蛋白质组学技术在提高水产品

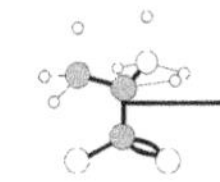

肉质的研究中的应用。

7.5.2 基于基因组学和蛋白质组学方法评价鱼类肉质

鱼类的选育一般针对具有较高遗传力的重要生理性状（如鲑鳟鱼的生长速度），并且这对水产养殖表型已经产生显著的影响。其他性状（如与肉质相关的性状）受到较少关注。肉质是一类高度复杂的性状，受到一系列生理过程的影响，并且每个生理过程可能受到多种激素的调节。深入研究肉质性状的生物学基础，将为生产特制的海产品提供重要的依据。

7.5.2.1 色泽

肌肉的色泽是一个复杂的性状并且很难从外部形态进行判断，这限制了传统的数量遗传学方法的应用。采用全同胞家系的表型鉴定，对育种值进行预测和选择，在一定程度上能缓解这个问题（Gjedrem 2000）。大多数鱼类的肌肉色泽的遗传力处于中低下水平。由于表型值预测需要对每个个体进行有效的表型鉴定，因此对肌肉色泽进行选育比较困难。利用银鲑鱼肌肉色泽相关的1个RAPD多态性标记，开发出与肌肉色泽性状关联的单位点SCAR标记（Oki206，GenBank accession AY661427），该标记可进一步用于辅助选育（Araneda et al. 2005，Lam et al. 2007）。肌肉类胡萝卜素含量与肌肉的感知和色泽特征的遗传相关性较低，表明色素沉积与可见的颜色并不是由相同的基因控制。很有可能鱼肉的颜色不单是受色素沉积的影响。

深入研究类胡萝卜素（如虾青素）与鲑鱼肌肉蛋白质之间的相互作用，有助于更好地了解鲑鱼肌肉类胡萝卜素的沉积。关于虾青素结合至鱼类肌肉机制的了解还很有限（Matthews et al. 2006）。对大西洋成年鲑的膜结合虾青素运输蛋白的蛋白质组学分析研究，发现一些应答膳食虾青素的表达上调的蛋白（Saha 2005）。虾青素从血浆至肌肉蛋白的运输机制还需进一步研究。

7.5.2.2 肌肉纹理（肌肉细胞特性方面）

纹理是反映肌肉品质最主要的指标之一。消费者对肌肉的纹理很敏感，并且纹理是肌肉进行物理切块的重要依据。由于常提及质地松软，业界一直在寻找评价鱼肉纹理的方法并且探究引起肌肉松软度的原因。肌肉的纹理取决于肌肉的化学组成和结构属性，特别是肌纤维和结缔组织蛋白。对于鱼类，肌肉细胞特性（纤维的数量和分布）已被证实能影响其纹理（Johnston 1999，Kiessling et al. 2006）。种内的比较发现，以平均肌纤维横截面面积来度量肌纤维的话，在烹饪过的鱼中，肌纤维随肌肉感官硬度的下降而增加（Hurling et al. 1996）。对于新鲜和熏制的大西洋鲑和淡水褐鲑鱼的研究表明：纤维越

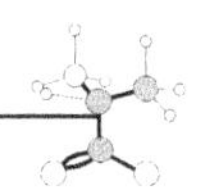

大（或纤维密度降低）肌肉硬度略微减小（Johnston et al. 2004，Bugeon et al. 2003）。

哺乳类和家畜的肌纤维数量由产前的一些因素决定（母体营养、生物活性物质、非生物因素、饲养和基因型），但鱼类还受到孵化后各种因素的影响（Johnston 2006）。尽管硬骨鱼类肌肉发生的细胞生物学与哺乳类相差甚远，但是生长调节相关的基因却高度保守（Watabe 2001）。河鲀和斑马鱼基因组序列为研究鱼类胚后肌肉生长的分子调节机制提供了新视角。斑马鱼的骨膜蛋白基因在肌纤维粘附于肌节和肌纤维的分化过程中起重要作用（Kudo et al. 2004）。在啮齿类动物中，骨膜蛋白基因（又名成骨细胞特异因子或 Osf 2）在肌肉再生过程中的表达显著上调（Goetsch et al. 2003）。该基因已被认为是提高猪肌肉生长速度的候选基因（Bílek et al. 2008），但这个基因在研究鱼类肌肉品质中的潜在价值还有待验证。

不同养殖条件下，肌肉生长的可塑性是决定肌肉品质的主要因素，特别是肌肉纹理和肉质加工性能（Johnston 1999），因此相关的研究对水产养殖具有重要的经济意义。另外，大部分硬骨鱼类属于变温动物并且是行体外受精，相比于哺乳类而言，环境因素对鱼类肌肉生长的影响更广泛（Johnston 2006）。

7.5.2.3 肌肉纹理（受死亡后降解的影响）

鱼类死亡后，肌肉纹理迅速发生改变。大量的文献报道，以剪切力作为评价指标，肌肉的纹理随捕获强度、屠宰方法、饮食要素、储存时间和储存温度的变化而变化（Sigholt et al. 1997，Johnston et al. 2002，Bencze-Røra et al. 2003，Bugeon et al. 2003，Espe et al. 2004）。尽管有大量的文献报道鱼被宰后储存过程中的降解方式，某些蛋白质的降解与肌肉硬度之间的关系还不是很清晰。

采用实验的方法和技术对大量的基因和蛋白质同时进行并行分析，将有益于对鱼类死亡后肌肉降解的详细特征的研究。基因组学和蛋白质组学技术可以同时分析多个基因或蛋白质，为我们研究生物化学系统提供复合的技术手段。最近，利用一个消减 cDNA 文库识别出在冰上储存的死亡虹鳟肌肉中表达上调的特异基因（Saito et al. 2006）。在分析的 200 条 cDNA 序列中，有 82 条与其他已识别的鱼类基因是同源的，如肌钙蛋白 I 和甘油醛 3－磷酸脱氢酶（GAPDH）。比较基因表达谱与斑点杂交的结果，证实了死亡后在冰上储存 3 小时肌肉 mRNA（编码肌钙蛋白 I 和 GAPDH）的含量比刚死亡时高。荧光定量 PCR 分析表明，在鱼保存在冰上至少 24 小时后肌肉组织的细胞还能合成肌钙蛋白 I 和 GAPDH，甚至也许在 48 小时后还有合成活性。这些结果表

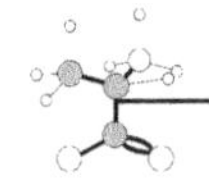

明，细胞水平的转录谱是冰鲜鱼类新鲜程度的敏感性指标。

其他采用蛋白质组分析海产品死亡后变化的研究（Kjærsgård and Jessen 2003，Martinez and Friis 2004，Morzel et al. 2000，Verrez-Bagnis et al. 2001，Schiavone et al. 2008），证实了海产品在储存和加工过程中蛋白质水解的复杂性。为评价鱼类死亡后肌肉的完整性，采用蛋白质组分析，研究2种不同的屠宰前处理（受限的或15 min剧烈肌肉运动）对鳟鱼肌肉蛋白质的影响（Morzel et al. 2006）。剧烈肌肉运动的条件下，非典型的肌间线蛋白（一种关键的细胞骨架蛋白）反复产生，表明这种屠宰前处理会影响鱼类死亡后肌肉完整性的结构。

冰冻是提高鱼类保质期的好方法，但储存过长、较高的冷冻温度及温度波动会对产品质量有负面效应，比如含有脂肪的鱼类发生腐败（Min and Ahn 2005）。冰冻也会降低蛋白质的溶解度（Saeed and Howell 2002）和影响肌肉的品质，如增加粘性和汁液减少（Mackie 1993）。近期，在虹鳟和鳕鱼中，识别到与冰冻阶段降解和氧化过程相关的蛋白质组变化（Kjærsgård et al. 2006a，b），不论是结构蛋白（MHC，肌动蛋白和原肌球蛋白）还是细胞质蛋白（肌酸激酶和烯醇酶）都被严重氧化。相反，核苷二磷酸激酶（NDPK）出现轻微羟基化现象，但其溶解度取决于冰冻温度，也许可以将其作为冰冻储存条件的生物标记。

近年来，这个领域越来越多的研究报道表明，能同时监测多种生化过程的基因组学、蛋白质组学和转录组学技术，适合于挖掘生化或代谢标记，这些标记适用于预测或监测不同养殖条件下和各种死亡前后处理手段的鱼类的肌肉品质。

7.5.2.4 营养价值和健康价值

鱼类的营养价值和健康价值主要与其肌肉的脂肪含量和脂肪酸成分有关。鱼类是人类n-3高不饱和脂肪酸（HUFA）唯一的主要食物来源。随着野生捕获量的下降，人类食物中人工养殖鱼类的比例逐渐升高。采用高能量饲料饲养的鱼类（如鳟鱼，鲑鱼），其内脏和肝脏的能量贮存升高，肌肉中升高的幅度小一些，对于过度肥胖的人群，一般不会选择水产品。通常来讲，养殖鱼类比野生鱼类的脂类含量更高。养殖鱼类肌肉中脂肪酸成分受到饲料中脂类成分的影响。养殖鱼类n-3 HUFA相对含量低于野生鱼类，但由于其较高的脂肪总量，其平均n-3 HUFA含量与野捕鱼类的基本相同。随着全球性鱼油供应不足，很有必要采用富含C18 PUFA而n-3 HUFA含量不及鱼油的植物油替代饲料用鱼油（Tocher 2003）。这引起消费者对人工养殖鱼类n-3 HUFA含量降

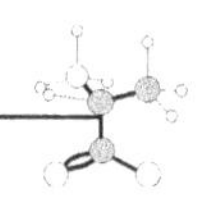

低的关注，因为这损害了鱼类的营养价值。

实验证据表明，海鱼需要饲喂 HUFA 的原因是其缺乏生物合成 HUFA 所需的关键酶，包括 D5 和 D6 脂肪酸脱饱和酶、脂肪酸延长酶（Tocher 2003）。比较海水鱼与淡水鱼脂肪酸脱饱和与延长途径关键因子的编码基因，将有助于深入了解鱼类调控合成 HUFA 的分子遗传机制。这个领域目前受到广泛的关注（Zheng et al. 2005，Tocher et al. 2006，Salem et al. 2007，Izquierdo et al. 2008，Leaver et al. 2008，Panserat et al. 2008a）。尽管对脂类的组成及代谢了解较多，但对脂滴形成所需的分子原件的了解仍然有限。Kadereit et al.（2008）报道了命名为脂肪诱导转录因子（FIT1 和 FIT2），它们对脂滴形成具有重要作用的保守基因家族的特征。利用吗啉基反义技术敲除斑马鱼 FIT2 基因，导致肠道和肝脏出现饲料诱导的脂滴累计的堵塞，揭示了 FIT2 对体内脂滴形成的重要作用。其他调节脂肪降解程度和模式的基因（如油脂 - 1a 和瘦素），在养殖动物包括鱼类中均有研究（Hanchuan et al. 2006，He et al. 2008）。

7.5.3 其他质量性状

多数消费者关注海产品的生产方法、动物的处理方式及健康状况和环境友好型生产系统及其可持续性，这些对选择何种方式进行海产品生产的影响越来越广泛（Harlizius et al. 2004）。“质量”的观念主要与水产养殖产品有关，但渔业可持续开发利用状况在海产品市场同样受到重视。这其中的主要原因是，产品供应链及其预期的调控变化的可持续性、鱼类养殖的健康标准以及对产品和肌肉品质都有影响的屠宰方式，受到的公众关注度越来越高。一般来说，水产动物的健康涉及人类活动哲学与道德的相互影响（Håstein et al. 2005）。尽管如此，生理健康及生物性压力指示因子是广为接受的健康指标。一个有效的健康管理程序必须覆盖水产养殖活动的各个环节，这包括鱼类健康状态的实时资料，识别并应对鱼类健康的风险，减少病原体的接触和传播，以及控制药物和化学物质的使用（Hill 2005）。

分子生物学技术可以很好地运用于鱼类的健康管理。PCR、实时 PCR 和基于核苷酸序列扩增（NASBA）技术能快速地检测、识别和定量微量的水生病原体，这对于有效地控制疾病非常重要，并可以减少抗生素和化学物质的使用。芯片技术为多维度筛选以及解释宿主与病原体间的相互作用提供了新的平台。重组 DNA 技术催生了大规模、低成本的疫苗生产。而且，在不久的将来，DNA 疫苗、蛋白质组学和口服疫苗必将促进有效鱼类疫苗的发展（Adams and Thompson 2006）。近期一些综述评价了基因组技术在促进鱼类健

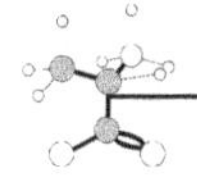

康和应激的研究中的应用和前景（Dios et al. 2008，Martin et al. 2008，Prunet et al. 2008）。

鱼类养殖对环境的影响同样备受关注，其主要是与过量投喂和次佳养分利用有关。鱼类对植酸磷（主要来自植物中的磷）的低效消化导致了水产养殖场对环境的磷污染。更复杂的是，植酸能螯合微量元素和蛋白质，从而导致鱼类不能充分吸收微量元素及蛋白质（Kaushik 2005）。目前解决这个问题的策略是，在鱼类的饲料中添加能把植酸降解为能直接被鱼类利用的无机磷的添加剂（Vielma et al. 2000）。转黑曲霉（*Aspergillus niger*）植酸酶基因的青鳉被用于检测利用植酸磷的能力，作为评价鱼类降解植酸的可行性和效率的模型（Hostetler et al. 2005）。细胞培养技术，包括转染、RT-PCR、Northern blot、Western blot 和酶活分析证明被转进去的基因所编码的蛋白能被表达并分泌，且鱼的生存和生长没有受到影响，表明类似的转基因方法在将来能被用于解决这个问题。

7.5.4　海产品安全

安全是食品质量最重要的议题。目前，基因组学在食品安全中的应用主要有两个方面，即食品成分的安全性评估（Ommen and Groten 2004）和对会引起食品腐败或威胁人类健康的微生物的检测（Abee et al. 2004）。食品安全性检测一般集中在危险物质的识别（食品是否对健康有负作用）和危险物质的特征（引起副作用的浓度）。采集合适的数据进行危险性分析往往成本较高并且耗时，还需要具体的动物毒性试验。基因组学技术提供了一个传统毒理学评价方法的备选方案。首先，高通量的特性意味着能以及时、高性价比的方式分析多个组织。另外，DNA 扩增能更快速且轻松地识别出生物性污染（如微生物或病毒）。再者，采用转录组学、蛋白质组学和宏基因组学，可以研究从基因表达到细胞功能一系列全景式的生物学应答。

7.5.4.1　海产品的健康危害

基于 DNA 分析技术在确保海产品安全中的应用和发展一直是研究热点。海产品安全主要的危害有：有毒物质（如环境化学污染物、生物毒素）、病毒、细菌和寄生虫。目前，相对于传统的毒理学参数（如形态变化、繁殖能力或性成熟的变化），利用 DNA 芯片检测基因表达来研究毒理能提供更加综合、更高灵敏度且更具特色的结果（Steinberg et al. 2008）。近些年，水产养殖和水生毒理学重要物种 EST 数据库的建立，促进了 DNA 芯片平台的发展，基于这些平台就能研究环境污染物和自然环境化合物应激下众多基因的即时

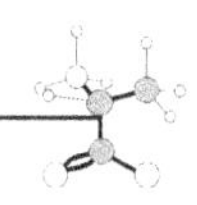

表达状况。另外，由于能同时监测多个具有代表性的毒理学标记基因和发现新的生物标记，表达谱分析对于生态毒理学筛查也是很有吸引力的选择方案（Battershill 2005，Heijne et al. 2005，Brul et al. 2006）

7.5.4.2 海产品的过敏性

海产品过敏是全世界公共健康的重要议题。尽管在公众舆论中有些被拔高，在鱼类和贝类消费与加工较多的地区（如斯堪的纳维亚和伊比利亚半岛的国家）的沿海居民中，海产品过敏最普遍存在。海产品主要的过敏原是鱼类中的 Ca^{2+} 结合蛋白和小清蛋白、贝类的原肌球蛋白（Pen a1）和虾类的精氨酸激酶（Pen a2；Lehrer et al. 2003）。这些蛋白质似乎是耐热、不易被化学降解和蛋白水解，因此，应付海产品过敏原只能是拒绝能使你过敏的海产品。

近期有研究报道，通过定点突变可获得低过敏性的鱼类小清蛋白和虾类精氨酸激酶，这种海产品过敏免疫疗法也许值得一试（Swoboda et al. 2002，Lehrer et al. 2003，Reese et al. 2005）。这些研究充分表明，通过修饰特异位点的氨基酸序列，可以显著降低变异抗原的反应（在某些情况下可大于99%）。

7.5.5 海产品的品种认定和溯源

由于食品安全事件以及公众对食品供应环节透明度的了解需求，在农产品业界，产品的溯源逐渐成为标准。在法律（CR-EC No. 2065/2001 and 104/2000）的规定下，欧洲市民可以知道鱼产品对应的种名、生产方式（养殖或野生）、鱼（野生的话）被捕获的区域或（养殖的话）养殖的区域。溯源系统在食品和饲料供应环节的应用已被写入欧洲食品法（法规号 178/2002/EC），其中第18条关于溯源的规定已于2005年1月1日生效。欧盟食品法把溯源定义为“追踪食品、饲料、生产食品的动物、食品或饲料添加剂的生产、加工和配送”（Martinez et al. 2007）

已被用于鱼类品种认定的基于分析蛋白质的传统分析技术包括电泳技术（如等电点聚焦或 SDS-PAGE）、色谱技术和免疫技术（如免疫扩散和 ELISA）。尽管大部分方法在特定情况下具有较好的可信度，但并不适用于常规的检测分析，因为蛋白质在动物死亡之后会丧失生物学活性，并且蛋白质的状态和特征与所处的细胞类型有关。另外，大多数蛋白质都是热不稳定的。因此如果产品经过剧烈的加工过程（如灭菌），必须依赖于针对 DNA 片段的技术，因为蛋白质和 DNA 差异巨大且呈现出不同的识别模式。鱼类和鱼产品的品种认定所采用的基于 PCR 的方法以及蛋白质组分析的应用可参考有关最

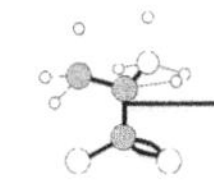

新和最全的综述（Martinez and Friis 2004，Martinez et al. 2005，Gil 2007）。生物条形码技术在渔业中应用的可参见第一章。

很明显，为了确保品种认定方法学特别是基于DNA相关技术的有效性，必需有质粒这种可量化的参考材料。在欧盟SEAFOODPlus项目框架中，一个质粒标准池作为鱼类品种认定的DNA技术的参考材料。一项由12家具有鱼类遗传鉴定资质和为企业提供品种认定分析的中心和研究所参与的试验，已对这些标准进行验证。这些进展已提交专利申请。同时，包括从53种经济鱼类获取到的700条DNA序列的动态DNA数据在互联网上可免费使用（http://www.azti.es/DNA_ database）。市场上也出现了一些鱼类品种认定的商业试剂盒并且检查项目符合现场筛查的目的（Gil 2007）。

尽管还处于发展初期，mRNA或蛋白质水平的全景式基因表达谱分析无疑为我们提供了研究基因调控的重要工具，而这些基因调控是研究海产品生产、运输安全等相关问题的基础。

7.6 基因组学技术在宿主—病原体相互作用研究中的应用

7.6.1 鱼类的宿主—病原体相互作用

对于高密度养殖的鱼类（如欧洲海鲈、金头鲷、大西洋鲑或塞内加尔鳎），疾病频发且引起巨大的经济损失，这主要是由病原感染引起的死亡，或者是染病动物的传播。大多数硬骨鱼类病原体隶属于病毒、细菌和寄生虫，并且具有物种特异性（表7.1）。

表7.1 养殖鱼类中常见的病原生物

Table 7.1 Most common infection agents encountered in farmed finfish

Infection	Type	Host	Publication
Vibrio anguillarum	Bacterial	Marine fish, salmonids and anadromous species	Samuelsen et al. (2006) Lopez-Castejon et al. (2007) Sepulcre et al. (2007) Boesen et al. (1999)
Aeromonas salmonicida	Bacterial	Salmonids in fresh and seawater	Emmerich and Weibel (1894)
Renibacterium salmoninarum	Bacterial	Salmonids in fresh and seawater	Fryer and Sanders (1981)

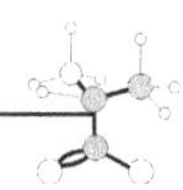

续上表

Table 7.1 (continued)

Yersinia ruckeri	Bacterial	Salmonids in freshwater	Rucker (1966)
Edwardsiella ictaluri	Bacterial	Catfish, particulary channel catfish *Ictalurus punctatus*	Hawke et al. (1981)
Photobacterium damselae piscicida	Bacterial	Marine fish, e.g. yellowtail *Seriola quinqueradiata*	Toranzo et al. (1991)
Nodavirus (small viruses with a simple architecture)	Virus	Marine fish (also detected in some freshwater fish e.g. *Acipenser sp.* and *Poecilia reticulata*)	Frerichs et al. (1996) Hegde et al. (2003) Athanassopoulou et al. (2004)
Monogenean (flatworms) e.g. *Diplectanum aequans*	Parasite	All fish species	Buchmann and Lindenstrom (2002) and Faliex et al. (2008)

英文注释:*Vibrio Anguillarum*,鳗弧菌;*Aeromonas Almonicida*,鲑嗜水气单胞菌;*Renibacterium Almoninarum*,鲑肾杆菌;*Yersinia Ruckeri*,耶尔森氏菌;*Edwardsiella Ictaluri*,鮰爱德华氏菌;*Photobacterium Amselae Iscicida*,发光菌;Nodavirus,诺达病毒;Monogenean,单殖吸虫

7.6.2 宿主免疫反应的转录组特征

7.6.2.1 EST 分析识别宿主免疫反应相关的基因

EST 分析是基因挖掘和识别最快速的方法之一，为研究基因表达谱、可变剪切或多聚腺苷酸化差异分析等提供了有用数据，同时也可以识别 1 类分子标记。此外，EST 是比较作图和发展芯片技术的基础。大多数已发表的数据都来源于 EST 项目（如 Gong et al. 1994，Douglas et al. 1999，Karsi et al. 2002，Sarropoulou et al. 2005a，Bai et al. 2007，Li et al. 2007；Marine Genomics Europe 2004－2008），采用未感染的组织获得基本的基因目录。但是，为了识别免疫反应的基因转录本，必需从感染组织的 cDNA 文库获取 EST 信息。识别和分离鱼类免疫反应相关的转录本在鲤鱼（Kono et al. 2003，Sakai et al. 2005）和欧洲海鲈（Sarropoulou et al. 2009）均有报道。后来，用 EST 分析基因表达水平，EST 来自 6 个鳗弧菌感染组织的 cDNA 文库（肝脏、脾脏、头肾、腹腔渗出组织、鳃和肠道）和 4 个诺达病毒感染组织的 cDNA 文库（头肾、脾脏、脑和肝脏）。分离感染之后的差异表达基因并采用实时 PCR 进行分析，证实了其中一些基因可作为鱼类感染细菌和病毒的生物标记。总之，不同硬骨鱼类日益丰富的序列信息，以及相应的功能信息，为我们更好地了解硬骨鱼类免疫相关机制及进化提供了重要基础。

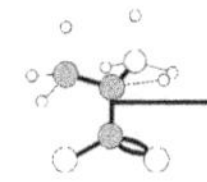

7.6.2.2 芯片分析识别宿主免疫反应相关的基因

DNA 芯片可以高通量地分析基因表达谱（见 7.2.5 功能基因组学技术），却较少被应用于识别养殖鱼类免疫反应相关的基因。有文献报道，采用 cDNA 芯片研究革兰氏阴性细菌急性感染鲶鱼的免疫反应相关的基因表达谱（Peatman et al. 2007），结果显示鲶鱼里大部分典型急性期反应蛋白连同一组假定硬骨鱼急性期反应物一起表达上调。对海鲈的一项类似研究（E. Sarropoulou，私人通讯）显示，铁代谢相关基因在免疫反应过程中被显著诱导，表明有限的自由铁可能会抑制细菌的生长同时预防金属诱导的细胞损伤。随后 Peatman et al.（2008）采用寡核苷酸芯片开展 MHC I 通路相关的转录本分析。该研究识别了 131 个差异表达基因，其中 103 个属于独特的基因。有趣的是，一个公认的管家基因 β-actin 经常作为 qPCR 的内参，也是属于差异表达的基因，这说明该基因在免疫反应的条件下可能发挥某些功能。在该研究中，作者还报道了在感染之后某些信号通路的应激反应，以及两种近亲物种蓝鲶鱼和斑点叉尾鮰存在显著的差异，这些差异主要集中在 MHC I 通路上的转录本。近期，应用芯片研究黄尾（*Seriola quinquerdiata*）在免疫刺激剂如刀豆蛋白 A（ConA）和脂多糖（LPS）作用下的免疫反应特征（Darawiroj et al. 2008）。类似的研究在大西洋鲑（Ewart et al. 2005，Martin et al. 2006）和虹鳟（Tilton et al. 2005，Gerwick et al. 2007）中也有报道。MacKenzie et al.（2008）研究显示，采用芯片基因表达分析可以区分虹鳟对两种不同的免疫因子，如病毒（IHNV）和细菌细胞壁成分的免疫反应。

7.6.3.2 采用 Real-Time PCR 筛选疾病诊断的候选标记

Kleppe et al.（1971）首次介绍了 PCR 技术，之后 Mullis and Faloona（1987）将 PCR 用于检测特定的转录本。Higuchi et al.（1993）首次将 PCR 用于定量研究。目前，实时定量 PCR（qPCR）被广泛运用于人类疾病诊断和各种生物学系统的表达研究（Bustin et al. 2005）。qPCR 的精确性取决与多种因素诸如 RNA 模板的质量、聚合酶类型、引物、采用的参考基因和数据分析。qPCR 已被用于重要经济鱼类免疫反应相关基因的研究。例如，qPCR 分析阐明了抗菌肽铁调素在金头鲷感染细菌之后天然免疫过程中的作用（Cuesta et al. 2008）。Raida and Buchmann（2007）通过 qPCR 研究发现，鳟鱼天然免疫和特异性免疫基因属于温度调节型。qPCR 同样可用于研究病毒和细菌相关疾病的致病机制、预防和处理（Samuelsen et al. 2006）。

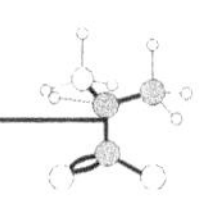

7.6.3 遗传连锁图、RH 作图和物理图谱如何阐明鱼—病原体的相互作用

除了生长和抗逆，大多数鱼类养殖的重要经济性状还包括免疫反应和宿主与病原体之间的相互作用。在畜禽动物中，已识别出 QTL 对应的基因组区域，在某些情况下 QTL 的分子多态性也被识别出来（Stear et al. 2001，Andersson and Georges 2004）。但是，由于缺少适用于水产生物的工具，限制了前沿技术（如 QTL 作图和 MAS）的应用。少数关于水产生物 QTL 的研究，其目标都是提高生长速度（Moghadam et al. 2007，Wang et al. 2008）和抗病力（Moen et al. 2007）。对于 QTL 的检测，分子生物学工具如遗传图谱、基因组图谱和足够的基因组信息（如大规模 ESTs 信息或 BAC - end 序列）是必不可少的，目前，硬骨鱼类和水产动物这方面的研究正如火如荼地进行。鲑科、鲤科、丽鱼科、狼鲈科、牙鲆科和叉尾鮰科鱼类的基因组图谱已有报道（Kocher and Kole 2008），同时分子标记（Bouza et al. 2007，Sanetra and Meyer 2008）、BAC 文库（Matsuda et al. 2001，Whitaker et al. 2006，Wang et al. 2008）、基因表达数据（Ewart et al. 2005，Sarropoulou et al. 2005b，Tilton et al. 2005，Gunnarsson et al. 2007，Darawiroj et al. 2008，Darias et al. 2008，Peatman et al. 2008）和作图平台（Gilbey et al. 2006，Houston et al. 2008）也有报道。图谱和多态性分子标记是更好地了解宿主与病原体之间相互作用以及宿主基因组水平免疫反应的必要工具，并且可以阐明与宿主抗性、耐受性和易感性相关的 QTL 基因。候选基因作图能显著减少 QTL 筛查的区域。对叉尾鮰 EST 的分析表明，大约有 10% 的 EST 含有短串联重复序列（STR；Serapion et al. 2004a，b）。但是，这些包含 STR 的 EST 均不是与免疫相关的转录本或者间接与免疫反应相关。因此，免疫相关转录本的作图将有利于挖掘与免疫应答相关 QTL 相联系的标记。候选标记的筛选可以通过表达分析手段（如芯片技术、qPCR 筛选或 cDNA 文库测序）来实现。

7.6.4 贝类宿主—寄生虫的相互作用

7.6.4.1 采用分子生物学技术改进诊断方法

目前，分子生物学提供了大量简单易行的诊断工具，从而替代传统耗时且样品需求较多的方法。PCR 为贝类疾病诊断提供许多新方法，这些方法特异性强、灵敏性高而且结果可靠。PCR 已被用于检测弧菌（Hill et al. 1991，Brauns et al. 1991）、病毒（Desenclos et al. 1991，Batista et al. 2007）、利斯

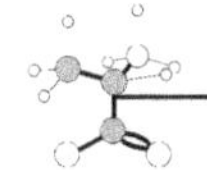

特菌（Jeyasekaran and Karunasagar 1996）、包那米虫（Cochennec et al. 2000）、沙门菌（Dupray et al. 1997）、帕金虫（Reece et al. 1997）、鞭毛虫（Graczyk et al. 1999）、马尔太虫（Le Roux et al. 1999）和隐孢子虫（Gomez-Bautista et al. 2000）。以 PCR 作为部分步骤的更复杂方法也被用于贝类病原体的检测，如 RFLP（Buchrieser et al. 1995，Gomez-Bautista et al. 2000，Hine et al. 2001，Le Chevalier et al. 2003，Abollo et al. 2006）、酶联免疫检测（ELISA；Gonzalez et al. 1999，Schwab et al. 2001，Elandalloussi et al. 2004）；或者是直接对 PCR 进行改进的方法，如多重 PCR（Shangkuan et al. 1995，Brasher et al. 1998，Penna et al. 2001）和实时定量 PCR（Blackstone et al. 2003，Campbell and Wright 2003，Audemard et al. 2006）。多重 PCR 可在 1 次 PCR 反应中检测多种病原体，实时 PCR 则可以进行精确定量。基于分子生物学技术的诊断方法已非常流行。近期，多重实时定量 PCR 已被用于即时检测和定量多种病原生物（Panicker et al. 2004，Nordstrom et al. 2007）。高通量技术的发展将在未来实现快速并且同步检测大量样本的病原。

7.6.4.2 双壳类的分子免疫

双壳类面对环境中大量与其共同生存的病原和寄生虫。同其他无脊椎生物一样，双壳类缺乏适应性免疫系统（Zinkernagel et al. 1996），仅依靠先天免疫抵御病原入侵（Bachère et al. 2004）。双壳类体内防疫机制可分为细胞免疫和体液免疫机制，很明显这两者是相互关联且跟主要的免疫活性血细胞紧密联系（Cheng 1981，Hine 1999）。为了增加对双壳类免疫反应的了解，基因组学技术已用于相关研究和某些种类的免疫相关基因也有报道（Gueguen et al. 2003，Tanguy et al. 2008）。

宿主的防御反应主要分为 3 个步骤：①病原识别；②诱导先天免疫信号转导通路的激活；③效应因子生成的启动。识别阶段主要是通过受体系统和可以识别保守的病原体相关分子模式（PAMP；Janeway and Medzhitov 2002）的分子，来鉴别自身的和外源的物质。PAMP 是微生物特有的，并非由宿主产生，因此与微生物的适应性无关（Nürnberger et al. 2004）。尽管如此，牡蛎防疫机制的激活不仅能由 PAMP 启动，还能由在马青格损伤模型（Janeway and Medzhitov 2002，Matzinger 2002，Montagnani et al. 2007）中定义的损伤相关分子模式（DAMP）来启动。宿主识别 PAMP 的分子称为模式识别分子（PRM）或模式识别受体（PRR；Medzhitov and Janeway 2002，Janeway and Medzhitov 2002）。在双壳类中，已鉴定了一些 PRR，包括肽聚糖识别蛋白（PGRP；Su et al. 2007，Ni et al. 2007，Itoh and Takahashi 2008）、一些碳水化合物结合凝集素（Kang et al. 2006，Wang et al. 2007a，Yamaura et al. 2008）、

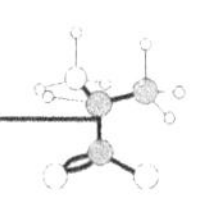

甲壳素结合凝集素（Badariotti et al. 2007）和 LPS 结合凝集素（Gonzalez et al. 2005a，2007，Ni et al. 2007，Bettencourt et al. 2007）。Toll 样受体（Tanguy et al. 2004；Qiu et al. 2007a），以及含 C1q 结构域蛋白质（Zhang et al. 2008a）。PRR 和 PAMP 的相互作用启动了免疫机制。

太平洋牡蛎有 6 个与 Rel/NF－κB 通路相关基因的特征支持信号通路保守性的观点（Gueguen et al. 2003，Escoubas et al. 1999，Montagnani et al. 2004，2008）。在其他双壳类，Toll 受体、MyD88 和 Rel 等组分的特征也有报道（Tanguy et al. 2004，Qiu et al. 2007a，b，Wu et al. 2007，Bettencourt et al. 2007）。双壳类与昆虫 Rel/NF-κB 通路的同源性表明，该调节通路参与先天免疫基因的调节作用（Lemaitre et al. 1995，Silverman and Maniatis 2001）。类似地，TGF-β 或 TGF-β 通路可能参与诱导防御系统的激活（Lelong et al. 2007）。最后，免疫反应的第 3 阶段是具有抗微生物活性的效应因子表达的诱导。这些效应因子主要由免疫活性细胞血细胞产生，也有一些是由上皮组织（如鳃和外套膜的上皮组织）产生，这些组织构成双壳类的第一道防线。在这些效应因子中，蛋白酶抑制剂和抗菌肽均有研究报道，抑制剂以微生物蛋白酶作为靶位点，从而防止宿主感染（Labreuche et al. 2006a，b）。在双壳类中，某些效应因子属于 α2－巨球蛋白（Gueguen et al. 2003，Ma et al. 2005）、丝氨酸蛋白酶抑制剂（serpin；Gueguen et al. 2003，Tanguy et al. 2004）和金属蛋白酶抑制剂（TIMP；Montagnani et al. 2001）。抗菌肽首次在贻贝中发现（Charlet et al. 1996，Hubert et al. 1996），其包括 4 个家族：defensin（Hubert et al. 1996，Mitta et al. 1999b），myticin（Mitta et al. 1999a），mytilin（Mitta et al. 2000b）和 mytimicin（Mitta et al. 2000a）。最近，在其他双壳类中也分离出抗菌肽（Seo et al. 2005，Gueguen et al. 2006，Zhao et al. 2007，Gestal et al. 2007，Bettencourt et al. 2007），表明抗菌肽广泛地存在于生物界的各个门类。在感染应答时，免疫效应因子的放大化主要是与转录或转录后的调节有关，同时也能激活造血作用，从而增肌血细胞数量（Tirape et al. 2007）。

7.6.4.3 帕金虫感染的免疫反应

对双壳类—帕金虫相互作用分子机制的了解仍然很有限。双壳类中描述较清楚的是由凝集素诱导的机制。双壳类靠凝集素识别感染源，并以此诱导防御机制。这种识别具有高度特异性，只是识别帕金虫的感染（Kim et al. 2006，2008）。比如，奥尔森帕金虫（Bulgakov et al. 2004，Kang et al. 2006，Kim et al. 2006，2008）和海水帕金虫（Gauthier et al. 2004，Tasumi and Vasta 2007）在蛤仔和牡蛎中不是由同一种凝集素识别。最近发现一种新的半乳凝集素，可识别特异的原虫期的帕金虫，海水帕金虫原虫是寄生虫的有毒阶段

(Tasumi and Vasta 2007)。

胞外蛋白酶，特别是丝氨酸蛋白酶，在帕金虫的致病和致毒阶段发挥重要作用（La Peyre et al. 1996，Faisal et al. 1999，Tall et al. 1999）。双壳类血浆中存在的蛋白酶抑制剂，应该对帕金虫防御起一定作用（Xue et al. 2006）。北美牡蛎血浆蛋白酶抑制剂活性比太平洋牡蛎的低，因此其更易感帕金虫（Faisal et al. 1998）。另外，有报道认为疾病暴发强度与蛋白酶抑制剂活性成负相关关系（Oliver et al. 2000）。

贝类对帕金虫的另外一种防疫机制是对寄生虫细胞的包裹（Navas et al. 1992，Montes et al. 1995b，Sagristà et al. 1995；图7.3）。这种机制是炎症反应，在蛤类如欧洲蛤仔（*R. decussatus*）和菲律宾蛤仔（*R. philippinarum*）中都有报道，奥尔森帕金虫诱导这两种蛤仔产生特异的细胞反应（Montes et al. 1995a，b），该反应过程中颗粒白细胞被招募并渗透到被感染的组织。它们合成一种轻微糖化多肽，以极化的方式释放并在帕金虫原虫的周围形成胶囊。这种肽在未感染帕金虫或感染其他微生物如细菌或真菌的蛤仔中未表达（Montes et al. 1995b）。由于帕金虫细胞壁可抵抗蛋白质水解，因此这个过程致死帕金虫的比例很低（Montes et al. 1996），但包裹作用能有效地阻断原虫的扩散（Montes et al. 1995a，Rodriguez and Navas 1995）。炎症反应也能消除导致宿主死亡的血窦（Montes et al. 1995a）。

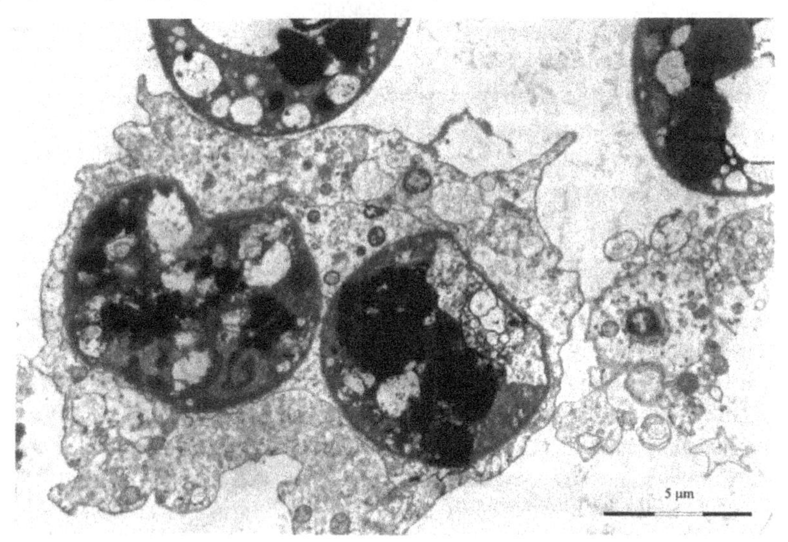

Fig. 7.3 *Perkinsus olseni* trophozoites upon phagocytosis by a clam (*Ruditapes decussatus*) hemocyte. (Figure credit: R. Leite)

图7.3　欧洲蛤仔（*R. decussatus*）对奥尔森帕金虫原虫的吞噬作用

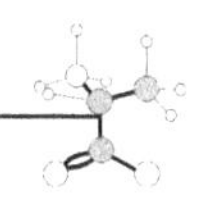

7.6.4.4 弧菌感染的免疫反应

作为滤食性动物，双壳类不断地接触到自然存在于近海微生物群落里面的致病菌和条件致病菌。在这些细菌中，弧菌是海洋环境中最常见的（Potasman et al. 2002）。作为共生菌，弧菌一般被认为是引起双壳类（特别是长牡蛎）死亡的条件致病菌（Paillard et al. 2004）。双壳类血细胞具备溶菌酶活性相关的氧化和非氧化杀菌系统（Hine 1999）。致病性弧菌 Vibrio 会激发长牡蛎血细胞的氧化爆发（Lambert et al. 2003，Labreuche et al. 2006a，b）。而且，双壳类的血细胞是流动的，具有较强的凝结潜力和趋化活性（Prieur et al. 1990，Canesi et al. 2002），并且细胞质能自发扩展（伪足）增强吸附能力。不同的双壳类以及不同的细菌感染时，血细胞的移动模式有所差异（Howland and Cheng 1982，Kumazawa and Morimoto 1992，Fawcett and Tripp 1994）。研究证实，弧菌可诱导贻贝伪足损伤和细胞变圆（Nottage and Birkbeck 1990，Lane and Birkbeck 1999），导致蛤仔血细胞粘附能力降低（Choquet et al. 2003）。弧菌的胞外产物影响长牡蛎的吞噬作用和血细胞的粘附（Labreuche et al. 2006a）。弧菌感染的蛤仔溶菌酶活性升高（Allam et al. 2006）。因此，细菌表面配体和可溶性血淋巴组分等因子，以及细菌对血细胞应答相关的特定信号通路的激活能力，均与弧菌在海洋双壳类中存留有关（Pruzzo et al. 2005）。

双壳类—弧菌相互作用以及相关的免疫应答的分子机制目前还知之甚少。感染致病弧菌的牡蛎血细胞的 EST 文库已构建好（Gueguen et al. 2003）。1142 条 EST 序列有 55 条序列属于免疫基因。在这些 EST 中，组织金属蛋白酶抑制剂（*Cg-Timp*）的丰度最高。这个基因在血细胞中特异性表达（Montagnani et al. 2001），并且其 mRNA 浓度受到弧菌产生的分泌或排泄分子诱导（Montagnani et al. 2007）。另外，扇贝丝氨酸蛋白酶抑制剂和丝氨酸蛋白酶受鳗弧菌的诱导（Zhu et al. 2006，2007）。从这个文库还识别到与 Rel/NF - κB 信号转导通路分子同源的 4 个 cDNA 以及 2 个之前已识别的基因，即 *oIKK* 和 *Cg-Rel*（Escoubas et al. 1999，Montagnani et al. 2004）。*Cg-MyD88* 作为这个通路上最重要的接头分子，在细菌感染的条件下表达上调（Tirape et al. 2007）而 *oIKK* 和 *Cg-Rel* 则无显著变化（Escoubas et al. 1999，Montagnani et al. 2004）。而且，扇贝 Toll 受体基因在 LPS 的处理下表达上调，表明该通路参与对弧菌的免疫反应（Qiu et al. 2007a）。LPS 结合蛋白在免疫系统的激活过程中起着重要作用。牡蛎的 1 种 BPI 蛋白（*Cg* - BPI）已被识别（Gonzalez et al. 2007a）。在大多数组织的上皮细胞中 *Cg* - BPI 属于组成型表达，在受细菌感染后的成年牡蛎以及发育阶段，血细胞中的 *Cg*-BPI 表达受诱导（Gonzalez et

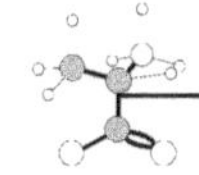

al. 2007）。另外，牡蛎在受弧菌感染之后，抗氧化酶 *Cg*-SOD（特别是在血细胞中产生的）表达明显下调（Gonzalez et al. 2005），而同时活性氧（ROS）产量升高（Labreuche et al. 2006b）。这些结果与其他研究结果相矛盾，因为其他结果显示细菌感染之后血细胞 ROS 产量降低（Lambert et al. 2003）。Lambert et al.（2006b）指出 ROS 产量升高会导致氧化应激，并使弧菌避开宿主的细胞应答。在扇贝中，过氧化氢酶家族的一员（包括参与分解过氧化氢而消除 ROS 的酶）已被识别出来，并且在弧菌感染之后表达上调（Li et al. 2008）。关于抗菌肽，对细菌感染之后的表达调控研究（Mitta et al. 1999a，Mitta et al. 2000a，b，Zhao et al. 2007；Gonzalez et al. 2007，Gestal et al. 2007）表明，抗菌肽对弧菌可能有作用。因此，弧菌引发的免疫应答效应因子的特殊功能还需进一步研究。研究免疫系统中诱导型和组成型基因的作用，并且阐明双壳类是如何区分病原菌和各种共生微生物，是很有意义的研究工作。

7.6.4.5　转录组技术的应用

迄今，大多数阐述双壳类—寄生虫相互作用的研究都是针对太平洋牡蛎的。目前，大多数实验室和研究项目广泛开展多种双壳类 cDNA 文库构建、EST 测序和基因组测序，从而大大增加了基因组资源。双壳类基因组学的重要进展主要由一些大型研究联盟推动的，如欧洲海洋基因组学联盟（www.marine-genomics-europe.org）和美国海洋基因组学联盟（www.marinegenomics.org），他们的主要目标是通过大量的 EST 和 BAC 文库测序，来识别与细胞和生化过程相关的重要基因。基于这种策略，细菌感染的太平洋牡蛎（Gueguen et al. 2003）和帕金虫感染的马尼拉蛤（Kang et al. 2006）中，免疫防御相关的血细胞基因已被识别。结束于 2008 年的欧洲海洋基因组学项目留下了大量双壳类 EST 数据，主要来自太平洋牡蛎、欧洲蛤仔、菲律宾蛤仔、紫贻贝和深海贻贝（Tanguy et al. 2008）。2008 年 8 月，19 个物种的 EST 数据已提交至公共数据库（www.ncbi.nlm.nih.gov）。其中，最多的是太平洋牡蛎，有29 018条 EST，加利福尼亚贻贝和美洲牡蛎分别有 23 871 和 14 560 条 EST。因此，最有代表性的物种是牡蛎、贻贝、蛤仔和扇贝，他们都是养殖种类。同时，两种淡水贻贝南非斑马贻贝（*Dreissena rostriformis bugensis*）和斑马纹贻贝（*Dreissena polymorpha*）的 EST 数据也有收集，虽然它们不是经济种类，但是在许多国家它们是属于入侵种，从而引发越来越多的问题，收集它们的 EST 是为制定恰当的解决方案提供有用的信息。EST 项目不仅可以开展基因注释，同时也有助于构建物理连锁群和进行比较作图，

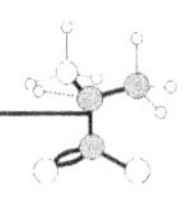

以及可变剪切分析和基因复制、基因芯片开发。开发包含感染了帕金虫和弧菌的太平洋牡蛎和美洲牡蛎基因的 cDNA 芯片，即可在宿主—病原体条件下对差异表达基因的特征进行研究（Jenny et al. 2006）。在欧洲海洋基因组学网络框架里，包括牡蛎（太平洋牡蛎）、贻贝和蛤仔基因的芯片正在研发。

同样，差减文库在研究宿主—寄生虫相互作用中可以提供基因差异表达显著相关的信息。该技术（Diatchenko et al. 1999）为研究疾病双方提供有力并且可靠的工具，可以对宿主或寄生虫的基因调控进行细致的阐述。基于该技术，感染弧菌的蛤仔（Gestal et al. 2007）、感染帕金虫（Tanguy et al. 2004）和弧菌（Jenny et al. 2006）的牡蛎以及感染帕金虫的蛤仔血细胞（Ascenso et al. 2007）中的一系列差异表达基因才有可能被鉴定出了。除了以上提到的技术，能对宿主或寄生虫的转录组采集连续且大量信息的其他可靠方法将在不久显示出强大的应用价值（如焦磷酸测序、SAGE 和 MPSS）。

7.6.4.6 总结

基因组学与其他技术结合，可以为了解宿主—寄生虫相互作用和控制双壳类疾病提供有力的方法和工具。除了基于宿主分子标记的品系选育，如选育抗尼氏单孢子虫的太平洋牡蛎品系（Ford and Haskin 1987），全球范围内都致力于开展宿主和病原生物的基因组测序，从而了解防御和感染机制。即使一些重要的工具，如双壳类细胞系、大规模高通量基因组学和芯片技术还有所欠缺，但一些可用的数据已经为了解宿主防御相关的免疫系统和生理过程提供诸多重要的线索。

参考文献

Abee T, Van Schaik W, Siezen RJ (2004) Impact of genomics on microbial food safety. Trends Biotechnol 22:653 – 660

Abollo E, Casas SM, Ceschia G et al (2006) Differential diagnosis of *Perkinsus* species by polymerase chain reaction-restriction fragment length polymorphism assay. Mol Cell Probes 20:323 – 329

Adams A, Thompson KD (2006) Biotechnology offers revolution to fish health management. Trends Biotechnol 24:201 – 205

Allam B, Paillard C, Auffret M et al (2006) Effects of the pathogenic *Vibrio tapetis* on defence factors of susceptible and non-susceptible bivalve species: II. Cellular and biochemical changes following in vivo challenge. Fish Shellfish Immunol 20:384 – 397

Allendorf FW, England PR, Luikart G et al (2008) Genetic effects of harvest on wild animal populations. Trends Ecol Evol 23:327 – 337

Anderson EC, Garza JC (2006) The power of single-nucleotide polymorphisms for large-scale parentage inference. Genetics 172:2567 – 2582

Andersson L, Georges M (2004) Domestic-animal genomics: deciphering the genetics of complex traits. Nat Rev Genet 5:202 – 212

Araneda C, Neira R, Iturra P (2005) Identification of a dominant SCAR marker associated with colour traits in Coho salmon (*Oncorhynchus kisutch*). Aquaculture 247:67 – 73

Arkush KD, Giese AR, Mendonca HL et al (2002) Resistance to three pathogens in the endangered winter-run Chinook salmon (*Oncorhynchus tshawytscha*): effects of inbreeding and major histocompatibility complex genotypes. Can J Fish Aquat Sci 59:966 – 975

Artamonova VS (2007) Genetic markers in population studies of Atlantic salmon *Salmo salar* L.: Analysis of DNA sequences. Russ J Genet 43:341 – 353

Ascenso RMT, Leite RB, Afonso R et al (2007) Suppression-subtractive hybridization: A rapid and inexpensive detection methodology for up-regulated *Perkinsus olseni* genes. Afr J Biochem Res 3:24 – 28

Athanassopoulou F, Billinis C, Prapas T (2004) Important disease conditions of newly cultured species in intensive freshwater farms in Greece: first incidence of nodavirus infection in *Acipenser* sp. Dis Aquat Organ 60:247 – 252

Audemard C, Ragone Calvo LM, Paynter KT et al (2006) Real-time PCR investigation of parasite ecology: in situ determination of oyster parasite *Perkinsus marinus* transmission dynamics in lower Chesapeake Bay. Parasitology 132:827 – 842

Bachère E, Gueguen Y, Gonzalez M et al (2004) Insights into the anti-microbial defense of marine invertebrates: the penaeid shrimps and the oyster *Crassostrea gigas*. Immunol Rev 198:149 – 168

Badariotti F, Lelong C, Dubos MP et al (2007) Characterization of chitinase-like proteins (Cg-Clp1 and Cg-Clp2) involved in immune defence of the mollusc *Crassostrea gigas*. FEBS J 274: 3646 – 3654

Bai J, Solberg C, Fernandes JM et al (2007) Profiling of maternal and developmental-stage specific mRNA transcripts in Atlantic halibut *Hippoglossus hippoglossus*. Gene 386:202 – 210

Baker CS, Cooke JG, Lavery S et al (2007) Estimating the number of whales entering trade using DNA profiling and capture-recapture analysis of market products. Mol Ecol 16:2617 – 2626

Barrett RDH, Rogers SM, Schluter D (2008) Natural selection on a major armor gene in threespine stickleback. Science 322:255 – 257

Batista FM, Arzul I, Pepin JF et al (2007) Detection of ostreid herpesvirus 1 DNA by PCR in bivalve molluscs: a critical review. J Virol Methods 139:1 – 11

Battershill JM (2005) Toxicogenomics: regulatory perspective on current position. Hum Exp

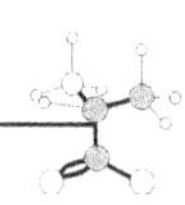

Toxicol 24:35 - 40

Beaumont MA, Nichols RA (1996) Evaluating loci for use in the genetic analysis of population structure. Proc R Soc B-Biol Sci 263:1619 - 1626

Beaumont MA (2005) Adaptation and speciation: what can F-st tell us? . Trends Ecol Evol 20:435 - 440

Bencze-Røra AM, Regost C, Lampe J (2003) Liquid holding capacity, texture and fatty acid profile of smoked fillets of Atlantic salmon fed diets containing fish oil or soybean oil. Food Res Int 36:231 - 239

Bernatchez L, Landry C (2003) MHC studies in nonmodel vertebrates: what have we learned about natural selection in 15 years? . J Evolution Biol 16:363 - 377

Bettencourt R, Roch P, Stefanni S et al (2007) Deep sea immunity: Unveiling immune constituents from the hydrothermal vent mussel *Bathymodiolus azoricus*. Mar Environ Res 64:108 - 127

Bierne N, Tsitrone A, David P (2000) An inbreeding model of associative overdominance during a population bottleneck. Genetics 155:1981 - 1990

Bílek K, Knoll A, Stratil A et al (2008) Analysis of mRNA expression of CNN3, DCN, FBN2, POSTN, SPARC and YWHAQ genes in porcine foetal and adult skeletal muscles. Czech J Anim Sci 53:181 - 186

Birney E, Stamatoyannopoulos JA, Dutta A et al (2007) Identification and analysis of functional elements in 1% of the human genome by the ENCODE pilot project. Nature 447:799 - 816

Blackstone GM, Nordstrom JL, Vickery MC et al (2003) Detection of pathogenic *Vibrio parahaemolyticus* in oyster enrichments by real time PCR. J Microbiol Methods 53:149 - 155

Blanco G, Borrell YJ, Bernardo D (2007) The use of microsatellites for optimizing broodstocks in a hatchery of gilthead seabream (*Sparus aurata* L.). Aquaculture 272:S246

Boesen HT, Pedersen K, Larsen JL et al (1999) *Vibrio anguillarum* resistance to rainbow trout (*Oncorhynchus mykiss*) serum: role of O-antigen structure of lipopolysaccharide. Infect Immun 67: 294 - 301

Bonin A (2008) Population genomics: a new generation of genome scans to bridge the gap with functional genomics. Mol Ecol 17:3583 - 3584

Borrell YJ, Alvarez J, Vazquez E et al (2004) Applying microsatellites to the management of farmed turbot stocks (*Scophthalmus maximus* L.) in hatcheries. Aquaculture 241:133 - 150

Bouck A, Vision T (2007) The molecular ecologist's guide to expressed sequence tags. Mol Ecol 16:907 - 924

Boudry P, Collet B, Cornette F et al (2002) High variance in reproductive success of the Pacific oyster (*Crassostrea gigas*, Thunberg) revealed by microsatellite-based parentage analysis of multifactorial crosses. Aquaculture 204:283 - 296

Boudry P, Dégremont L, Haffray P (2008) The genetic basis of summer mortality in Pacific

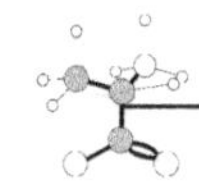

oys-ter spat and potential for improving survival by selective breeding in France. In: Samain JF, Mc-Combie H (eds) Summer mortality of Pacific oyster *Crassostrea gigas*-The morest project, Quae edn. Versailles, France

Boughman JW (2001) Divergent sexual selection enhances reproductive isolation in sticklebacks. Nature 411: 944 – 948

Bourre JM (2005) Dietary omega-3 fatty acids and psychiatry: mood, behaviour, stress, depression, dementia and aging. J Nutr Health Aging 9: 31 – 38

Boutet I, Tanguy A, Moraga D (2004) Response of the Pacific oyster *Crassostrea gigas* to hydrocarbon contamination under experimental conditions. Gene 329: 147 – 157

Bouza C, Hermida M, Pardo BG et al (2007) A microsatellite genetic map of the turbot (*Scophthalmus maximus*). Genetics 177: 2457 – 2467

Bramble L, Anderson RS (1997) Modulation of *Crassostrea virginica* hemocyte reactive oxygen species production by *Listonella anguillarum*. Dev Comp Immunol 21: 337 – 348

Brasher CW, DePaola A, Jones DD et al (1998) Detection of microbial pathogens in shellfish with multiplex PCR. Curr Microbiol 37: 101 – 107

Brauns LA, Hudson MC, Oliver JD (1991) Use of the polymerase chain reaction in detection of culturable and nonculturable *Vibrio vulnificus* cells. Appl Environ Microbiol 57: 2651 – 2655

Brenner S, Williams SR, Vermaas EH et al (2000) In vitro cloning of complex mixtures of DNA onmicrobeads: physical separation of differentially expressed cDNAs. Proc Natl Acad Sci USA 97: 1665 – 1670

Brul S, Schuren F, Montijn R et al (2006) The impact of functional genomics on microbiological food quality and safety. Int J Food Microbiol 112: 195 – 199

Buchmann K, Lindenstrom T (2002) Interactions between monogenean parasites and their fish hosts. Int J Parasitol 32: 309 – 319

Buchrieser C, Gangar VV, Murphree RL et al (1995) Multiple *Vibrio vulnificus* strains in oysters as demonstrated by clamped homogeneous electric field gel electrophoresis. Appl Environ Microbiol 61: 1163 – 1168

Bugeon J, Lefevre F, Fauconneau B (2003) Fillet texture and muscle structure in brown trout (*Salmo trutta*) subjected to long-term exercise. Aquac Res 34: 1287 – 1295

Bulgakov AA, Park KI, Choi KS et al (2004) Purification and characterisation of a lectin isolated from the Manila clam *Ruditapes philippinarum* in Korea. Fish Shellfish Immunol 16: 487 – 499

Burnett KG, Bain L, Baldwin WS et al (2007) *Fundulus* as the premier teleost model in environmental biology: Opportunities for new insights using genomics. Comp Biochem Physiol D-Genomics Proteomics 2: 257 – 286

Bustin SA, Benes V, Nolan T et al (2005) Quantitative real-time RT-PCR - a perspective. J Mol Endocrinol 34: 597 – 601

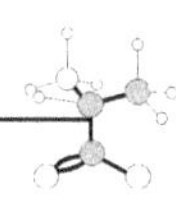

Calder PC (2008) Polyunsaturated fatty acids, inflammatory processes and inflammatory bowel diseases. Mol Nutr Food Res 52:885 – 897

Camara M, Evans F, Langdson CJ (2008) Parental relatedness and survival of Pacific oysters from a naturalized population. J Shellfish Res 27:323 – 336

Campana SE, Thorrold SR (2001) Otoliths, increments, and elements: keys to a comprehensive understanding of fish populations? Can J Fish Aquat Sci 58:30 – 38

Campbell MS, Wright AC (2003) Real-time PCR analysis of *Vibrio vulnificus* from oysters. Appl Environ Microbiol 69:7137 – 7144

Campbell D, Bernatchez L (2004) Generic scan using AFLP markers as a means to assess the role of directional selection in the divergence of sympatric whitefish ecotypes. Mol Biol Evol 21:945 – 956

Canesi L, Gallo G, Gavioli M et al (2002) Bacteria-hemocyte interactions and phagocytosis in marine bivalves. Microsc Res Tech 57:469 – 476

Carvalho GR, Hauser L (1998) Advances in the molecular analysis of fish population structure. Ital J Zool 65:21 – 33

Case RAJ, Hutchinson WF, Hauser L et al (2005) Macro-and micro-geographic variation in pan-tophysin (Pan I) allele frequencies in NE Atlantic cod *Gadus morhua*. Mar Ecol-Prog Ser 301:267 – 278

Castaño-Sanches C, Fuji K, Ozaki A et al (2007) High-density linkage map of the Japanese flounder, *Paralichthys olivaceus*. Aquaculture 272:S248

Castro J, Bouza C, Presa P et al (2004) Potential sources of error in parentage assessment of turbot (*Scophthalmus maximus*) using microsatellite loci. Aquaculture 242:119 – 135

Castro J, Pino A, Hermida M et al (2006) A microsatellite marker tool for parentage analy-sis in Senegal sole (*Solea senegalensis*): Genotyping errors, null alleles and conformance to theoretical assumptions. Aquaculture 261:1194 – 1203

Castro J, Pino A, Hermida M et al (2007) A microsatellite marker tool for parentage assessment in gilthead seabream (*Sparus aurata*). Aquaculture 272:S210 – S216

Cerda J, Mercade J, Lozano JJ et al (2008) Genomic resources for a commercial flatfish, the Senegalese sole (*Solea senegalensis*): EST sequencing, oligo microarray design, and development of the bioinformatic platform Soleamold. BMC Genomics 9:508

Charlet M, Chernysh S, Philippe H et al (1996) Innate immunity: isolation of several cysteine-rich antimicrobial peptides from the blood of a mollusc, *Mytilus edulis*. J Biol Chem 271:21808 – 21813

Cheng T (1981) Bivalves. In: Ratcliffe NA, Rowley A (eds) Invertebrate blood cells. Academic Press, London

Chistiakov DA, Hellemans B, Volckaert FAM (2006) Microsatellites and their genomic distri-

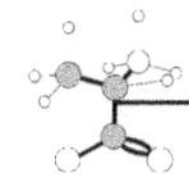

bu-tion, evolution, function and applications: A review with special reference to fish genetics. Aquaculture 255:1 – 29

Choquet G, Soudant P, Lambert C et al (2003) Reduction of adhesion properties of *Ruditapes philippinarum* hemocytes exposed to *Vibrio tapetis*. Dis Aquat Organ 57:109 – 116

Christiansen FB, Frydenberg O (1974) Geographical patterns of 4 polymorphisms in *Zoarces viviparus* as evidence of selection. Genetics 77:765 – 770

Clark AG, Hubisz MJ, Bustamante CD et al (2005) Ascertainment bias in studies of human genome-wide polymorphism. Genome Res 15:1496 – 1502

Cnaani A, Hallerman EM, Ron M et al (2003) Detection of a chromosomal region with two quan-titative trait loci, affecting cold tolerance and fish size, in an F2 tilapia hybrid. Aquaculture 223:117 – 128

Cnaani A, Zilberman N, Tinman S et al (2004) Genome-scan analysis for quantitative trait loci in an F-2 tilapia hybrid. Mol Genet Genomics 272:162 – 172

Cnaani A, Hulata G (2008) Tilapias. In: Kocher TD, Kole C (eds) Genome mapping and genomics in animals, vol 2. Springer-Heidelberg, Berlin

Cochennec N, Le Roux F, Berthe F et al (2000) Detection of *Bonamia ostreae* based on small subunit ribosomal probe. J Invertebr Pathol 76:26 – 32

Colborn J, Crabtree RE, Shaklee JB et al (2001) The evolutionary enigma of bonefishes (*Albula* spp.): Cryptic species and ancient separations in a globally distributed shorefish. Evolution 55:807 – 820

Coltman DW (2008) Molecular ecological approaches to studying the evolutionary impact of selective harvesting in wildlife. Mol Ecol 17:221 – 235

Cossins AR, Crawford DL (2005) Opinion-fish as models for environmental genomics. Nat Rev Genet 6:324 – 333

Cossins A, Fraser J, Hughes M et al (2006) Post-genomic approaches to understanding the mechanisms of environmentally induced phenotypic plasticity. J Exp Biol 209:2328 – 2336

Cowen RK, Paris CB, Srinivasan A (2006) Scaling of connectivity in marine populations. Science 311:522 – 527

Coyne JA, Orr HA (2004) Speciation. Sinauer Associates, MA

Cuesta A, Meseguer J, Esteban MA (2008) The antimicrobial peptide hepcidin exerts an important role in the innate immunity against bacteria in the bony fish gilthead seabream. Mol Immunol 45:2333 – 2342

Cunningham C, Hikima JI, Jenny MJ et al (2006) New resources for marine genomics: Bacterial artificial chromosome libraries for the eastern and pacific oysters (*Crassostrea virginica* and *C. gigas*). Mar Biotechnol 8:521 – 533

Curole JP, Hedgecock D (2005) High frequency of SNPs in the Pacific oyster genome. http:

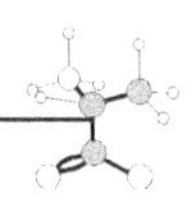

//intl-pag. org/13/abstracts/PAG13_W026. html

Cushing DH (1969) Regularity of spawning season of some fishes. Journal du conseil international de l'exploitation de la mer 33:81 – 92

Dalvit C, De Marchi M, Cassandro M (2007) Genetic traceability of livestock products: A review. Meat Sci 77:437 – 449

Darawiroj D, Kondo H, Hirono I et al (2008) Immune-related gene expression profiling of yellowtail (*Seriola quinqueradiata*) kidney cells stimulated with ConA and LPS using microarray analysis. Fish Shellfish Immunol 24:260 – 266

Darias MJ, Zambonino-Infante JL, Hugot K et al (2008) Gene expression patterns during the lar-val development of European sea bass (*Dicentrarchus labrax*) by microarray analysis. Mar Biotechnol 10:416 – 428

David E, Boudry R, Degremont L et al (2007) Genetic polymorphism of glutamine synthetase and delta-9 desaturase in families of Pacific oyster *Crassostrea gigas* and susceptibility to summer mortality. J Exp Mar Biol Ecol 349:272 – 283

Davis MB, Shaw RG, Etterson JR (2005) Evolutionary responses to climate change. Ecology 86:1704 – 1714

Delaney JR (2007) NEPTUNE: Transforming ocean and earth sciences with distributed subma-rine sensor networks wired to next generation internet. In: 2007 Symposium on Underwater Technology and Workshop on Scientific Use of Submarine Cables and Related Technologies

Delgado CL, Wada N, Rosegrant MW et al (2003) Outlook for fish to 2020: meeting global demand. In: Food Policy Report-International Food Policy Research Institute WorldFish Center, Malaysia, Washington, DC

Derelle E, Ferraz C, Rombauts S et al (2006) Genome analysis of the smallest free-living eukaryote *Ostreococcus tauri* unveils many unique features. Proc Natl Acad Sci USA 103: 11647 – 11652

Derome N, Bernatchez L (2006) The transcriptomics of ecological convergence between two limnetic coregonine fishes (Salmonidae). Mol Biol Evol 23:2370 – 2378

Derome N, Duchesne P, Bernatchez L (2006) Parallelism in gene transcription among sympatric lake whitefish ecotypes (*Coregonus clupeaformis* Mitchill). Mol Ecol 15:1239 – 1250

De-Santis C, Jerry DR (2007) Candidate growth genes in finfish-Where should we be looking? . Aquaculture 272:22 – 38

Desenclos JC, Klontz KC, Wilder MH et al (1991) A multistate outbreak of hepatitis A caused by the consumption of raw oysters. Am J Public Health 81:1268 – 1272

Diatchenko L, Lukyanov S, Lau YF et al (1999) Suppression subtractive hybridization: a versatile method for identifying differentially expressed genes. Methods Enzymol 303:349 – 380

Dios S, Novoa B, Buonocore F et al (2008) Genomic resources for immunology and disease of

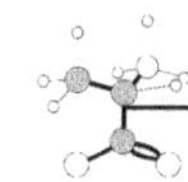

salmonid and non-salmonid fish. Rev Fish Sci 16:119 – 132

Dondero F, Piacentini L, Marsano F et al (2006) Gene transcription profiling in pollutant exposed mussels (*Mytilus* spp.) using a new low-density oligonucleotide microarray. Gene 376:24 – 36

Douglas SE, Gallant JW, Bullerwell CE et al (1999) Winter flounder expressed sequence tags: Establishment of an EST database and identification of novel fish genes. Mar Biotechnol 1:458 – 464

Dulvy NK, Sadovy Y, Reynolds JD (2003) Extinction vulnerability in marine populations. Fish Fish 4:25 – 64

Dulvy NK, Rogers SI, Jennings S et al (2008) Climate change and deepening of the North Sea fish assemblage: a biotic indicator of warming seas. J Appl Ecol 45:1029 – 1039

Dupray E, Caprais MP, Derrien A et al (1997) Salmonella DNA persistence in natural seawaters using PCR analysis. J Appl Microbiol 82:507 – 510

Duvernell DD, Lindmeier JB, Faust KE (2008) Relative influences of historical and contemporary forces shaping the distribution of genetic variation in the Atlantic killifish, *Fundulus heteroclitus*. Mol Ecol 17:1344 – 1360

Elandalloussi LM, Leite RM, Afonso R et al (2004) Development of a PCR-ELISA assay for diagnosis of *Perkinsus marinus* and *Perkinsus atlanticus* infections in bivalve molluscs. Mol Cell Probes 18:89 – 96

Ellegren H (2004) Microsatellites: simple sequences with complex evolution. Nat Rev Genet 5:435 – 445

Emmerich R, Weibel E (1894) Ueber eine durch bakterien erzengte seuche unter den forellen. Arch Hyg Bakteriol 21:1 – 21

Endler JA (1986) Natural selection in the wild. Princeton University Press, Princeton

Escoubas J-M, Briant L, Montagnani C et al (1999) Oyster IKK-like protein shares structural and functional properties with its mammalian homologues. FEBS Lett 453:293 – 298

Espe M, Ruohonen K, Bjørnevik M et al (2004) Interactions between ice storage time, collagen composition, gaping and textural properties in farmed salmon muscle harvested at different times of the year. Aquaculture 240:489 – 504

Ewart KV, Belanger JC, Williams J et al (2005) Identification of genes differentially expressed in Atlantic salmon (*Salmo salar*) in response to infection by *Aeromonas salmonicida* using cDNA microarray technology. Dev Comp Immunol 29:333 – 347

Faisal M, MacIntyre EA, Adham KG et al (1998) Evidence for the presence of protease inhibitors in Eastern (*Crassostrea virginica*) and Pacific (*Crassostrea gigas*) oysters. Comp Biochem Physiol B-Biochem Mol Biol 121:161 – 168

Faisal M, Schafhauser DY, Garreis KA et al (1999) Isolation and characterization of *Perkinsus*

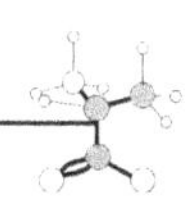

marinus proteases using bacitracin-sepharose affinity chromatography. Comp Biochem Physiol B-Biochem Mol Biol 123:417 – 426

Falch E, Rustad T, Jonsdottir R et al (2006) Geographical and seasonal differences in lipid com-position and relative weight of by-products from gadiform species. J Food Compos Anal 19:727 – 736

Faliex E, Da Silva C, Simon G et al (2008) Dynamic expression of immune response genes in the sea bass, *Dicentrarchus labrax*, experimentally infected with the monogenean *Diplectanum ae-quans*. Fish Shellfish Immunol 24:759 – 767

Fawcett LB, Tripp MR (1994) Chemotaxis of *Mercenaria mercenaria* hemocytes to bacteria in vitro. J Invertebr Pathol 63:275 – 284

Feder ME, Mitchell-Olds T (2003) Evolutionary and ecological functional genomics. Nat Rev Genet 4:651 – 657

Fernandes JMO, Mackenzie MG, Elgar G et al (2005) A genomic approach to reveal novel genes associated with myotube formation in the model teleost, *Takifugu rubripes*. Physiol Genomics 22:327 – 338

Ferraresso S, Vitulo N, Mininni AN et al (2008) Development and validation of a gene expres-sion oligo microarray for the gilthead sea bream (Sparus aurata). BMC Geno-mics 9:580

Fishback AG, Danzmann RG, Ferguson MM et al (2002) Estimates of genetic parameters and genotype by environment interactions for growth traits of rainbow trout (*Oncorhynchus mykiss*) as inferred using molecular pedigrees. Aquaculture 206:137 – 150

Ford SE, Haskin HH (1987) Infection and mortality patterns in strains of oysters *Crassostrea virginica* selected for resistance to the parasite *Haplosporidium nelsoni* (MSX). J Parasitol 73:368 – 376

Franch R, Louro B, Tsalavouta M et al (2006) A genetic linkage map of the hermaphrodite teleost fish *Sparus aurata*. Genetics 174:851 – 861

Frerichs GN, Rodger HD, Peric Z (1996) Cell culture isolation of piscine neuropathy nodavir-us from juvenile sea bass, *Dicentrarchus labrax*. J Gen Virol 77:2067 – 2071

Fryer JL, Sanders JE (1981) Bacterial kidney disease of salmonid fish. Annu Rev Microbiol 35:273 – 298

Gahr SA, Vallejo RL, Weber GM et al (2008) Effects of short-term growth hormone treatment on liver and muscle transcriptomes in rainbow trout (*Oncorhynchus mykiss*). Physiol Genomics 32: 380 – 392

Galindo HM, Olson DB, Palumbi SR (2006) Seascape genetics: A coupled oceanographic-ge-netic model predicts population structure of Caribbean corals. Curr Biol 16:1622 – 1626

Gallardo JA, Garcia X, Lhorente JP et al (2004) Inbreeding and inbreeding depression of fe-male reproductive traits in two populations of Coho salmon selected using BLUP predictors of breed-

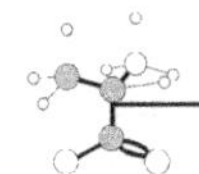

ing values. Aquaculture 234:111 – 122

Gauthier JD, Jenkins JA, La Peyre JF (2004) Flow cytometric analysis of lectin binding to in vitro-cultured *Perkinsus marinus* surface carbohydrates. J Parasitol 90:446 – 454

Geisler R, Rauch GJ, Baier H et al (1999) A radiation hybrid map of the zebrafish genome. Nat Genet 23:86 – 89

Gerber S, Mariette S, Streiff R et al (2000) Comparison of microsatellites and amplified fragment length polymorphism markers for parentage analysis. Mol Ecol 9:1037 – 1048

Gerwick L, Corley-Smith G, Bayne CJ (2007) Gene transcript changes in individual rainbow trout livers following an inflammatory stimulus. Fish Shellfish Immunol 22:157 – 171

Gestal C, Costa M, Figueras A et al (2007) Analysis of differentially expressed genes in response to bacterial stimulation in hemocytes of the carpet-shell clam *Ruditapes decussatus*: identification of new antimicrobial peptides. Gene 406:134 – 143

Gharbi K, Gautier A, Danzmann RG et al (2006) A linkage map for brown trout (*Salmo trutta*): chromosome homeologies and comparative genome organization with other salmonid fish. Genetics 172:2405 – 2419

Gienapp P, Teplitsky C, Alho JS et al (2008) Climate change and evolution: disentangling environmental and genetic responses. Mol Ecol 17:167 – 178

Gil LA (2007) PCR-based methods for fish and fishery products authentication. Trends Food Sci Technol 18:558 – 566

Gilbey J, Verspoor E, Mo TA et al (2006) Identification of genetic markers associated with *Gyrodactylus salaris* resistance in Atlantic salmon *Salmo salar*. Dis Aquat Organ 71:119 – 129

Gjedrem T (2000) Genetic improvement of cold-water fish species. Aquac Res 31:25 – 33

Goetsch SC, Hawke TJ, Gallardo TD et al (2003) Transcriptional profiling and regulation of the extracellular matrix during muscle regeneration. Physiol Genomics 14:261 – 271

Gomez-Bautista M, Ortega-Mora LM, Tabares E et al (2000) Detection of infectious *Cryptosporidium parvum* oocysts in mussels (*Mytilus galloprovincialis*) and cockles (*Cerastoderma edule*). Appl Environ Microbiol 66:1866 – 1870

Gong Z, Hu Z, Gong ZQ et al (1994) Bulk isolation and identification of fish genes by cDNA clone tagging. Mol Mar Biol Biotechnol 3:243 – 251

Gonzalez I, Garcia T, Fernandez A et al (1999) Rapid enumeration of *Escherichia coli* in oysters by a quantitative PCR-ELISA. J Appl Microbiol 86:231 – 236

Gonzalez M, Romestand B, Fievet J et al (2005) Evidence in oyster of a plasma extracellular superoxide dismutase which binds LPS. Biochem Biophys Res Commun 338:1089 – 1097

Gonzalez M, Gueguen Y, Destoumieux-Garzon D et al (2007a) Evidence of a bactericidal perme-ability increasing protein in an invertebrate, the *Crassostrea gigas* Cg-BPI. Proc Natl Acad Sci USA 104:17759 – 17764

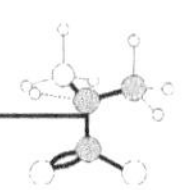

Gonzalez M, Gueguen Y, Desserre G et al (2007b) Molecular characterization of two isoforms of defensin from hemocytes of the oyster *Crassostrea gigas*. Dev Comp Immunol 31:332 – 339

Gracey AY (2007) Interpreting physiological responses to environmental change through gene-expression profiling. J Exp Biol 210:1584 – 1592

Gracey AY, Fraser EJ, Li W et al (2004) Coping with cold: An integrative, multitissue analysis of the transcriptome of a poikilothermic vertebrate. Proc Natl Acad Sci USA 101:16970 – 16975

Graczyk TK, Thompson RC, Fayer R et al (1999) *Giardia duodenalis* cysts of genotype A recovered from clams in the Chesapeake Bay subestuary, Rhode River. Am J Trop Med Hyg 61:526 – 529

Gregory SG, Sekhon M, Schein J et al (2002) A physical map of the mouse genome. Nature 418:743 – 750

Grimholt U, Larsen S, Nordmo R et al (2003) MHC polymorphisms and disease resistance in Atlantic salmon (*Salmo salar*); facing pathogens with single expressed major histocompatibil-ity class I and class II loci. Immunogenetics 55:210 – 219

Gross R, Lulla P, Paaver T (2007) Genetic variability and differentiation of rainbow trout (*Oncorhynchus mykiss*) strains in Northern and Eastern Europe. Aquaculture 272:139 – S146

Gueguen Y, Cadoret JP, Flament D et al (2003) Immune gene discovery by expressed sequence tags generated from hemocytes of the bacteria-challenged oyster, *Crassostrea gigas*. Gene 303:139 – 145

Gueguen Y, Herpin A, Aumelas A et al (2006) Characterization of a Defensin from the oys-ter *Crassostrea gigas*: Recombinant production, folding, solution structure, antimicrobial activities, and gene expression. J Biol Chem 281:313 – 323

Guelinckx J, Maes J, Geysen B et al (2008) Estuarine recruitment of a marine goby reconstructed with an isotopic clock. Oecologia 157:41 – 52

Guerard F, Sellos D, Le Ga Y (2005) Fish and shellfish upgrading, traceability. Mar Biotechnol 96:127 – 163

Guinand B, Lemaire C, Bonhomme F (2004) How to detect polymorphisms undergoing selection in marine fishes? A review of methods and case studies, including flatfishes. J Sea Res 51:167 – 182

Gunnarsson L, Kristiansson E, Forlin L et al (2007) Sensitive and robust gene expression changes in fish exposed to estrogen: a microarray approach. BMC Genomics 8:149

Gupta PK (2008) Ultrafast and low-cost DNA-sequencing methods for applied genomics research. Proc Natl Acad Sci USA 78:91 – 102

Hanchuan D, Liangqi L, Guang D (2006) Molecular cloning of the obese gene from *Cyprinus carpio* and its expression in *Escherichia coli*. Front Biol China 1:50 – 55

Hansen MM, Kenchington E, Nielsen EE (2001) Assigning individual fish to populations u-

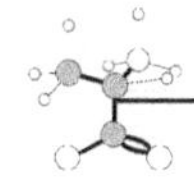

sing microsatellite DNA markers: Methods and applications. Fish Fish 2:93 – 112

Hanski IA, Gilpin ME (1997) Metapopulation biology: ecology, genetics and evolution. Academic Press, New York

Hanski IA, Gaggiotti OE (2004) Ecology, genetics, and evolution of metapopulations. Elsevier Academic Press, San Diego

Harlizius B, van Wijk R, Merks JWM (2004) Genomics for food safety and sustainable animal production. J Biotechnol 113:33 – 42

Håstein T, Hill BJ, Berthe F et al (2001) Traceability of aquatic animals. Rev Sci Tech 20: 564 – 583

Håstein T, Scarfe AD, Lund VL (2005) Science-based assessment of welfare: aquatic animals. Rev Sci Tech 24:529 – 547

Hauser L, Adcock GJ, Smith PJ et al (2002) Loss of microsatellite diversity and low effective population size in an overexploited population of New Zealand snapper (*Pagrus auratus*). Proc Nat Acad Sci USA 99:11742 – 11747

Hawke JP, McWhorter AC, Steigerwalt AG et al (1981) *Edwardsiella ictaluri* sp. nov., the causative agent of enteric septicemia of catfish. Int J Syst Bacteriol 31:396 – 400

Hayes B, Sonesson AK, Gjerde B (2005) Evaluation of three strategies using DNA markers for traceability in aquaculture species. Aquaculture 250:70 – 81

Hayes B, He J, Moen T et al (2006) Use of molecular markers to maximise diversity of flounder populations for aquaculture breeding programs. Aquaculture 255:573 – 578

Hayes B, Lærdahl JK, Lien S et al (2007) An extensive resource of single nucleotide polymorphism markers associated with Atlantic salmon (*Salmo salar*) expressed sequences. Aquaculture 265:82 – 90

He XP, Xu XW, Zhao SH et al (2008) Investigation of Lpin1 as a candidate gene for fat deposition in pigs. Mol Biol Rep DOI 10.1007/s11033 – 008 – 9294 – 4

Hedgecock D (1994) Does variance in reproductive success limit effective population sizes of marine organisms? In: Beaumont AR (ed) Genetics and evolution of aquatic organisms. Chapman and Hall, London

Hedgecock D, Li G, Hubert S et al (2004) Widespread null alleles and poor cross-species amplifi-cation of microsatellite DNA loci cloned from the Pacific oyster, *Crassostrea gigas*. J Shellfish Res 23:379 – 385

Hedgecock D, Davis J (2007) Heterosis for yield and crossbreeding of the Pacific oyster *Crassostrea gigas*. Aquaculture 272:S17 – S29

Hedgecock D, Lin JZ, DeCola S (2007) Transcriptomic analysis of growth heterosis in larval Pacific oysters (*Crassostrea gigas*). Proc Natl Acad Sci USA 104:2313 – 2318

Hegde A, Teh HC, Lam TJ et al (2003) Nodavirus infection in freshwater ornamental fish,

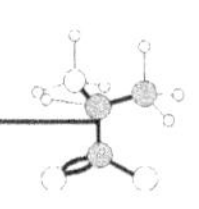

guppy, *Poicelia reticulate*-comparative characterization and pathogenicity studies. Arch Virol 148: 575 – 586

Heijne WHM, Kienhuis AS, van Ommen B et al (2005) Systems toxicology: applications of toxicogenomics, transcriptomics, proteomics and metabolomics in toxicology. Expert Rev Proteomics 2:767 – 780

Hemmer-Hansen J, Nielsen EE, Frydenberg J et al (2007) Adaptive divergence in a high gene flow environment: Hsc70 variation in the European flounder (*Platichthys flesus* L.). Heredity 99: 592 – 600

Hendriks IE, Duarte CM, Heip CHR (2006) Biodiversity research still grounded. Science 312:1715 Hendriks IE, Duarte CM (2008) Allocation of effort and imbalances in biodiversity research. J Exp Mar Biol Ecol 360:15 – 20

Herbinger CM, Doyle RW, Pitman ER (1995) DNA fingerprint based analysis of paternal and maternal effects on offspring growth and survival in communally reared rainbow trout. Aquaculture 137:245 – 256

Herlin M, Taggart JB, Mcandrew BJ et al (2007) Parentage allocation in a complex situation: A large commercial Atlantic cod (*Gadus morhua*) mass spawning tank. Aquaculture 272: 195 – 203

Herlin M, Delghandi M, Wesmajervi M (2008) Analysis of the parental contribution to a group of fry from a single day of spawning from a commercial Atlantic cod (*Gadus morhua*) breeding tank. Aquaculture 274:218 – 224

Higuchi R, Fockler C, Dollinger G et al (1993) Kinetic PCR analysis: real-time monitoring of DNA amplification reactions. Biotechnology 11:1026 – 1030

Hill WE, Keasler SP, Trucksess MW et al (1991) Polymerase chain reaction identification of *Vibrio vulnificus* in artificially contaminated oysters. Appl Environ Microbiol 57:707 – 711

Hill JA, Kiessling A, Devlin RH (2000) Coho salmon (*Oncorhynchus kisutch*) transgenic for a growth hormone gene construct exhibit increased rates of muscle hyperplasia and detectable levels of differential gene expression. Can J Fish Aquat Sci 57:939 – 950

Hill BJ (2005) The need for effective disease control in international aquaculture. Dev Biol 121:3 – 12

Hine P (1999) The inter-relationships of bivalve haemocytes. Fish Shellfish Immunol 9:367 – 385 Hine PM, Cochennec-Laureau N, Berthe FC (2001) *Bonamia exitiosus* n. sp. (Haplosporidia)

infecting flat oysters *Ostrea chilensis* in New Zealand. Dis Aquat Organ 47:63 – 72

Hitte C, Kirkness EF, Ostrander EA et al (2008) Survey sequencing and radiation hybrid mapping to construct comparative maps. Methods Mol Biol 422:65 – 77

Hjort J (1914) Fluctuations in the great fisheries of Northern Europe. Rapp Proc-Verb Réun

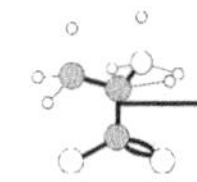

Cons Int Expl Mer 20:1 – 228

Holloway AJ, van Laar RK, Tothill RW et al (2002) Options available-from start to finish-for obtaining data from DNA microarrays II. Nat Genet 32 Suppl:481 – 489

Hostetler HA, Collodi P, Devlin RH et al (2005) Improved phytate phosphorus utilization by Japanese medaka transgenic for the *Aspergillus niger* phytase gene. Zebrafish 2:19 – 31

Houston RD, Gheyas A, Hamilton A et al (2008) Detection and confirmation of a major QTL affecting resistance to infectious pancreatic necrosis (IPN) in Atlantic salmon (*Salmo salar*). Dev Biol 132:199 – 204

Howland K, Cheng TC (1982) Identification of bacterial chemoattractants for oyster (*Crassostrea virginica*) hemocytes. J Invertebr Pathol 89:123 – 132

Hubert F, Noel T, Roch P (1996) A member of the arthropod defensin family from edible Mediterranean mussels (*Mytilus galloprovincialis*). Eur J Biochem 240:302 – 306

Hubert S, Hedgecock D (2004) Linkage maps of microsatellite DNA markers for the Pacific oyster *Crassostrea gigas*. Genetics 168:351 – 362

Hukriede NA, Joly L, Tsang M et al (1999) Radiation hybrid mapping of the zebrafish genome. Proc Natl Acad Sci USA 6:9745 – 9750

Hurling R, Rodell JB, Hunt HD (1996) Fibre diameter and fish texture. J Texture Studies 27:679 – 685

Huvet A, Herpin A, Degremont L et al (2004) The identification of genes from the oyster *Crassostrea gigas* that are differentially expressed in progeny exhibiting opposed susceptibility to summer mortality. Gene 343:211 – 220

Huvet A, Jeffroy F, Fabioux C et al (2008) Association among growth, food consumption-related traits and *amylase* gene polymorphism in the Pacific oyster *Crassostrea gigas*. Anim Genet 39:662 – 665

Itoh N, Takahashi KG (2008) Distribution of multiple peptidoglycan recognition proteins in the tissues of Pacific oyster, *Crassostrea gigas*. Comp Biochem Physiol B-Biochem Mol Biol 150:409 – 417

Izquierdo MS, Robaina L, Juárez E et al (2008) Regulation of growth, fatty acid composition and delta 6 desaturase expression by dietary lipids in gilthead seabream larvae (*Sparus aurata*). Fish Physiol Biochem 34:117 – 127

Janeway CA, Medzhitov R (2002) Innate immune recognition. Annu Rev Immunol 20:197 – 216

Jenny MJ, Warr GW, Ringwood AH et al (2006) Regulation of metallothionein genes in the American oyster (*Crassostrea virginica*): ontogeny and differential expression in response to different stressors. Gene 379:156 – 165

Jenny MJ, Chapman RW, Mancia A (2007) A cDNA microarray for *Crassostrea virginica* and

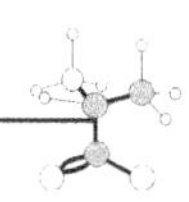

C. gigas. Mar Biotechnol 9:577 - 591

Jeyasekaran G, Karunasagar I (1996) Incidence of *Listeria* spp. in tropical fish. Int J Food Microbiol 31:333 - 340

Johnston IA (1999) Muscle development and growth: potential implications for flesh quality in fish. Aquaculture 177:99 - 115

Johnston IA, Manthri S, Alderson R et al (2002) Effects of dietary protein level on muscle cellularity and flesh quality in Atlantic salmon with particular reference to gaping. Aquaculture 210: 259 - 283

Johnston IA, Alderson R, Sandham C et al (2004) Muscle fibre density in relation to the colour and texture of smoked Atlantic salmon (*Salmo salar* L.). Aquaculture 189:335 - 349

Johnston IA (2006) Environment and plasticity of myogenesis in teleost fish. J Exp Biol 209: 2249 - 2264

Jones GP, Planes S, Thorrold SR (2005) Coral reef fish larvae settle close to home. Curr Biol 15:1314 - 1318

Joost S, Bonin A, Bruford W et al (2007) A spatial analysis method (SAM) to detect candidate loci for selection: towards a landscape genomics approach to adaptation. Mol Ecol 16:3955 - 3969

Jordal A-EO, Torstensen BE, Tsoi S et al (2005) Dietary rapeseed oil affects the expression of genes involved in hepatic lipid metabolism in Atlantic salmon (*Salmo salar* L.). J Nutr 135:2355 - 2361

Jørgensen HBH, Hansen MM, Bekkevold D et al (2005) Marine landscapes and population genetic structure of herring (*Clupea harengus* L.) in the Baltic Sea. Mol Ecol 14:3219 - 3234

Jørgensen HBH, Pertoldi C, Hansen MM et al (2008) Genetic and environmental correlates of morphological variation in a marine fish: the case of Baltic Sea herring (*Clupea harengus*). Can J Fish Aquat Sci 65:389 - 400

Kadereit B, Kumar P, Wang WJ et al (2008) Evolutionarily conserved gene family important for fat storage. Proc Natl Acad Sci USA 105:84 - 99

Kalbe M, Kurtz J (2006) Local differences in immunocompetence reflect resistance of sticklebacks against the eye fluke *Diplostomum pseudospathaceum*. Parasitology 132:105 - 116

Kang YS, Kim YM, Park KI et al (2006) Analysis of EST and lectin expressions in hemocytes of Manila clams (*Ruditapes philippinarum*) (Bivalvia: Mollusca) infected with *Perkinsus olseni*. Dev Comp Immunol 30:1119 - 1131

Kang JH, Kim WJ, Lee WJ (2008) Genetic linkage map of olive flounder, *Paralichthys olivaceus*. Int J Biol Sci 4:143 - 149

Karlsson S, Renshaw MA, Rexroad CE et al (2008) PCR primers for 100 microsatellites in red drum (*Sciaenops ocellatus*). Mol Ecol Resour 8:393 - 398

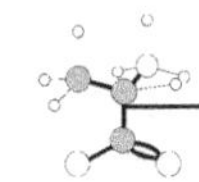

Karsi A, Cao D, Li P et al (2002) Transcriptome analysis of channel catfish (*Ictalurus punctatus*): initial analysis of gene expression and microsatellite-containing cDNAs in the skin. Gene 285: 157 – 168

Kaushik SJ (2005) Besoins et apport en phosphore chez les poissons. INRA Prod Anim 18: 203 – 208

Kawecki TJ, Ebert D (2004) Conceptual issues in local adaptation. Ecol Lett 7: 1225 – 1241

Khoo SK, Ozaki A, Nakamura F et al (2004) Identification of a novel chromosomal region as-so-ciated with infectious hematopoietic necrosis (IHN) resistance in rainbow trout. Fish Pathol 39: 95 – 102

Kiessling A, Ruohonen K, Bjørnevik M (2006) Muscle fibre growth and quality in fish. Arch Tierz Dummerstorf 49: 137 – 146

Kim YM, Park K-I, Choi K-S et al (2006) Lectin from the Manila clam *Ruditapes philippinarum* is induced upon infection with the protozoan parasite *Perkinsus olseni*. J Biol Chem 281: 26854 – 26864

Kim JY, Kim YM, Cho SK et al (2008) Noble tandem-repeat galectin of Manila clam *Ruditapes philippinarum* is induced upon infection with the protozoan parasite *Perkinsus olseni*. Dev Comp Immunol 32: 1131 – 1141

King M (1995) Fisheries biology, assessment and management. Fishing News Books, Oxford

Kingsley DM, Zhu BL, Osoegawa KJ et al (2004) New genomic tools for molecular studies of evolutionary change in three-spined sticklebacks. Behaviour 141: 1331 – 1344

Kjærsgård IVH, Jessen F (2003) Proteome analysis elucidating post-mortem changes in cod (*Gadus morhua*) muscle proteins. J Agric Food Chem 51: 3985 – 3991

Kjærsgård IVH, Nørrelykke MR, Jessen F (2006a) Changes in cod muscle proteins during frozen storage revealed by proteome analysis and multivariate data analysis. Proteomics 6: 1606 – 1618

Kjærsgård IVH, Nørrelykke MR, Baron CP et al (2006b) Identification of carbonylated protein in frozen rainbow trout (*Oncorhynchus mykiss*) fillets and development of protein oxidation during frozen storage. J Agric Food Chem 54: 9437 – 9446

Kleppe K, Ohtsuka E, Kleppe R et al (1971) Studies on polynucleotides. XCVI. Repair replications of short synthetic DNA's as catalyzed by DNA polymerases. J Mol Biol 56: 341 – 361

Kocher TD, Kole C (2008) Genome mapping and genomics in fishes and aquatic animals. Volume 2. Springer-Heidelberg, Berlin

Koehn RK, Milkman R, Mitton JB (1976) Population genetics of marine pelecypods. 4. Selection, migration and genetic differentiationn in blue mussels *Mytilus edulis*. Evolution 30: 2 – 32

Koehn RK, Immermann FW (1981) Biochemical studies of aminopeptidase polymorphism in *Mytilus edulis*. 1. Dependence of enzyme activity on season, tissue, and genotype. Biochem Genet 19: 1115 – 1142

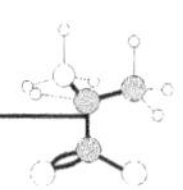

Koljonen ML, Tahtinen J, Saisa M et al (2002) Maintenance of genetic diversity of Atlantic salmon (*Salmo salar*) by captive breeding programmes and the geographic. Aquaculture 212:69 – 92

Koljonen ML, Pella JJ, Masuda M (2005) Classical individual assignments versus mixture mod-eling to estimate stock proportions in Atlantic salmon (*Salmo salar*) catches from DNA micro-satellite data. Can J Fish Aquat Sci 62:2143 – 2158

Kono T, Ponpompisit A, Sakai M (2003) The analysis of expressed genes in head kidney of common carp *Cyprinus carpio* L. stimulated with peptidoglycan. Aquaculture 235:37 – 52

Koskinen MT, Hirvonen H, Landry PA et al (2004) The benefits of increasing the number of microsatellites utilized in genetic population studies: an empirical perspective. Hereditas 141:61 – 67

Kudo H, Amizuka N, Araki K et al (2004) Zebrafish periostin is required for the adhesion of muscle fiber bundles to the myoseptum and for the differentiation of muscle fibers. Dev Biol 267: 473 – 487

Kuhl H, Beck A, Wozniak G, Canario AVM et al (2010) The European sea bass *Dicentrarchus labrax* genome puzzle: comparative BAC-mapping and low coverage shotgun sequencing. BMC Genetics (in press)

Kumazawa N, Morimoto N (1992) Chemotactic activity of hemocytes derived from a brackish-water clam, *Corbicula japonica*, to *Vibrio parahaemolyticus* and *Escherichia coli* strains. J Vet Med Sci 5:851 – 855

Kwok C, Korn RM, Davis ME et al (1998) Characterization of whole genome radiation hybrid mapping resources for non-mammalian vertebrates. Nucleic Acids Res 26:3562 – 3566

La Peyre JF, Yarnall HÁ, Faisal M (1996) Contribution of *Perkinsus marinus* extracellular products in the infection of eastern oysters (*Crassostrea virginica*). J Invertebr Pathol 68:312 – 313

Labreuche Y, Soudant P, Gonçalves M et al (2006a) Effects of extracellular products from the pathogenic *Vibrio aestuarianus* strain 01/32 on lethality and cellular immune responses of the oyster *Crassostrea gigas*. Dev Comp Immunol 30:367 – 379

Labreuche Y, Lambert C, Soudant P et al (2006b) Cellular and molecular hemocyte responses of the Pacific oyster, *Crassostrea gigas*, following bacterial infection with Vibrio *aestuarianus* strain 01/32. Microbes Infect 8:2715 – 2724

Lallias D, Arzul I, Heurtebise S et al (2008) Bonamia-ostreae induced mortalities in one-year old European flat oysters *Ostrea edulis*: experimental infection by cohabitation challenge. Aquat Living Resour 21:423 – 439

Lam N, Araneda C, Díaz NF et al (2007) Physical mapping of SCAR-RAPD markers associ-a-ted to spawning date and flesh color traits in cultivated Coho salmon (*Oncorhynchus kisutch*). Aquaculture 272:S282

Lambert C, Soudant P, Choquet G et al (2003) Measurement of *Crassostrea gigas* hemocyte oxida-tive metabolism by flow cytometry and the inhibiting capacity of pathogenic vibrios. Fish Shellfish Immunol 15:225 – 240

Lane E, Birkbeck TH (1999) Toxicity to bacteria towards haemocytes of *Mytilus edulis*. Aquat Living Resour 12:343 – 350

Lang P, Langdon CJ, Camara MD (2008) Predicting the resistance of adult pacific oysters (*Crassostrea gigas*) to summer mortality. J Shellfish Res 27:470

Langefors J, Lohm M, Grahn O (2001) Association between major histocompatibility complex class IIB alleles and resistance to *Aeromonas salmonicida* in Atlantic salmon. Proc R Soc B-Biol Sci 268:479 – 485

Larsen PF, Nielsen EE, Williams TD et al (2007) Adaptive differences in gene expression in European flounder (*Platichthys flesus*). Mol Ecol 16:4674 – 4683

Larsson LC, Laikre L, Palm S et al (2007) Concordance of allozyme and microsatellite dif-fe-rentiation in a marine fish, but evidence of selection at a microsatellite locus. Mol Ecol 16:1135 – 1147

Launey S, Barre M, Gerard A et al (2001) Population bottleneck and effective size in Bonamia ostreae-resistant populations of *Ostrea edulis* as inferred by microsatellite markers. Genet Res 78: 259 – 270

Le Chevalier P, Le Boulay C, Paillard C (2003) Characterization by restriction fragment length polymorphism and plasmid profiling of *Vibrio tapetis* strains. J Basic Microbiol 43:414 – 422

Le Roux F, Audemard C, Barnaud A et al (1999) DNA probes as potential tools for the detection of *Marteilia refringens*. Mar Biotechnol 1:588 – 597

Leaver MJ, Bautista JM, Björnsson BT et al (2008) Towards fish lipid nutrigenomics: current state and prospects for fin-fish aquaculture. Rev Fish Sci 16:73 – 94

Lehrer SB, Ayuso R, Reese G (2003) Seafood allergy and allergens: a review. Mar Biotechnol 5:339 – 348

Lelong C, Badariotti F, Le Quéré H et al (2007) Cg-TGF-b, a TGF-b/activin homologue in the Pacific oyster *Crassostrea gigas*, is involved in immunity against Gram-negative microbial infection. Dev Comp Immunol 31:30 – 38

Lemaire C, Allegrucci G, Naciri M et al (2000) Do discrepancies between microsatellite and allozyme variation reveal differential selection between sea and lagoon in the sea bass (*Dicentrarchus labrax*)? . Mol Ecol 9:457 – 467

Lemaitre B, Kromer-Metzger E, Michaut L et al (1995) A recessive mutation, immune deficiency (imd), defines two distinct control pathways in the *Drosophila* host defense. Proc Natl Acad Sci USA 92:9465 – 9469

Levesque HM, Shears MA, Fletcher GL et al (2008) Myogenesis and muscle metabolism in ju-

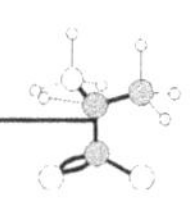

venile Atlantic salmon (*Salmo salar*) made transgenic for growth hormone. J Exp Biol 211:128 –137

Li G, Hubert S, Bucklin K et al (2003a) Characterization of 79 microsatellite DNA markers in the Pacific oyster *Crassostrea gigas*. Mol Ecol Notes 3:228 – 232

Li X, Field C, Doyle R (2003b) Estimation of additive genetic variance components in aquaculture populations selectively pedigreed by DNA fingerprinting. Biom J 45:61 – 72

Li Q, Kijima A (2006) Microsatellite analysis of gynogenetic families in the Pacific oyster, *Crassostrea gigas*. J Exp Mar Biol Ecol 331:1 – 8

Li P, Peatman E, Wang S et al (2007) Towards the ictalurid catfish transcriptome: generation and analysis of 31,215 catfish ESTs. BMC Genomics 8:177

Li C, Ni D, Song L et al (2008) Molecular cloning and characterization of a catalase gene from Zhikong scallop *Chlamys farreri*. Fish Shellfish Immunol 24:26 – 34

Liu N, Chen L, Wang S et al (2005a) Comparison of single-nucleotide polymorphisms and microsatellites in inference of population structure. BMC Genet 6:S26

Liu YG, Chen SL, Li BF et al (2005b) Analysis of genetic variation in selected stocks of hatchery flounder, *Paralichthys olivaceus*, using AFLP markers. Biochem Syst Ecol 33:993 – 1005

Liu XD, Liu X, Guo X et al (2006) A preliminary genetic linkage map of the pacific abalone *Haliotis discus hannai* Ino. Mar Biotechnol 8:386 – 397

Liu XD, Liu X, Zhang GF (2007) Identification of quantitative trait loci for growth-related traits in the Pacific abalone *Haliotis discus hannai Ino*. Aquac Res 38:789 – 797

Liu Z, Li RW, Waldbieser GC (2008) Utilization of microarray technology for functional genomics in ictalurid catfish. J Fish Biol 72:2377 – 2390

Lohm J, Grahn M, Langefors A et al (2002) Experimental evidence for major histocompatibility complex allele-specific resistance to a bacterial infection. Proc R Soc B-Biol Sci 269:2029 – 2033

Lopez-Castejon G, Sepulcre MP, Roca FJ et al (2007) The type II interleukin-1 receptor (IL-1RII) of the bony fish gilthead seabream *Sparus aurata* is strongly induced after infection and tightly regulated at transcriptional and post-transcriptional levels. Mol Immunol 44:2772 – 2780

Lucassen M, Koschnick N, Eckerle L et al (2006) Mitochondrial mechanisms of cold adaptation in cod (*Gadus morhua* L.) populations from different climatic zones. J Exp Biol 209:2462 – 2471

Luikart G, England PR, Tallmon D et al (2003) The power and promise of population genomics: From genotyping to genome typing. Nat Rev Genet 4:981 – 994

Lynch M, Walch B (1998) Genetics and analysis of quantitative traits. Sinauer Associates Inc, Sunderland, Massachusetts

Ma H, Mai K, Xu W et al (2005) Molecular cloning of a2-macroglobulin in sea scallop *Chla-*

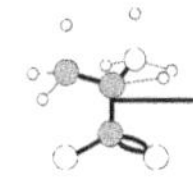

mys farreri (Bivalvia, Mollusca). Fish Shellfish Immunol 18:345 – 349

MacKenzie S, Balasch JC, Novoa B et al (2008) Comparative analysis of the acute response of the trout, *O. mykiss*, head kidney to in vivo challenge with virulent and attenuated infectious hematopoietic necrosis virus and LPS-induced inflammation. BMC Genomics 9:141

Mackie IM (1993) The effects of freezing on flesh proteins. Food Rev Int 9:575 – 610

Maes GE, Pujolar JM, Hellemans B et al (2006) Evidence for isolation by time in the European eel (*Anguilla anguilla* L.). Mol Ecol 15:2095 – 2107

Mäkinen HS, Cano JM, Merilä J (2008) Identifying footprints of directional and balancing selec-tion in marine and freshwater three-spined stickleback (*Gasterosteus aculeatus*) populations. Mol Ecol 17:3565 – 3582

Malek RL, Sajadi H, Abraham J et al (2004) The effects of temperature reduction on gene expres-sion and oxidative stress in skeletal muscle from adult zebrafish. Comp Biochem Physiol C-Toxicol Pharmacol 138:363 – 373

Manel S, Gaggiotti OE, Waples RS (2005) Assignment methods: matching biological questions with appropriate techniques. Trends Ecol Evol 20:136 – 142

Margulies M, Egholm M, Altman WE et al (2005) Genome sequencing in microfabricated high-density picolitre reactors. Nature 437:376 – 380

Marteinsdottir G, Begg GA (2002) Essential relationships incorporating the influence of age, size and condition on variables required for estimation of reproductive potential in Atlantic cod *Gadus morhua*. Mar Ecol-Prog Ser 235:235 – 256

Martin SAM, Cash P, Blaney S et al (2001) Proteome analysis of rainbow trout (*Oncorhynchus mykiss*) liver proteins during short term starvation. Fish Physiol Biochem 24:259 – 270

Martin SAM, Vilhelmsson O, Médale F et al (2003) Proteomic sensitivity to dietary manipulations in rainbow trout. Biochim Biophys Acta 1651:17 – 29

Martin SAM, Blaney SC, Houlihan DF et al (2006) Transcriptome response following administra-tion of a live bacterial vaccine in Atlantic salmon (*Salmo salar*). Mol Immunol 43:1900 – 1911

Martin SAM, Collet B, Mackenzie S et al (2008) Genomic tools for examining immune gene function in salmonid fish. Rev Fish Sci 16:112 – 118

Martinez I, Friis TJ (2004) Application of proteome analysis to seafood authentication. Proteomics 4:347 – 354

Martinez I, James D, Loréal H (2005) Application of modern analytical techniques to ensure seafood safety and authenticity. FAO Fish Tech Paper 455, FAO, Rome

Martinez I, Šližyte · R, Daukšas E (2007) High resolution two-dimensional electrophoresis as a tool to differentiate wild from farmed cod (*Gadus morhua*) and to assess the protein composition of klipfish. Food Chem 102:504 – 510

Martinez V, di Giovanni S (2007) Breeding programmes of scallops: effect of self-fertilization

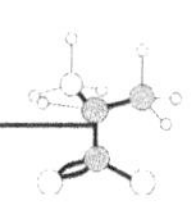

when estimating genetic parameters. Aquaculture 272:S287

Matsuda M, Kawato N, Asakawa S et al (2001) Construction of a BAC library derived from the inbred Hd-rR strain of the teleost fish, *Oryzias latipes*. Genes Genet Syst 76:61 - 63

Matthews SJ, Ross NW, Lall SP et al (2006) Astaxanthin binding protein in Atlantic salmon. Comp Biochem Physiol B-Biochem Mol Biol 144:206 - 214

Matzinger P (2002) The danger model: A renewed sense of self. Science 296:301 - 305

McAvoy ES, Wood AR, Gardeur JN (2008) Development and evaluation of microsatellite mark-ers for identification of individual Greenshell™ mussels (*Perna canaliculus*) in a selective breeding programme. Aquaculture 274:41 - 48

McDonald JH, Verrelli BC, Geyer LB (1996) Lack of geographic variation in anonymous nuclear polymorphisms in the American oyster, *Crassostrea virginica*. Mol Biol Evol 13:1114 - 1118

McDonald GJ, Danzmann RG, Ferguson MM (2004) Relatedness determination in the absence of pedigree information in three cultured strains of rainbow trout. Aquaculture 233:65 - 78

McKay SD, Schnabel RD, Murdoch BM et al (2007) Construction of bovine whole-genome radiation hybrid and linkage maps using high-throughput genotyping. Anim Genet 38:120 - 125

McKinnon JS, Rundle HD (2002) Speciation in nature: the threespine stickleback model systems. Trends Ecol Evol 17:480 - 488

Medzhitov R, Janeway CA Jr (2002) Decoding the patterns of self and nonself by the innate immune system. Science 296:298 - 300

Meyers BC, Scalabrin S, Morgante M (2004) Mapping and sequencing complex genomes: let's get physical! . Nat Rev Genet 5:578 - 588

Milinski M, Griffiths S, Wegner KM (2005) Mate choice decisions of stickleback females predictably modified by MHC peptide ligands. Proc Natl Acad Sci USA 102:4414 - 4418

Milkman R, Koehn RK (1977) Temporal variation in relationship between size, numbers, and allele frequency in a population of *Mytilus edulis*. Evolution 31:103 - 115

Min B, Ahn DU (2005) Mechanism of lipid peroxidation in meat and meat products: a review. Food Sci Biotechnol 14:152 - 163

Mitta G, Vandenbulcke F, Hubert F et al (1999a) Mussel defensins are synthesised and processed in granulocytes then released into the plasma after bacterial challenge. J Cell Sci 112:4233 - 4242

Mitta G, Hubert F, Noël T et al (1999b) Myticin, a novel cysteine-rich antimicrobial peptide iso-lated from haemocytes and plasma of the mussel *Mytilus galloprovincialis*. Eur J Biochem 265: 71 - 78

Mitta G, Vandenbulcke F, Hubert F et al (2000a) Involvement of mytilins in mussel antimicrobial defense. J Biol Chem 275:12954 - 12962

Mitta G, Vandenbulcke F, Roch P (2000b) Original involvement of antimicrobial peptides in

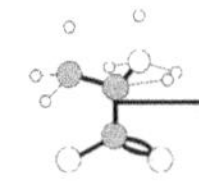

mussel innate immunity. FEBS Lett 486:185 – 190

Moen T, Agresti JJ, Cnaani A et al (2004) A genome scan of a four-way tilapia cross supports the existence of a quantitative trait locus for cold tolerance on linkage group 23. Aquac Res 35:893 – 904

Moen T, Sonesson AK, Hayes B et al (2007) Mapping of a quantitative trait locus for resistance against infectious salmon anaemia in Atlantic salmon (*Salmo Salar*): comparing survival analysis with analysis on affected/resistant data. BMC Genet 8:53

Moen T, Hayes B, Baranski M et al (2008a) A linkage map of the Atlantic salmon (*Salmo salar*) based on EST-derived SNP markers. BMC Genomics 9:223

Moen T, Hayes B, Nilsen F et al (2008b) Identification and characterisation of novel SNP markers in Atlantic cod: Evidence for directional selection. BMC Genet 9:18

Moghadam HK, Poissant J, Fotherby H et al (2007) Quantitative trait loci for body weight, con-dition factor and age at sexual maturation in Arctic charr (*Salvelinus alpinus*): comparative analysis with rainbow trout (*Oncorhynchus mykiss*) and Atlantic salmon (*Salmo salar*). Mol Genet Genomics 277:647 – 661

Mommsen TP (2001) Paradigms of growth in fish. Comp Biochem Phys B-Biochem Mol Biol 129:207 – 219

Montagnani C, Le Roux F, Berthe F et al (2001) Cg-TIMP, an inducible tissue inhibitor of metal-loproteinase from the Pacific oyster *Crassostrea gigas* with a potential role in wound healing and defense mechanisms. FEBS Lett 500:64 – 70

Montagnani C, Kappler C, Reichhart JM et al (2004) Cg-Rel, the first Rel/NF-kB homolog characterized in a mollusk, the Pacific oyster *Crassostrea gigas*. FEBS Lett 561:75 – 82

Montagnani C, Avarre JC, de Lorgeril J et al (2007) First evidence of the activation of Cg-timp, an immune response component of Pacific oysters, through a damage-associated molecular pattern pathway. Dev Comp Immunol 31:1 – 11

Montagnani C, Labreuche Y, Escoubas JM (2008) Cg-IkB, a new member of the IkB pro-tein family characterized in the Pacific oyster *Crassostrea gigas*. Dev Comp Immunol 32:182 – 190

Montes JF, Durfort M, Garcia-Valero J (1995a) Cellular defense mechanisms of the clam *Tapes semidecussatus* against infection by the protozoan parasite *Perkinsus* sp. Cell Tissue Res 279:529 – 538

Montes JF, Durfort M, Garcia-Valero J (1995b) Characterization and localization of an Mr 225 kDa polypeptide specifically involved in the defence mechanisms of the clam *Tapes semidecussatus*. Cell Tissue Res 280:27 – 37

Montes JF, Dufort M, Garcia-Valero J (1996) When the venerid clam *Tapes decussatus* is parasitized by the protozoan *Perkinsus* sp. it synthesizes a defensive polypeptide that is closely related to p225. Dis Aquat Organ 26:149 – 157

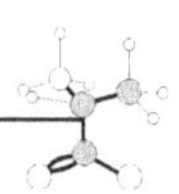

Morin PA, Luikart G, Wayne RK et al (2003) SNPs' in ecology, evolution and conservation. Trends Ecol Evol 19:208 – 216

Morzel M, Verrez-Bagnis V, Arendt EK et al (2000) Use of two-dimensional electrophoresis to evaluate proteolysis in salmon *Salmo salar* muscle as affected by a lactic fermentation. J Agric Food Chem 48:239 – 244

Morzel M, Chambon C, Lefèvre F et al (2006) Modifications of trout (*Oncorhynchus mykiss*) muscle proteins by pre-slaughter activity. J Agric Food Chem 54:2997 – 3001

Mullis K, Faloona F (1987) Specific synthesis of DNA in vitro via a polymerase-catalyzed chain reaction. Methods Enzymol 155:335 – 350

Naciri-Graven Y, Launey S, Lebayon N et al (2000) Influence of parentage upon growth in *Ostrea edulis*: evidence for inbreeding depression. Genet Res 46:159 – 168

Navas JI, Castillo MC, Vera P et al (1992) Principal parasites observed in clams, *Ruditapes decus-satus* (L.), *Ruditapes philippinarum* (Adams et Reeve), *Venerupis pullastra* (Montagu) and *Venerupis aureus* (Gmelin), from the Huelva coast (S. W. Spain). Aquaculture 107:193 – 199

Ni D, Song L, Wu L et al (2007) Molecular cloning and mRNA expression of peptidoglycan recognition protein (PGRP) gene in bay scallop (*Argopecten irradians*, Lamarck 1819). Dev Comp Immunol 31:548 – 558

Nielsen EE, Hansen MM, Schmidt C et al (2001) Determining the population origin of individual cod in the Northeast Atlantic. Nature 413:272

Nielsen EE, Hansen MM, Ruzzante DE et al (2003) Evidence of a hybrid-zone in Atlantic cod (*Gadus morhua*) in the Baltic and the Danish Belt Sea, revealed by individual admixture analysis. Mol Ecol 12:1497 – 1508

Nielsen EE, Hansen MM, Meldrup D (2006) Evidence of microsatellite hitch-hiking selection in Atlantic cod (*Gadus morhua* L.): Implications for inferring population structure in non-model organisms. Mol Ecol 15:3219 – 3229

Nielsen EE, MacKenzie BR, Magnussen E et al (2007) Historical analysis of Pan I in Atlantic cod (*Gadus morhua*): temporal stability of allele frequencies in the southeastern part of the species distribution. Can J Fish Aquat Sci 64:1448 – 1455

Nielsen EE, Hansen MM (2008) Waking the dead: the value of population genetic analyses of historical samples. Fish Fish 9:450 – 461

Nikula R, Strelkov P, Vainola R (2008) A broad transition zone between an inner Baltic hybrid swarm and a pure North Sea subspecies of *Macoma balthica* (Mollusca, Bivalvia). Mol Ecol 17:1505 – 1522

Nordstrom JL, Vickery MC, Blackstone GM et al (2007) Development of a multiplex real-time PCR assay with an internal amplification control for the detection of total and pathogenic *Vibrio parahaemolyticus* bacteria in oysters. Appl Environ Microbiol 73:5840 – 5847

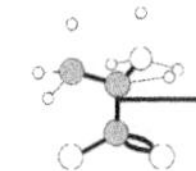

Norris AT, Bradley DG, Cunningham EP (1999) Microsatellite genetic variation between and within farmed and wild Atlantic salmon (*Salmo salar*) populations. Aquaculture 180:247 – 264

Norris AT, Bradley DG, Cunningham EP (2000) Parentage and relatedness determination in farmed Atlantic salmon (*Salmo salar*) using microsatellite markers. Aquaculture 182:73 – 83

Nottage AS, Birkbeck TH (1990) Interactions between different strains of *Vibrio alginolyticus* and hemolymph fractions from adult *Mytilus edulis*. J Invertebr Pathol 56:15 – 19

Nürnberger T, Brunner F, Kemmerling B et al (2004) Innate immunity in plants and animals: striking similarities and obvious differences. Immunol Rev 198:249 – 266

Nygaard V, Liu F, Holden M et al (2008) Validation of oligoarrays for quantitative exploration of the transcriptome. BMC Genomics 9:258

Oleksiak MF, Churchill GA, Crawford DL (2002) Variation in gene expression within and among natural populations. Nat Genet 32:261 – 266

Oliver JL, Gaffney PM, Allen SK et al (2000) Protease inhibitory activity in selectively bred families of eastern oysters. J Aquat Anim Health 12:136 – 145

Ommen B, Groten JP (2004) Nutrigenomics in efficacy and safety evaluation of food components. World Rev Nutr Diet 93:134 – 152

Orr HA (2005) The genetic theory of adaptation: a brief history. Nat Rev genet 6:119 – 127

Ozaki A, Sakamoto T, Khoo S et al (2001) Quantitative trait loci (QTLs) associated with resistance/susceptibility to infectious pancreatic necrosis virus (IPNV) in rainbow trout (*Oncorhynchus mykiss*). Mol Genet Genomics 265:23 – 31

Paillard C, Le Roux F, Borrego JJ (2004) Bacterial disease in marine bivalves, a review of recente studies: Trends and evolution. Aquat Living Resour 17:477 – 498

Palumbi SR (2004) Marine reserves and ocean neighborhoods: The spatial scale of marine populations and their management. Annu Rev Environ Resour 29:31 – 68

Pampoulie C, Jorundsdottir TD, Steinarsson A et al (2006) Genetic comparison of experimental farmed strains and wild Icelandic populations of Atlantic cod (*Gadus morhua* L.). Aquaculture 261:556 – 564

Panicker G, Vickery MC, Bej AK (2004) Multiplex PCR detection of clinical and environmental strains of *Vibrio vulnificus* in shellfish. Can J Microbiol 50:911 – 922

Panserat S, Ducasse-Cabanot S, Plagnes-Juan E et al (2008a) Dietary fat level modifies the expression of hepatic genes in juvenile rainbow trout (*Oncorhynchus mykiss*) as revealed by microarray analysis. Aquaculture 275:235 – 241

Panserat S, Kolditz C, Richard N et al (2008b) Hepatic gene expression profiles in juvenile rainbow trout (*Oncorhynchus mykiss*) fed fishmeal or fish oil-free diets. Br J Nutr 100:953 – 967

Park CC, Ahn S, Bloom JS et al (2008) Fine mapping of regulatory loci for mammalian gene expression using radiation hybrids. Nat Genet 40:421 – 429

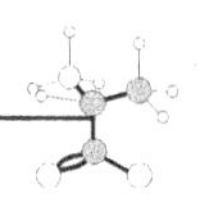

Pauly D, Christensen V, Dalsgaard J et al (1998) Fishing down marine food webs. Science 279:860 – 863

Peatman E, Baoprasertkul P, Terhune J et al (2007) Expression analysis of the acute phase response in channel catfish (*Ictalurus punctatus*) after infection with a Gram-negative bacterium. Dev Comp Immunol 31:1183 – 1196

Peatman E, Terhune J, Baoprasertkul P et al (2008) Microarray analysis of gene expression in the blue catfish liver reveals early activation of the MHC class I pathway after infection with *Edwardsiella ictaluri*. Mol Immunol 45:553 – 566

Peichel CL, Nereng KS, Ohgi KA et al (2001) The genetic architecture of divergence between threespine stickleback species. Nature 414:901 – 905

Pella J, Masuda M (2001) Bayesian methods for analysis of stock mixtures from genetic characters. Fish Bull 99:151 – 167

Penna MS, Khan M, French RA (2001) Development of a multiplex PCR for the detection of *Haplosporidium nelsoni*, *Haplosporidium costale* and *Perkinsus marinus* in the eastern oyster (*Crassostrea virginica*, Gmelin, 1971). Mol Cell Probes 15:385 – 390

Pieniak Z, Verbeke W, Brunsø K et al (2006) Consumer knowledge and interest in information about fish. In: Luten JB, Jacobsen C, Bekaert K, Sæbø A, Oehlenschläger J (eds) Seafood Research from Fish to Dish. Wageningen Academic Publishers, Wageningen

Place SP, O'Donnell MJ, Hofmann GE (2008) Gene expression in the intertidal mussel *Mytilus californianus*: physiological response to environmental factors on a biogeographic scale. Mar Ecol-Prog Ser 356:1 – 14

Pogson GH, Fevolden SE (2003) Natural selection and the genetic differentiation of coastal and Arctic populations of the Atlantic cod in northern Norway: a test involving nucleotide sequence variation at the pantophysin (PanI) locus. Mol Ecol 12:63 – 74

Pollock DD, Bergman A, Feldman MW et al (1998) Microsatellite behavior with range constraints: Parameter estimation and improved distances for use in phylogenetic reconstruction. Theor Popul Biol 53:256 – 271

Porta J, Porta JM, Martinez-Rodriguez G et al (2006) Genetic structure and genetic relatedness of a hatchery stock of Senegal sole (*Solea senegalensis*) inferred by microsatellites. Aquaculture 251:46 – 55

Potasman I, Paz A, Odeh M (2002) Infectious outbreaks associated with bivalve shellfish consumption: a worldwide perspective. Clin Infect Dis 35:921 – 928

Prieur G, Mevel G, Nicolas JL et al (1990) Interactions between bivalve molluscs and bacteria in the marine environment. Oceanogr Mar Biol Annu Rev 28:277 – 352

Primmer CR, Koskinen MT, Piironen J (2000) The one that did not get away: individual assignment using microsatellite data detects a case of fishing competition fraud. Proc R Soc B-Biol Sci

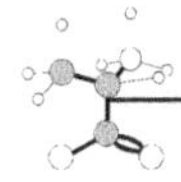

267:1699 – 1704

Prudence M, Moal J, Boudry P et al (2006) An amylase gene polymorphism is associated with growth differences in the Pacific cupped oyster *Crassostrea gigas*. Anim Genet 37:348 – 351

Prunet P, Cairns MT, Winberg S et al (2008) Functional genomics of stress responses in fish. Rev Fish Sci 16:157 – 166

Pruzzo C, Gallo G, Canesi L (2005) Persistence of vibrios in marine bivalves: the role of interactions with haemolymph components. Environ Microbiol 7:761 – 772

Purcell MK, Nichols KM, Winton JR et al (2006) Comprehensive gene expression profiling fol-lowing DNA vaccination of rainbow trout against infectious hematopoietic necrosis virus. Mol Immunol 43:2089 – 2106

Qiu L, Song L, Xu W et al (2007a) Molecular cloning and expression of a Toll receptor gene homologue from Zhikong Scallop, *Chlamys farreri*. Fish Shellfish Immunol 22:451 – 466

Qiu L, Song L, Xu W et al (2007b) Identification and characterization of a myeloid differentiation factor 88 (MyD88) cDNA from Zhikong scallop *Chlamys farreri*. Fish Shellfish Immunol 23: 614 – 623

Raeymaekers JAM, Van Houdt JKJ, Larmuseau MHD et al (2007) Divergent selection as revealed by PST and QTL-based FST in three-spined stickleback (*Gasterosteus aculeatus*) populations along a coastal-inland gradient. Mol Ecol 16:891 – 905

Raida MK, Buchmann K (2007) Temperature-dependent expression of immune-relevant genes in rainbow trout following *Yersinia ruckeri* vaccination. Dis Aquat Organ 77:41 – 52

Rasmussen RS, Morrissey MT (2008) DNA-based methods for the identification of commercial fish and seafood species. Compr Rev Food Sci Food Saf 7:280 – 295

Raymond M, Vaanto RL, Thomas F et al (1997) Heterozygote deficiency in the mussel *Mytilus edulis* species complex revisited. Mar Ecol-Prog Ser 156:225 – 237

Reece KS, Bushek D, Graves JE (1997) Molecular markers for population genetic analysis of *Perkinsus marinus*. Mol Mar Biol Biotechnol 6:197 – 206

Reese G, Viebranz J, Leong-Kee SM et al (2005) Reduced allergenic potency of VR9-1, a mutant of the major shrimp allergen Pen a1 (Tropomyosin). J Immunol 175:8354 – 8364

Reid DP, Szanto A, Glebe B et al (2005) QTL for body weight and condition factor in Atlantic salmon (*Salmo salar*): Comparative analysis with rainbow trout (*Oncorhynchus mykiss*) and Arctic charr (*Salvelinus alpinus*). Heredity 94:166 – 172

Reid DP, Smith CA, Rommens M et al (2007) A genetic linkage map of Atlantic halibut (*Hippoglossus hippoglossus* L.). Genetics 77:1193 – 1205

Rengmark AH, Slettan A, Skaala O et al (2006) Genetic variability in wild and farmed Atlantic salmon (*Salmo salar*) strains estimated by SNP and microsatellites. Aquaculture 253:229 – 237

Rescan PY, Montfort J, Ralliere C et al (2007) Dynamic gene expression in fish muscle during

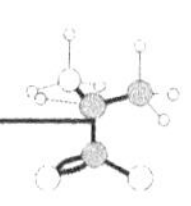

recovery growth induced by a fasting-refeeding schedule. BMC Genomics 8:438

Reusch TBH, Wood TE (2007) Molecular ecology of global change. Mol Ecol 19: 3973 - 3992

Rexroad CE, Rodriguez MF, Coulibaly I et al (2005) Comparative mapping of expressed sequence tags containing microsatellites in rainbow trout (*Oncorhynchus mykiss*). BMC Genomics 6: 54

Rise ML, Jones SR, Brown GD et al (2004) Microarray analyses identify molecular biomarkers of Atlantic salmon macrophage and hematopoietic kidney response to *Piscirickettsia salmonis* infection. Physiol Genomics 20:21 - 35

Rise ML, Douglas SE, Sakhrani D et al (2006) Multiple microarray platforms utilized for hepatic gene expression profiling of GH transgenic Coho salmon with and without ration restriction. J Mol Endocrinol 37:259 - 282

Roberge C, Einum S, Guderley H et al (2006) Rapid parallel evolutionary changes of gene transcription profiles in farmed Atlantic salmon. Mol Ecol 15:9 - 20

Roberge C, Normandeau E, Einum S et al (2008) Genetic consequences of interbreeding between farmed and wild Atlantic salmon: insights from the transcriptome. Mol Ecol 17:314 - 324

Rodriguez F, Navas JI (1995) A comparison of gill and hemolymph assays for the thioglycolate diagnosis of *Perkinsus atlanticus* (Apicomplexa, Perkinsea) in clams, *Ruditapes decussatus*, (L.) and *Ruditapes philipinarum* (Adams et Reeve). Aquaculture 132:145 - 152

Rodriguez MF, LaPatra S, Williams S et al (2004) Genetic markers associated with resistance to infectious hematopoietic necrosis in rainbow trout and steelhead trout (*Oncorhynchus mykiss*) backcrosses. Aquaculture 241:93 - 115

Roessig JM, Woodley CM, Cech JJ et al (2004) Effects of global climate change on marine and estuarine fishes and fisheries. Rev Fish Biol Fisher 14:251 - 275

Rogers SM, Bernatchez L (2005) Integrating QTL mapping and genomic scans towards the charac-terization of candidate loci under parallel directional selection in the lake whitefish (*Coregonus clupeaformis*). Mol Ecol 14:351 - 361

Rogers SM, Bernatchez L (2007) The genetic architecture of ecological speciation and the associ-ation with signatures of selection in natural lake whitefish (*Coregonus* sp. Salmonidae) species pairs. Mol Biol Evol 24:1423 - 1438

Rogers SM, Isabel N, Bernatchez L (2007) Linkage maps of the dwarf and normal lake whitefish (*Coregonus clupeaformis*) species complex and their hybrids reveal the genetic architecture of population divergence. Genetics 175:1 - 24

Rooker JR, Bremer JRA, Block BA et al (2007) Life history and stock structure of Atlantic bluefin tuna (*Thunnus thynnus*). Rev Fish Sci 15:263 - 310

Rucker RR (1966) Redmouth disease of rainbow trout (*Salmo gairdneri*). Bull Off Int

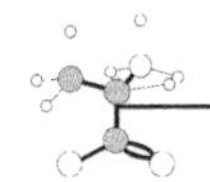

Epizoot 65:825 – 830

Ruzzante DE, Mariani S, Bekkevold D et al (2006) Biocomplexity in a highly migratory pelagic marine fish, Atlantic herring. Proc R Soc B-Biol Sci 273:1459 – 1464

Ryynanen HJ, Primmer CR (2004) Distribution of genetic variation in the growth hormone 1 gene in Atlantic salmon (*Salmo salar*) populations from Europe and North America. Mol Ecol 13: 3857 – 3869

Saeed S, Howell NK (2002) Effect of lipid oxidation and frozen storage on muscle proteins of Atlantic mackerel (*Scomber scombrus*). J Sci Food Agric 82:579 – 586

Sagristà E, Durfort M, Azevedo C (1995) *Perkinsus* sp. (Phylum Apicomplexa) in Mediterranean clam *Ruditapes semidecussatus*: ultrastructural observations of the cellular response of the host. Aquaculture 132:153 – 160

Saha MR (2005) The role of muscle proteins in the retention of carotenoid in Atlantic salmon flesh. PhD Thesis Dalhousie University, Canada

Saito M, Higuichi T, Suzuki H et al (2006) Post-mortem changes in gene expression of the muscle tissue of rainbow trout, *Oncorhynchus mykiss*. J Agric Food Chem 54:9417 – 9421

Sakai M, Kono T, Savan R (2005) Identification of expressed genes in carp (*Cyprinus carpio*) head kidney cells after in vitro treatment with immunostimulants. Dev Biol 121:45 – 51

Salem M, Kenney PB, Rexroad IIICE et al (2006) Microarray gene expression analysis in atrophying rainbow trout muscle: a unique nonmammalian muscle degradation model. Physiol Genomics 28:33 – 45

Salem M, Silverstein J, Rexroad IIICE et al (2007) Effect of starvation on global gene expression and proteolysis in rainbow trout (*Oncorhynchus mykiss*). BMC Genomics 8:328

Samuelsen OB, Nerland AH, Jorgensen T et al (2006) Viral and bacterial diseases of Atlantic cod *Gadus morhua*, their prophylaxis and treatment: a review. Dis Aquat Organ 71:239 – 254

Sanetra M, Meyer A (2008) A microsatellite-based genetic linkage map of the cichlid fish, *Astatotilapia burtoni* and a comparison of genetic architectures among rapidly speciating cichlids. Genetics doi:10.1534/genetics.108.089367

Sarropoulou E, Power DM, Magoulas A et al (2005a) Comparative analysis and characterization of expressed sequence tags in gilthead sea bream (*Sparus aurata*) liver and embryos. Aquaculture 243:69 – 81

Sarropoulou E, Kotoulas G, Power DM et al (2005b) Gene expression profiling of gilthead sea bream during early development and detection of stress-related genes by the application of cDNA microarray technology. Physiol Genomics 23:182 – 191

Sarropoulou E, Franch R, Louro B et al (2007) A gene-based radiation hybrid map of the gilthead sea bream *Sparus aurata* refines and exploits conserved synteny with *Tetraodon nigroviridis*. BMC Genomics 8:44

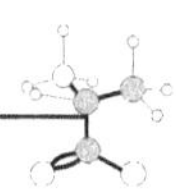

Sarropoulou E, Nousdili D, Magoulas A et al (2008) Linking the genomes of nonmodel teleosts through comparative genomics. Mar Biotechnol 10:227 – 233

Sarropoulou E, Sepulcre P, Poisa-Beiro L et al (2009) Profiling of infection specific mRNA transcripts of the European sea bass *Dicentrarchus labrax.* BMC Genomics 10:157

Sauvage C, Bierne N, Lapègue S et al (2007) Single-nucleotide polymorphisms and their relationship to codon usage bias in the Pacific oyster *Crassostrea gigas.* Gene 406:13 – 22

Scheffer M, Carpenter S, de Young B (2005) Cascading effects of overfishing marine systems. Trends Ecol Evol 20:579 – 581

Schena M, Shalon D, Davis RW et al (1995) Quantitative monitoring of gene expression patterns with a complementary DNA microarray. Science 270:467 – 470

Schiavone R, Zilli L, Storelli C et al (2008) Identification by proteome analysis of muscle proteins in sea bream (*Sparus aurata*). Eur Food Res Technol 227:1403 – 1410

Schlötterer C (2002) A microsatellite-based multilocus screen for the identification of local selective sweeps. Genetics 160:753 – 763

Schlötterer C (2004) The evolution of molecular markers-just a matter of fashion? . Nat Rev Genet 5:63 – 69

Schluter D (1995) Adaptive radiation in sticklebacks: Trade-offs in feeding performance and growth. Ecology 76:82 – 90

Schwab KJ, Neill FH, Le Guyader F et al (2001) Development of a reverse transcription-PCR-DNA enzyme immunoassay for detection of Norwalk-like viruses and hepatitis A virus in stool and shellfish. Appl Environ Microbiol 67:742 – 749

Sekino M, Hara M, Taniguchi N (2002) Loss of microsatellite and mitochondrial DNA variation in hatchery strains of Japanese flounder *Paralichthys olivaceus.* Aquaculture 213: 101 – 122
Sekino M, Hara M (2007) Linkage maps for the Pacific abalone (genus *Haliotis*) based on mi-crosatellite DNA markers. Genetics 175:945 – 958

Senger F, Priat C, Hitte C et al (2006) The first radiation hybrid map of a perch-like fish: the gilthead seabream (*Sparus aurata* L). Genomics 87:793 – 800

Seo J-K, Crawford JM, Stone KL et al (2005) Purification of a novel arthropod defensin from the American oyster, *Crassostrea virginica.* Biochem Biophys Res Commun 338:1998 – 2004

Sepulcre MP, Sarropoulou E, Kotoulas G et al (2007) *Vibrio anguillarum* evades the immune response of the bony fish sea bass (*Dicentrarchus labrax* L.) through the inhibition of leukocyte respiratory burst and down-regulation of apoptotic caspases. Mol Immunol 44:3751 – 3757

Serapion J, Kucuktas H, Feng J et al (2004a) Bioinformatic mining of type I microsatellites from expressed sequence tags of channel catfish (*Ictalurus punctatus*). Mar Biotechnol 6:364 – 377

Serapion J, Waldbieser GC, Wolters W et al (2004b) Development of type I markers in channel catfish through intron sequencing. Anim Genet 35:463 – 466

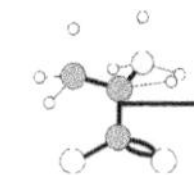

Shangkuan YH, Show YS, Wang TM (1995) Multiplex polymerase chain reaction to detect toxigenic *Vibrio cholerae* and to biotype *Vibrio cholerae* O1. J Appl Bacteriol 79: 264 – 273

Shapiro MD, Bell MA, Kingsley DM (2006) Parallel genetic origins of pelvic reduction in vertebrates. Proc Natl Acad Sci USA 103: 13753 – 13758

Sick K (1965) Haemoglobin polymorphism of cod in the Baltic and the Danish Belt Sea. Hereditas 54: 19 – 48

Sigholt T, Erikson U, Rustad T et al (1997) Handling stress and storage temperature affect meat quality of farmed-raised Atlantic salmon (*Salmo salar*). J Food Sci 62: 898 – 905

Silverman N, Maniatis T (2001) NF-kB signaling pathways in mammalian and insect innate immunity. Genes Dev 15: 2321 – 2342

Sinclair M (1988) Marine populations: An essay on population regulation and speciation. University of Washington Press, Seattle

Smith CT, Antonovich A, Templin WD et al (2007) Impacts of marker class bias relative to locus-specific variability on population inferences in Chinook salmon: A comparison of single-nucleotide polymorphisms with short tandem repeats and allozymes. Trans Am Fish Soc 136: 1674 – 1687

Smith CT, Seeb LW (2008) Number of alleles as a predictor of the relative assignment accuracy of short tandem repeat (STR) and single-nucleotide-polymorphism (SNP) baselines for chum salmon. Trans Am Fish Soc 137: 751 – 762

Solé-Cava AM, Thorpe JP (1991) High levels of genetic variation in natural populations of ma-rine lower invertebrates. Biol J Linnean Soc 44: 65 – 80

Somorjai IML, Danzmann RG, Ferguson MM (2003) Distribution of temperature tolerance quan-titative trait loci in Arctic charr (*Salvelinus alpinus*) and inferred homologies in rainbow trout (*Oncorhynchus mykiss*). Genetics 165: 1443 – 1456

Sonesson AK (2005) A combination of walk-back and optimum contribution selection in fish: a simulation study. Genet Sel Evol 37: 587 – 599

Sørensen JG, Kristensen TN, Loeschcke V (2003) The evolutionary and ecological role of heat shock proteins. Ecol Lett 6: 1025 – 1037

St-Cyr J, Derome N, Bernatchez L (2008) The transcriptomics of life-history trade-offs in whitefish species pairs (*Coregonus* sp.). Mol Ecol 17: 1850 – 1870

Stear MJ, Bishop SC, Mallard BA et al (2001) The sustainability, feasibility and desirability of breeding livestock for disease resistance. Res Vet Sci 71: 1 – 7

Steinberg CEW, Stürzenbaum SR, Menzel R (2008) Genes and environment-striking the fine balance between sophisticated biomonitoring and true functional environmental genomics. Sci Total Environ 400: 142 – 161

Stemshorn KC, Nolte AW, Tautz D (2005) A genetic map of *Cottus gobio* (Pisces, Teleostei) based on microsatellites can be linked to the physical map of *Tetraodon nigroviridis*. J Evol Biol 18:

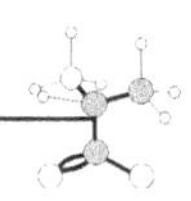

1619 - 1624

Storz JF (2005) Using genome scans of DNA polymorphism to infer adaptive population divergence. Mol Ecol 14:671 - 688

Su J, Ni D, Song L et al (2007) Molecular cloning and characterization of a short type peptidoglycan recognition protein (CfPGRP-S1) cDNA from Zhikong scallop *Chlamys farreri*. Fish Shellfish Immunol 23:646 - 656

Sultan M, Schulz MH, Richard H et al (2008) A global view of gene activity and alternative splicing by deep sequencing of the human transcriptome. Science 321:956 - 960

Swoboda I, Bugajska-Schretter A, Verdino P et al (2002) Recombinant carp parvalbumin, the major cross-reactive fish allergen: a tool for diagnosis and therapy of fish allergy. J Immunol 168: 4576 - 4584

Symonds JE, Bowman S (2007) Atlantic cod genomics and broodstock development in Canada. Aquaculture 272:S313

Takezaki N, Nei M (1996) Genetic distances and reconstruction of phylogenetic trees from microsatellite DNA. Genetics 144:389 - 399

Tall BD, La Peyre JF, Bier JW et al (1999) *Perkinsus marinus* extracellular protease modulates survival of *Vibrio vulnificus* in Eastern oyster (*Crassostrea virginica*) hemocytes. Appl Environ Microbiol 65:4261 - 4263

Tanguy A, Guo X, Ford SE (2004) Discovery of genes expressed in response to *Perkinsus marinus* challenge in Eastern (*Crassostrea virginica*) and Pacific (*C. gigas*) oysters. Gene 338:121 - 131

Tanguy A, Boutet I, Laroche J et al (2005) Molecular identification and expression study of dif-ferentially regulated genes in the Pacific oyster *Crassostrea gigas* in response to pesticide exposure. FEBS J 272:390 - 403

Tanguy A, Bierne N, Saavedra C et al (2008) Increasing genomic information in bivalves through new EST collections in four species: development of new genetic markers for environmental studies and genome evolution. Gene 408:27 - 36

Tao WJ, Boulding EG (2003) Associations between single-nucleotide polymorphisms in candidate genes and growth rate in Arctic charr (*Salvelinus alpinus* L.). Heredity 91:60 - 69

Taris N, Ernande B, McCombie H et al (2006) Phenotypic and genetic consequences of size selection at the larval stage in the Pacific oyster (*Crassostrea gigas*). J Exp Mar Biol Ecol 333:147 - 158

Tasumi S, Vasta GR (2007) A galectin of unique domain organization from hemocytes of the Eastern oyster (*Crassostrea virginica*) is a receptor for the protistan parasite *Perkinsus marinus*. J Immunol 179:3086 - 3098

Tilton SC, Gerwick LG, Hendricks JD et al (2005) Use of a rainbow trout oligonucleotide microar-ray to determine transcriptional patterns in aflatoxin B1-induced hepatocellular carcinoma compared to adjacent liver. Toxicol Sci 88:319 – 330

Tirape A, Bacque C, Brizard R et al (2007) Expression of immune-related genes in the oyster *Crassostrea gigas* during ontogenesis. Dev Comp Immunol 31:859 – 873

Tocher DR (2003) Metabolism and functions of lipids and fatty acids in teleost fish. Rev Fish Sci 11:107 – 184

Tocher DR, Zheng XZ, Schlechtriem C et al (2006) Highly unsaturated fatty acid synthesis in marine fish: Cloning, functional characterization, and nutritional regulation of fatty acyl delta-6 desaturase of Atlantic cod (*Gadus morhua* L.). Lipids 41:1003 – 1016

Toranzo A, Barreiro S, Casa JF et al (1991) *Pasteurellosis* in cultured gilthead seabream (*Sparus aurata*): first report in Spain. Aquaculture 99:1 – 15

Trudel M, Tremblay A, Schetagne R et al (2001) Why are dwarf fish so small? An energetic analysis of polymorphism in lake whitefish (*Coregonus clupeaformis*). Can J Fish Aquat Sci 58:394 – 405

Vandeputte M, Kocour M, Mauger S et al (2004) Heritability estimates for growth-related traits using microsatellite parentage assignment in juvenile common carp (*Cyprinus carpio* L.). Aquaculture 235:223 – 236

Vandeputte M, Mauger S, Dupont-Nivet M (2006) An evaluation of allowing for mismatches as a way to manage genotyping errors in parentage assignment by exclusion. Mol Ecol Notes 6:265 – 267

Vasemägi A, Nilsson J, Primmer CR (2005) Expressed sequence tag-linked microsatellites as a source of gene-associated polymorphisms for detecting signatures of divergent selection in Atlantic salmon (*Salmo salar* L.). Mol Biol Evol 22:1067 – 1076

Velculescu VE, Zhang L, Vogelstein B et al (1995) Serial analysis of gene expression. Science 270:484 – 487

Venier P, De Pittà C, Pallavicini A et al (2006) Development of mussel mRNA profiling: can gene expression trends reveal coastal water pollution?. Mutat Res 602:121 – 134

Verrez-Bagnis V, Ladrat C, Morzel M et al (2001) Protein changes in post-mortem sea bass *D. labrax* muscle monitored by one-and two-dimensional electrophoresis. Electrophoresis 22:1539 – 1544

Vielma J, Makinen T, Ekholm P et al (2000) Influence of dietary soy and phytase levels on performance and body composition of large rainbow trout (*Oncorhynchus mykiss*) and algal availability of phosphorus load. Aquaculture 183:349 – 362

Vilhelmsson OT, Martin SAM, Médale F et al (2004) Dietary plant-protein substitution affects hepatic metabolism in rainbow trout (*Oncorhynchus mykiss*). Br J Nutr 92:71 – 80

Volckaert FAM, Batargias C, Canário A et al (2008) European sea bass. In: Kocher TD, Kole

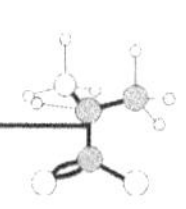

C (eds) Genome mapping and genomics in animals Volume 2. Springer-Heidelberg, Berlin

Walter MA, Spillett DJ, Thomas P et al (1994) A method for constructing radiation hybrid maps of whole genomes. Nat Genet 7:22 – 28

Wang CM, Zhu ZY, Lo LC et al (2007) A microsatellite linkage map of Barramundi, *Lates calcarifer*. Genetics 175:907 – 915

Wang H, Song L, Li C et al (2007) Cloning and characterization of a novel C-type lectin from Zhikong scallop *Chlamys farreri*. Mol Immunol 44:722 – 731

Wang CM, Lo LC, Feng F et al (2008) Construction of a BAC library and mapping BAC clones to the linkage map of Barramundi, *Lates calcarifer*. BMC Genomics 9:139

Waples RS (1998) Separating the wheat from the chaff: Patterns of genetic differentiation in high gene flow species. J Hered 89:438 – 450

Waples RS, Gaggiotti O (2006) What is a population? An empirical evaluation of some genetic methods for identifying the number of gene pools and their degree of connectivity. Mol Ecol 15: 1419 – 1439

Ward RD, Woodwark M, Skibinski DOF (1994) A comparison of genetic diversity levels in marine, freshwater, and anadromous fishes. J Fish Biol 44:213 – 232

Was A, Wenne R (2002) Genetic differentiation in hatchery and wild sea trout (*Salmo trutta*) in the Southern baltic at microsatellite loci. Aquaculture 204:493 – 506

Watabe S (2001) Myogenic regulatory factors. In: Johnston A (ed) Muscle development and growth. Academic Press, San Diego

Wegner KM, Kalbe M, Reusch TBH (2007) Innate versus adaptive immunity in sticklebacks: evidence for trade-offs from a selection experiment. Evol Ecol 21:473 – 483

Wenne R, Boudry P, Hemmer-Hansen J et al (2007) What role for genomics in fisheries management and aquaculture?. Aquat Living Resour 20:241 – 255

Werbeke W, Vermeir I, Brunsø K (2007) Consumer evaluation of fish quality as basis for fish market segmentation. Food Qual Prefer 18:651 – 661

Whitaker HA, McAndrew BJ, Taggart JB (2006) Construction and characterization of a BAC library for the European sea bass *Dicentrarchus labrax*. Anim Genet 37:526

Williams TD, Gensberg K, Minchin SD et al (2003) A DNA expression array to detect toxic stress response in European flounder (*Platichthys flesus*). Aquat Toxicol 65:141 – 157

Withler RE, Candy JR, Beacham TD et al (2004) Forensic DNA analysis of Pacific salmonid samples for species and stock identification. Environ Biol Fishes 69:275 – 285

Wu X, Xiong X, Xie L et al (2007) Pf-Rel, a Rel/nuclear factor-kappaB homolog identified from the pearl oyster, *Pinctada fucata*. Acta Biochim Biophys Sin 39:533 – 539

Xue Q-G, Waldrop GL, Schey KL et al (2006) A novel slow-tight binding serine protease inhibitor from Eastern oyster (*Crassostrea virginica*) plasma inhibits perkinsin, the major extracel-lu-

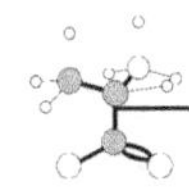

lar protease of the oyster protozoan parasite *Perkinsus marinus*. Comp Biochem Physiol B-Biochem Mol Biol 145:16 – 26

Yamaura K, Takahashi KG, Suzuki T (2008) Identification and tissue expression analysis of C-type lectin and galectin in the Pacific oyster, *Crassostrea gigas*. Comp Biochem Physiol B-Biochem Mol Biol 149:168 – 175

Yooseph S, Sutton G, Rusch DB et al (2007) The Sorcerer II global ocean sampling expedition: Expanding the universe of protein families. PLoS Biol 5:432 – 466

Yu ZN, Guo XM (2006) Identification and mapping of disease-resistance QTLs in the Eastern oyster, *Crassostrea virginica* Gmelin. Aquaculture 254:160 – 170

Yu H, Li QI (2008) Exploiting EST databases for the development and characterization of EST-SSRs in the Pacific oyster (*Crassostrea gigas*). J Hered 99:208 – 214

Zane L, Bargelloni L, Patarnello T (2002) Strategies for microsatellite isolation: a review. Mol Ecol 11:1 – 16

Zhang H, Song L, Li C et al (2008a) A novel C1q-domain-containing protein from Zhikong scallop *Chlamys farreri* with lipopolysaccharide binding activity. Fish Shellfish Immunol 25:281 – 289

Zhang Y, Zhang XJ, Scheuring CF et al (2008b) Construction and characterization of two bacte-rial artificial chromosome libraries of Zhikong scallop, *Chlamys farreri* Jones et Preston, and identification of BAC clones containing the genes involved in its innate immune system. Mar Biotechnol 10:358 – 365

Zhao J, Song L, Li C et al (2007) Molecular cloning, expression of a big defensin gene from bay scallop *Argopecten irradians* and the antimicrobial activity of its recombinant protein. Mol Immunol 44:360 – 368

Zheng XZ, Tocher DR, Dickson CA et al (2005) Highly unsaturated fatty acid synthesis in verte-brates: New insights with the cloning and characterization of a delta 6 desaturase of Atlantic salmon. Lipids 40:13 – 24

Zhu L, Song L, Chang Y et al (2006) Molecular cloning, characterization and expression of a novel serine proteinase inhibitor gene in bay scallops (*Argopecten irradians*, Lamarck 1819). Fish Shellfish Immunol 20:320 – 331

Zhu L, Song L, Zhao J et al (2007) Molecular cloning, characterization and expression of a serine protease with clip-domain homologue from scallop *Chlamys farreri*. Fish Shellfish Immunol 22:556 – 566

Zinkernagel RM, Bachmann MF, Kundig TM et al (1996) On immunological memory. Annu Rev Immunol 14:333 – 367

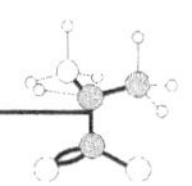

第8章　海洋生物技术

摘要：海洋环境有着丰富的生物多样性以及大量新的未被发现的生物代谢途径，这使得基于海洋环境基因的生物技术（有时称作“蓝色生物技术”）在现阶段还未被有效利用，但已展现出了巨大的应用潜力。人们对海洋生物基因组的不断了解，将对海洋生物技术领域产生重大的影响。基因组测序和重要宏基因组资源开发利用的出现，为解决海洋新陈代谢多样性的问题提供了新途径，也大大促进了海洋生物技术转化为实际产品。本章对海洋生物技术领域进行简要描述，并从已经面世的产品入手，分析如何获取基因资源，以及探讨它们将会给未来的生物技术带来怎样的影响。

8.1　海洋生物技术概览

海洋生物技术的一个简单定义是：利用海洋生物或它们的化合物提供产品或服务。在本章中，除了食品制造方面的应用外，还涉及海洋生物技术在其他方面的应用。我们不仅简要介绍一下海洋生物技术概论，还重点强调基因组学和生物技术的联合会给人类未来面临的挑战提供强有力的解决方案。

很多理由让我们相信，海洋生物技术在近几年内将会成为一个激动人心、硕果累累的研究领域。生命起源于海洋，已经有了38亿年的历史，比陆地上的生物有着更高的多样性。比如，动物界的36个门中有14个只存在于海洋中，而陆地特有的仅为1个门（Gray 1997）。系统发育多样性结合漫长的进化历史，产生的代谢途径极富多样性，又由于海洋环境极大的多样性和特殊性（比如海底热泉高温且含特殊化学成分，潮间带则在水环境与陆地环境间保持动态平衡），这些多样的环境反过来进一步增加了生物代谢的多样性。海水中盐份很高，包括溴化物和碘离子都能被海洋生物的代谢过程所利用。比起陆地生物，人们对海洋生物的新陈代谢知之甚少。我们相信，基因组学研究是揭开其神秘面纱、发现其应用价值的强有力工具。

海洋生物技术是一个相对较新的概念，仍然处于起步阶段。20世纪50年代，通过分析海绵（*Tethya crypta*）的基因首次发现药物，成为50年代开发抗病毒药物的新模式，并在AZT（葛兰素威康制药公司生产的一种抗艾滋病药物，它能够防止艾滋病毒的母婴传播）和Acyclovir（一种治疗疱疹的药物，由葛兰素史克公司生产）的发现中起到重要的作用（Newman and Cragg 2004，Leary et al. 2009）。60年代，人们描述了第一个海洋来源的抗生素[2,3,4-tribromo-5(1′ hydroxy,2′,4′-dibromo phenyl)pyrrole]，它来源于海洋细菌*Pseudomonas bromoutilis*（Burkholder et al. 1966）。海洋生物技术仍停留在初级发展阶段。虽然海洋覆盖了地球上大部分的面积，并且存在许多新颖的海洋生物代谢产物，但来源于海洋的收益并不高。2006年医药市场产值约6500亿美元，尽管27%的药物为生物来源，但来自于海洋天然产物的药物尚不足0.5%。同样，在500亿美元的酶市场也显示出了相似的比率（Leary et al. 2009）。并且，到目前为止，在15 000种注册的海洋产品中只有Prialt[R]和Yondelis[R]两种药物，另外有大约50种正处于研发的不同阶段（Newman and Cragg 2004，www.marinebiotech.org/pipeline.html）。对基因组信息更高效的应用将能够改变这种情况，能为生物技术提供新型酶类，为疾病治疗和健康状况监控提供新的途径，进一步提高水产品产量，为工业生产材料和加工提供新的资源等。例如，从珊瑚中提取到的一种化合物被用于抗炎症药物（Mayer et al. 1998）、从海藻中获得新的抗癌药物（Fuller et al. 1992，1994）和可降解泄漏石油（Head et al. 2006）的海洋细菌以及其他海洋生物（Amador et al. 2003），都是很好的例子。我们认为，关于化学品及其安全使用的欧洲共同体条约（REACH），能够显著增加人们对海洋生物技术的关注度（http://ec.europa.eu/environment/chemicals/reach/reach_intro.htm），这将提高人们对寻找新酶和新代谢途径的关注，为我们带来一个更高效和更少污染的化工业。

8.2 基因组学如何影响海洋生物技术

到目前为止，海洋基因资源的开发利用严重滞后。1973年至2007年间的“象征性”调研，仅形成135个相关专利（Leary et al. 2009）。造成这个现象的原因之一是目前对海洋化合物的研究要么靠盲目的、大规模的筛选，要么以生理生态学的知识为导向进行挖掘（Sennett 2001）。对基因组学认识的不断加深，有助于我们进一步了解海洋生物在海洋环境中的复杂代谢过程的遗传基础，辅助我们获得更多新的发现，从而研发出新产品。

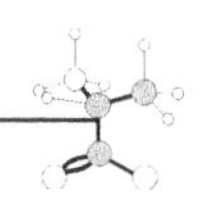

许多海洋微生物都难以培养，使得它们很难用于生产。那些目前难以培养或无法培养的生物或群落，可以通过宏基因组了解它们的生理生化机制，这些信息是其他方法难以获得的（见下文）。事实上，宏基因组和适当的筛选方法相结合，将会成为现代生物技术的有力工具。对生物催化酶的大量需求，催生了大规模、特异性的基于构建海洋宏基因组文库的筛查技术，并得到广泛的应用，比如对酯酶的寻找。去年，通过构建南中国海和北极圈沉积物的宏基因组文库，已经发现并筛选了四种酯酶（Jeon et al. 2009，Chu et al. 2008）。像酯酶这样的水解酶是很好的生物催化剂，因为它们在工业上有着广泛的用途，比如对底物的特异性选择以及对空间结构的选择性。

基因组学知识的增长能够促进新基因的发现，使我们更容易从蛋白质找到核苷酸序列。例如，如果一个人们感兴趣的生物活性物质被找到，且对应的蛋白质或酶已被纯化，利用已知的基因组去寻找相关的基因就是很简单的事情了。已知功能的基因与来自其他生物的基因比对，可以发现功能相似的基因。这样，对于人们感兴趣的蛋白，可以扩大其编码基因的多样性。比如，最近 Marsic 等利用以上方法，从 *Thermococcus thioreducens*（一种从 Rainbow 深海热泉分离得到的古细菌）中发现了一种新的 B 型 DNA 聚合酶，并对它进行了克隆、表达、纯化和性质鉴定。在一系列 PCR 反应条件下，这个新酶比许多常用的聚合酶表现得更快、更稳定和更准确（Marsic et al. 2008）。相反地，如果某个基因某种生物缺失，就不要在该生物中寻找这种生物活性物质，或者，如果该生物活性是必须的，就寻找具有类似活性的替代酶。

8.3 通过微生物群落宏基因组、单个物种的全基因组及数据挖掘来扩充基因资源

过去的十年里，由于测序技术的进步和由此带来的基因组项目的增多，可供生物技术利用的基因资源呈指数性增长。然而，这些数据均是以人类及模式生物（包括人、大鼠、小鼠、水稻）以及与之相关的病原菌基因组为主。尽管如此，基因组学革命已直接影响到在进化中有特殊地位或有生物技术应用潜质的一些海洋物种和生物群落。

基因组规模的信息最初来源于可培养的细菌，即已获得纯培养的菌种。但实际上大部分的细菌和古细菌是不可培养的（Amann et al. 1995，Rappé and Giovannoni 2003）。同样，对海洋病毒的研究发现，大部分海洋病毒基因在现有数据库中仍找不到相关联的信息（Breitbart et al. 2002，Angly et al. 2006），

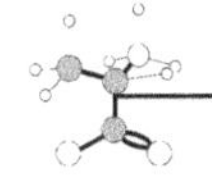

当然它们大多仍没有被开发。从各种各样的海洋环境中得到几乎无限的微生物基因资源是一件激动人心的事情。它主要依赖于两种互补的方法：宏基因组学和单细胞基因组学。而第三种则是建立高通量的方法，来筛选和培养以前难以培养的微生物（占绝大多数），这种方法被大大地忽略了，尽管在过去的十年中有一些有趣的、开拓性的结果被报道（Giovannoni and Stingl 2007）。

大多数宏基因组项目是基于环境 DNA 文库的构建，如马尾藻海（Sargasso Sea）项目（Venter et al. 2004）。为了更全面的理解，减少随机性带来的误差，应当更加合理地规划微生物多样性调查。人们正在建立更有效地评估海洋环境样本的采样方法和策略（Quince et al. 2008）。尽管宏基因组有明显的局限性，但还是引起了人们很大的兴趣，相信将来通过高通量测序可以得到生物圈中大多数基因资源，包括那些稀有物种的基因。单是一个全球海洋调查（Sorcerer II Global Ocean Survey）的宏基因组计划，就几乎将已知的蛋白数据库数据量翻倍（Rusch et al. 2007）。这极大地改变了人们对底泥、深海热泉等海洋环境中生物的认识（Sogin et al. 2006，Huber et al. 2007，Quince et al. 2008）。在这些环境下要想发掘更大的基因宝库，尤其是那些稀有物种和种系，仍需要改良挖掘基因资源的策略。

尽管人们对挖掘新的生物信息学方法进行了很多的努力，但从复杂的群落宏基因中组装、分离出单个微生物基因组仍然是一件困难的任务。这使得单细胞基因组策略发展了起来。这种技术先分离出单细胞，然后用 φ29 DNA 合成酶实现多重置换扩增的方法扩增整个基因组（Dean et al. 2002）。这为那些非可培养的物种或可获得 DNA 量非常少的物种，打开了另一个窗口。

无论是单个物种还是对宏基因组的测序项目，都是一种获取新的特定基因序列的途径。如何将这些新的基因资源充分利用起来是现代基因组学面临的主要挑战。传统的结合电脑分析和功能筛选的方法，在生物技术应用中仍然具有里程碑式的意义。如今，大量复杂数据产生时，我们一方面会很快淹没在数据海洋中，但却仍然十分渴求我们想要的数据。未来，亟需发展新的多学科交叉方法，来高效拓展我们对基因组学的认识。

8.3.1　全基因组

过去很长一段时间，人们认为测定典型物种的全基因组，就能够得到这个物种的大部分基因信息，几十个有代表性的细菌和古细菌基因组能够代表微生物界大多数的基因。但事实显然不是这样。对致病菌，比如大肠杆菌（*E. coli*）和无乳链球菌（*Streptococcus agalactiae*）的比较基因组学发现，即使

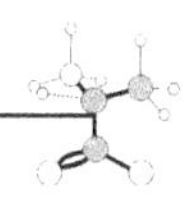

是同一物种基因组间也会存在差异（Bielaszewska et al. 2007）。这些研究证明，用单一个体基因组信息来描述一个物种是不可行的，人们需要一个群体来描述基因组（泛基因组）。这个群体量需要多少，因不同的物种而异。因此，一些重要的病原菌、菌株的基因组测序正快速增加。例如，霍乱弧菌已有2个全基因组和14个部分基因组序列，2年内可能会增长到50多个。到现在为止，所有人们感兴趣的可用于生物技术的海洋物种基因组信息，都是来自单克隆培养细胞。在这个过程中，一部分（也有可能是大多数!）变异被忽略了，其相应的基因很难通过基因分析和信息挖掘的方法得到。这就是为什么仍然需要功能筛查和信息挖掘相结合，来发掘得到这些变异的菌株信息。在不远的将来，对于人们感兴趣的有生物技术应用价值的物种，像针对病原菌一样，对不同菌株都进行基因组测序的项目会得到广泛实施。这在真菌 *Penicillium chrysogenum* 项目中已经实行，其2个基因组序列已经完成（www. genomesonline. org/gold. cgi）。

截至2008年8月，已进行全基因组测序的微生物大多为细菌（684个在NCBI Entrez Genome数据库；696个在Genomesonline数据库），其次是古细菌（52～53个）。虽然真菌在许多海洋环境中都存在，但在全基因组测序物种列表中仍然很少。不同生活域的海洋与陆地已测基因组的物种数量比率差异相差很大（表8.1）。

表8.1　三个领域的海洋物种的基因组测序情况

（来源于 www. genomesonline. org 并加以修改和编辑，2008年8月）

Table 8.1 Status of genome sequencing projects in the three domains of life and in marine species (compiled from genomesonline.org, and modified, August 2008)

	Bacteria			Archaea			Eukarya		
Genome status	All	Marine	Ratio (%)	All	Marine	Ratio (%)	All	Marine	Ratio (%)
Complete and published	690	74	11	50	19	38	94	6	6.4
Complete	13	1	7.7	1	1	–	28	1	–
Incomplete	1573	184	11.7	69	10	14.5	547	2	–
Targeted	302	8	2.6	22	6	27	6	?	–
ALL	2578	267	10.4	142	36	25	675	9	1.3

英文注释：Bacteria，真细菌；Archaea，古细菌；Eukarya，真核生物

从表中可以清楚看出，海洋真核生物测序项目所占的比例很小，细菌相对多一些（有10%是海洋物种），而海洋来源的古细菌测序比例最大。虽然古细菌测序项目总量较低，但已完成并发表的海洋物种序列所占比例高达

38%，准备测序的占25%。从全球来看，尽管可用的基因组数据总量上升，但现阶段海洋物种的增长速度仍较为缓慢。这并不符合人们对海洋物种基础研究（尤其进化和发育方面）以及海洋生物技术两方面都有巨大热情的现状。另外，对于有工业利用价值的极端微生物的测序项目还不到4%。

对已经证明有生物技术应用价值或有应用潜力的全基因组进行测序，发现不同的基因簇，其中 *Pyrococci* 和 *Thermococci* 这两种古细菌十分有代表性。这两个属的许多种中都发现了（人们感兴趣的 DNA 相关酶（如 DNA 合成酶和连接酶）以及大量有价值的降解淀粉、纤维素和水解酶（见综述 Egorova and Antranikian 2007），用于制造生物燃料以及可降解角蛋白与朊病毒蛋白等顽固污渍）（Tsiroulnikov et al. 2004）。到20世纪90年代末，所有从这些物种中得到的耐热性酶都是基于功能筛查的生化方法。继几个全基因组信息发表后（Kawarabayasi et al. 1998，Maeder et al. 1999，Cohen et al. 2003），功能筛查的方法就和基因组信息发掘结合起来了。在一些案例中，最开始的研究只是通过计算机寻找目的基因，然后再与生化特征相结合，最终取得成功。在 *Pyrococcus abyssi* 中发现耐热性腈水解酶就是一个很好例子。

近来，石油危机迫使人们寻找可替代化石燃料的新物质。最近一篇“生物能源基因组”的综述（Rubin 2008）列出了一些已经有全基因组序列或是正在进行测序的物种，这些物种是生物降解菌或是燃料生产者。表8.2提到了海洋基因组对这个大项目的小贡献。但这并不意味着海洋物种在这个大项目贡献很少，我们已经例举了在极端嗜热古菌中存在的热稳定性淀粉酶、纤维素酶和木聚糖酶等，产生这个现象是因为人们在筛选时对陆地物种有着强烈的偏好性（见下文）。

8.3.2 宏基因组不断增长的贡献

在《生物技术杂志》（*Biotechnological Journal*）发表的题为《宏基因组：通往自然多样性的无穷道路》的综述中提出一个共识，即宏基因组对生物技术有着潜在的推动作用（Langer et al. 2006）。海洋宏基因组应用的技术和方法大多由土壤宏基因组研究发展而来，是土壤细菌中 DNA 的提取、消化技术（Torsvik 1980），构建环境 DNA 文库技术（Pace et al. 1986）以及海洋浮游生物环境 DNA 文库概念（Schmidt et al. 1991）几方面的结合。宏基因组这个词由 Handelsman 提出，用以描述在指定生境下所有生物基因序列的总和（Handelsman et al. 1998）。这种新技术在生物技术应用上的潜能很快就受到重视

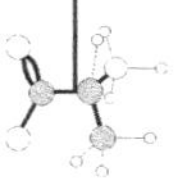

表8.2 与生物技术有关的海洋微生物基因组

Table 8.2 Marine microbial genomes of biotechnological interest

	Organism	Size(kb)	Orf s	NCBI Ref	Institution	Targets, type of enzymes, type of applications References (genomes; biotech) Archae
Archaea	*Aciduliprofundum boonei T469*	2973		NZ ABSD00000000	JCVI, Univ Portland	Incomplete (proteases) (unpublished)
	*Aeropyrum pernix K*1	1669	1700	NC_000854	NITE	Proteases (Kawarabayasi et al. 1999)
	Hyperthermus butylicus DSM 5456	1667	1602	NC_008818	Eipdauros Biotech.	Peptidases (Brügger et al. 2007)
	*Nitrosopumilus maritimus SCM*1	1645	1795	NC_010085	JGI	* (unpublished)
	Pyrobaculum aerophilum IM2	2222	2605	NC_003364	UCLA, Caltech	Glycoside-hydrolases, proteases (Fitz-Gibbon et al. 2002)
	Pyrococcus abyssi GE5	1765	1896	NC_000868	Genoscope	DNA processing; starch, cellulose, xylan degrading; proteases (Cohen et al. 2003)
	Pyrococcus furiosus JCM 8422	1908	2125	NC_003413	Univ. Utah/Maryland	DNA processing; starch, cellulose, chitin degrading; proteases (Vanfossen et al. 2008, Jenney and Adams 2008)
	Pyrococcus horikoshii OT3	1738	1955	NC_000961	NITE, Univ Tokyo	DNA processing; starch degrading; proteases, hydrogenases (Kawarabayasi et al. 1998)
	Pyrodictium abyssi DSM 6198	*	*	*	JCVI, GBM-MGSP	Starch, xylan degrading (unpublished);
	Pyrolobus fumarii	1850	2000	*	Celera, Diversa	Thersmostability/pressure; organic solutes (unpublished); (Gonçalves et al. 2008)
	*Staphylothermus marinus F*1	1570	1570	NC_009033	JGI	Starch processing, protease, DNA processing (unpublished)
	Sulfolobus solfataricus P2	2292	2977	NC_002754	CBR	Starch, cellulose degrading, proteases (She et al. 2001)
	*Thermococcus kodakaraensis KOD*1	2088	2306	NC_006624	Kwansei Gakuin Univ	DNA processing; starch degrading enzymes; hydrohenases (Fukui et al. 2005, Kanai et al. 2005)
	Thermococcus barophilus MP	2059*	*	NZ ABSF00000000	JCVI, Prokarya	Proteases, starch degrading, pressure (unpublished)
	*Thermococcus onnurineus NA*1	*	*	*	KORDI	DNA polymerase, Proteases, starch processing (unpublished); (Lim et al. 2007)
	Thermococcus gammatolerans EJ3	*	*	*	IGM	DNA processing, proteases (unpublished)

英文注释：Protease，蛋白酶；Peptidase，肽酶；Glycoside hydrolase，糖苷水解酶；Starch，淀粉；Cellulose，纤维素；Xylan，木聚糖；Chitin，几丁质；Hydrogenase，氢化酶；Thermostability，热稳定性，耐热性；Solute，溶质；Hydrogenase，氢化酶

Table 8.2 (continued)

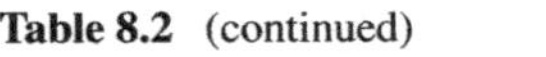

Organism	Size(kb)	Orf s	NCBI Ref	Institution	Targets, type of enzymes, type of applications References (genomes; biotech) Archae
*Alcanivorax borkumensis SK*2	3120	2755	NC_008260	Bielefeld Univ	Oil hydrocarbon degradation (Schneiker et al. 2006)
Alteromonas macleodii DSMZ 17117	4236	4163	NZ_AAOD00000000	JCVI	Exopolysaccharide production (Ivars-Martinez et al. 2008)
*Aquifex aeolicus VF*5	1551	1529	NC_000918	Diversa, Univ Illinois	Glucan branching, molecular biology (Deckert et al. 1998)
Bacillus halodurans C-125	4202	4066	NC 002570	JAMSTEC JCVI	Starch, pullulan and xylan degrading (Takami et al. 2000)
Dehalobium chlorocoercia DF-1	*	*		Epidauros Biotech.	PCB dechlorination (unpublished)
*Desulfotalea psychrophila ESV*54	3523	3116	NC_006138	JGI	Psychrophilic enzymes (Rabus et al. 2004)
*Marinoboacter hydrocarbonoclasticus VT*8	4326	3858	NC_008740	JVI	Recalcitrant hydrocarbon sources degrading (unpublished)
*Marinitoga piezophila KA*3	2000	*	*	JGI	Piezophilic thermostable enzymes (unpublished)
*Marinitoga camini MV*1075	*	*	*	JAMSTEC	Thermostable proteases (unpublished)
*Nitratiruptor sp SB*155-2	1877	1843	NC_009662	MPI, MPI-MM	Virulence factors (Nakagawa et al. 2007)
*Rhodhopirellula baltica SH*1	7145	7325	NC_005027	JCVI	Sulfatases (Glockner et al. 2003)
*Thermotoga maritima MSB*8	1860	1858	NC_000853	JCVI	Thermostable glycosides hydrolases (Nelson et al. 1999, Vanfossen et al. 2008)
*Vibrio fischeri ES*114	4478*	4061	NZ_ABIH00000000	CNRS-SBR	Luminescence (Ruby et al. 2005)
Zobellia galactanovorans Dsij	*	*	*		Sulfatases, sulfotransferases, food industry (unpublished)

* indicates unpublished

英文注释：Exopolysaccharide，胞外多糖；Glucan，葡聚糖；Dechlorination，脱氯作用；Psychrophilic，嗜冷的；Recalcitrant，顽强的；Hydrocarbon，碳氢化合物；Thermostable，热稳定的；Virulence，毒性；Sulfatase，硫酸酯酶；Glycoside，糖苷；Hydrolase，水解酶；Luminescence，荧光；Sulfotransferase，磺基转移酶

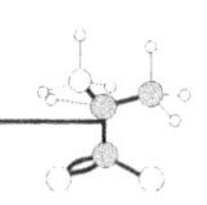

(Short et al. 1997)。海洋宏基因组最初是为了调查海洋中的微生物多样性，以此来了解未被认知的在地球化学循环中起关键作用的物种。到21世纪初，人们就已经能够结合高通量测序技术，大规模地应用这项技术了，比如C. Venter和他的团队在马尾藻海宏基因组调查和全球海洋样品收集（Sorcerer II/GOS）考察（Rusch et al. 2007）中就应用了该技术。

宏基因组一个最大的优势是可以打破培养技术的限制。当微生物样品来自高度复杂的群落或宏有机体系时（通常是一种真核生物及其附属的微生物群落），宏基因组技术就特别有用，比如在海绵中绝大多数微生物群落都是无法培养的。

地球上的海洋环境是十分多样的，对它们的研究和开发为生物技术提供了全新的基因资源。宏基因组结合异源表达技术、合适的高通量筛查流程和定点突变技术，成功地被一些研究团队、中小企业及大型企业（Diversa、Genencor、Degussa、Henkel等）应用于发现新的酶或天然产物。表8.3例举了近年来宏基因组学的突出贡献。

虽然上面提及的Sorcerer II/GOS项目只采集海洋表面的微生物，但对宏基因组仍然具有里程碑的意义，因为它在数据库中预测了600万种蛋白，几乎是现有蛋白数据库的2倍（Yooseph et al. 2007）。该计划的实际应用价值无法估计，还需要很多年才能达到预计的开发目标。其中一个首要目标就是在预测蛋白数量增加的情况下，加快了解新蛋白家族的发现速率。从已有的信息中他们得出结论，现有蛋白样品库还远没有达到饱和，还有许多的蛋白家族等待人们去发现。但这是个有争议的结论，Koonin（Koonin 2007）认为不仅要比较Sorcerer II/GOS数据库的蛋白家族数量与序列数量的增长量，还要考虑同源蛋白的增长速度。人们对几个海洋区域的病毒基因组（virome）（给定栖息地的微生物群落量的病毒部分）研究也揭示了全新的蛋白家族多样性（Breitbart et al. 2002，2004，Angly et al. 2006）。关于未知功能的古细菌病毒蛋白基因组的一项研究，揭示出海洋病毒是发现新的折叠方法和新的结构，进而发现新蛋白的宝库（Vestergaard et al. 2008）。

随着宏基因组的发展，微生物多样性的发掘越来越快。有工业应用价值的生物催化剂将会呈指数式增长。近来如腈水解酶、脂肪酶和酯酶的发现，已经证实了这一观点（Robertson et al. 2004，Bertram et al. 2008），至少在功能筛查效率足够的前提下可以保证这一点。需要注意的是，与全球海洋宏基因组调查相比，这些结果是直接针对工业化应用的，有着鲜明的生物技术目标。

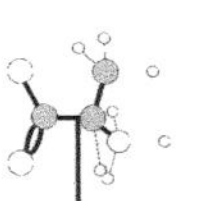

表8.3 对海洋宏基因组和生物技术的贡献

Table 8.3 Contributions to marine metagenomics and biotechnology

	Authors	Origin of sample(s)	Screening method	Main targets	Industry	Reference
Methods	Rondon et al. (2000)	Soils	Function based	Enzymes, antibacterial, hemolytic activity	– –	
	Henne et al. (2000)	Soils (meadow, field)	Function based	Lipases, esterases	Patent –	
	Uchiyama et al. (2005)	Groundwater	SIGEX	Method, aromatic hydrocarbon degradation		
	Kalyuzhanya et al. (2008)	Lake sediment	C_1 substrates	Group diversity/function		
Diversity	Breitbart et al. (2002)	Surface seawater	–	Viral diversity	–	
	Venter et al. (2004)	Sargasso Sea surface	–	Microbial diversity	–	
	Rusch et al. (2007)	Atlantic, Pacific Oceans	–	Microbial diversity	–	
	Yooseph et al. (2007)	Atlantic, Pacific Oceans	–	Microbial diversity		

英文注释：Meadow，草地，牧场；Sargasso Sea，马尾藻海（在西印度群岛东北）；Hemolytic，溶血的；Lipase，脂肪酶；Esterase，酯酶；Aromatic，芳香族的；Hydrocarbon，碳氢化合物；Viral，滤过性毒菌引起的；Microbial，由细菌引起的

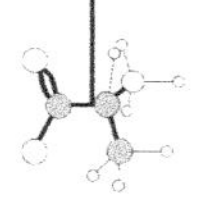

Table 8.3 (continued)

Biotechnology

Authors	Origin of sample(s)	Screening method	Main targets	Industry	Reference
Schirmer et al. (2005)	Sponge *Discodermia*	Sequenced/PCR	Polyketide synthases (PKS)	Kosan[a]	
Fieseler et al. (2007)	20 sponge species	Sequenced/PCR	Identification of pharma relevant PKS genes	Kosan	
Chu et al. (2008)	Surface seawater, China	Function based	Lipases, esterases	?	
Short et al. (1997)	Diverse incl. plankton	Function based	Hydrolases (notably thermostables)	Diversa[b]	WO9704077
Short (1999)	Whale bone, picoplankton	Function based	Hydrolases	Diversa	US5958672
Weiner et al. (2007)	Pacific Ocean picoplankton	Function based	Oxidative enzymes (epoxidases, P450) :	Diversa	US2007231820
Lee et al. (2008)	Deep-sea sediment	Function based	Fibrinolytic Metalloprotease	Patent	WO2008056840
Robertson et al. (2004)	>600 biotopes	Function based	Nitrilases (137 novel enzymes)	Diversa	
Bertram et al. (2008)	Information missing	Function based	Lipases, esterases (350 novel enzymes)	Verenium	

[a]Kosan Biosciences is now a subsidiary of Bristol-Myers;
[b]Diversa is currently a branch of Verenium. Patent data were compiled from the European Patent Office (http://ep.espacenet.com/). In August 2008, a query with metagenome as keyword led to 8 entries, most of them dedicated to soil metagenomes and cow rumen.

[a] Kosan生命科学现在是Bristol Myers公司的子公司；

[b] Diversa现在是Verenium公司的一个分公司。专利数据来自欧洲专利局(http://epespacenet.com/)。在2008年8月，以宏基因组为关键词的查询有8条结果，大部分是针对土壤宏基因组和牛瘤胃的。

英文注释：Sponge，海绵；Discodermia，二氟亚甲基；Plankton，浮游生物；Picoplankton，超微型浮游生物；Biotope，群落生境；Polyketide，聚酮化合物；Synthases，合成酶；Pharma，制药公司；Lipase，脂肪酶；Esterase，酯酶；Hydrolase，水解酶；Thermostable，热稳定的；Oxidative，氧化的；Epoxidase，环氧酶；Fibrinolytic，溶解纤维蛋白的；Metalloprotease，金属蛋白酶；Nitrilase，腈水解酶

对于海洋极端生物开发来说有两个主要的困难。第一，筛查方法是第一局限（序列为基础还是功能筛查为基础）。第二，在极端环境下生物量较低，这意味着克隆所需的 DNA 含量非常低，需要后续加入如全基因组扩增的步骤。以基因序列为基础的筛查受到已知基因家族的限制，它无法鉴定宏基因组数据库如 CAMERA（Seshadri et al. 2007，http://camera.calit2.net）中不存在的一些全新基因。并且，许多在宏基因组库中的基因无法用常规工程菌株如 *E. coli* 表达，因此无法用活性筛查的方法检测到。为了突破这些局限性，最近人们发明了一种叫做底物诱导基因表达筛查（substrate-induced gene expression screening，SIGEX）的新方法（Uchiyama et al. 2005）。该方法用带有 GFP 蛋白的各种底物诱导克隆菌中相应基因的表达，然后用荧光活化细胞分拣器（fluorescence-activated cell sorting，FACS）筛选具有该基因的克隆。这项高通量的方法成功地从宏基因库中筛选到可被芳香族烃化诱导表达的基因。但这种方法仍然存在一些局限性，比如基因在克隆载体中的方向或基因中存在终止子，导致插入大片段产出率很低（De Lorenzo 2005，Yun and Ryu 2005）。

通过在环境中添加底物，然后富集可降解这些底物的细菌，从而寻找各种天然生物催化酶。同样，也可以通过从大量数据中分类 DNA 信息，来富集我们感兴趣的 DNA，从而减少测序量。去年已经用稳定性同位素探针进行了大量类似工作。为了在湖底沉积物中寻找可以利用单碳化合物（无 C－C 键）的微生物，通过 ^{13}C 标记 C_1 化合物后可以从微生物群落中提取 DNA 并分离含有标记的目的基因（Kalyuzhnaya et al. 2008）。虽然这种技术的开发主要是用于分析微生物群落中的不同化合物，但也可以为复杂群落宏基因组富集所要测序的 DNA，为生物技术研究做贡献。

8.4 海洋生物技术在发现天然产物、新药物和白色生物技术中的应用

在海洋生物中发现了许多天然产物和酶类。这里给出了细菌、古细菌、海洋真菌、后生动物、病毒和藻类的例子。但这些产物绝不代表在海洋环境中的所有发现，这里只是列出几个例子而已，用以呈现该物种的潜在价值。没有提到的物种，譬如变形虫（amoebozoa）、有孔虫（rhizaria）和古虫界（excavates）可能蕴含着还未发现的重要代谢多样性。

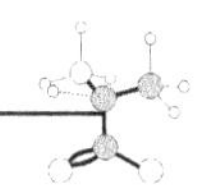

8.4.1 病毒

海洋病毒可能代表了地球上最大的未被开发的生物技术资源。平均每毫升海水含有 10^6 个病毒，而整个海洋的海水中估计有 10^{30} 个病毒，这是一个有待研究和利用的巨大资源（Suttle 2007）。过去，酶和天然产物的发现都是基于可培养的方法，这些方法对于需要合适宿主的病毒来说难以实现。因此，不依赖培养的宏基因组学方法打开了海洋病毒研究的大门。没有一个基因是所有病毒所共有的，海洋病毒序列的随机测序揭示了基因的多样性，这在以前是无法实现的（Angly et al. 2006）。

过去，病毒被认为是一种结构简单、可以自我复制的基因。巨大病毒的发现完全颠覆了这种观念，病毒可以用自己编码的蛋白控制复杂的代谢通路（Raoult et al. 2004，Wilson et al. 2005，la Scola et al. 2008）。比如，分子生物学家发现的许多有价值的蛋白都是病毒来源的：T4 噬菌体（可感染 *E. coli*）的 DNA 连接酶、多聚核苷酸激酶、DNA 聚合酶等都已经商品化，用于克隆技术；从 T7、phi6 和 SP6 中提取的 RNA 聚合酶已得到广泛应用；禽成髓细胞性白血病毒和 Moloney 小鼠白血病毒中的反转录酶（RNA 指导下的 DNA 聚合酶，现阶段只发现于病毒）是以 RNA 模板制备 cDNA 的标准酶。虽然上述这些酶类没有一个是来源于海洋病毒的，但是这些例子证明病毒中确实存在着许多有用、高效的可供开发的酶。现在对海洋病毒进行深入研究的报道很少，但已经发现了几个具有生物技术应用潜力的病毒（Allen and Wilson 2008）。比如，球形病毒（*coccolithovirus*）EhV－86 有一个基因几乎编码了合成神经酰胺所需的所有酶，包括一个转运酶、一种延长蛋白、一个磷酸酶和三个脱饱和酶（Wilson et al. 2005，Han et al. 2006）。神经酰胺是质膜的组成部分，常作为化妆品的抗衰老组分。这个新发现不仅有学术上的意义，还有商业开发的潜力。同样，在这个病毒中还发现了可用于生物催化的脂肪酶和酯酶（也有一些核酸酶和蛋白酶；Allen et al. 2006b）。仅对病毒基因组多样性进行最简单的调查就会发现，我们对这些有机体如何行使功能仍知之甚少：80% 海洋来源的病毒序列是独特的，没有已知功能的序列与之相对应（Suttle 2005）。显然，这些新基因对于病毒来说是有用的，只是我们还不了解它们的功能以及它们之间的关系（Yin and Fischer 2008）。由于这些基因的功能尚未知晓，我们预言，挖掘新的生化代谢途径和反应可以推动生物技术的更多应用。

自从内含肽（inteins）被发现以来，引起了人们越来越多的关注（Gogarten and Hilario 2006）。内含肽是一类可以自我剪切的蛋白，它可以切断

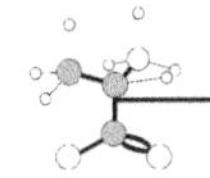

一部分蛋白序列，然后把剩余部分用肽键连接起来。它在蛋白表达、合成、纯化和标记等生物技术领域有着重要的应用（Perler 2002）。内含肽在所有生物界均有发现，但最近发现其在海洋病毒中十分普遍。在赤潮异弯藻病毒（*Heterosigma akashiwo*）（Nagasaki et al. 2005）、球星病毒一些亚种（Allen et al. 2006a，Goodwin et al. 2006）以及微型病毒（Ogata et al. 2005）中均有发现，与马尾藻海宏基因组数据库中内含肽的序列有一定的相似性。

8.4.2 古细菌和细菌

细菌和近期发现的古细菌，包括海洋来源的原核生物，是生物技术中重要的有机体。极端嗜热细菌和古菌的全基因组测序为人们打开了发现新酶的大门。一些深海极端嗜热古菌的耐热性酶，已经在分子生物学中得到了广泛应用。从生物质发酵生产乙醇和生物燃料，到有机化学、生物产氢等领域，这些酶都有应用潜力。

红小梨形菌（*Rhodopirellula baltica*）和放线菌（*Zobellia galactanovorans*）的开发就是很好的例子。前者大量聚集，形成在沿海出现的“海雪”，它能够高效地降解有机物。后者通常与大型藻类共生，能够降解细胞壁。全基因组测序发现，在这两个物种中发现的硫酸酯酶（Glockner et al. 2003）和磺基转移酶（未发表结果）数量特别多。基于这些发现，功能基因组计划出现了，旨在探讨各种不同底物及与之相对应的酶。

8.4.3 藻类

藻类，包括蓝藻细菌（*cyanobacteria*），是一个多样性丰富的类群，囊括了泛植物界（*archaeplastida*）、囊胞藻界（*chromalveolates*）和真细菌界（*eubacteria*），因此有着相当高的代谢多样性。但应用于生物技术的藻类基因却寥寥无几，一个成功例子是红藻（*Chondrus crispus*）的己糖氧化酶基因。当把己糖氧化酶加入食品中时，它可以减少美拉德反应（Maillard reaction），从而不会过度染色，并能增加面筋的强度（Hansen and Stougaard 1997），通过异源高效表达，该酶已经有商业化产品。多不饱和脂肪酸也是海洋生物技术和海洋基因组的很好例子，因为只有海洋是这种重要脂类的稳定来源。浮游植物是一个很重要的资源，如 EPA 和 DHA，主要来源于浮游植物。藻类中有关脂肪酸合成的基因引起了人们很大的兴趣。将硅藻 *Thalassiosira* 的相关基因转入农作物中，可以产生更长链的多不饱和脂肪酸（Tonon et al. 2004a，b，2005）。另一个藻类产物应用的例子是从掌状海带纯化的海带多糖，用来刺激烟草产

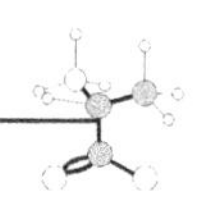

生防御反应（Klarzynski et al. 2000），这是Goëmar公司生产农药昆布素（Iodus）的基础，该产品可以部分地替代商业用谷物杀虫剂。

实际上，最近吸引人们眼球的一个领域是生物燃料，而海洋基因组可以在这方面一显身手。全球生物燃料的投资在近几年大幅度上升，超过了10亿美元，预计已经到达峰值。相比陆地植物，微藻显出一些优越性，因为单细胞藻类在合适环境下的光合效率远超过高等植物。海洋微藻主要有两方面的优势：他们的开发与淡水资源、食物供应不冲突；相比高等植物，微藻可以在水塘或光合作用反应器中培养，通过调控进入的营养物可以方便地调控它们的生长代谢。另外，这些微藻的养殖需要 CO_2 和氮、磷等营养元素，可以利用这一特点进行减少温室气体排放和治理污水等方面的探索。在光合反应器中和自然池塘环境下养殖微藻有着重要区别，前者无论以人工光或自然光为光源，产量都比后者要高出许多，但成本也较后者高。实际上，自然环境下高密度培养有着较强的竞争优势，但缺点是环境相对复杂，所以难以控制。除了高盐、高pH等极端条件培养条件之外，一般极易受本地优势物种的入侵以及一些草食性动物的污染。陆生植物每平方米每天只能产生1 g生物质，而微藻的产量可达10～30 g，因此藻类的培养可以在不与食品行业竞争的情况下基本满足生物燃料的需求，从而节约耕地与淡水资源。但如果要使藻类燃料的价格更具竞争力，人们还需要克服一些技术上的难题（Cadoret and Bernard 2008）。

除了生物燃料外，藻类还可以用在以下四个方面：制造生物柴油、产氢气、产酒精和作为发酵生物质（Chisti 2007）。

8.4.4 藻类用于生产生物柴油

一些藻类尤其是微藻储存脂类作为能量储备。美国国家可再生资源实验室（National Renewable Energy Laboratory）筛查了3000种候选藻类（Sheehan et al. 1998），筛选出含油量占干重比例达75%的一些品种（Schenk et al. 2008）。通常，大部分最佳候选者的产油率为40%（Rodolfi et al. 2009）。为了生产生物柴油，利用诱导产油的条件培养藻类，然后收集藻体，将油脂提取出来后转化为燃料。通过理论计算，每公顷每年产超过100m³的生物柴油产量是可行的，比油棕榈树产量高出10～50m³（Chisti 2007）。为了达到这样的目标，我们需要加深对不同藻类油脂合成生化途径及油脂生理代谢的了解。硅藻（*Phaeodactylum tricornutum*）中油脂含量可以达到干重的31%（Sheehan et al. 1998），它的基因组测序已经完成（Bowler et al. 2008）。因此通过对这

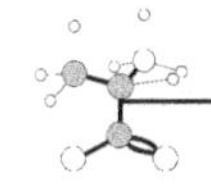

个物种的基因组进行研究，可以帮助我们了解产油藻类的生化机理。

8.4.5 藻类用于生产酒精

生产生物燃料的另一种方法是利用蓝藻光合作用在生物体内生产酒精。其可行性已经被 Deng and Coleman（1999）证实。他们将运动发酵单胞菌（*Zymomonas mobilis*）的丙酮酸脱羧酶和乙醇脱氢酶转入聚球藻（*Synechococcus* sp. PCC 7942），该转基因菌株就可以合成酒精，并将其分泌到培养基中。据说这套系统已经可以进行商业化生产，Algenol 和 BioFields 公司计划投资 1 亿美元建设这套系统（www.algenolbiofuels.com）。该公司还声称这项技术现在也适用于在海水中生长的蓝藻。这是个有趣的先例，它暗示终有一天，海洋生物和"陆地基因"的结合能够带来更有前景的产品（反之亦然）。

8.4.6 藻类用于生产氢气

许多蓝藻和一些绿藻具有能产氢气的酶（Tamagnini et al. 2002，Schütz et al. 2004，Melis and Happe 2001），这样光合作用的能量可以直接转化为高效有用的能源。并且，因为氢气可以直接释放，使得该能源的提取十分便捷。对绿藻的研究目前主要集中在莱茵衣藻（*Chlamydomonas reinhardtii*），它有一种高效、但对氧气敏感的产氢酶（Hankamer et al. 2007）。Melis 等（2000）通过降低硫磺的摄入量，使细菌在光合作用和厌氧产氢两种状态下循环，已经在一定程度上克服了它对氧气敏感的缺点。人们开始驯化菌株使它拥有更高的产气量，比如 Surzycki et al.（2007）的工作。再比如 Polle et al.（2003）筛选具有更小捕光复合体的菌株，Kruse et al.（2005）筛选更高淀粉含量的菌株来优化产气。在蓝藻中也可以产氢气，它通常是固氮酶固氮过程的副产物，氢气常常被一种氢化酶氧化（见综述 Tamagnini et al. 2002，Sakurai and Masukawa 2007）。同绿藻一样，蓝藻产氢效率也受到氧气的抑制，因为其固氮酶也对氧气敏感。

8.4.7 藻类用于生物质发酵

培养或收集微藻和大型海藻用于生物质发酵也是可行的。它既可以用来生产酒精，也可以用来产生燃气。需要着重考虑的是，该藻是否能达到高生长速率以及具有高比例的易发酵化合物（比如淀粉和脂类）。开发高淀粉含量的莱茵衣藻就是一个很好的例子（Kruse et al. 2005）。

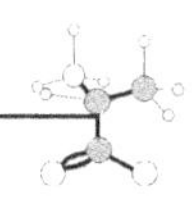

8.4.8 海洋基因组和藻类燃料

海洋基因组如何有助于生物燃料产品的开发？有一些藻类的全基因组已经完成测序，包括一些真核藻类，如绿藻门的莱茵衣藻、绿色鞭毛藻（*Ostreococcus lucimarinus*）和一种海洋真核微藻（*Ostreococcus tauri*）、红藻（*Cyanidioschyzon merolae*）、硅藻（*Thalassiosira pseudonana*）和褐指藻（*Phaeodactylum tricornutum*）（Bowler et al. 2008），以及一些蓝藻细菌（如 *Synechococcus* spp）（Six et al. 2007）。这些基因组序列可以提供多方面的信息，比如 *Ostreococcus tauri* 作为全世界已知最小的自养型真核生物，让我们进一步了解了小基因组，这样有助于我们探索生物所需要的最小基因组，人们就可以删除对生长无用的基因，提高生长效率，避免浪费与生长无关的资源。另一个例子是嗜极红藻（*Cyanidioschyzon merolae*），用非常小的基因组适应了高温、低 pH 的环境。聚球藻属（*Synechococcus*）中不同种或不同菌株对光照强度有不同的适应性，它们的基因组信息对了解其捕光天线的结构和功能带来新的启示（Six et al. 2007）。

用微藻作为生物燃料来源是个相对新的概念，所用的菌株多为野生型（Sheehan et al. 1998），因此还有很大的改进空间。改进方法可以是传统方法，也可以用基因修饰的方法。如果知道了菌株的基因组信息，对于指导传统方法和优化基因修饰的现代方法都可以起到很大的帮助。

表 8.4 海洋基因组学可以发挥关键作用的研究领域

Table 8.4 Research areas where marine genomics can play a key role

Research area	Comment
Increasing growth rates	Growth rates vary enormously between algal species and strains
Reducing photo-synthetic antenna size	Reducing antennae size would reduce internal shading and diminish photoinhibition allowing higher densities in culture
More efficient inorganic carbon uptake	Allowing for high inorganic carbon concentration in cultures will be a technological challenge; therefore traits such as C4 like mechanisms, high activity of carbonic anhydrase, or bicarbonate pumps could be utilized
Wider range of tolerance to CO_2	It is important to tolerate variable CO_2 concentrations and pH. Extremophiles such as *Galderia sulfuraria* and *Cyanidioschyzon merolae* can provide insights

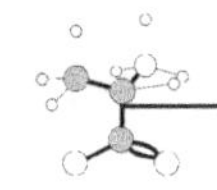

续上表

Table 8.4 (continued)

Increasing resistance to photoinhibition	High rates of photosynthesis need to be achieved even at high light intensities. Different strains of cyanobacteria with different light adaptation can guide the search for relevant genes
Increasing tolerance to oxidative stress	Intensive cultures causing increased concentrations of O_2, inducing photorespiration and pseudocyclic photophosphorylation that need to be compensated by efficient antioxidative systems
Increasing thermotolerance	Temperature can be a problem in cultures; *C. merolae* and *G. sulfuraria* could be important thermophilic models
Effective channelling of photosynthetates	Important lessons can be learned from "minimal organisms" such as *Ostreococcus tauri*

英文注释：Photo-synthetic，光合的；Tolerance，耐受性；Photoinhibition，光抑制作用；Oxidative，氧化的；Thermotolerance，耐热性；Algal，海藻的；Bicarbonate，碳酸氢盐；Extremophile，极端微生物；Photorespiration，光呼吸作用；Pseudocyclic Photophosphorylation，假环式性光合磷酸化；Antioxidative，抗氧化的；Thermophilic，嗜热的

8.4.9 藻类细胞工厂

转化藻类形成所谓"绿色细胞工厂"是个快速发展的领域，这项技术除了应用于生物燃料外，在特殊化合物、高附加值化合物、食品添加剂和生物医药等方面也能有广泛的应用（Rosenberg et al. 2008）。现在药物分子仍以天然产物提取为主，由于找到该蛋白对应的基因是可行的，那么就可以将这些基因导入到培养细胞中，作为"细胞工厂"的一部分，指导细胞合成人们需要的产品。这种通过可控细胞系统表达高附加值分子的技术策略非常有前景，不同来源的数据都显示它有数百亿美元的市场价值（Gasdaska et al. 2003，Schmidt 2004）。现有的表达系统有细菌、酵母、动物细胞和陆生植物，将它们在基因层面做了改造，以便更好地适应胰岛素、生长激素、单克隆抗体和其他药物蛋白的生产，在生产成本、产品安全性、提取纯化方便程度、产物是否易降解等方面，它们各有优劣（Leon-Banares et al. 2004，Walker et al. 2005，Cadoret et al. 2008）。大部分表达宿主都是非海洋来源的（如衣藻 *Chlamydomonas*），稳定的海洋微藻转化系统的出现只是个时间问题。目前，藻类系统操作的稳定性是主要的限制因素，但随着对基因组和发酵系统的了解，这个困难将会很快被攻克。目前已经可以成功地转化绿藻、红藻和不等鞭毛藻（Cadoret et al. 2008），在淡水藻类（Rosenberg et al. 2008，Dawson et al. 1997）和海水藻类（Geng et al. 2003，Teng et al. 2002）中，已经可以稳定地表达外源蛋白。同源重组转化在红微藻中（*Cyanidioschyzon merolae*）已有展示（Minoda et al. 2004）。在不等鞭毛体中，报道转化成功的藻

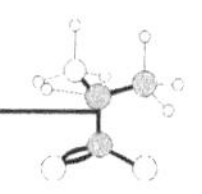

类有硅藻(Poulsen et al. 2006, Kroth 2007)和巨藻(Jiang et al. 2002, 2003)。

8.4.10 海洋真菌

陆地真菌是抗生素等生物分子的重要来源。如今,人们对海洋真菌丰富的抗肿瘤、抗菌、抗寄生虫、抗炎症和抗病毒的活性分子关注程度逐渐提高(见综述 Bhadury et al. 200, Raghukumar 2008)。比如桔青霉菌(*Penicillium citrinum*)、链孢霉菌(*Fusarium* sp)和蒙塔涅梨孢假壳菌(*Apiospora montagnei*)发现了具有潜在抗肿瘤活性的生物碱,这些菌株都是从海藻中分离出来的(Tsuda et al. 2004, Ebel 2006, Klemke et al. 2004);从另一株海洋真菌壳二孢菌(*Ascochyta salicorniae*)中则提取到了具有抗疟疾的活性物质(Osterhage et al. 2000)。另一个在奶酪和乳制品中也存在的海洋真菌汉德巴利氏酵母(*Debaryomyces hansenii*)(Dujon et al. 2004)已被测序,人们正在研究它的木糖代谢途径(Sampaio et al. 2004)。

8.4.11 后生动物

荧光蛋白的发现、描述与应用在海洋生物技术中得到很好的应用。Osamu Shimomura、Martin Chalfie 和 Roger Y. Tsien 因为对绿色荧光蛋白的研究而获得 2008 年诺贝尔化学奖。荧光蛋白最早发现于多管水母属(*Aequorea victoria*)中(Shimomura et al. 1962),这个蛋白的基因由 Prasher et al. (1985)克隆,Chalfie et al. (1994)将其用于基因在大肠杆菌和线虫中的表达研究。后来在水螅虫、珊瑚虫和桡足类海洋生物中也发现了荧光蛋白。这些荧光蛋白经过改造,可以发出不同颜色的光,产生了不同的商业产品(Mocz 2007)。

海绵及与其共生的微生物虽然合成很多有潜力的生物活性物质,但由于生物量小,成为制约其发展的瓶颈。下面几个技术的应用都有助于解决原料供给的瓶颈问题,如异源表达在宏基因组中发现的基因,在无海绵存在的条件下培养海绵共生微生物,以及建立海绵的细胞系(Wijffels 2008)。

海洋无脊椎动物来源的抗菌肽最近倍受关注,比如鲎素具有抗巴西利什曼原虫(*Leishmania braziliensis*)和克氏锥虫(*Trypanosoma cruzi*)的活性(Löfgren et al. 2008),从玻璃海鞘((Fedders et al. 2008)和贻贝(Mitta et al. 2000)中分离的多肽也具有很强的抗菌活性。

8.4.12 总结

显然,海洋环境具有巨大的生物技术开发潜能。正如这篇综述所展示的,

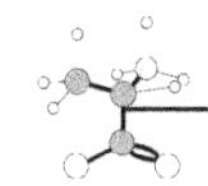

这是个有待利用的资源。虽然目前开发得很少,但成效斐然。目前海洋基因组的开发是与现代社会更加关注绿色生活、降低碳排放、提高技术利用率的理念相符合的,暗示着海洋生物技术应用的美好前景。地球上70%的面积是水,很明显,海洋能够更长久地承载人们对地球资源的需求,而且这种潜力已经显现出来了。为了在呈上升趋势的海洋研究中找到新的机会,我们需要增加科研资金,更需要海洋科学家们同心协力。自然已经教会我们未雨绸缪,对于生物技术学家,海洋环境已经展示了它丰富的功能多样性,一定能不断地满足我们的需求。我们必须以极大的热情,随时准备全力以赴。

参考文献

Allen MJ,Schroeder DC,Donkin A et al (2006a) Genome comparison of two Coccolithoviruses. Virol J 3:15

Allen MJ,Schroeder DC,Holden MT et al (2006b) Evolutionary history of the Coccolithoviridae. Mol Biol Evol 23:86 – 92

Allen MJ, Wilson WH (2008) Aquatic virus diversity accessed through omic techniques: a route map to function. Curr Opin Microbiol 11:226 – 232

Amador ML,Jimeno J,Paz-Ares L et al (2003) Progress in the development and acquisition of anticancer agents from marine sources. Ann Oncol 14:1607 – 1615

Amann RI,Ludwig W,Schleifer K – H (1995) Phylogenetic identification and in situ detection of individual microbial cells without cultivation. Microbiol Rev 59:143 – 169

Angly FE,Felts B,Breitbart M et al (2006) The marine viromes of four oceanic regions. PLoS Biol 4:e368

Bertram M,Hildebrandt P,Weiner D et al (2008) Characterization of lipases and esterases from metagenomes for lipid modification. J Am Oil Chem Soc 85:47 – 53

Bhadury P,Mohammad BT,Wright PC (2006) The current status of natural products from marine fungi and their potential as anti – infective agents. J Ind Microbiol Biotechnol 33:325 – 337

Bielaszewska M,Dobrindt U,Gärtner J et al (2007) Aspects of genome plasticity in pathogenic *Escherichia coli*. Int J Med Microbiol 297:625 – 639

Bowler C,Allen AE,Badger JH et al (2008) The *Phaeodactylum* genome reveals the evolutionary history of diatom genomes. Nature 456:239 – 244

Breitbart M,Felts B,Kelley S et al (2004) Diversity and population structure of a near-shore marine-sediment viral community. Proc Biol Sci 271:565 – 574

Breitbart M,Salamon P,Andresen B et al (2002) Genomic analysis of uncultured marine viral communities. Proc Natl Acad Sci USA 99:14250 – 14255

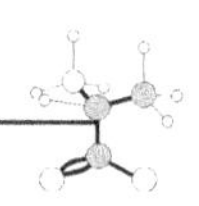

Brügger K, Chen L, Stark M et al (2007) The genome of *Hyperthermus butylicus*: a sulfur-reducing, peptide fermenting, neutrophilic Crenarchaeote growing up to 108 degrees C. Archaea 2: 127 - 135

Burkholder PR, Pfister RM, Leitz FH (1966) Production of a pyrrole antibiotic by a marine bacterium. Appl Environ Microbiol 14:649 - 653

Cadoret J - P, Bardor M, Lerouge P et al (2008) Les microalgues: Usines cellulaires productrices de molécules commerciales recombinants. Med Sci 24:375 - 382

Cadoret J - P, Bernard O (2008) La production de biocarburant lipidique avec des microalgues: promesses et défis. J Soc Biol 202:201 - 211

Chalfie M, Tu Y, Euskirchen G et al (1994) Green fluorescent protein as a marker for gene expression. Science 263:802 - 805

Chisti Y (2007) Biodiesel from microalgae. Biotechnol Adv 25:294 - 306

Chu X, He H, Guo C et al (2008) Identification of two novel esterases from a marine metagenomic library derived from South China Sea. Appl Microbiol Biotechnol 80:615 - 625

Cohen GN, Barbe V, Flament D et al (2003) An integrated analysis of the genome of the hyperthermophilic archaeon *Pyrococcus abyssi*. Mol Microbiol 47:1495 - 1512

Dawson HN, Burlingame R, Cannons AC (1997) Stable transformation of *Chlorella*: Rescue of nitrate reductase-deficient mutants with the nitrate reductase gene. Curr Microbiol 35:356 - 362

De Lorenzo V (2005) Problems with metagenomic screenings. Nat Biotech 23:1045 - 1046

Dean FB, Hosono S, Fang L et al (2002) Comprehensive human genome amplification using multiple displacement amplification. Proc Natl Acad Sci USA 99:5261 - 5266

Deckert G, Warren PV, Gaasterland T et al (1998) The complete genome of the hyperthermophilic bacterium *Aquifex aeolicus*. Nature 392:353 - 358

Deng M - D, Coleman JR (1999) Ethanol synthesis by genetic engineering in Cyanobacteria. Appl Environ Microbiol 65:523 - 528

Dujon B, Sherman D, Fischer G et al (2004) Genome evolution in yeasts. Nature 430:35 - 44

Ebel R (2006) Secondary metabolites from marine derived fungi. In: Proksch P, Müller WEG (eds) Frontiers in marine biotechnology. Horizon Bioscience, England, pp 73 - 143

Egorova K, Antranikian G (2007) Biotechnology. In: Garrett RA, Klenk HP (eds) Archaea: evolution, physiology, and molecular biology. Blackwell, Malden

Fedders H, Michalek M, Grötzinger J et al (2008) An exceptional salt-tolerant antimicrobial peptide derived from a novel gene family of haemocytes of the marine invertebrate *Ciona intestinalis*. Biochem J 416:65 - 75

Fieseler L, Hentschel U, Grozdanov L et al (2007) Widespread occurrence and genomic context of unusually small polyketide synthase genes in microbial consortia associated with marine sponges. Appl Environ Microbiol 73:2144 - 2155

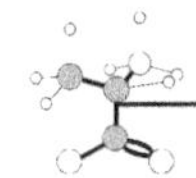

Fitz-Gibbon ST, Ladner H, Kim UJ et al (2002) Genome sequence of the hyperthermophilic crenarchaeon *Pyrobaculum aerophilum*. Proc Natl Acad Sci USA 99:984 – 989

Fukui T, Atomi H, Kanai T et al (2005) Complete genome sequence of the hyperthermophilic archaeon *Thermococcus kodakaraensis* KOD1 and comparison with *Pyrococcus* genomes. Genome Res 15:352 – 363

Fuller RW, Cardellina JH II, Jurek J et al (1994) Isolation and structure/activity features of halomon-related antitumor monoterpenes from the red alga *Portieria hornemannii*. J Med Chem 37: 4407 – 4411

Fuller RW, Cardellina II JH, Kato Y et al (1992) A pentahalogenated monoterpene from the red alga *Portieria hornemannii* produces a novel cytotoxicity profile against a diverse panel of human tumor cell lines. J Med Chem 35:3007 – 3011

Gasdaska JR, Spencer D, Dickey L (2003) Advantages of therapeutic protein production in the aquatic plant *Lemna*. Bioprocess J Mar/Apr

Geng DG, Wang YQ, Wang P et al (2003) Stable expression of hepatitis B surface antigen gene in *Dunaliella salina* (Chlorophyta). J Appl Phycol 15:451 – 456

Ghedin E, Claverie JM (2005) Mimivirus relatives in the Sargasso sea. Virol J 2:62

Giovannoni S, Stingl U (2007) The importance of culturing bacterioplankton in the 'omics' age. Nat Rev Microbiol 5:820 – 826

Glockner FO, Kube M, Bauer M et al (2003) Complete genome sequence of the marine planctomycete *Pirellula sp.* strain 1. Proc Natl Acad Sci USA 100:8298 – 8303

Gogarten JP, Hilario E. (2006) Inteins, introns, and homing endonucleases: recent revelations about the life cycle of parasitic genetic elements. BMC Evol Biol 6:94

Gonçalves LG, Lamosa P, Huber R et al (2008) Di – myo-inositol phosphate and novel UDP – sugars accumulate in the extreme hyperthermophile *Pyrolobus fumarii*. Extremophiles 12:383 – 389

Goodwin TJ, Butler MI, Poulter RT (2006) Multiple, non-allelic, intein-coding sequences in eukaryotic RNA polymerase genes. BMC Biol 4:38

Gray JS (1997) Marine biodiversity: patterns, threats and conservation needs. Biodiv Conserv 6:153 – 175

Han G, Gable K, Yan L et al (2006) Expression of a novel marine viral single-chain serine palmi – toyltransferase and construction of yeast and mammalian single-chain chimera. J Biol Chem 281:39935 – 39942

Handelsman J, Rondon MR, Brady SF et al (1998) Molecular biological access to the chemistry of unknown soil microbes: a new frontier for natural products. Chem Biol 5: R245 – R249

Hankamer B, Lehr F, Rupprecht J et al (2007) Photosynthetic biomass and H_2 production by green algae: from bioengineering to bioreactor scale-up. Physiol Plant 131:10 – 21

Hansen OC, Stougaard P (1997) Hexose oxidase from the red alga *Chondrus crispus*: purifica-

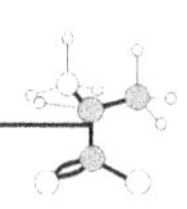

tion, molecular cloning, and expression in *Pichia pastoris*. J Biol Chem 272:11581 – 11587

Head IM, Jones DM, Röling WFM (2006) Marine microorganisms make a meal of oil. Nat Rev Microbiol 4:173 – 182

Henne A, Schmitz RA, Bomeke M et al (2000) Screening of environmental DNA libraries for the presence of genes conferring lipolytic activity on *Escherichia coli*. Appl Environ Microbiol 66: 3113 – 3116

Huber JA, Mark WDB, Morrison HG et al (2007) Microbial population structures in the deep marine biosphere. Science 318:97 – 100

Ivars – Martinez E, Martin-Cuadrado A – B, D'Auria G et al (2008) Comparative genomics of two ecotypes of the marine planktonic copiotroph *Alteromonas macleodii* suggests alternative lifestyles associated with different kinds of particulate organic matter. ISME J 2:1194 – 1212

Jeon JH, Kim JT, Kang SG et al (2009) Characterization and its potential application of two esterases derived from the arctic sediment metagenome. Mar Biotechnol 11:307 – 311

Jiang P, Qin S, Tseng CK (2002) Expression of hepatitis B surface antigen gene (HBsAg) in *Laminaria japonica* (Laminariales Phaeophyta). Chin Sci Bull 47:1438 – 1440

Jiang P, Qin S, Tseng CK (2003) Expression of the lacZ reporter gene in sporophytes of the sea-weed *Laminaria japonica* (Phaeophyceae) by gametophyte-targeted transformation. Plant Cell Rep 21:1211 – 1216

Kalyuzhnaya MG, Lapidus A, Ivanova N et al (2008) High-resolution metagenomics targets specific functional types in complex microbial communities. Nat Biotechnol 26:1029 – 1034

Kanai T, Imanaka H, Nakajima A et al (2005) Continuous hydrogen production by the hyper-ther-mophilic archaeon, Thermococcus kodakaraensis KOD1. J Biotechnol 116: 271 – 282

Kawarabayasi Y, Hino Y, Horikawa H et al (1999) Complete genome sequence of an aerobic hyper-thermophilic crenarchaeon, Aeropyrum pernix K1. DNA Res 6:145 – 152

Kawarabayasi Y, Sawada M, Horikawa H et al (1998) Complete sequence and gene organization of the genome of a hyper- thermophilic archaebacterium, *Pyrococcus horikoshii* OT3. DNA Res 5:147 – 155

Klarzynski O, Plesse B, Joubert J – M et al (2000) Linear β – 1,3 glucans are elicitors of defense responses in tobacco. Plant Physiol 124:1027 – 1038

Klemke C, Kehraus S, Wright AD et al (2004) New secondary metabolites from the endophytic fungus *Apiospora montagnei*. J Nat Prod 67:1058 – 1063

Koonin EV (2007) Metagenomic sorcery and the expanding protein universe. Nat Biotechnol 25:540 – 542

Kroth PG (2007) Genetic transformation: a tool to study protein targeting in diatoms. Methods Mol Biol 390:257 – 267

Kruse O, Rupprecht J, Bader K – P et al (2005) Improved photobiological H_2 production in

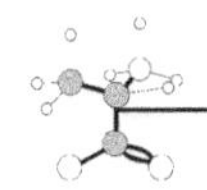

engineered green algal cells. J Biol Chem 280:34170 – 34177

La Scola B, Desnues C, Pagnier I et al (2008) The virophage as a unique parasite of the giant mimivirus. Nature 455:100 – 104

Langer M, Gabor EM, Liebeton K et al (2006) Metagenomics: an inexhaustible access to nature's diversity. Biotechnol J 1:815 – 821

Leary D, Vierros M, Hamon G et al (2009) Marine genetic resources: A review of scientific and commercial interest. Mar Policy 33:183 – 194

Lee S – H, Lee D – G, Jeon J – H et al (2008) Fibrinolytic metalloprotease and composition comprising the same. WO2008056840

Leon-Banares R, Gonzalez-Ballester D, Galvan A et al (2004) Transgenic microalgae as green cell-factories. Trends Biotechnol 22:45 – 52

Lim JK, Lee HS, Kim YJ et al (2007) Critical factors to high thermostability of an alpha-amylase from hyperthermophilic archaeon *Thermococcus onnurineus* NA1. J Microbiol Biotechnol 17:1242 – 1248

Læfgren SE, Milettib LC, Steindel M et al (2008) Trypanocidal and leishmanicidal activities of different antimicrobial peptides (AMPs) isolated from aquatic animals. Exp Parasitol 118:197 – 202

Maeder DL, Weiss RB, Dunn DM et al (1999) Divergence of the hyperthermophilic archaea *Pyrococcus furiosus* and *P. horikoshii* inferred from complete genomic sequences. Genetics 152:1299 – 1305

Marsic D, Flaman JM, Ng JD (2008) New DNA polymerase from the hyperthermophilic marine archaeon *Thermococcus thioreducens*. Extremophiles 12:775 – 788

Mayer AMS, Jacobson PB, Fenical W et al (1998) Pharmacological characterization of the pseu – dopterosins: novel anti – inflammatory natural products isolated from the Caribbean soft coral, *Pseudopterogorgia elisabethae*. Life Sci 62:PL401 – PL407

Melis A, Zhang L, Forestier M et al (2000) Sustained photobiological hydrogen gas production upon reversible inactivation of oxygen evolution in the green alga *Chlamydomonas reinhardtii*. Plant Physiol 122:127 – 136

Melis A, Happe T (2001) Hydrogen production. Green algae as a source of energy. Plant Physiol 127:740 – 748

Minoda A, Rei Sakagami R, Yagisawa F et al (2004) Improvement of ulture conditions and evi – dence for nuclear transformation by homologous recombination in a red alga, *Cyanidioschyzon merolae* 10D. Plant Cell Physiol 45:667 – 671

Mitta G, Vandenbulcke F, Roch P (2000) Original involvement of antimicrobial peptides in mussel innate immunity. FEBS Lett 486:185 – 190

Mocz G (2007) Fluorescent proteins and their use in marine biosciences, biotechnology, and

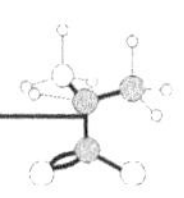

proteomics. Mar Biotech 9:305 – 328

Mueller P, Egorova K, Vorgias CE et al (2006) Cloning, overexpression, and characterization of a thermoactive nitrilase from the hyperthermophilic archaeon *Pyrococcus abyssi*. Protein Exp Purif 47:672 – 681

Nagasaki K, Shirai Y, Tomaru Y et al (2005) Algal viruses with distinct intraspecies host specificities include identical intein elements. Appl Environ Microbiol 71:3599 – 3607

Nakagawa S, Takaki Y, Shimamura S et al (2007) Deep – sea vent epsilon-proteobacterial genomes provide insights into emergence of pathogens. Proc Natl Acad Sci USA 104:12146 – 12150

Nelson KE, Clayton RA, Gill SR, et al (1999) Evidence for lateral gene transfer between Archaea and bacteria from genome sequence of Thermotoga maritima. Nature 399:323 – 329

Newman DJ, Cragg GM (2004) Marine natural products and related compounds in clinical and advanced preclinical trials. J Nat Prod 67:1216 – 1238

Ogata H, Raoult D, Claverie JM (2005) A new example of viral intein in Mimivirus. Virol J 2:8 Osterhage C, Kaminsky R, Konig GM et al (2000) Ascosalipyrrolidonone A, an antimicrobial alkaloid from the obligate marine fungus *Ascochyta salicorniae*. J Org Chem 65:6412 – 6417

Pace NR, Stahl DA, Lane DJ et al (1986) The analysis of natural microbial populations by ribosomal RNA. Adv Microbiol Ecol 9:1 – 55

Perler FB (2002) InBase: the intein database. Nucleic Acids Res 30:383 – 384

Polle JEW, Kanakagiri SD, Melis A (2003) tla1, a DNA insertional transformant of the green alga *Chlamydomonas reinhardtii* with a truncated light-harvesting chlorophyll antenna size. Planta 217:49 – 59

Poulsen N, Chesley PM, Kröger N (2006) Molecular genetic manipulation of the diatom *Thalassiosira pseudonana* (Bacillariophyceae). J Phycol 42:1059 – 1065

Prasher D, McCann RO, Cormier MJ (1985) Cloning and expression of the cDNA coding foe aequorin, a bioluminescent calcium-binding protein. Biochem Biophys Res Com 126:1259 – 1268

Quince C, Curtis TP, Sloan WT (2008) The rational exploration of microbial diversity. ISME J 2:997 – 1006

Rabus R, Ruepp A, Frickey T et al (2004) The genome of *Desulfotalea psychrophila*, a sulfate-reducing bacterium from permanently cold Arctic sediments. Environ Microbiol 6:887 – 902

Raghukumar C (2008) Marine fungal biotechnology: an ecological perspective. Fungal Divers 31:19 – 35

Raoult D, Audic S, Robert C et al (2004) The 1. 2 – megabase genome sequence of Mimivirus. Science 306:1344 – 1350

Rappé MS, Giovannoni SJ (2003) The uncultured microbial majority. Annu Rev Microbiol 57:369 – 394

Robertson DE, Chaplin JA, DeSantis G et al (2004) Exploring nitrilase sequence space for en-

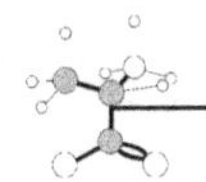

antioselective catalysis. Appl Environ Microbiol 70:2429 – 2436

Rodolfi L, Zittelli GC, Bassi N et al (2009) Microalgae for oil: strain selection, induction of lipid synthesis and outdoor mass cultivation in a low-cost photobioreactor. Biotechnol Bioeng 102: 100 – 112

Rondon MR, August PR, Bettermann AD et al (2000) Cloning the soil metagenome: a strategy for accessing the genetic and functional diversity of uncultured microorganisms. Appl Environ Microbiol 66:2541 – 2547

Rosenberg JN, Oyler GA, Wilkinson L et al (2008) A green light for engineered algae: redirecting metabolism to fuel a biotechnology revolution. Curr Opin Biotechnol 19:430 – 436

Rubin EM (2008) Genomics of cellulosic biofuels. Nature 454:841 – 845

Ruby EG, Urbanowski M, Campbell J et al (2005) Complete genome sequence of *Vibrio fis – cheri*: a symbiotic bacterium with pathogenic congeners. Proc Natl Acad Sci USA 102:3004 – 3009

Rusch DB, Halpern AL, Sutton G et al (2007) The Sorcerer II global ocean sampling expedition: northwest Atlantic through eastern tropical Pacific. PLoS Biol 5:e77

Sakurai H, Masukawa H (2007) Promoting R & D in photobiological hydrogen production utilizing mariculture-raised cyanobacteria. Mar Biotechnol 9:128 – 145

Sampaio FC, Torre P, Passos FML et al (2004) Xylose metabolism in *Debaryomyces hansenii* UFV – 170. Effect of the specific oxygen uptake rate. Biotechnol Prog 20:1641 – 1650

Schenk P, Thomas – Hall S, Stephens E et al (2008) Second generation biofuels: high-efficiency microalgae for biodiesel production. BioEnerg Res 1:20 – 43

Schirmer A, Gadkari R, Reeves CD et al (2005) Metagenomic analysis reveals diverse polyketide synthase gene clusters in microorganisms associated with the marine sponge *Discodermia dissoluta*. Appl Environ Microbiol 71:4840 – 4849

Schmidt FR (2004) Recombinant expression systems in the pharmaceutical industry. Microbiol Biotechnol 65:363 – 372

Schmidt TM, DeLong EF, Pace NR (1991) Analysis of a marine picoplankton community by 16S rRNA gene cloning and sequencing. J Bacteriol 173:4371 – 4378

Schneiker S, Martins SVA, Bartels D et al (2006) Genome sequence of the ubiquitous hydrocarbon-degrading marine bacterium *Alcanivorax borkumensis*. Nat Biotechnol 24:997 – 1004

Schütz K, Happe T, Troshina O et al (2004) Cyanobacterial H_2 production-a comparative analysis. Planta 218:350 – 359

Sennett SH (2001) Marine chemical ecology: application in marine biomedical prospecting. In: McClintock JB, Baker BJ (eds) Marine chemical ecology. CRC Press, Boca Ratton, FL, pp 523 – 542

Seshadri R, Kravitz SA, Smarr L et al (2007) CAMERA: a community resource for metagenomics. PLoS Biol 5:S18 – S21

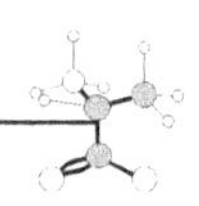

She Q, Singh RK, Confalonieri F et al (2001) The complete genome of the crenarchaeon *Sulfolobus solfataricus* P2. Proc Natl Acad Sci USA 98:7835 - 7840

Sheehan J, Dunahay T, Benemann J et al (1998) A look back at the US Department of Energy's aquatic species program: Biodiesel from Algae. US Report NREL/TP - 580 - 24190 Golden, US Department of Energy:323

Shimomura O, Johnson FH, Saiga Y (1962) Extraction, purification and properties of aequorin, a bioluminescent protein from the luminous hydromedusan *Aequorea*. J Cell Comp Physiol 59: 223 - 239

Short JM, Marss B, Stein JL (1997) Screening methods for enzymes and enzyme kits. WO9704077 (A1)

Short JM (1999) Protein activity screening of clones having DNA from uncultivated microorgan-isms. US5958672

Six C, Thomas J - C, Garczarek L et al (2007) Diversity and evolution of phycobilisomes in marine *Synechococcus* spp: a comparative genomics study. Genome Biol 8: R259

Sogin ML, Morrison HG, Huber JA et al (2006) Microbial diversity in the deep sea and the underexplored "rare biosphere". Proc Natl Acad Sci USA 103:12115 - 12120

Surzycki R, Cournac L, Peltier G et al (2007) Potential for hydrogen production with inducible chloroplast gene expression in *Chlamydomonas*. Proc Natl Acad Sci USA 104:17548 - 17553

Suttle CA (2005) Viruses in the sea. Nature 437:356 - 361

Suttle CA (2007) Marine viruses - major players in the global ecosystem. Nat Rev Microbiol 5:801 - 812

Takami H, Nakasone K, Takaki Y et al (2000) Complete genome sequence of the alkaliphilic bac-terium *Bacillus halodurans* and genomic sequence comparison with *Bacillus subtilis*. Nucleic Acids Res 28:4317 - 4331

Tamagnini P, Axelsson R, Lindberg P et al (2002) Hydrogenases and hydrogen metabolism of Cyanobacteria. Microbiol Mol Biol Rev 66:1 - 20

Teng C, Qin S, Liu J et al (2002) Transient expression of lacZ in bombarded unicellular green alga *Haematococcus pluvialis*. J Appl Phycol 14:497 - 500

Tonon T, Harvey D, Qing R et al (2004a) Identification of a fatty acid 11 - desaturase from the microalga *Thalassiosira pseudonana*. FEBS Lett 563:28 - 34

Tonon T, Qing R, Harvey D et al (2005) Identification of a long-chain polyunsaturated fatty acid acyl-coenzyme A synthetase from the diatom *Thalassiosira pseudonana*. Plant Physiol 138:402 - 408

Tonon T, Sayanova O, Michaelson LV et al (2004b) Fatty acid desaturases from the microalga *Thalassiosira pseudonana*. FEBS J 272:3401 - 3412

Torsvik V (1980) Isolation of bacterial DNA from soil. Soil Biol Biochem 12:15 - 21

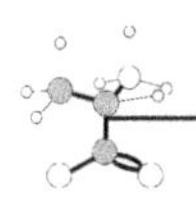

Tsiroulnikov K, Rezai H, Bonch-Osmolovskaya E et al (2004) Hydrolysis of the amyloid prion protein and nonpathogenic meat and bone meal by anaerobic thermophilic prokaryotes and streptomyces subspecies. J Agric Food Chem 52:6353 – 6360

Tsuda M, Kasai Y, Komatsu K et al (2004) Citrinadin A, a novel pentacyclic alkaloid from marine-derived fungus *Penicillium citrinum*. Org Lett 6:3087 – 3089

Uchiyama T, Abe T, Ikemura T et al (2005) Substrate-induced gene-expression screening of environmental metagenome libraries for isolation of catabolic genes. Nat Biotech 23:88 – 93

VanFossen AL, Lewis DL, Nichols JD et al (2008) Polysaccharide degradation and synthesis by extremely thermophilic anaerobes. Ann NY Acad Sci 1125:322 – 337

Venter JC, Remington K, Heidelberg JF et al (2004) Environmental genome shotgun sequencing of the Sargasso Sea. Science 304:66 – 74

Vestergaard G, Aramayo R, Basta T et al (2008) Structure of the *Acidianus* filamentous virus 3 and comparative genomics of related archaeal lipothrixviruses. J Virol 82:371 – 381

Walker TL, Collet C, Purton S (2005) Algal transgenics in the genomic era. J Phycol 41:1077 – 1093

Weiner D, Short JM, Hitchman T et al (2007) P450 enzymes, nucleic acids encoding them and methods of making and using them. US2007231820(A1)

Wijffels RH (2008) Potential of sponges and microalgae for marine biotechnology. Trends Biotechnol 26:26 – 31

Wilson WH, Schroeder DC, Allen MJ et al (2005) Complete genome sequence and lytic phase transcription profile of a Coccolithovirus. Science 309:1090 – 1092

Yin Y, Fischer D (2008) Identification and investigation of ORFans in the viral world. BMC Genomics 9:24

Yooseph S, Sutton G, Rusch DB et al (2007) The Sorcerer II Global Ocean Sampling expedition: expanding the universe of protein families. PLoS Biol 5:S56 – S90

Yun J, Ryu S (2005) Screening for novel enzymes from metagenome and SIGEX, as a way to improve it. Microb Cell Fact 4:8

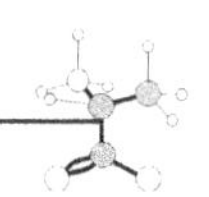

第9章　实践指南：基因组学技术及其在海洋生物学方面的应用

摘要：近年来，运用于基因组和后基因组研究的高通量技术对海洋科学产生了深远影响。如今，大规模平行DNA测序和杂交技术不仅能够鉴定基因库，甚至还能分析物种内的基因调控网络。高通量的测序技术产生的海量数据，需要强大的生物信息平台，用于这些数据的存储及分析。生物信息学依赖有效的数据分析算法、易于使用的分析工具和应用程序，以及强大的硬件设备以支持基因组层面的分析。

本章将首先简要介绍与功能及结构基因组学最相关的一些生物信息学主题，接着介绍做一个基因组分析项目的步骤（诸如测序及数据管理），后续内容将向读者呈现应用于海洋基因组学分析的不同生物信息技术手段。

第一部分聚焦于数据的产生，不仅介绍经典的Sanger法和鸟枪法基因组测序技术，亦对当前新一代测序技术做一个简短概述。第二部分简要介绍基于生物信息学应用的数据管理的相关概念。第三部分主要介绍基因组序列分析的基本原理，并阐述EST聚类和组装、基因预测、基因功能注释和分类以及全基因组注释等方面的内容。第四部分对利用微阵列杂交技术进行转录组数据分析进行综述。本章最后简要介绍微阵列技术以及用于图像处理、数据标准化、显著性检验和聚类分析的最先进方法。

9.1　序列数据产生

在遗传学研究早期，科学家们一次只能研究几个基因。现在微生物基因组分析快速发展，自1995年第一个被完整测序的微生物——流感嗜血杆菌（*Haemophilus influenzae*）的全基因组序列（Fleischmann et al. 1995）被公布以来，成百上千的微生物基因组已经被测序。截至2008年1月，约700个全基因组已经发表，另外基因组在线数据库（Genomes OnLine Database，GOLD；

http://www.genomesonline.org）还列出了正在进行的3 250个基因组测序项目。2007 年 9 月的 GOLD 中（Liolios et al. 2008），记录的古细菌和细菌测序项目是 1 950 个，而新的测序技术平台如焦磷酸测序的出现又极大促进了新的微生物基因组测序项目的开展。GOLD 网站公布的 134 个项目使用 454 技术平台测序，这是全基因组鸟枪测序的一种新技术平台（Liolios et al. 2008）。

9.1.1　经典基因组测序技术

虽然有多种测序技术，但 Sanger et al.（1977）等发明的 DNA 序列末端终止法测序技术，仍然是基因测序的基础技术，尽管该方法已经使用超过 25 年。

9.1.1.1　Sanger 法

经典 DNA 测序是通过双脱氧链终止法得以实现（Sanger et al. 1977）。双脱氧链终止法又称 Sanger 法，利用与待测序 DNA 序列互补的短 DNA 引物、DNA 合成酶和四种核苷酸来实现一段单链 DNA 序列的复制。测序时混合有常规的脱氧核苷酸（dNTP）和特殊的双脱氧核苷酸（ddNTP）。由于 ddNTP 缺乏延伸所需要的 3′—OH 基团，当 ddNTP 掺入到 DNA 链的末端时，就会阻止后续的核苷酸结合，进而导致整条 DNA 链的合成反应终止。另外，不同的荧光基团标签结合在四种不同的 ddNTP 上，作为检测哪种 ddNTP 结合上去的标记（Prober et al. 1987）。经过重复的合成循环，就产生了一系列长度不等的以 ddNTP 为 3′端的 DNA 片段，其末端都有一个可供后续检测的荧光标记。

扩增的 DNA 可以通过凝胶电泳按长度大小分开。早期的 Sanger 测序技术使用垂直聚丙烯酰胺凝胶电泳的方法进行分离，胶块需要手工操作制备。现在，毛细管电泳测序仪能够让 DNA 片段在一个毛细管中电泳来实现，然后通过一台紫外激光器来检测聚合反应中掺入的荧光标记。最新的仪器（如 ABI 3730XL），可以同时分析 96 个样本，并用荧光标记代替 Sanger 法中同位素标记，使 DNA 测序准确性进一步提高。DNA 片段上带有 4 种不同荧光基团的双脱氧核苷酸（dd NTP）的碱基，在通过毛细管时，被激光激发出不同颜色的荧光，被激光检测系统识别，并直接翻译成 DNA 序列。

9.1.1.2　鸟枪法技术

Sanger 测序法可产生长达 1 000 个 bp 的序列数据。为了测定全基因组，需要采用基于此方法的特殊策略。对于细菌基因组和小的真核生物全基因组

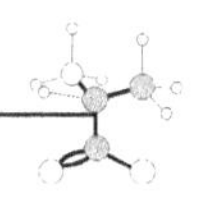

测序，一般选择全基因组鸟枪测序技术。该种技术将基因组 DNA 序列片段化，形成特定长度的 DNA 片段，再把这些片段克隆到测序载体后转化到大肠杆菌（*E. coli*）中。随机选取产生的重组克隆并进行测序，再根据测序的序列间重叠关系将序列拼接成连续区段（重叠群 contig）。对于更大的基因组，可采用分层鸟枪法（Green 2002），即基因组被切割成大的片段，然后将大的片段克隆到细菌人工染色体（BAC）中，通过 PCR 或高强度的杂交实验按最小重叠度对 BAC 进行排序，然后对选择的亚克隆（最小重叠度）通过鸟枪法对每个 DNA 分子单独测序（Kaiser et al. 2003）。

9.1.1.3 细菌基因组组装和加工

在人类基因组计划实施的过程中，研究人员开发了多种生物信息学工具。例如碱基荧光信号读取和序列峰图优化剪接软件 PHRED（Ewing et al. 1998），DNA 序列组装程序 PHRAP（Green1996）和 ACP3（Huang and Madan 1999）以及基因组序列加工程序 Consed（Gordon te al. 1998）。Kaiser et al.（2003）在比勒费尔德大学的生物信息学平台上开发出一种全基因组鸟枪法测序的优化方法，该方法有机地结合了高通量鸟枪测序数据的快速产生和序列加工及后续大片段插入的 BAC 或 Fosmid 克隆文库的验证过程。这种方法的目标是在时间和费用都合算的情况下，产出高质量的细菌基因组序列。测序策略主要由两步构成：①通过 Sanger 测序法产生高通量的鸟枪读取片段（建议至少 8X 的测序深度）；②手工的序列法和拼接步骤，使用 BAC 或 Fosmid 文库的末端序列信息，通过引物步查法来完成序列读长的拼接。

为监控第一步（即高通量鸟枪法测序阶段），一种被称为 SAMS（序列分析和管理系统）的生物信息学工具被开发出来（Bekel et al. 2009）。SAMS 处理原始序列数据（如 scf 文件）的步骤如下：首先，标准化步骤涉及使用软件 PHRED（Ewing et al. 1998）实现碱基读取和质量剪切（参见 9.3.1 节）；接着，进行 BLAST（Altschul et al. 1990）去除载体序列（另见 9.3.1 节）。为确定项目的整体进度，需对给定的序列数据的子集进行装配。通过使用 CAP3 和 PHRAP 组装工具产生的数据，对 gaps 数目与测序读长数目的关系绘制成一个类似于 Lander-Waterman（Lander and Waterman 1988）的统计分析图。

第二步涉及序列编辑和手工装配检验，需要使用 Consed 软件包（Gordon et al. 1998）。为了拼接和优化现有的重叠群，可使用 Autofinish 软件（Gordon et al. 2001）。使用 BACCardI 工具可以进行重叠群的拼接（Bartels et al. 2005），还可以自动生成 BAC 和 Fosmid 图谱。

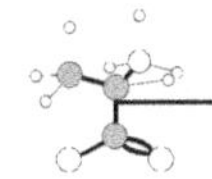

这种基于生物信息学分析流程的全基因组鸟枪法测序装配的优化方案，已成功地应用于不同的细菌全基因组计划中，包括农业、环境和生物技术相关细菌的基因组研究，如可降解石油的海洋嗜油菌（Schneiker et al. 2006）、植物病菌（Gartemann et al. 2008，Thieme et al. 2005，Vorhölter et al. 2008）、植物益生菌（Krause et al. 2006）和生物技术相关细菌（Schneiker et al. 2007）。

高通量鸟枪测序步骤现在逐渐被高通量平行测序技术所取代，不过对于全基因组测序项目，序列的装配和加工仍然是个难点。

9.1.2　第二代测序技术

基于 Sanger 法的测序技术不能无限地改良。克隆偏差和很难对含有很强二级结构的基因组区段进行测序（“强制终止”），制约了基于 Sanger 法测序的序列组装质量。一些非常有前景的方法是大规模并行化测序。用于体外克隆扩增的乳液 PCR 法被用于焦磷酸测序技术（Margulies et al. 2005，由 454 Life Sciences 商业化，后被 Roche 收购）、Dolony 测序法（Shendure et al. 2005）和 SOLiD 系统（由 Agencourt 公司开发，后被美国 Applied Biosystems 公司并购）。由 Illumina 公司发布的单个碱基延伸系统使用了“桥式 PCR”的方法。Roche/454 公司的 454 GS20 测序平台被证明可以有效地测序细菌的全基因组（Goldberg et al. 2006），同时，还可以通过消除宏基因组中的各种克隆问题来测序环境样本。目前有三种可进行大规模平行 DNA 测序的平台，它们是：Roche/454 GS FLX 系统（标准和钛系列），该系统相比 GS20 系统，测序读长更长；Illumina/Solexa 的 Genome Analyser 和 Applied Biosystems 的 SOLiD™ 系统。表格 9.1 比较了这三种大规模平行化测序技术以及在上面 9.1.1 章节部分提到的使用 Sanger 测序法和毛细管电泳的第一代测序技术。

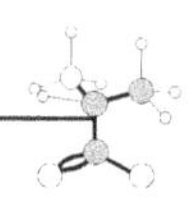

表 9.1　基因组测序技术的比较（第二代测序技术参照文献 Millar et al. 2008）

Table 9.1 Genome sequencing technologies comparison (next generation technologies according to Millar et al. 2008)

Genome sequencing technologies				
	Second or Next generation			First generation
Company	Roche®	Illumina®	Applied Biosystems®	Applied Biosystems®
Machine	Genome Sequencer FLX (standard series)	Genome analyser	SOLiD gene sequencer	3730xl DNA analyser
	Emulsion PCR of bead anchored oligos and pyrosequencing using light emission	Solid-phase -anchored oligo bridge amplification and sequencing with reversible dNTP terminators	Paired-end oligo cloning. Emulsion PCR of bead-anchored oligos. Fluorescent oligo ligation and detection	Sanger dideoxy chemistry and capillary array electrophoresis-based DNA analyser
Read length	~250 bp	~50 bp	~35 bp	~900 bp
Number of reads/run	400,000	40,000,000	85,000,000 (mate-pair run)	96
Raw data	100 MB/run/7.5 h	1 GB/run/ 67–91 h	1 GB/run/4 days for fragment library, 8 days for paired library	1 MB/day and system
Future prospects	500 bases reads; 1 GB per run (Titanium series, since October 2008 on the market)	> 6 GB per run	50 bp single read; 6 Gb/run	Up to 1,100

英文注释：Genome Sequencing Technologies，基因组测序技术；Second or Next Generation，第二代或新一代；First Generation，第一代；Read Length，能读取的碱基的个数；Number of Reads/run，每次测序能读取序列片段数；Raw Data，每次产生的下机数据；Future Prospects，未来前景

大规模平行 DNA 测序技术减少了基因组测序所耗费的时间和成本。以下段落将简要介绍每种方法。读者如想了解更多关于第二代测序方法的信息，可以阅读综述（如 Mardis 2008，Millar et al. 2008），里面对测序技术的最新领域做了详细的介绍。

9.1.2.1　焦磷酸测序（或称 454 测序）

美国 454 Life Sciences 公司开发了一种可扩展的平行测序系统—454 焦磷

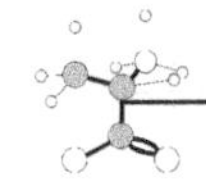

酸测序系统，该测序方法的详细介绍见 Margulies and colleagues（2005）。简单地说，454 的初代产品 GS20，在 4 小时内可测出 2500 万个碱基，准确率可达到 99% 以上，该技术无需文库构建，没有克隆误差。这种机器对固定在每个微珠上一条单链分子在乳液中形成的油包小液滴（Dressman et al. 2003）进行 DNA 扩增，（被称为乳液聚合酶链式反应，简称乳液 PCR；Nakano et al. 2003），并使用了全新的光学纤维玻片。使用的技术包括焦磷酸测序和一台用于合成测序的设备，在测序中使用了优化了载体支撑物和微滴容量的焦磷酸测序步骤（Ronaghi et al. 1998，Margulies et al. 2005）。在焦磷酸测序中，DNA 合成酶每延伸一个碱基就会产生一份焦磷酸，它引发了一系列级联反应，最终由萤火虫荧光素酶处理生成光信号，光信号的强度与合成的核苷酸数目成正比（Mardis 2008）。

关于 454 系统的使用、通量、准确率和稳健性方面的更多细节见 Margulies 等在 2005 年发表的文章。自从和罗氏诊断公司合作，454 焦磷酸测序方法经历了快速发展时期，于 2005 年推出 GS20 测序仪，于 2007 年推出 GS FLX 测序仪（见表 9.1）。新的 GS FLX 系统钛系列在每 10 小时的测序反应中，可产生高达 100 万的序列（片段）和 1GB 的数据。

关于 454 测序技术和 Roche GS FLX 测序仪的更多细节信息和最近发展情况，可访问网址 http://www. 454. com。

9.1.2.2 Illumina 测序技术

Illumina 测序技术通过可逆的终止化学反应，实现对数百万的序列片段进行大规模平行测序（Ju et al. 2006）。这项技术依赖于测序时将基因组 DNA 的随机片段附着到光学透明的玻璃表面（即 Flow cell），和随后的固相阶段扩增（也称为桥式扩增；Adams et al. 1997，Fedurco et al. 2006），可形成含有大于 5 000万个簇的超高密度测序 flow cell，每个簇包含数千份相同模板的拷贝。然后利用四种带有可洗脱荧光基团的特殊荧光标记 dNTP，通过可逆性终止反应对高密度 DNA 序列簇测序，产生 1GB 左右的原始数据（表 9.1）。这种新的测序方法保证了很高的准确度和真正意义的单个碱基依次测序，减少了由特定序列组成引起的测序错误，并能测通单核苷重复片段和短重复序列。

当第一轮测序完成后，通过双端测序模式（Paired-End Module），引导互补链原位再生，保证第二轮从模板链的反向测序 36bp 以上。这种双端测序方法提高了数据产量，可产生 3GB 以上的双端测序数据。

有关 Illumina 测序技术和 Illumina Genome Analyser 的更详细信息和最新发展情况请参阅 Illumina 网页：http://www. illumina. com。Illumina 测序平台包括

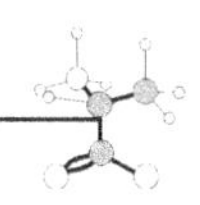

来源于 Solexa、Lynx 和 Manteia SA 公司的融合技术。

9.1.2.3 SOLiD™系统

Applied Biosystems（ABI）开发的 SOLiD 系统（支持寡核苷酸的连接和检测系统）是一个遗传学分析平台，能够对连接在磁珠上的克隆扩增 DNA 片段进行大规模的并行测序。SOLiD 测序方法是以染料标记的寡核苷酸进行连续的连接反应为基础。该公司声称其技术可提供无可匹敌的准确度、超高通量和应用灵活性。SOLiD 系统采用双碱基编码技术，即检测每个碱基两次的一项专有机制。SOLiD 系统单次运行可以产生大于 1 GB 的序列数据（见表 9.1）。在系统可拓展性发面，增加磁珠产量、改善微珠富集方法、提高微珠密度以及优化软件，能使平台轻松升级到更高的通量。双碱基编码算法能够过滤掉原始数据中的测序错误，这为测序提供了内置式的纠错机制。

有关 SOLiD 测序技术和 ABI SOLID 系统的更详细信息和最新发展可访问 Applied Biosystems 公司网页：https://products.appliedbiosystems.com/index.cfm 或 http://solid.appliedbiosystems.com

9.1.3 其他新的高级 DNA 测序方法

2007 年 8 月，美国国立卫生研究院（NIH）发布新闻宣布数个新的课题基金，以促进下一代测序技术的发展（http://www.genome.gov）。Shendure et al.（2008）对 DNA 测序策略，包括微电泳和质谱分析法等方法进行了综述。这里基于 Shendure et al.（2004，2008）两篇文献对三种不同的 DNA 测序方法，即聚合酶克隆测序、杂交法测序和纳米孔测序，进行概述。

9.1.3.1 开源式“聚合酶克隆测序”系统

使用“聚合酶克隆循环合成测序”方法，使得能够以每百万碱基测序中错误低于一个的精确度，对一个大肠杆菌（*Escherichia coli*）进化品系重测序（Shendure et al. 2005）。聚合酶克隆重测序涉及在 1 微米磁珠上进行乳液聚合酶链式反应，对短的 DNA 片段进行扩增（见 9.1.2 节焦磷酸测序部分），然后这些磁珠被固定在聚丙烯酰胺凝胶上，并进行另一种酶法测序（被称为“连接法测序”）。在每个连接循环后，通过四色（例如红色表示腺嘌呤；绿色，胞嘧啶；蓝色，鸟嘌呤；黄色，胸腺嘧啶）荧光显微镜检测每个磁珠释放的荧光波长。“聚合酶克隆测序”方法是开源式的平台，已经被许可用于 Applied Biosystems/Agencourt 的进一步发展。

9.1.3.2 杂交法测序

和本章其他已描述的测序方法比较，杂交测序是一种将目的 DNA 差异杂

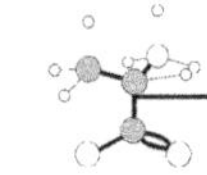

交到一系列固化的寡核苷酸探针的非酶测序法。这种技术最近被成功应用在重测序中。

一项研究使用 Affymetrix 芯片对两种酿酒酵母（*Saccharomyces cerevisae*）品系进行重测序（Gresham et al. 2006），检测到二个品系之间大约30 000个已知单核苷酸多态性位点（覆盖90%以上）。另一项研究通过使用 NimbleGen 重测序芯片对 5 个大肠杆菌（*Escherichia coli*）品系进行研究（Herring et al. 2006），以检测在适当添加甘油的培养基上生长时，发生有利于适应性生长的自发性突变。

9.1.3.3 纳米孔测序技术

安捷伦科技公司和一些学术研究团队正在开发纳米孔测序技术。这种方法的目标是运用电场，对通过小孔道或纳米孔上的单个碱基直接进行测序，不需要扩增。这项技术还在开发阶段，对单个碱基的检测精度还不够。

9.1.4 结论

DNA 测序技术发展很快。因此我们对现有可利用的测序技术进行的概述，主要集中在已经开发并且广泛应用的技术。尽管如此，我们也提到新出现的技术，虽然这些技术在未来几年离成为“标准”测序技术还很远。

无论使用何种测序技术，开发测序产生的巨量数据分析的生物信息学工具仍然是未来几年的关键问题。

采用何种测序技术，同样取决于课题可用的经费和从事研究的类型（*de novo* 测序、重测序、宏基因组、序列标签基因表达等）。将二代测序技术的高覆盖度和 Sanger 测序的低覆盖度结合在一起的混合策略常常是最佳方案，这样一方面能够保证足够的测序深度，另一方面有利于大片段的组装和 Scaffold 的装配完成。

9.2 生物信息学应用数据管理

新的测序技术不断发展，产生了大量的序列数据，这些序列数据又可衍生新的数据。处理这些海量数据，不仅需要有大量的计算资源，同时还需要强大的数据存储能力。这一节将对筹备基因组测序项目时需要重点考虑的问题做一个概述。

9.2.1 数据建模和存储

对于经常被访问的数据，结构化和组织良好的数据存储系统是非常必要

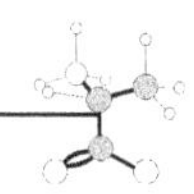

的。这不仅涉及开发形式化的数据描述模型、关联的元数据和需要存储的现成实体关系，同时也要评估将会产生的量和需要的资源存储配置。

在传统的方案中，核酸和蛋白序列可以很容易保存为按顺序排列的只包含序列数据的文本文档。通常创建元数据（如索引）以便更快地访问单条记录。但是，这种方法虽然对某种类型的数据非常有效，但对其他信息（如基因组注释数据）却不可行。当众多的关系存在于数据中，关系型数据库管理系统（Codd 1990）或其他结构化存储可能更合适。

对于经常更新和变化的数据，集中式存储系统能够更好地让数据分布式拷贝在每个本地系统上。但是，这样的中央存储模式，当面临很多的子系统对存储的数据进行存取使用时，很容易遇到瓶颈（例如在许多计算机访问一个中央序列数据库的集群环境下）。因此，任何解决方案，我们都有必要考虑其所有潜在的缺点和困难。

对于那些不经常访问的数据，我们可能更容易按需求重新创建它们，而不需要全部存储。还有一个选择是提取相关的细节，仅存储原始数据的一个子集，用于后续的数据处理。

另一个必须考虑的重要方面是数据的可用性。这不仅包括制定一个访问控制方案，通过这个规定谁能访问数据，如何访问数据；还包括制定数据的存储时间，关键是要确定数据需要存储的时间，到期后应该怎么处理。可能的解决方案是要么删除数据，要么将数据转移到别的存储单元进行长期归档。

9.2.2 数据访问

数据访问可以通过多种手段来实现。数据可能仅仅被本地安装的应用程序处理，或者被非本地用户通过网站或通过其他应用程序界面（例如网络服务器）访问。

传统的数据访问是由安装在计算机系统独立的应用程序来实现，对本地存储的数据（如序列数据库或其他像微阵列数据等）资源进行处理或提取所需信息。

虽然这种方法很简单，但还是存在几个缺点：无论是软件本身还是相关数据对用户都有一定的维护费用要求，例如当出现新版本的软件时就需要更新，或与经常变化的数据库相关的软件需要维护。当涉及在不同的系统架构和操作系统的环境中使用这些软件时，维护将成为一个日益严重的问题。

在这样的一个典型环境中，不同操作系统和软件能够通过基于网络的应用程序被使用，用于不同研究人员之间的跨平台合作。这样他们使用集中存

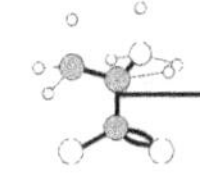

储的数据从事共同的项目。此外，当发布相关工具不合适（例如由于许可限制）或不实际时（在大型或频繁更新的数据上操作的工具），通常使用网络界面为公众提供新开发的工具。

网络服务为访问信息提供一套标准化方法，或通过基于XML语言进行信息交换。作为一个访问远程资源的简单途径，网络服务在生物信息学领域越来越流行，它可以执行计算量较大的运算而无需提供必要的硬件设备，或访问大型的序列数据库而不需要在本地存储这些数据（图9.1）。相对于基于网络的界面，网络服务可以很容易地被融入到其他应用并扩展它们的功能。因为网络服务可以被有网络连接的任何系统访问，相对于本地安装的工具，其结合了基于网络应用的优点和较低的维护成本。大量不同的公开提供的网络服务，让研究者简便地使用各种工具和信息，帮助他们分析和评价他们的数据。例如BioMOBY系统（Wilkinson and Links 2002）整合了大量的数据资源和数据分析软件，可通过网络服务的形式进行访问。

9.2.3 常用的文件格式

为了满足生物信息学家的需求，便于研究者数据共享多种多样的文件格式被开发出来。

由NCBI首先使用的FASTA格式，已经成为原始数据交换的标准格式。一个FASTA文件一般包含一条或者数条序列信息，序列文件的第一行是由大于号“>”开头的任意文字说明，用于序列标记。从第二行开始为可读格式的核苷酸或氨基酸序列本身，使用IUBMB/IUPAC标准代码（http://www.chem.qmul.ac.uk/iupac/jcbn/）确定的单个编码符号。

其他常用文件格式包括EMBL和GenBank格式。两种格式不只是包含简单的序列数据，通常可用于储存（比如一个基因组带有相应注释信息的核苷酸或蛋白质序列；见9.3.5.1节）。

在生物信息学领域，一些模块（如BioJava或BioPerl；Mangalam 2002）被开发出来，在最常用的编程语言中运用，从而能够很简单方便地访问和处理以上格式的数据。

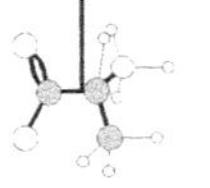

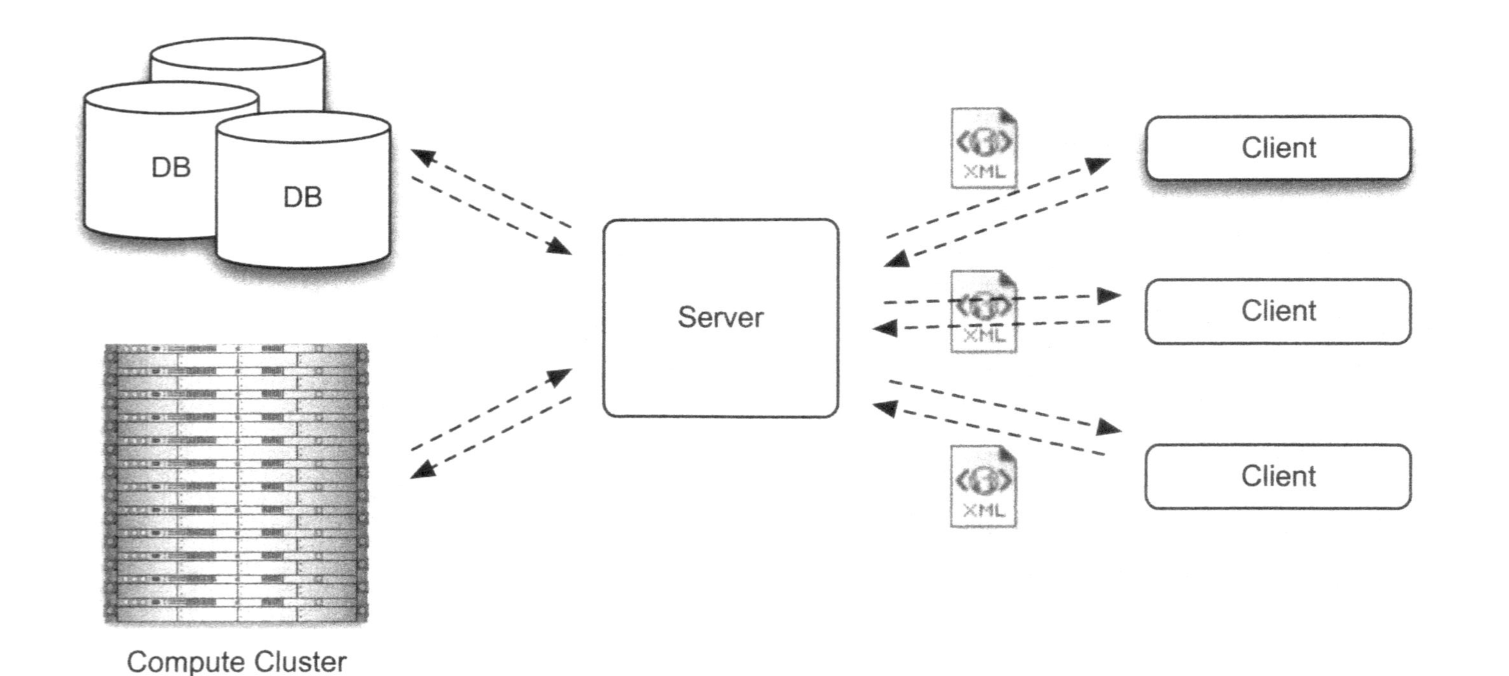

Fig. 9.1 The Web Services server provides access to stored data or compute resources. Clients can access these resources by exchanging XML-based messages with the Web Services server

图9.1 网络服务器提供存储数据或计算资源的访问。用户可通过与网络服务器进行基于XML的信息交换来获取这些资源

英文注释：DB：Database，数据库；Server，服务器；Compute Cluster，计算集群；Client，用户

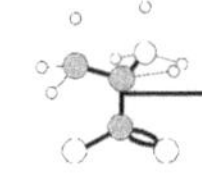

9.3 DNA 序列分析

在本节中，我们将向海洋生物研究者介绍一些可用于海洋生物基因组分析的生物信息学工具。首先，我们将介绍真核生物 EST（表达序列标签）分析，而序列数据产生那一节已经介绍过细菌基因组测序。接下来我们会关注基因预测，这是新测序的基因组注释分析的第一步，下一步对基因或组装的 EST 功能和作用进行预测和分析。我们使用多种方法和信息资源（如 InterPro）进行序列注释，接着介绍比较基因组学和功能分类，目的是鉴定序列特定区域的功能。在本节末尾我们将介绍主要的公共序列数据库和其他资源，以期为科学界提供尽可能多的序列和注释信息。

9.3.1 EST 分析

各种基因组项目产生了成千上万的表达序列标签（EST）或者鸟枪片段。EST 是通过反转录 mRNA 到 cDNA，然后测序得到的序列。EST 能够快速实惠地鉴定编码蛋白，并在某个细胞阶段或细胞种类中表达的 DNA 片段（数百个核苷酸）。经过数步分析操作，可获得高质量的序列及序列内容的总体情况。EST 序列的分析包括：①前处理；②聚类和装配；③功能分析。

9.3.1.1 基本概念

EST 前处理：将测序获得的原始 EST 序列以序列跟踪文件的形式存储，而后转化为 FASTA 格式文件，并去除（Trimming）序列两端的低质量序列（质量剪切）。

质量剪切之后，通过 BLAST 对特定的数据库（包含载体序列），移除残存在序列里克隆载体的序列，仅保留目的基因转录本的序列（载体剪切）。

EST 聚类和装配：为了减少冗余，利用一些聚类工具，通过 DNA 序列水平上的比较对 EST 序列进行分组（聚类；例如 TGICL 是美国基因组研究所 TIGR 开发的聚类工具；Pertea et al. 2003）。聚类需要选择不同的参数，包括高得分片段对（HSP）的长度、非匹配序列长度和 HSP 的相似度值（参见图 9.2）。

通过 MagaBLAST 对所有的 EST 进行同源性分析（Zhang et al. 2000），来决定 EST 所属的簇，接着这些簇被组装成暂定一致序列（TC）。这一步通过生物信息学软件（如 CAP3）来实现（Huang and Madan 1999）。可以将数个 EST 文库进行聚类和组装，将在不同文库中表达（即在不同组织或不同条件下表达）的基因组装成一个 TC 集。图 9.3 是将 EST 聚类和组装成 TC 的图解。最后生成的 TC 可以用于后续的功能分析。

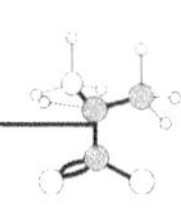

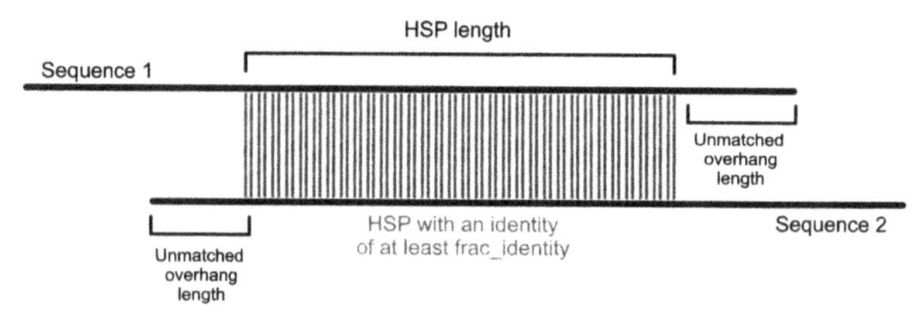

Fig. 9.2 This figure shows the overlap of two EST sequences. For the clustering procedure, the high-scoring segment pair (HSP) length, the length of the unmatched overhang and the percentage of the identity of the sequences have to be defined

图9.2　此图显示两个EST序列间的重叠。在聚类过程中，需要确定高得分片段对（HSP）的长度、非配对区域的长度和序列一致性的百分比。

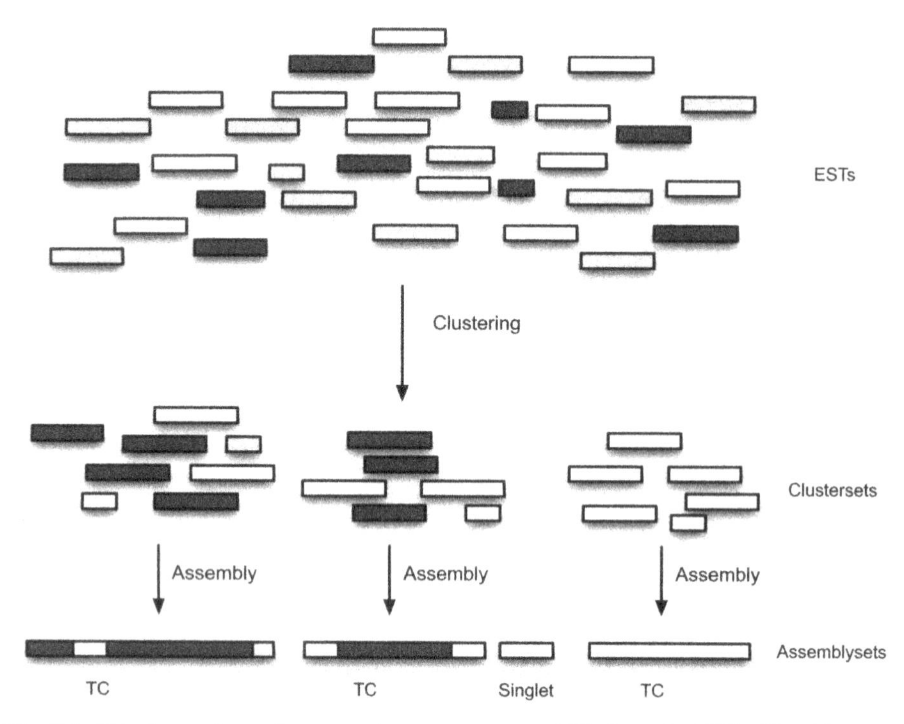

Fig. 9.3 This figure shows the processing steps to create TCs from the ESTs. The ESTs are clustered according to the similarity of their nucleotide sequence. The clusters are assembled afterwards to produce TCs plus singletons. The different colours indicate the different EST libraries

图9.3　此图显示EST聚类装配生成暂定一致序列的步骤。EST通过核苷酸序列的相似性聚类成簇。随后这些簇经组装生成暂定一致序列和单体。不同的颜色表示不同的EST文库

英文注释：Singleton，指不能组装的单一Read；Clustering，聚类；Assembly，组装；Cluster-set，聚类集；Assembly-set，组装集；Singlet，单体；TC，暂定一致序列

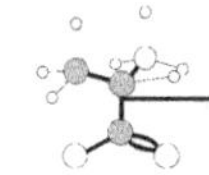

功能分析：功能分析是一种预测某一基因功能的分析技术。预测工具大都基于同源性比对（如 Blast 或 InterProScan），搜索现有的数据库（如 SwissProt 或 NR，NCBI 的非冗余蛋白数据库）查找具有相似结构的基因。然后，基于相似基因有相似功能的假设，对这些基因进行功能注释。

9.3.1.2 SAMS：一种 EST 分析工具

序列分析和管理系统（Sequence Analysis and Management System，SAMS）是一个包含数据后台的生物信息分析平台。它提供了一个环境，可以使用多种分析软件来处理基因组片段或 EST 序列，并以网页界面的形式展示结果（Bekel et al. 2009）。SAMS EST 分析主要包括四个主要部分：前处理、聚类、装配和功能分析。主体软件的参数可调，也可以整合新的工具。由于功能分析是关键部分，SAMS 中植入了几个图形和交互设置，例如 KEGG 图（京都基因和基因组百科全书，见 9.3.4.4）用于显示组装发现的基因，或表达分析工具（SteN-Northern Blot）用于分析差异表达基因。序列和注释数据可以通过多种文件格式导出。SAMS 使用面向对象的后端和一个关系型数据库管理系统 MySQL，与 Bielefeld 大学生物信息学资源机构开发的对象关系映射相连接。

SAMS 可通过网页浏览器下载（https://sams.cebitel.uni-bielefeld.de/cgi-bin/Sams_login.cgi? cookie_test=1）。使用者需要登录名和密码。项目访问有一定的用户限制，以满足数据保密的需求。密码可通过登陆页面设定。

用 SAMS 系统分析 EST 序列比较方便快捷。首先要选定一个序列文库，用户可创建或指定一个文库用来储存序列数据。接着通过导入功能导入序列。这需要几步连续的操作：第一步，用户选择本地存储的文件上传到 SAMS 服务器。SAMS 支持多种数据格式，推荐使用原始数据文件。第二步，上传原始的峰图文件，使用软件 PHRED 进行质量过滤，用户可以选择需要的质量值（例如 PHRED 13 表示碱基识别的准确率为 95%，即碱基识别错误的概率为 5%；PHRED 20 表示碱基识别的准确率为 99%，即碱基识别错误的概率为 1%）。最后，如果序列中仍然含有载体序列，可以使用整合在系统中的载体过滤选项进行去除。

经过上面几步之后，长度小于 50 bp 的序列被去除；当序列导入完成后，这些序列将在 SAMS 系统上以 EST 的形式列出。接着用户可以进入聚类步骤，以减少序列的冗余，这一步建议使用标准参数，但也可根据需要进行调整。标准的聚类参数设置由 TIGR 默认（HSP 长度：40，identity：0.95，unmatched overhang length：20）。SAMS 使用了一个聚类组件，是 TIGR 聚类方法的类 TGICL 方法。聚类之后，使用软件 CAP3 将序列装配成暂时一致性序列

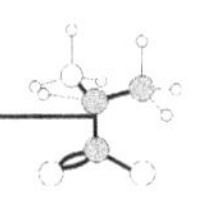

(TC)。SAMS系统为检验不同参数设置的影响，可能会不止一次地对同一个序列数据进行聚类和装配。当用户对聚类结果满意时，再执行装配的步骤，得到TC和singleton（指不能组装的单一序列），这些装配结果存储在数据库中。

SAMS功能分析：这步操作的目的是对装配得到的TC进行功能注释。装配过程完成之后，可以使用名为Metanor（Goesmann et al. 2005）的自动注释流程执行注释步骤。这个流程包含了几个用于分析TC、EST和singleton的生物信息学工具，使用BLAST（Basic Local Alignment Search Tool，基于局部比对算法的搜索工具）来查找目标序列与不同数据库间的同源性，这些数据库如NT（Non-redundant nucleotide database from NCBI，NCBI非冗余核酸数据库）、NR（Non-redundant protein database from NCBI，NCBI非冗余蛋白数据库）、KEGG、KOG/COG（clusters of euKaryotic Orthologous Groups or Clusters of Orthologous Groups，参见9.3.4.4节）和SwissProt数据库（见9.3.5.2节）。此外，使用InterProScan（见9.3.3.3节）来查找所有的6种阅读框。这些结果以“observations”的形式存储在数据库中，可供进一步使用。

所有的软件运行完之后，会产生一个一致性的注释结果，含有EC编号、基因名称、KOG/COG类别和推测的功能。可通过注释对话窗对TC、EST和singleton进行手工注释，该对话窗展示了之前分析软件运行的结果。所有注释的结果都会备份，之前的注释结果也会在几秒内被重新保存。所有的序列、注释结果和软件分析的结果都可以通过SAMS网页界面以不同的文件格式导出。

SteN-Statistical electronic Northern Blot：SteN是利用EST数据进行基因表达分析的工具。几种不同文库被装配成TC后，使用SteN可以比较每个文库中的EST数目，根据不同文库中EST的组成对TC进行过滤，这样就可以鉴定在某个文库中特异性表达的基因。需要用来源于不同时间点或组织的cDNA文库的EST集进行这样的分析。经过TC过滤之后，将对表达数据进行统计学评估，结果列表显示。

SAMS的备选系统ESTexplorer：ESTexplorer是可以代替SAMS的系统（Nagaraj et al. 2007），可通过下面的网址免费获得用于学术研究：http://estexplorer.els.mq.edu.au/estexplorer/main_page.php。

数据上传之后，载体剪切、重复序列屏蔽（repeat masking）和组装这些步骤可采用下面的软件：SeqClean（Chen et al. 2007）、RepeatMasker（http://www.repeatmasker.org）和CAP3。使用BLAST和Blast2GO（Conesa et al.

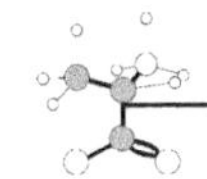

2005）进行基因水平的注释，使用 ESTSCAN（Iseli et al. 1999）、InterProScan（见 9.3.3.3 节）和 KOBAS（Wu et al. 2006）进行蛋白水平的注释。流程运行结束后，结果将会被保存，并在一周内可供下载，一周后所有的文件将会被删除。这个系统没有使用通过登录名和密码来识别用户的用户管理系统，而是用户通过系统提供的 ID 号（例如 John_123）来获取结果。因此我们需要特别注意，这样的系统存在安全风险。如果一个用户猜对了另一个用户的 ID，那么就能获得其所有的数据结果。此外，可上传和分析的数据量是受限制的。表 9.2 对 SAMS（Bekel et al. 2009）和 ESTexplorer（Nagaraj et al. 2007）这 2 种应用系统进行了比较。

表 9.2　SAMS 与 ESTexplorer 的比较

Table 9.2　Comparison SAMS vs. ESTexplorer

SAMS vs. ESTexplorer		
Function\system	SAMS	ESTexplorer
User authentification	Yes	No
Permanent data storage	Yes	No
Clustering and assembly	Yes	Yes
Automatic annotation	Yes	Yes
Manual annotation	Yes	No
Export of sequences and annotations	Yes/yes	Yes/no
Expression analysis	Yes	No

英文注释：Function\system，功能\系统；User Authentification，用户认证；Permanent Data Storage，永久性数据存贮；Clustering and Assembly，聚类和装配；Annotation，注释；Automatic Annotation，自动注释；Manual annotation，人工注释；Export of Sequences，序列输出；Expression Analysis，表达分析

在欧洲海洋基因组网（MGE），SAMS 已被成功用于分析 44 个 EST 项目，包括墨角藻（*Fucus serratus* 和 *Fucus vesiculosus*）、欧洲鲈（*Dicentrarchus labrax*）、富硒海洋球石藻（*Emiliania huxleyi*）和纹壶藻（*Balanus amphitrite*）。

9.3.2　基因预测

从原始的基因组序列中预测 tRNA、rRNA 和蛋白编码基因，是对新测序基因组注释的关键步骤之一。这一节主要介绍自动识别基因组序列中的蛋白编码基因（编码序列，CDS）的计算策略和已有的软件。对于基因预测，可采用两种不同的方法，基于内在特征或基于相似性。基于内在特征的方法分析基因组的序列特征用以区分编码和非编码区域，可找出编码和非编码序列的不同组成特征，这种差异主要是由于 CDS 区密码子使用偏好导致，而这种

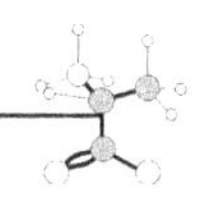

使用偏好可以优化蛋白质合成时的翻译效率（Gouy and Gautier 1982）。

基于内在特征的基因查找方法通常引用统计学模型，以模拟短寡核苷酸序列在编码和非编码区域的出现频率。通常采用具有马尔可夫性质（Durbin et al. 1998）的生成模型来描绘序列构成。这种马尔可夫性质包括在核苷酸（Delcher et al. 1999，Larsen and Krogh 2003）或密码子（Badger and Olsen 1999）上有固定顺序的马尔可夫链，以及其他特性（例如核糖体结合位点或者相邻基因的重叠区域），当使用隐马尔可夫模型（HMM）描述基因上下游时，同样可以被整合成概率模型（Delcher et al. 1999）。

基于相似性的方法是通过搜索与参考序列具有显著性相似的 DNA 片段来预测基因，参考序列包括近缘物种的基因组、系统发生树相关蛋白质或基因转录本。多种基于相似性的预测方法能够根据同义替换率区别保守的编码区和保守的非编码区（Badger and Olsen，Moore and Lake 2003，Nekrutenko et al. 2003）。这是因为为了确保编码蛋白的氨基酸序列，编码序列比保守的非编码区具有更多的同义突变（即不改变氨基酸的突变）。

9.3.2.1　原核生物的基因预测

在原核生物（细菌和古细菌）中，蛋白编码基因通常是开放阅读框（ORF），一段由起始密码子开始、终止密码子结束、中间没有终止密码子的序列。因此，在原核生物基因组中预测蛋白编码基因存在一种二级分类问题，即从大多数非编码 ORF（在体内不转录和翻译的基因组区域）中区分出编码序列。

对于原核生物，许多基因预测软件可以使用，通过基于内在特征或将内在特征与相似性结合在一起的方法，基因组序列进行对初步的基因预测。这些方法能实现很高的准确性，但对短基因预测（即没有典型序列特性和翻译起始位点的基因预测）仍然很困难。短基因很难预测是因为它们的序列信息太少，难以进行鉴定（Skovgaard et al. 2001，Larsen and Krogh 2003，Ou et al. 2004）。另外，由于表达相关的密码子用法（McHardy et al. 2004b），前导链和后随链相关的偏差或不同细菌之间的基因转移（称为水平基因转移；Smith et al. 1992）等原因，所以同一个物种的基因可能具有不同的序列特性。在这种情况下，内在特征的方法可能很难识别不同的基因类型。这个问题已经通过引入一个能识别具有“非典型性”序列组成的新模型或在预测阶段之前进行无监督 CDS 发掘等方案解决（Lukashin and Borodovsky 1998，Krause et al. 2007）。

在基于内在特征的基因预测软件中，Glimmer（Delcher et al. 1999，2007）和 Genemark（Lukashin and Borodovsky 1998）被广泛使用。Glimmer 在

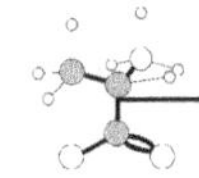

本地安装快速简单，平均可以找到99%功能清晰的“确定”基因。然而，Glimmer-2被报道会产生较高的假阳性预测（McHardy et al. 2004a，Krause et al. 2007）。最新版本Glimmer-3，已经解决了这个问题，它是通过选择一套在处理后期重叠度最大的、分数最高的预测来解决的。Genemark，使用隐马尔可夫模型（Besemer et al. 2001，Besemer and Borodovsky 2005）也实现了很高的预测准确度，灵敏度可高达99%，特异性为93%左右（Delcher et al. 2007），这个软件可以通过网络界面来简便操作。

通常，基于相似性的基因预测方法比基于内在特征的预测方法慢，并且安装更复杂，但预测结果更加可信。此外，在采用二者结合的方法时，相似性的预测能够为基因组特异性内在特征模型提供可靠的初始训练集。譬如基因预测软件GISMO，将基于Pfam搜索保守的蛋白家族成员的方法和一种内在特征预测方法结合在一起，能够鉴定出高达99%的具有已知功能的基因（Krause et al. 2007）。GISMO也能够得到非常可靠的预测结果，其特异性超过94%，在鉴定短基因和从富含GC中预测基因方面，能够达到高准确度。为进行基于内在特征的基因鉴定，GISMO采用了一个支持向量机，能够以无监督的方式学习不同基因的序列特征。

研究人员已经开发出多个软件，用来预测翻译的起始位点，譬如软件GS-Finder（Ou et al. 2004）、RBS-Finder（Suzek et al. 2001）和Tico（Tech and Meinicke 2006），这些软件都可以达到90%以上的准确率。但是，可用于性能评估实验验证的翻译起始位点的数目有限，无法普遍达到这样的精度。

在线资源REGANOR（Linke et al. 2006）和RAST（Aziz et al. 2008）提供了简单的方法，通过网络界面来自动预测原核生物基因组中的tRNA、rRNA和蛋白编码基因。通过结合基因预测软件Glimmer-2（Delcher et al. 1999）和CRITICA（Badger and Olsen 1999），REGANOR对鉴定具有已知功能的“确定”基因的灵敏度达到98%，特异性为95%（McHardy et al. 2004a）。软件RAST从另外一个方面结合了不同的资源，来鉴定蛋白编码基因，这些资源包括Glimmer-2预测、保守蛋白家族（FIGfams）的同源性查找和与大的蛋白数据库的BLAST比对。另外，RAST会对所鉴定的基因进行自动的功能注释。REGANOR和RAST都使用了tRNAscan-SE（Lowe and Eddy 1997）和Search-ForRNAs（Niels Larsen et al.，未发表），程序用来自动鉴定初始基因组序列中的tRNA和rRNA。软件REGANOR可通过以下网址下载https://www.cebitec.uni-bielefeld.de/groups/brf/software/reganor/，RAST可通过http://rast.nmpdr.org/下载。

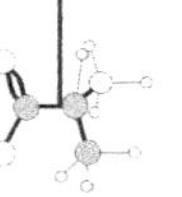

表9.3 现有的细菌基因预测软件。基因预测采用的策略可基于内在特征（I）和基于序列相似性（S），可通过下载获取（D）或通过网络界面在线访问（O）。

Table 9.3 Existing bacterial gene finding software. The implemented strategy is depicted as I (intrinsic) and S (similarity based), the availability as D (available for download) or O (accessible online via web-interface)

Gene finding programs					
Program	I	S	Comments	A	URL (in 12.2008)
Critica (Badger and Olsen 1999)	+	+	Identifies genes based on synonymous substitution rate	D	http://www.ttaxus.com/software.html
EasyGene (Larsen and Krogh 2003)	+	+	Employs HMMs, training set derived by BLAST search	O	http://www.cbs.dtu.dk/services/EasyGene/
GeneMark.hmm/S (Lukashin and Borodovsky 1998, Delcher et al. 1999)	+	–	Uses HMMs	O	http://exon.gatech.edu/GeneMark/
Gismo (Krause et al. 2007)	+	+	Combines Pfam search with SVM	D	http://www.cebitec.uni-bielefeld.de/brf/gismo/gismo.html
Glimmer-3 (Delcher et al. 1999)	+	–	Employs interpolated context model. Uses dynamic programming to reduce overlapping genes and to refine gene starts	D	http://www.cbcb.umd.edu/software/glimmer/
MetaGene (Noguchi et al. 2006)	+		Employs pre-trained, GC content dependent codon-usage model. Applicable also for metagenomic fragments and draft genomes	D+O	http://metagene.cb.k.u-tokyo.ac.jp
Rast (Aziz et al. 2008)	+	+	Combines Glimmer-2 with homology searches	O	http://rast.nmpdr.org/
Reganor (McHardy et al. 2004b)	+	+	Combines Glimmer-2 and Critica	O	https://www.cebitec.uni-bielefeld.de/groups/brf/software/reganor/
RescueNet (Suzek et al. 2001)	+	+	Uses SOM	D	http://bioinf.nuigalway.ie/RescueNet/
SearchForRNAs (Niels Larsen et al., unpublished)			Identifies rRNA genes		Available from author upon request.
tRNAscan-SE (Lowe and Eddy 1997)			Predicts tRNA genes using covariance models	D+O	http://lowelab.ucsc.edu/tRNAscan-SE/
ZCurve (Guo et al. 2003)	+	–	Employs LDA. Performs Z-transformation of DNA	D	http://tubic.tju.edu.cn/Zcurve_B/

SVM: Support Vector Machine; SOM: Self-Organizing Map

英文注释：I: Intrinsic，基于内在特征；S: Similarity based，基于序列相似性； A: Availability，可获得性; D: Download,下载；O: Online,在线。

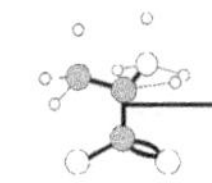

总之，在线基因预测软件 REGANOR 和 RAST 可提供高质量的基因预测结果，且操作简便。另外，开放的基因查找软件 GLIMMER 和 GISMO 可在本地安装，并可整合在已有的基因组注释流程中。Glimmer 操作简便并且省时。GISMO 运行较慢且维护较难，但基因鉴定准确度很高。tRNAscan-SE （Lowe and Eddy 1997）、SearchForRNAs 和 RNAmmer （Lagesen et al. 2007） 都可以自动鉴定 tRNA 和 rRNA（见表 9.3）。

9.3.2.2 真核生物的基因预测

虽然真核生物的基因预测软件在最近几年有所改进，但仍然很难对真核生物基因组中的基因结构进行预测。真核生物复杂的基因结构和少量染色体片段对应蛋白编码外显子，是主要的挑战（Mathe et al. 2002，Zhang 2002，Brent 2007）。相对于原核生物 90% 的基因组是用来编码蛋白质而言，真核生物的外显子嵌合在浩瀚的非编码 DNA 中。此外，真核生物的基因具有可变剪切和多腺苷酸化位点，并具有可变的转录和翻译起始位点。

目前，最可靠的基因结构预测可通过将基因转录本序列映射到相关基因组上，针对这项任务已开发了许多不同的程序出来，其中包括 EST_ GENOME（Mott 1997）、AAT （Huang et al. 1997）、SIM4 （Florea et al. 1998）、GENESEQUER （Usuka et al. 2000）、BLAT（Kent 2002） 和 GMAP （Wu and Watanabe 2005）。这种预测方法的主要缺点在于，低水平表达或只在某些特定组织或特定条件下表达的基因很可能检测不到。

双重或多重基因组基因预测算法，如 SGP2 （Parra et al. 2003）、SLAM（Alexandersson et al. 2003）、TWAIN （Majoros et al. 2005）、N-SCAN （Gross and Brent 2006） 和 TWINCAN（Korf et al. 2001），可分析进化相关的两个或多个物种的基因组间序列保守的模式。在没有测序的基因转录本情况下，如果近源物种的基因组可用，这些软件可以准确预测外显子。但是，双重或多重基因组基因预测算法对完整基因结构的预测准确度较低，会遗漏没有同源序列的基因，并经常将假基因误认为有功能的基因。最近有研究通过结合假基因识别比对软件 PPFINDER（van Baren and Brent 2006），解决了这些问题。如果没有近缘物种的基因组序列，通过与已知蛋白比对的基因预测方法可准确鉴定外显子的位置（Huang et al. 1997，Slater and Birney 2005）。大众化软件 GENEWISE（Birney et al. 2004）使用隐马尔可夫模型，融合了蛋白序列比对到基因组的模型和真核生物基因结构的模型。

为了补充基于比对的方法，也可通过评估序列组成特征的内在特征的算法预测基因。大部分基于这种算法的软件，如 GENSCAN（Burge and Karlin

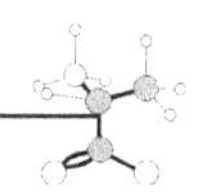

1997)、GENEMARK. hmm（Lomsadze et al. 2005）、GLIMMERHMM（Majoros et al. 2004）、FGENESH（Salamov and Solovyev 2000）和 AUGUSTUS（Stanke and Waack 2003），都采用隐马尔可夫模型（HMM）或广义隐马尔可夫模型（GHMM）区分内含子、外显子和基因间区。通过使用 HMM 或预测内含子、外显子、基因间区、起始和终止密码子、可变剪切位点和多聚腺苷酸化信号的 GHMM 模型，多种序列特征可以组合成一致的概率模型。对新基因组使用内在算法之前，需要对基因组特异的序列组成特征诸如密码子使用和可变剪切模式进行机器学习。虽然大多数基于内在特征的算法需要对已知基因进行监督训练，但软件 SNAP（Korf 2004）和 GENEMARK. hmm ES（Lomsadze et al. 2005）可以以无监督的方式发觉新测序基因组的序列特征。对于多数已经测序的真核生物基因组，预训练的内在模型已公开可用。通常内在基因识别程序用于外显子的鉴定非常有效，但是在重构全基因的外显子与内含子结构方面还存在很多困难。另外，由于预测结果的高假阳性率，没有外在证据（例如序列相似性）支持的预测结果并不可信。

多种真核生物基因识别算法可以利用不同类型的证据进行预测。例如 N-SCAN、N-SCAN-EST（Wei and Brent 2006）、TWINSCAN、AUGUSTUS 和 GENIE（Reese et al. 2000）等程序通过 HMM 整合内在和外在信息成单一的概率模型。曾用于拟南芥（*Arabidopsis thaliana*）注释的真核生物基因识别工具 EUGENE（Schiex et al. 2001），采用了类似的方法。在 EUGENE 软件中，通过有向无环图将内在和外在信息结合在一起，以图中最短的路径决定所分析 DNA 序列的最可能的基因结构。其他程序，如 JIGSAW（Allen and Salzberg 2005）、EXONHUNTER（Brejova et al. 2005）、GLEAN（Elsik et al. 2007）、EXOGEAN（Djebali et al. 2006）、EVIDENCEMODELER（Haas et al. 2008）和 AGUSTUS-any（Stanke et al. 2006），通过对不同来源的证据评估以模仿人类基因识别过程。内含子数目和大小越小，真核生物基因识别方法的准确性一般就会越高。使用比对算法对紧凑基因组的完整基因结构进行从头预测，准确度达 70%。对于含有高比例基因间区的哺乳动物基因组，因其复杂的基因结构和可变剪切位点，通常预测的准确度很低。

2005 年 EGASP 项目启动，通过人基因组编码区对已有的基因识别软件的准确性进行的评估（Guigo and Reese 2005，Guigo et al. 2006），最好的软件能够对将近 70% 的已注释基因正确地预测出至少一个转录本。但是，如果考虑选择性剪接变异体，准确率将降低到 40%～50%。在编码外显子水平上，评估的最好预测方法可以达到 80% 以上的灵敏度和特异性，接近在编码核苷酸

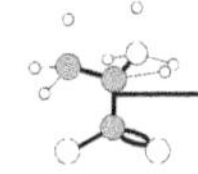

水平上的90%。依赖多种信息，尤其是基因转录本或蛋白序列的预测程序是最准确的，其次是依赖两个或多个基因组间的序列比较的预测方法。仅基于内在序列特征的预测方法准确率最低。被评估的几乎所有软件对非编码外显子的预测准确度都很低。

总之，现在最准确的真核基因识别方法，通常在鉴定外显子方面准确率很高，但对复杂基因组完整基因结构的预测仍然是很困难的。为了得到可能最准确的基因预测结果，建议将能够检测重复区域和假基因的程序和基因转录本、双重和多重基因组比对的方法以及基于序列内在特征的预测方法结合在一起。几种流行的基因识别算法可以通过公开的网络界面简便使用：AUGUSTUS（http://augustus. gobics. de/）、GENEMARK(malab. wustl. ebu/nscan)、TWINSCAN/NSCAN（http://mblab. wustl. edu/software/twinscan/）、GLIMMERM（http://www. tigr. org/tdb/glimmerm/glmrform. html）和GLIMMERHMM(ccb. jhu. edu/software/glimmerhmm)。

9.3.3 基因组注释及其他

在基因预测或EST序列组装完成之后，接下来的工作就是对这些基因或者EST序列的功能和作用进行预测和分析。几乎所有的计算机方法都是建立在基因序列、相应的氨基酸序列、蛋白质的二维和三维结构以及蛋白质可能功能之间的相互关系上。由于大量不同来源的基因序列及功能都已经被描述和公布，所以搜索相似性序列成为注释新基因和基因组的有效方法。

9.3.3.1 序列相似性介绍

在进行序列比较时，需要对序列间的相似程度和差异程度进行准确衡量。如果没有这样的衡量标准，计算机就无法实现对序列的自动化处理。在生物信息学中，“比对”（alignments）被用来描述序列之间的相似性，它是一个序列变为另外一个序列的必要步骤，每一个步骤可能是下面中的一个：①一个氨基酸或核苷酸的匹配或者不匹配；②一个氨基酸或核苷酸的插入；③一个氨基酸或核苷酸的删除。

通过对比对过程中的步骤进行打分，比如说对插入和缺失的罚分和对相同序列的得分，可以得到序列的最佳比对。最简单的方法是列出所有可能的比对，根据他们的得分来选择最佳比对。然而，随着序列长度的增加，可能的比对数目呈指数性增加。在比对过程中，每增加一个核苷酸碱基或者氨基酸，比对数目将增至3倍。

所以，计算序列相似性的软件主要过滤大部分比对，直到留下最佳比对，

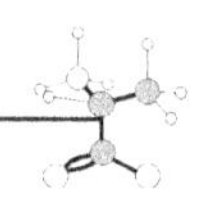

这取决于如何评估比对的质量。这就涉及到对不同步骤进行简单的打分，直到包括突变、重组和其他改变序列事件的高度复杂模型。1970 年，Needleman 和 Wunsch 提出了动态编程算法，其处理时间与两条序列长度的乘积呈线性关系，这使得两条长序列的比对成为可能。这样的全局比对在序列保守性高的时候效果比较好。1981 年，Smith 和 Waterman 提出了一种类似的方法用于局部比对，这种方法在序列中有部分不保守时也能得到比较好的结果（Smith and Waterman 1981）。

随着公布的序列不断增加，需要把这些序列收集起来储存在一个大的数据库里供公众使用。1982 年，GenBank 的第一个版本只包含 600 条记录，随后数据量便以指数级速度快速增长。到了 2008 年，已经达到了 8 千万条 DNA 序列。1988 年 Pearson 和 Lipman 发布的 FASTA 是第一个广泛使用的在数据库里搜索相似序列的程序。这个程序对 Needleman-Wunsch 算法进行了特别优化，使得在大数据库里搜索相似序列成为可能。1990 年，Altschul et al. 公布了 BLAST 的第一个版本，一个基于局部比对搜索工具，它提供了一种在数据库里基于局部比对搜索相似序列的快速算法（Altschul et al. 1990）。

虽然还有其他一些在大数据库里搜索相似序列的方法，但是 FASTA 和 BLAST 是两种最为广泛应用的工具，几乎所有大基因组的测序和注释都基于它们。

9.3.3.2 从基因注释到基因组注释

由于有大量各种可用的生物信息学工具和资源，基于相似性搜索的新基因分析和功能预测变得比较容易。很多机构都提供容易使用的网络界面和图形化的 BLAST 等工具进行单基因分析的服务。

对一个完整基因组的基因集或者 EST 序列集合的分析，则与单基因的分析有很多不同的地方。如果有几千条序列需要分析，手动地去运行所有必要的预测工具是不可能的。数据的管理和自动化分析是注释一个完整基因组和大的 EST 序列集合的关键。有一些整合的系统可以帮助生物学家方便地对大量序列进行注释、管理，利用不同工具进行分析并给出易理解的结果。根据所储存的数据信息，大部分的系统也提供一些高级功能，例如新陈代谢网络的重构、实验数据的整合和与其他基因组的比较分析。

2000 年，由 Sanger 中心建立并公布的 Artemis 是一个能够容易地进行基因组注释的系统（Rutherford et al. 2000）。它提供可视化界面，对本地序列数据进行注释。本地序列可以是简单的序列文件、表格文件或者是已经包含基因及注释信息的结构文件。只需要下载并解压就可以完成程序的安装，并且还能在本地计算机系统上整合多种功能预测工具。由于 Artemis 基于本地系统的

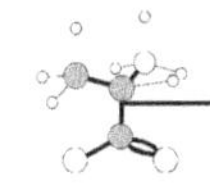

工具，它不允许多个用户在同一时间对同一个基因组进行分析。

GenDB（Meyer et al. 2003）和SAMS（见9.3.1.b）等注释系统解决了多用户使用的问题。这样的系统将基因序列和信息、分析和注释结果都放在中央关系数据库里，为用户提供网络界面而使他们不必安装本地软件。大的计算机集群和数据处理流程的支持，使得GenDB和SAMS成为基因组分析强有力的工具系统，而定义良好、开放的编程接口使开发者们能够容易地增加更多新功能。两个系统都支持大量序列的全自动分析，也能引导部分用户自己进行手动注释。

多数大型测序和生物信息中心都建立了自己的数据分析流程与系统，并允许其他研究机构访问。这些流程和系统多数是在大的真核生物基因组项目的背景下建立起来的。Ensembl就是这样一个很好的系统（Flicek et al. 2008），它储存了从酿酒酵母（*Saccharomyces cerevisiae*）到人类的各种各样已经注释的真核生物基因组数据，其网页界面能够允许从多种不同的水平（大到染色体，小到单碱基）来查看基因组信息。它还含有一个复杂的数据挖掘机制。

在这里介绍的系统仅仅是个例子。除此之外还有很多其他系统，并且每年都有新的改良系统出现。几乎每个大的测序项目都有自己的网站，介绍测序项目的背景信息，并构建了专业数据库。

9.3.3.3　蛋白质注释工具

一个蛋白的功能通常与其具有的独特特征或者基序（motif）是相关的。预测一个蛋白的功能也与确定它的亚细胞定位有关系。通常，在对蛋白序列进行注释的时候，最好能够利用多种方法和信息资源，而不是仅仅使用前面章节提到的单一同源搜索工具。

第一部分，我们将介绍InterPro和InterProScan。InterPro是一个蛋白功能和结构分析的整合数据库，而InterProScan是一个结合了来自InterPro数据库的蛋白特征识别功能的搜索工具，允许用户将未知序列与数据库进行比对来进行蛋白功能的预测。

第二部分，我们将介绍两种可靠的蛋白亚细胞定位的预测工具，即进行跨膜结构域和拓扑结构预测的TMHMM以及进行信号肽预测的SignalP。

（1）InterPro和InterProScan

InterPro（Mulder et al. 2007）是蛋白家族、结构域和功能位点的整合资源数据库。同一个家族的蛋白或者蛋白结构域通常具有进化上保守的区域。有的与蛋白功能有着直接关系，还有的对蛋白三维结构的维持有重要作用。识别出这样的序列相似区域对于确定一个蛋白家族或者结构域特征是非常重

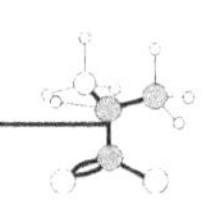

要的。通过比较特征区域，也可以将一个蛋白家族中的不同蛋白与其他家族中的蛋白区分开来。

InterPro 联盟目前由分布在欧洲和美国的 11 个成员数据库组成，包括 UniProt（见 9.3.5.2）、PROSITE、Pfam、PRINTS、ProDom、SMART、TIGRFAMs、PIRSF、SUPERFAMILY、Gene3D 和 PANTHER。成员数据库通过不同的内部特征（例如配制文件或者位置特异的打分机制）来构建（表 9.4）。除了 ProDom 数据库中的蛋白特征序列是从 UniProt 序列数据库中自动生成以外，其他成员数据库的蛋白特征都是通过手工分析得到的。访问这些成员数据库的主页，可以进一步了解它们采用的方法和标准的细节信息。

InterPro 整合了几个各有优缺点的蛋白特征数据库，提供了一个能进行蛋白功能和结构分析以及序列注释的整合工具。每一条 InterPro 记录都由人工创建而成，包含了一个或多个成员数据库中的一个或多个蛋白特征。预测一个蛋白的同一个结构域的两个特征将被归到同一条 InterPro 记录中。InterPro 对数据的整合还有其他方法，可以在 InterPro 网页的用户手册上了解更多信息。

表 9.4　成员数据库建立蛋白特征的两种策略

Table 9.4 Two categories of methods used by the member databases for building the protein signatures

Sequence-motif methods			Sequence cluster method
Regular expression and profiles	Motifs	Hidden Markov models or HMMs	Sequence clustering derived from the UniProtKB database
PROSITE	PRINTS	Pfam, SMART, TIGRFAMs, PIRSF, SUPERFAMILY, PANTHER, Gene3D	ProDom

英文注释：Regular Expression and Profiles，正则表达和特征谱；Sequence-motif Methods，基于序列基序的方法；Sequence Cluster Method，基于序列聚类的方法；HMMs，隐马尔科夫模型

InterPro 版本 17.0 有 16583 条记录，包括蛋白结构域、蛋白家族、转录后修饰、重复序列、活性位点、结合位点或保守位点。

InterPro 记录的例子详见链接：http://www.ebi.ac.uk/interpro/entry/ IPR000719

一个 InterPro 记录一般包含 8 方面内容：

- 匹配的蛋白（匹配的 UniProt、登录号、类型、特征）
- 在 InterPro 数据库里的关系（父子关系、包含/被包含）

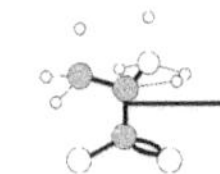

- InterPro 注释（摘要、结构和数据库链接）
- 蛋白分类
- 重叠的 InterPro 记录
- 示例蛋白（图形显示）
- 文献
- 补充阅读

匹配蛋白的展现形式有表格、简单图形、复杂图形（每一个成员数据库用不同颜色标出）或者 InterPro 蛋白结构域（对于多结构域的蛋白组织可以得到很好的图形化展示）。结构信息在视图的底部可以看到，它显示了 SCOP 和 CATH 结构域在 InterPro 蛋白序列上的映射，这个信息是基于 InterPro、UniProt 和 MSD（大分子结构数据库，http://www.ebi.ac.uk/msd/）的联合使用而产生的。InterPro 图形界面显示了序列中结构域的位置，它是通过在 PDB（蛋白数据库，见 9.3.5.4）链和来源于 MSD 数据的 UniProt 序列之间进行逐个残基映射得到的。只有 PDB 链才展示出不重叠区域。即使蛋白结构比序列更难确定，目前已经确定结构的蛋白家族也接近 2000 个。蛋白结构比序列更为保守，而且往往能揭示一些在序列水平上不易被发现的进化关系。蛋白结构对于研究蛋白功能、催化机理和蛋白间互作关系也是至关重要的。

那怎样利用已注释的蛋白（收录在 InterPro 数据库中）来研究新蛋白的功能呢？InterProScan 在这里就发挥作用了。InterProScan（Quevillon et al. 2005）是一款能对特征序列数据库进行搜索的工具（这与 BLAST 和 FASTA 对数据库中每条序列进行搜索来寻找未知序列的相似性不同，见 9.3.3.1）。InterProScan 允许输入的蛋白序列包括文本格式、FASTA 格式或者 UniProt 格式，核酸序列应该为文本格式、FASTA 格式或者 DDBJ/EMBL/GenBank 格式。InterProScan 可以通过网络在 EBI 网站（http://www.ebi.ac.uk/InterProScan/）上在线使用，也可以下载（ftp://ftp.ebi.ac.uk/pub/databases/interpro/iprscan/）后在本地计算机上安装使用。

2can 网站（EBI 的生物信息学教学资源）上的两个教程（http://www.ebi.ac.uk/2can/tutorials/）可以帮助用户快速使用 InterProScan 进行蛋白的特征发现和注释：

（2）TMHMM 和 SignalP

TMHMM 是用于鉴别跨膜蛋白，而 SignalP 的作用则是检测信号肽。跨膜蛋白是穿过双层脂膜的多肽链，他们通常参与物质转运和胞内外通迅等一系列细胞活动。大部分蛋白质组都包含 20%～25% 的跨膜蛋白。跨膜蛋白一般

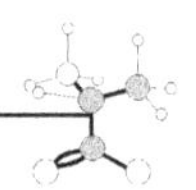

可分为两类：在任何膜中都会存在的 α 螺旋蛋白和主要存在于外膜（革兰氏阴性细菌的外膜，但也存在于线粒体和叶绿体的外膜中）的 β 桶状蛋白。由于可用的跨膜蛋白三维结构太少，对跨膜结构域的识别目前还是比较困难。截至2008 年7 月，在 PDB 数据库中接近52 000 个蛋白的三维结构中，跨膜蛋白的三维结构不到1 000 个。蛋白质前体的疏水区域常被误认为是跨膜区域，而脂肪族螺旋常常不会被识别为跨膜区域。目前，最可靠的跨膜结构域和拓扑结构预测软件是基于隐马尔可夫模型的 TMHMM 2.0（http://www.cbs.dtu.dk/services/TMHMM/），它能预测 α 螺旋的位置和方向（Sonnhammer et al. 1998，Krogh et al. 2001）。但是，由于信号肽也有的核心区域，跨膜结构域预测软件会把信号肽错误地识别为靠近蛋白氨基端的跨膜区域。

信号肽是引导蛋白运送出细胞质的短肽链。根据蛋白目的地的不同，也存在不同的信号肽。SignalP 3.0（http://www.cbs.dtu.dk/services/SignalP/）能够预测信号肽，进而提供蛋白的亚细胞定位的信息，同时还能预测原核生物和真核生物序列的信号肽内的剪切位点（Bendtsen et al. 2004，Emanuelsson et al. 2007）。然而，信号肽预测软件有时候也会将蛋白氨基末端的跨膜结构域错误地识别为信号肽。为了解决这个问题，Phobius 网站服务器（http://phobius.binf.ku.dk）应运而生，将跨膜结构和信号肽的预测结合起来对蛋白进行综合预测（Kall et al. 2007）。

表9.5 总结了这部分所提到的预测软件。G. Von Heijne 团队还开发了一些其他的预测软件，比如预测叶绿体转运肽存在及位置的软件 ChloroP（Emanuelsson et al. 1999），以及预测叶绿体转运肽、线粒体靶肽或者分泌途经信号肽的亚细胞定位的软件 TargetP（Emanuelsson et al. 2007）等。

表 9.5 能够可靠预测蛋白亚细胞定位的程序

Table 9.5 Programs that provide a reliable prediction of subcellular location

Prediction program	TMHMM 2.0 (Phobius server)	SignalP 3.0 (Phobius server)	ChloroP	TargetP
Subcellular location	Transmembrane helices	Signal peptide cleavage sites	Chloroplast transit peptide	Any N-terminal presequence
Organism	Prokaryotes and eukaryotes	Prokaryotes and eukaryotes	Plants	Eukaryotes

英文注释：Prediction Program，预测程序；Trans-membrane Helices，跨膜螺旋；Signal Peptide Cleavage Sites，信号肽剪切位点；Chloroplast Transit Peptide，叶绿体转运肽；Any N-terminal Pre-sequence：任何氨基末端前导肽，Organism，生物体；Prokaryote，原核生物；Eukaryote，真核生物；Plant，植物

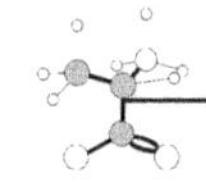

9.3.4 比较基因组学和功能分类

随着完全测序的真核生物和原核生物基因组数目的指数增长，越来越需要可以对基因组进行精确、一致和自动化的基因功能注释。

9.3.4.1 同源性和相似性

Chothia 等人认为，绝大多数蛋白是通过基因复制、重组和分化形成的（Chothia et al. 2003）。为了描述蛋白之间的进化关系，同源性、直系同源和旁系同源等术语常被使用。关于这点，同源性的意思是两个蛋白或序列拥有共同的祖先。蛋白或 DNA 序列之间的同源性可以基于序列的相似性推断，因为同源性的确实证据需要分析共同的祖先和所有中间形式（Reeck et al. 1987）。如果两个基因拥有几乎相同的 DNA 序列，很可能它们是同源的。然而，同源性不是序列相似性的唯一原因。短序列可能因为随机性而相似，或因两个基因都被选择结合到一个特定蛋白（例如转录因子），它们的序列也可能相似。同源性可以区分为两个主要类型。直系同源序列是由于物种形成事件而分开的同源序列。一个存在于祖先物种的基因，因祖先物种分化成两个物种，变为 2 个拷贝。这些来源于祖先的拷贝被命名为直系同源基因，通常具有相同或相似的功能。

在祖先基因组中因基因复制事件产生的同源基因被称为旁系同源。旁系同源基因可能由于选择压力减弱而发生突变，获得新的功能。图 9.4 展示了这三个术语和它们之间的相互关系。

基因组功能注释可以描述为鉴定序列数据特定区域的功能的过程，这些特定区域几乎缺少相应的信息（Overbeek et al. 2005）。通过对编码区进行功能注释，对整个基因组的了解将深化。最直接和可靠的方法是通过生物实验得到编码区域的功能，但这种途径很耗时间，且花费高，并不适用于新产生的巨量基因组数据。

具有相似序列的基因编码蛋白质通常拥有相似的三维结构。由于三维结构决定了蛋白质的功能，有假设认为，具有相似序列的基因编码的蛋白质功能也相似。这种基因序列和相应的基因产物间的联系，可以用来对未知基因进行功能注释。为此，将未知功能的核苷酸或氨基酸序列与具有已知确定功能的基因序列进行比较，如果能够找到相关的序列相似性，可以将已知基因的功能赋给未知基因。多种计算工具可以用于序列相似性的检测，基于局部比对的搜索工具 BLAST（Altschul et al. 1990）是其中最有名并且最广泛使用的工具之一。

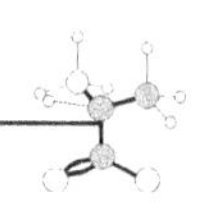

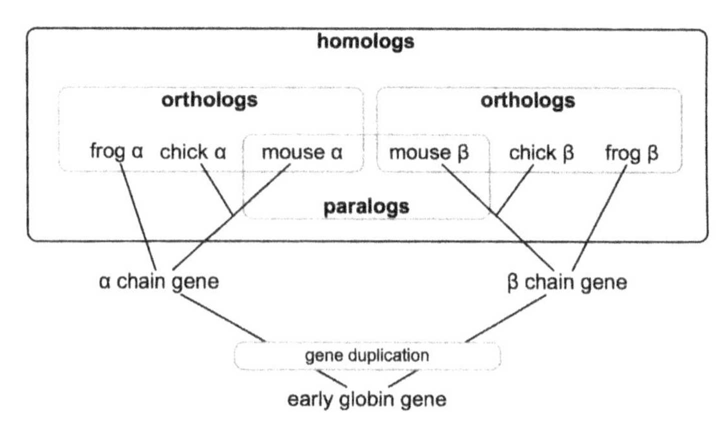

Fig. 9.4 Relation of the terms homology, orthology and paralogy. Genes that have a common ancestor are homologous, orthologous genes are separated by a speciation event. A gene duplication event generates two paralogous copies of a single gene. This figure is based on the figure at the following website: http://www.ncbi.nlm.nih.gov/Education/BLASTinfo/Orthology.html

图 9.4　同源性、直系同源和旁系同源的关系。拥有相同祖先的基因是同源的，直系同源基因通过物种形成事件而分离。一次基因复制事件产生一个基因的两个旁系拷贝。本图片是基于以下网址：http://www.ncbi.nlm.nih.gov/Education/BLASTinfo/Orthology.html

英文注释：Homolog，同源基因；Ortholog，直系同源基因；Paralog，旁系同源基因；Frog，青蛙；Mouse，小鼠；Chick，鸡；Gene Duplication，基因复制；Early Globin Gene，早期球蛋白基因

9.3.4.2　蛋白质结构域

蛋白质序列还可以进行保守结构域的细致分析。结构域可以被认为是蛋白质功能的构建模块。已经建立专业的数据库（如由 InterPro 联盟建立的数据库）进行蛋白质结构域的分析工作（见 9.3.3.3）。

序列分析和对数据库进行相似性搜索，为基因的功能注释提供了证据。如果检测到的序列相似性低或者证据是矛盾的，基因的功能注释将变得困难，这就需要人为干预，通常认为由专家人工进行的功能注释的质量是最好的。

9.3.4.3　使用基因聚类进行功能注释

如前所述，序列相似性可以用来推断基因的功能。结合基因组数据的高级分析和序列分析产生的结果会更加丰富。一个物种的基因在染色体上的排列顺序通常可以产生功能性相关基因的额外信息，尤其对原核生物。在细菌中，基因在染色体上的顺序（基因顺序）在进化分歧时间相近的物种中是保守的（Tamames et al. 1997），但是两个分化物种中基因顺序和组成会随着时间变化而发生变化。基因复制、丢失和横向基因转移可以改变基因组中的基因组成。易位、转座、倒位以及染色体断裂和融合会影响基因排列。

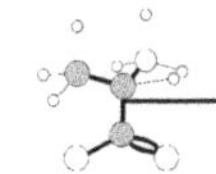

有趣的是，一些非常保守的基因组成和顺序可以在远缘物种中被发现。1966 年 Bauerle 和 Margolin 首次使用术语“基因簇”来描述鼠伤寒沙门氏菌（*Salmonella typhimurium*）中的色氨酸基因簇。随着更多物种的色氨酸基因簇被发现，Tatsuov et al.（1996）提出，虽然原核生物基因组的基因顺序通常不受选择的压力，但某些基因倾向于维持它们在染色体上的邻近位置。这种情况可以解释为相关基因的染色体邻近效应对物种具有进化优势。例如功能相关的基因倾向于相互邻近，如果它们在原核生物基因组中由同一个操纵子共同调控的话。如果它们的基因产物能相互作用，让其彼此邻近对细胞是有利的（Dandekar et al. 1998）。

系统发育上进化分歧物种中，它们的基因顺序保守的另外一个原因是横向基因转移事件的发生（Lawrence and Roth 1996）。利用基因所在染色体的邻近信息（Overbeek et al. 1999）来推断基因功能偶联，进而为功能注释提供附加证据是合理的。特别是如果未知功能基因在数个基因组的保守基因簇中找到，这些位置关联信息可能提示潜在的基因功能（Tamames et al. 1997）。

9.3.4.4 比较性分析的现有资源

各种各样的注释系统被开发出来，以支持基因组注释工作（见 9.3.3.2 节）。随着超过 1000 多种全基因组测序序列可以利用，比较基因组注释策略成为主要方法。这种策略通过整合近缘或所有可获得的现有基因组数据，来进行新基因组的注释工作。这里我们介绍一些支持此策略的注释系统、相关的内容和数据库。

MAGPIE：MAGPIE 是由 Gaasterland et al.（2000）开发的注释系统，提供图像化界面，注释者可以用来操作整个基因组，并且支持数据自动收集、分析和注释。这个系统使用内置 HERON 工具，基于相似性查找并进行自动注释。个体注释者的决策过程可以通过选择标准的模式，从高分匹配序列的描述中选出最优的注释。

ERGO：ERGO 基因组分析和挖掘软件包整合了基因组学、生物化学、高通量表达谱、遗传学和同行评议的学术期刊中的数据（Overbeek et al. 2003）。500 个不同完成程度的基因组已经被整合到了这个系统中。通过比较分析诸如共调控、基因融合事件、功能相关基因的染色体邻近效应等来进行基因的功能注释。数据库中包含了对来自生命三域（古菌、细菌和真核生物）基因组的细胞信号通路的重建。这个系统已不再免费使用，现在只能通过付费预约才可使用。

The SEED：SEED（Overbeek et al. 2004）系统能对指数增长的完整测序基

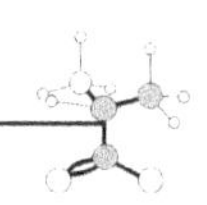

因组进行功能注释。它是商业 WIT（Overbeek et al. 2000）和 ERGO 注释系统的开放型继任系统。SEED 系统取代一次注释一个基因组的策略，而使用可同时注释所有可用基因组的策略。其中比较基因组分析支撑着有关的注释过程。

系统的数据库存储了公开可用的物种序列，可进行基因组特征穷举的两两对比（all-against-all）相似性矩阵的预计算，从而鉴定和可视化染色体簇和功能耦合的基因。另外，存储于数据库中预计算的相似性信息，有利于注释者找到直系同源基因，方便在所有物种中进行同源基因的一致注释。该系统还能在一个更高的层次上组织关联基因。SEED 系统首次引入子系统概念（Overbeek et al. 2005）。

KEGG 和 KOBAS：京都基因和基因组百科全书数据库（Kyoto Encyclopedia of Genes and Genomes，KEGG；Kanehisa and Goto 2000）是整合了各种生化通路中分子相互作用网络的数据库。这些数据库描述了酶类、化合物和相关通路的连接点。最近发布的 KOBAS（Mao et al. 2005）系统使用 KEGG 本体论（KEGG Ontology，KO）对大量基因进行自动注释。KO 是类似于基因本体论（Gene Ontology；Ashburner et al. 2000）的受控词汇，用于将基因关联到 KEGG 数据库中的代谢通路。KOBAS 是用 Python 语言写成的自动注释系统，它可以对一组序列进行完整通路的鉴定和注释。如果发现查询序列与已注释的基因有很高的序列相似性，系统就进行 KO 术语分配，并且将这些序列与 KEGG 图谱相关联。

STRING：STRING 数据库（von Mering et al. 2005）储存了基因间的基因组关联信息。两个基因间最相关的关联是在系统发生上分化的物种间具有保守的染色体邻近关系和基因融合事件。另外，数据库收集了微阵列实验中具有相似基因表达模式的信息和文献中共同表达的基因名称。这些综合的信息可以用来预测未知蛋白间的互作关系。这个数据库是预计算全球性资源用于研究基因间功能互作关系，可在线使用。

COG：COG 数据库（Clusters of Orthologous Groups of proteins，蛋白质直系同源簇；Tatusov et al. 2003）是由 Tatusov 和同事在 1997 年创立的。该数据库是基于基因间共同的起源，来对完整测序的物种中的基因进行分类。最初有 21 个全基因组被用来进行 all-against-all 序列比对，采用基因组特异的最优比对标准来比较所有的编码序列，从而构建直系同源簇。通过这些簇来进行 COG 分类。COG 分类作为受控的分层词表，可用于描述蛋白的功能。数据库创建初期，建立了 2091 COG 记录，覆盖了一个物种中多达 83% 的基因产物。随着新测序物种产生的基因序列，同源簇也在不断扩展。

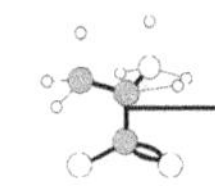

GO：基因本体论（Gene Ontology，GO）计划是为了能够使不同数据库中对基因产物的功能描述相一致而产生的合作项目（Ashburner et al. 2000）。基于生命三域中基因功能保守的事实，建立了一种共同的基因产物注释语言，从而使得基因组数据库间的交互使用变得简单。GO 数据库开发了三个结构层次的可控制词汇表（本体论），分别从生物过程、细胞组分和分子功能三方面描述基因产物，无物种相关性。基因本体论可以登陆 http://www.geneontology.org 网站查找，也可通过 AmiGO 本体论浏览器浏览现有的 GO 术语和它们之间的关系。

9.3.5 主要公共序列数据库和其他资源

这一节将介绍位于欧洲、日本、美国的三大主要公共序列数据库（参见表 9.6）。这些数据库和资源旨在为科学界提供全面的序列和注释信息。

在第一部分，我们首先介绍三个主流的核苷酸序列中心如何获得数据，接着描述向不同数据库中心提交不同类型数据所需要的相应步骤。第二部分，将集中讨论蛋白序列数据库 UniProt。最后，介绍一些对海洋生物学家有用的其他资源。

9.3.5.1 主要公共的核苷酸序列数据库

三大公共数据库中心分别位于欧洲、日本和美国（见表 9.6）。这些数据中心构成国际核苷酸序列数据库联盟（INSDC，www.insdc.org），已进行了超过 18 年的合作，每天相互交换新提交和更新的核苷酸序列和注释数据（Brunak et al. 2002）。这些数据库共同为公共领域提供全面的序列和注释信息。

表 9.6 主要的公共核苷酸序列数据库和它们的数据库中心

Table 9.6 The major public nucleotide sequences databases and their database centres

Major public database centres		
Europe	Japan	USA
EBI (European Bioinformatics Institute)	CIB (Center for Information Biology) at the NIG (National Institute of Genetics)	NCBI (National Center for Biotechnology Information)
Established in 1993 www.ebi.ac.uk EMBL-Bank created in 1981 (Cochrane et al. 2008)	Established in 1995 www.ddbj.nig.ac.jp DDBJ created in 1986 (Sugawara et al. 2008)	Established in 1988 www.ncbi.nlm.nih.gov GenBank created in 1982 (Benson et al. 2008)

英文注释：EBI，欧洲生物信息研究所；CIB，日本信息生物学中心；NCBI，美国国家生物技术信息中心。

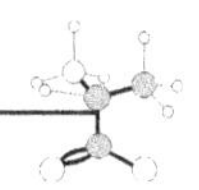

数据库联盟 INSDC 主要从事三类工作。首先，为数据生成者提供提交服务，让数据提交尽量简便，同时保留这些序列数据重要的生物学相关信息（诸如序列的生物学来源）和注释形式的功能解释。第二，开发相关结构和格式，用来精确简明地描述序列和注释数据。开发核心是用户的实用性，工具包括 INSDC 特性表定义文档和相关的词汇（例如以/country 和 /db_ xref 为限定符的词汇）。最后，通过和出版商合作，努力促进序列和注释数据的公共可用性。

每个中心都努力对提交的序列和注释信息进行质量控制，但是数据产生者仍然对其提交的序列内容具有编辑责任。面对持续增长的数据量，质量控制措施依赖于自动化确认过程。数据一旦提交后，接受的数据库会为提交序列提供一个唯一永久的登录号（譬如 AF123456），以方便后续的登录识别。主要的分子生物学杂志在文章发表之前，都需要作者提供数据库登录号。

登录号前缀由提交的数据库和数据类型决定（例如向 DDBJ 直接提交、EMBL 的基因组项目、GenBank 的 EST 以及向 JPO、EPO 或 USPTO 提交的专利）。目前使用的全部前缀编码可以在这里找到：http://www.ddbj.nig.ac.jp/sub/prefix.html.

注意 GI 号（例如 GI：26117688）是 Genbank 内部标识符，不是初始的 INSDC 登录号。为了处理 GenInfo 标识符或 GI，可使用 NCBI Entrez 网页提取对应的初始 INSDC 检索号（例子 GI：26117688 对应的标识符为 U00089）。

（1）CIB 的 DDBJ 数据库。日本国立遗传研究所（National Institute of Genetics，简称 NIG）在 1986 年创建了日本 DNA 数据库（DDBJ；Sugawara et al. 2008），2001 年建立了信息生物学与日本 DNA 数据库中心（Center for Information Biology and DNA Data Bank of Japan，简称 CIB－DDBJ）。该中心主要负责收集日本实验室提交的数据。

在 DDBJ 网站上定期更新的 DDBJ/EMBL/GenBank 特征表格文件中，可以找到 DDBJ 平面文件格式的示例条目。

网站上有几个有用的 DDBJ 注释的例子（例如 rRNA、EST、微卫星、转座子）。可通过多种方式获取 DDBJ 核苷酸数据库中的数据（见表 9.7）。

（2）EBI 的 EMBL 数据库。1981 年位于德国海德堡的欧洲分子生物学实验室（European Molecular Biology Laboratory，简称 EMBL）创建了 EMBL－Bank 核酸序列数据库（Cochrane et al. 2008）。1993 年，它被转移到英国剑桥附近的茵格斯顿欧洲生物信息研究所（EBI）分部，与基因组研究院（如桑格研究所）距离更近。目前 EMBL-Bank 成为 EBI 的蛋白质和核苷酸数据组

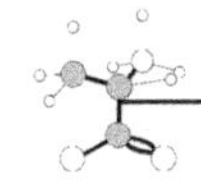

（PANDA）的一部分。

表 9.7　DDBJ 数据库不同的数据检索方式

Table 9.7 The different ways to retrieve data at DDBJ

Sequence retrieval at DDBJ	
Getentry	Data retrieval of nucleotide sequences mainly by accession numbers
ARSA	All-round Retrieval of Sequence and Annotation. Search of sequence libraries such as DDBJ and UniProt, sequence-related libraries such as PROSITE and Pfam, protein 3D structures, and metabolic pathways
SRS	Sequence Retrieval System offering term search
GIB	Genome information broker or data retrieval and comparative analysis system for completed genomes

英文注释：DDBJ，数据检索，包括 Getentry、ARSA、SRS 和 GIB 等方式，其中 Getentry 是指通过登录号来检索；ARSA 是全面序列注释检索系统，是高速关键词检索工具；SRS 是能提供词搜索的序列检索系统，GIB 是基因组信息数据库或全基因组数据检索和比较分析系统

在 EMBL 网站上定期更新的 DDBJ/EMBL/GenBank 特征表格文件中，可以找到 EMBL 平面文件格式的示例条目。

EMBL 网站上还有几个其他有用的 EMBL 注释的例子。

可通过多种方式获取 EMBL 核苷酸序列数据库中的数据（见表 9.8）

表 9.8　EMBL 数据库不同的数据检索方式

Table 9.8 The different ways to retrieve data at EMBL

Sequence retrieval at EMBL	
Simple sequence retrieval (embl fetch)	Sequence retrieval by accession number
SRS	Query all databases by term search, including EMBL-Bank standard, EST, STS and GSS data
EMBL sequence version archive	Repository of all current and historical EMBL entries
Browse data by geography	Geographical origin of sequenced samples
FTP server	Complete latest EMBL release, completed genomes, contigs, WGS sequences, patent sequences, etc.
Genomes	Access to completed genomes
Genome reviews	Genome annotation of Archaea, Bacteria, bacteriophages and selected Eukaryota
Ensembl genome browser	Annotation of large eukaryotic genomes
Integr8	Proteome analysis information

英文注释：Browse data by geography，按照测序样本的地理来源浏览数据；Ensembl genome browser，Ensembl 基因组浏览器，提供大型真核生物基因组的注释信息检索

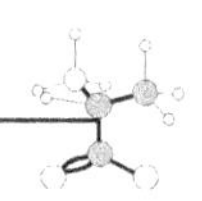

(3) NCBI 的 GenBank 数据库。1982 年，美国洛斯阿拉莫斯国家实验室（Los Alamos National Laboratory，简称 LANL）创建了 GenBank 核苷酸序列数据库（Benson et al. 2008）。该数据库于 1993 年归属于美国马里兰州贝塞斯达的美国国家生物技术信息中心（National Center for Biotechnology Information，简称 NCBI）管理。

在 NCBI 网站上定期更新的 DDBJ/EMBL/GenBank 特征表格文件中，可以找到 GenBank 平面文件格式的示例条目。

在 BankIt 网站上，可以找到几个有用的 GenBank 平面文件格式的注释例子。

可通过多种方式获取 GenBank 核苷酸序列数据库中的数据（见表 9.9）

表 9.9 GenBank 数据库不同的数据检索方式

Table 9.9 The different ways to retrieve data at GenBank

Sequence retrieval at GenBank	
Entrez nucleotide browser	Searches for sequences in Genbank, RefSeq and PDB
dbEST searching	Expressed sequence tags from one of the 6 major organism groups (Archea, Bacteria, Eukaryota, Viruses, Viroids, and Plasmids)
dbSTS searching	Sequence tagged sites
dbGSS searching	Genome survey sequences
FTP	Full release and daily updates of GenBank
Genomes	Views provided for genomes from one of the 6 major organism groups (Archea, Bacteria, Eukaryota, Viruses, Viroids, and Plasmids)

英文注释：STS：Sequence tagged sites，序列标签位点，是已知核苷酸序列的 DNA 片段，是基因组中任何单拷贝的短 DNA 序列，长度在 100 ～ 500bp 之间；GSS：Genome Survey Sequences，基因组序列，是基因组 DNA 克隆的一次性部分测序得到的序列

(4) 核苷酸序列提交。如何提交数据给核苷酸序列数据库？

如果你需要提交一条或几条序列（通常小的实验室是这种情况），推荐使用三大核苷酸数据库网站中的直接提交步骤（见表 9.10）。对于时间长或复杂的提交任务，需要提前联系好数据库管理方以获得帮助。

如果你需要提交大量的序列（通常大的测序中心是这种情况），这三个核苷酸数据库网站都提供了批量提交程序，例如 DDBJ 上的大规模提交系统 MSS（mass submission system）。

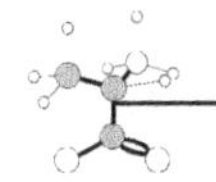

表 9.10　主要公共数据库中心的数据提交工具

Table 9.10 Submission tools available at the major public database centres

Direct submissions of DNA sequences			
	DDBJ (CIB)	EMBL (EBI)	GenBank (NCBI)
Submission tool	Sakura[1]	Webin[2]	BankIt[3] or Sequin[4]

[1] http://sakura.ddbj.nig.ac.jp/top-e.html
[2] http://www.ebi.ac.uk/embl/Submission/webin.html
[3] http://www.ncbi.nlm.nih.gov/BankIt/index.html
[4] http://www.ncbi.nlm.nih.gov/Sequin/index.html

注：在向公共数据库提交数据时，可参照数据提交工具的说明，准备相关的数据格式。

如果要向 NCBI 提交表达序列标签（Expressed Sequence Tags，简称 EST）、基因搜寻序列（Genome Survey Sequence，简称 GSS）和序列标签位点（Sequence Tagged Sites，简称 STS）之类的数据，可以仔细浏览 NCBI 网站上的相关细节信息。

对于任何数据的提交，建议数据提交者仔细研究所要提交的数据库（DDBJ、EMBL 或 GenBank）网站上更新的数据提交说明信息。三大数据库可通过下面的邮箱进行相关咨询：

DDBJ：ddbjsub@ ddbj. nig. ac. jp

EMBL：datasubs@ ebi. ac. uk

GenBank：gb-sub@ ncbi. nlm. nih. gov

（5）第三方注释数据库。数据库联盟 INSDC 中的第三方注释数据库（Third Party Annotation，简称 TPA；Cochrane et al. 2006），创建于 2002 年，以便从提交者那里接收核苷酸序列的高质量注释。这些核苷酸序列不是提交者自己创造的，这些注释信息的质量是以实验或推理分析来支持的。

TPA 数据分为两类：

• 实验证据类型（譬如 BK000016）

• 推理证据类型，源分子或其产物不是通过直接实验得到的（例如 BK000554）

更多关于 TPA 数据的信息可访问 DDBJ、EMBL 或 GenBank 的网站。

9.3.5.2　主要的公共蛋白质序列数据库：UniProt

美国的 Margaret Oakley Dayhoff 博士（1925 – 1983）是将计算机运用于化学和生物学的生物信息学领域的先驱，他创造了“蛋白质序列和结构地图集”（Atlas of Protein Sequence and Structure），从 1965 到 1978 年间由美国国家生物

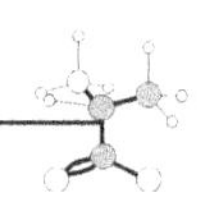

医学研究基金会（National Biomedical Research Foundation，简称 NBRF）发表。1984 年，NBRF 启动了“蛋白质信息资源”（Protein Information Resource，简称 PIR）计划，成为第一个蛋白质序列信息资源，蛋白质序列数据库 PIR 也因此而诞生。2004 年底，在 PIR 的基础上产生了名为 PIR-PSD 的蛋白质序列数据库。

Swiss-Prot 是 Amos Bairoch 于 1986 年在瑞士日内瓦创建的一个蛋白质序列注释数据库，其第一条记录是人细胞色素 c 蛋白（http://www.expasy.ch/uniprot/P99999）。从 1998 年起，瑞士生物信息研究所（Swiss Institute of Bioinformatics，简称 SIB）开始维护 Swiss-Prot，同时还维护名为蛋白质分析专家系统（Expert Protein Analysis System，ExPASy）的蛋白质组服务器（http://www.expasy.org），这个服务器 15 年来一直致力于对蛋白质序列和结构以及双向聚丙烯酰胺凝胶电泳的分析。

1996 年，欧洲生物信息研究所（EBI）建立了 TrEMBL 来应对日益增加的序列数据。TrEMBL 是一个计算机注释的蛋白质序列数据库，其中包含翻译 DDBJ/EMBL/GenBank 核酸数据库中的所有编码序列（coding sequences，CDS）而得到的蛋白质序列，这些蛋白在 Swiss-Prot 中还找不到。Swiss-Prot 和 TrEMBL 的建立是在 SIB 和 EBI 的共同合作下完成的。

2002 年，三大研究所（EBI、PIR 和 SIB）整合了他们的资源和专长，创建了通用蛋白质资源（Universal Protein Resource，简称 UniProt）联盟（Consortium 2008），提供一个优质和全面的蛋白质序列和功能注释数据库（http://www.uniprot.org）。

（1）UniProt 数据库的组织结构（表 9.11）。UniProt 知识库（UniProt Knowledgebase，简称 UniProtKB）由两个数据库构成，包括 UniProtKB/Swiss-Prot 和 UniProtKB/TrEMBL，前者是一个结合了实验结果和计算机辅助分析的高质量手工注释的非冗余蛋白质序列数据库，后者是一个高品质的计算机自动注释数据库。TrEMBL 条目可以经过手工注释，整合到 Swiss-Prot 中，并保留独特的登录号。

UniProtKB 包含了几乎所有可用的蛋白序列，但不包括以下几个：

- 大部分非生殖细胞系的免疫球蛋白和 T 细胞受体
- 合成的蛋白序列
- 大部分专利保护的序列
- 核酸序列翻译过来的小片段（< 8 个氨基酸）
- 假基因

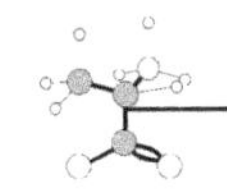

- 融合或截断蛋白
- 非真正的蛋白（如有足够的证据表明不太可能存在的蛋白）

上述的前五种蛋白在 UniProtKB/TrEMBL 建立的过程中可以自动被鉴定出来，后面两种由数据管理员手工识别（例如对基因组序列进行基因预测时，一些被错误预测为能编码蛋白的序列）并去除。

所有这七种类型的序列被收录在 UniParc 中，并给出了它们不被 UniProt-KB 包含的原因（见下文）。

表 9.11　UniProt 数据库的组织结构

Table 9.11 Organisation of the UniProt databases

The universal protein resource			
UniProtKB: protein knowledgebase	UniRef: sequence clusters	UniMES	UniParc
UniProtKB/Swiss-Prot (manually curated annotation) UniProtKB/TrEMBL (automatic annotation)	UniRef100 UniRef90 (at least 90% sequence identity) UniRef50 (at least 50% sequence identity)	Metagenomic and environmental sample sequences	UniProt archive

英文注释：UniProt：Universal Protein Resource，通用蛋白质资源，提供了一个有关蛋白质序列和功能注释的稳定、综合、可免费访问的中心资源；UniProtKB：UniProt 知识库；UniRef：UniProt 参考资料库；UniMES：宏基因组和环境微生物序列数据库；UniParc：UniProt 档案

更多关于 UniProtKB 的信息详见 UniProt 网站。

UniProt 参考聚类（UniProt Reference Clusters，简称 UniRef）将 UniProtKB 中相近的蛋白序列聚类成单个记录，以提高检索速度。UniRef90 和 UniRef50 分别代表序列一致性至少为 90% 和 50% 的序列集。如需更多的 UniRef 相关信息，请参见 UniProt 网站。

UniProt 宏基因组和环境微生物序列数据库（简称 UniMES）包含了全球海洋抽样考察队（Global Ocean Sampling Expedition，GOS）最初提交到国际核酸序列数据库（INSDC）中的数据。可以在 FTP 站点上下载 FASTA 格式的 UniMES 序列数据和“UniMES matches to InterPro methods”文件。

UniProt 归档（UniProt Archive，简称 UniParc）是用于包含所有公开蛋白质序列的非冗余数据库。UniParc 对每条唯一的序列只存一次，并给出一个稳定唯一的识别号（UPI），以便识别来自不同数据库的同一蛋白。一经确定，

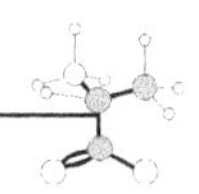

所有的 UPI 不会被删除、改变或者重新分配。UniParc 中每一条记录的基本信息包括标识符、序列、循环冗余校验码、来源数据库中的检索号、版本号和时间标识。如果 UniParc 中的记录没有收录在 UniProtKB 中，将会提供 UniProtKB 不收录的原因（譬如因为专利原因）。

更多 UniParc 的相关信息请见 UniProt 网站。

（2）数据的发布、提交和获取。UniProt 每三周更新一次，一年生成三次主要版本。第一个主要版本于 2003 年 12 月发布，最新版本（2008 年 2 月 26 日发布的第 13 版本）收录了 5 751 608 条记录，包括：

- 356194 条 UniProtKB/Swiss-Prot 记录，涉及 11290 个物种（版本 55）
- 5395414 条 UniProtKB/TrEMBL 记录，涉及 1552882 个物种（版本 38）。

当研究人员有新的蛋白序列需要提交时，可以通过 EBI 网站上的 SPIN 工具直接提交到 UniProtKB。

在 UniProt 的网站（http://www.uniprot.org）上，通过一个蛋白索取号，可以搜索到该蛋白在 UniRef 里的记录以及在 UniParc 里的唯一识别号等信息。

9.3.5.3 RefSeq

参考序列数据库（RefSeq；http://www.ncbi.nlm.nih.gov/RefSeq/）是 NCBI 建立的转录组、蛋白组和基因组区域的非冗余集合（Pruitt et al. 2007）。虽然 GeneBank 中包含了提交的任何物种（超过 250000 种不同的生物）的序列，但是 RefSeq 却只收录数据比较充足的主要物种（截止 2008 年 1 月发布的版本 27 中包含了 5000 种左右的不同生物）。每一条 RefSeq 记录只包含来自一个物种的一条单一序列，这些记录包含了各种各样的注释状态，既有已经注释的（已被验证或评审过）也有未注释的（推断、模拟、预测、暂时的或者全基因组鸟枪测序的序列）。

RefSeq 数据库可有多种访问方式。可以直接在网站上查询，也可以通过 Gene、Entrez、PubMed 和 Map Viewer 等 NCBI 资源提供的链接进行访问。

RefSeq 索引号是由两个字符接着一个下划线作为前缀，如 NP 010000。更多关于 RefSeq 索引号格式的信息请见 RefSeq 网站。

表 9.12 展示了 RefSeq 和 UniProt 的不同特征。

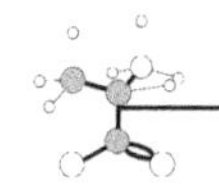

表 9.12　RefSeq 和 UniProt 的比较

Table 9.12 Comparison RefSeq vs UniProt

RefSeq	UniProt	
Coding sequences from the NCBI's set of genomic, transcript and protein reference sequences. Only the entries with validated or reviewed status are annotated entries	UniProtKB/Swiss-Prot: manually annotated protein sequences mostly derived from TrEMBL	UniProtKB/TrEMBL: automatically annotated protein sequences derived from coding sequences in nucleotide sequence database
Release 27, January 11, 2008: 4,426,609 protein entries and 4,926 organisms	Release 13, February 26, 2008: 356,194 entries (release 55) and 11,290 different species	Release 13, February 26, 2008: 5,395,414 entries (release 38) and 1,552,882 different species
Bi-monthly release	Major release 3 times per year, minor release every 3 weeks	
Limited to major organisms	All organisms	
Exclusive NCBI database	Produced by EBI, PIR, and SIB	
Protein and nucleotide data	Protein data only	

In Table 9.12 you can find different characteristics of RefSeq versus UniProt.

注：本表主要是对 RefSeq 和 UniProt 这两个数据库在数据类型、收录的数据量、涉及的物种、数据释放时间等进行了比较；UniProtKB/Swiss-Prot：主要来自 TrEMBL 的人工注释的蛋白序列；UniProtKB/TrEMBL：对核酸数据库中的编码序列进行计算机自动注释得到的蛋白序列

9.3.5.4　其他资源

下面将介绍一些包含更多特定数据的其他资源。

（1）物种特异的数据库。下面的数据库只包含特定一个物种或者基因组的信息，这里只列出了部分数据库，这些数据库收录的数据类型和数据存储方式都大不相同。《核酸研究》（Nucleic Acids Research）每年 1 月份都会公布这些数据库的最新版本：

- FlyBase：果蝇基因和基因组数据库(http://www.flybase.org)
- GDB：人基因组数据库(http://www.gdb.org/)
- MGI：小鼠基因组数据库(http://www.informatics.jax.org)
- SGD：酵母基因组数据库(http://www.yeastgenome.org)
- TAIR：拟南芥基因组数据库(http://www.arabidopsis.org)
- WormBase：秀丽隐杆线虫和相关线虫的遗传学数据库(http://www.wormbase.org)
- ZFIN：斑马鱼模式生物数据库(http://www.zfin.org)

（2）海洋生物数据库。近年来，用于基因组和后基因组研究的现代高通量技术也已经应用到海洋科学领域。如今大规模平行 DNA 测序或杂交方法，

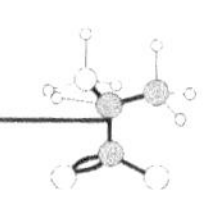

不仅能够鉴定出一个物种的基因集，还能识别基因调控网络。以下是最近的海洋基因组项目和数据库的部分列表：

• Moore 基金会海洋微生物基因组测序项目（http://www.moore.org/microgenome/）。该项目启动于 2004 年 4 月，更多详情请见 https://research.venterinstitute.org/moore/。

• Megx.net（http://www.megx.net）海洋生态基因组数据库，为海洋细菌和宏基因组学的全基因组分析提供了专业数据库和工具。

• 美国南卡罗来纳州查尔斯顿的海洋基因组项目（http://www.marinegenomics.org/）。这个项目自动化处理、维护、存储和分析 35 种不同海洋生物的 EST 和 16S RNA 序列以及从基因芯片实验得到的数据。

• 欧洲海洋基因组网（Marine Genomics Europe，简称 MGE；http://www.marine-genomics-europe.org/），致力于发展、利用和传播高通量技术在海洋生物研究上的应用。德国比勒费尔德大学的生物信息学平台已经为海洋生物基因组学欧洲共同体创建了一个生物信息学门户网站（http://www.cebitec.uni-bielefeld.de/brf/cooperations/mge.html），专门提供所有数据集和各种软件工具（如 GenDB、SAMS 和 EMMA 等）的中央访问点。

（3）wwPDB。1971 年，位于美国长岛的布鲁克海文国家实验室建立了蛋白质数据库（Protein Data Bank，简称 PDB），旨在存储经实验验证的生物大分子的三维结构。1974 年还只有 12 个有原子坐标的分子结构，但如今 PDB 已拥有超过 5 万个具有原子坐标和相关信息的结构，这些结构经过了 X 射线晶体、核磁共振（NMR）或电子显微镜技术的确定（Henrick et al. 2008）。

蛋白结构的重要性：虽然蛋白质的结构要比蛋白质的序列更难确定，但是目前结构已知的蛋白家族有 2000 左右。相比序列而言，蛋白结构显得更加保守，通常能揭示出很多在序列水平上无法探究的进化关系。结构信息对于研究一个蛋白的功能、催化机理和蛋白互作是非常重要的。近 10 年来存储的结构数据大量增加，再加上结构基因组学项目的不断兴起，也将产生大量的三维分子结构，这些都迫切需要一个全球性的蛋白质数据库来统一存储这些大分子结构数据并向公众开放。

全球蛋白质数据库（worldwide Protein Data Bank，简称 wwPDB）就是在这样的背景下产生的。2003 年，由美国的 RCSB PDB、欧洲 EBI 的大分子结构数据库（macromolecular structure database，简称 MSD）和日本大阪大学的蛋白质数据库（Protein Data Bank Japan，简称 PDBj），联合建立（Berman et al. 2003）。2006 年，美国的 BioMagResBank 也加入了 wwPDB。这些 wwPDB 站

点（表9.13）负责蛋白结构数据的存储、处理和分配。它们支持对每一个蛋白大分子结构建立唯一的标准化档案，并且每周更新一次。

表9.13　生物大分子三维结构公共数据库中心

Table 9.13 Public database centres for biological macromolecular 3D structures

wwwPDB data access sites			
BMRB: http://www.bmrb.wisc.edu USA	MSD-EBI: http://www.ebi.ac.uk/msd Europe	PDBj: http://www.pdbj.org Japan	RCSB-PDB: http://www.pdb.org USA

注：wwwPDB data access sites：wwwPDB 数据访问网站。

9.4　基于高通量技术的转录组分析

现代高通量测序技术产生了大量的生物序列数据。对海洋生物而言，这些数据可能由全微生物基因组或由前文描述的海洋真核生物大型 EST 文库构成。可利用的结构基因组学数据，进一步分析鉴定基因和其他序列的功能。通过对现有的基因组数据进行内在序列特征和序列比较的分析，能够初步了解新序列的功能。同时还需要定量实验来推断出未知功能基因的新假设以及对序列功能的预测验证。

分子生物学中心法则：蛋白质表达的信息流从 DNA 通过媒介信使 RNA（mRNA）到最终产物蛋白。基因表达调控发生在多个阶段，从 DNA、转录调控水平到氨基酸翻译水平和翻译后修饰。基因表达决定于机体的内在状态和环境条件。调控网络的控制在细胞内有复杂的信号传导网络介导。通常认为在特定实验条件下，对基因表达的定量测定可以推断基因的功能。多数科学家使用的一种方法，是通过基因之间的常见的调控模式来推断基因功能。通常，具有相似代谢通路或者其他共同功能的基因，在转录分析实验中往往表现出相似的基因表达模式。这种方法已被命名为“连接数法”（guilt-by-association；Quackenbush 2003）。

转录组学是功能基因组学中相对较新的领域，其目的是度量 mRNA 分子的表达丰度。有数种方法可以用来定量 mRNA 的丰度，包括实时定量反转录 PCR（qRT-PCR；Iizuka et al. 1994）、基因表达系列分析（Serial Analysis of Gene Expression，简称 SAGE；Velculescu et al. 1995）和微阵列。以下将对这些方法进行简单讨论。

实时定量反转录 PCR：是一种能对低丰度 mRNA 进行检测和定量的高度

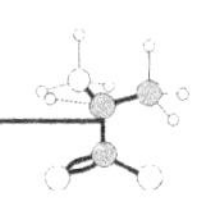

灵敏方法。在这种方法中，通过反转录酶将mRNA样本反转录成cDNA。然后使用基因特异的引物，通过标准的PCR技术对目标基因的cDNA进行扩增。基因特异性引物可以保证只扩增目标基因。样本cDNA的扩增通过荧光染料来指示，从而反映合成的DNA量。数种染料可供选择，最简便且最常用的是双链结合染料SYBR Green I。这种染料结合到双链DNA（dsDNA）上后会释放荧光，而荧光强度的增加和合成的dsDNA数量相关，因此可以对已经合成的PCR产物进行定量。最初的cDNA量可以通过荧光达到一个初始背景水平的时间来决定，这个背景水平被称为交点（crossing point，简称CP）。CP值越早到达，表明初始cDNA量越高。这种技术的不足就是低通量，一次实验只能检测一个基因的转录水平。但是，大规模商业化qPCR系统可以在一次qPCR运行中进行32～384个反应。

对这个技术所有方面的详细描述（诸如实验平台、试剂和数据标准化方面）超出了本章的范围，但是对qPCR相关内容的综述，包括更新的qPCR相关的文献列表，可以访问网址http://www.gene-quantification.info。目前关于qPCR数据的生物信息学分析方面，适用于全面分析。qPCR数据的开放软件非常多，CAmpER系统是其中的一个例子，可以从德国比勒费尔德大学的生物技术中心（http://www.cebitec.uni-bielefeld.de）免费下载使用。

基因表达系列分析（SAGE）和其他基于测序的方法：基因表达系列分析是对mRNA定量分析的高通量方法，该方法基于对cDNA的短片段（即标签）测序。相比于EST测序，这些标签仅是全长cDNA的一个短片段（11－25 bp）。

首先以生物素标记的Oligo（dT）引物结合在mRNA的5′末端的poly-A尾巴上，将mRNA反转录成双链cDNA，产生的cDNA在引物位点结合到链霉亲和素磁接上，接着被锚定限制性内切酶NlaIII酶切，得到较短的cDNA片段。

结合了cDNA的磁珠被分成两部分，cDNA分别接上接头（linker）A或B。连有接头的cDNA用标签酶BsmFI进行酶切，将cDNA从附着的磁珠上剪切下来，标签酶BsmFI具有结合特定的核苷酸序列（CATG）和剪切结合位点上游11 bpDNA序列的特性。A和B两部分产生的标签（tags）进一步被连接成双标签（di-tags），然后用接头A和B特异的PCR引物进行PCR扩增

当扩增到足够的水平，使用NlaIII去除接头A和B，然后将双标签随机串联成长链的标签，接着被克隆至质粒载体，通过Sanger测序获得序列。DNA载体上还留有特异的剪切位点，作为双标签体之间的分割。

如果研究物种的基因组序列是已知的（无论是EST或全基因组），可以将

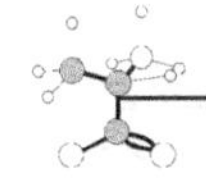

测序得到的序列标签比对到更长的序列上，从而进行差异基因表达的定量分析。相似序列标签来源于唯一的 cDNA 片段，标签数目可以认为与来源 cDNA 片段的初始量相关。但因为 PCR 扩增的原因，并不是呈线性相关。

如果所研究物种还没有可获得的序列信息，也可以对序列标签进行定量分析，因为标签序列是来源 mRNA 几乎唯一的标识符。然而，将 11 bp 长的短标签序列与可获得的数据库进行序列比较时，会产生大量的假阳性比对结果

SAGE 技术最初的问题是建库时需要含 poly-A 的 RNA 样本量较大，将近 2. 5 ~5 μg。针对这个问题，1999 年 Datson et al. 开发了 MicroSAGE 技术，使用链霉青霉素（strepatividin）涂层管代替磁珠。从而只需要 1 ng 的 RNA，大约相当于十万个哺乳动物细胞的重量。

最初的 SAGE 技术存在的另外一个缺点是标签长度很短。针对这个问题，研究人员发布了几种改良方法，通过使用不同的限制性酶来提高序列标签的长度。已经报道的有能产生 20 bp 标签的 LongSAGE 技术（Saha et al. 2002）和 26 bp 标签的 SuperSAGE 技术（ Matsumura et al. 2003），这些技术产生的较长标签可以显著提高对重要数据库的比对结果。

此外，基于双末端双标签的基因鉴定方法（简称 GIS-PET；Ng et al. 2005）的开发是另外一个技术改进。相比于 SAGE，PET 来源于 mRNA 的3′和5′端。加工 mRNA 的特异加帽结构的引物被添加到 3′末端，这样双末端双标签（paired di-tags）就含有同一个 mRNA 双末端的序列，进而可以用来识别准确的转录边界。

结合二代测序技术，PET 技术得到了进一步发展，产生通量更高的 MS-PET 技术（Ng et al. 2006）。该技术可以产生平均长度为 40 bp 的双标签（di-tag），因此，两个双标签就相当于第一代 454 测序仪的一个测序读长。

在这里需要说明的是，SAGE 和衍生的方法涉及了非常复杂的实验室标准性操作流程。二代测序技术增加的通量，使得测序的瓶颈变为对测序技术的适应。此外，需要指出的是，这些操作并不适用于原核生物，因为这些操作是基于 mRNA 的 5′加帽和 poly-A 尾巴结构，而这些结构只在真核生物中发现，原核生物中没有。

鸟枪转录组方法也可用于转录分析。从原理上讲，这个方法类似于前面描述的 EST 方法。将 mRNA 逆转录成 cDNA，再将 cDNA 片段化并测序。可使用高通量测序方法代替 Sanger 测序，测序产生的 read 可以比对到基因组进行定量分析。这个方法也适用于原核生物，并能实现对完整转录本的高覆盖度。

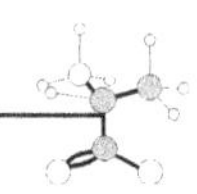

尽管该方法显示了很高的潜能，但目前仅被用于有限的研究，仍需要开发出可用于对鸟枪转录组数据进行统计分析的明确指导方法。

9.4.1 微阵列技术的基本原理

微阵列可以并行检测成千上万个基因对应的 mRNA 丰度（Lipshutz et al. 1995，Schena et al. 1995），因此，微阵列被认为是高通量技术（Lipshutz et al. 1999，Miron and Nadon 2006，Küster et al. 2007）。微阵列技术在 20 世纪 90 年代末期发展迅速，并因其广泛应用和相对好的性价比迎来功能基因组学的转折点，多种不同的微阵列技术和方法以及数据评估的统计方法被开发出来。

虽然微阵列平台在具体细节上存在许多技术差异，但所有微阵列平台的通用原则却非常明确。确定核苷酸序列的 DNA 分子被固定在载体表面（载体通常是镀膜玻璃），相同类型的分子聚在载体表面的同一区域（这些区域被称为特征或斑点）并且按网格排列。现今技术可以达到每 cm^2 一万个以上斑点的密度。为了测定 mRNA 丰度，mRNA 被抽提并且反转录成 cDNA 后，将 cDNA 标上荧光标记以便于 DNA 拷贝数的定量。一些操作技术可以直接使用 mRNA，不需要反转录。含标记分子的溶液（又称为目标分子）流入微阵列的载体表面，在平行杂交过程中，标记的单链 RNA 或 cDNA 分子和微阵列载体表面上的互补链杂交结合。通过一个检测装置来检测结合到特定点的目标分子的近似数量。通常是使用激光扫描仪生成可视图像。激光扫描仪通过一定波长的激光辐射激活荧光染料，释放出一定发射波长的光，产生的图像经过图像分析软件处理从而转变成强度测量值。

许多公司商业化生产微阵列。他们采用的技术在许多细节方面有差别，例如使用的阵列基板和产生探针序列的过程。斑点微阵列通过机械点样仪固定少量 DNA 溶液在基板上，其他公司（如 Agilent）用喷涂技术固定 DNA，Affymetrix 和 NimbleGen 阵列使用的是光刻技术，直接在基板上将核苷酸合成长的寡聚核苷酸。最重要的参数之一是固定或合成的探针序列的长度，当用 cDNA 点样，序列长度可以从数百个碱基到短的寡核苷酸。商业化的微阵列通常使用短的寡聚核苷酸序列。例如 Affymetrix 微阵列对每个基因有数种不同的 25 bp 长度的探针序列。结合几种不同的探针进行综合检测是必要的，可以弥补靶标 DNA 的非特异性结合。其他含有 50 ~ 80 bp 长度的寡核苷酸微阵列，特异性更高。各个微阵列平台另外一个重要的差异是它们支持的通道数。双色微阵列支持在一个阵列中对两个实验条件下的样品进行直接比较，单通道

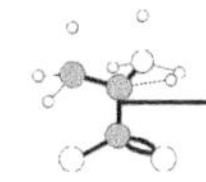

阵列可以用于单个条件的研究或者通过多个阵列进行比较研究。另外，针对单个阵列的设计，商业阵列供应商们有不同的生产成本和设计时间以及支持不同的斑点数。

9.4.1.1 可变性和重复性

和其他生物学实验一样，微阵列实验有一定程度的测量变异和测量误差。微阵列实验所产生的误差可以分为技术误差和生物误差（或说系统性误差）。

技术偏差对微阵列实验的结果影响很大。这种影响程度取决于所使用的技术和微阵列平台，可以通过同一个实验室和不同的实验室间的重复实验，来观察和衡量这种影响。主要原因是实验标准操作流程使用时的差异。更进一步的技术偏差产生于微阵列生产过程（如斑点大小和密度的差异；Bammler et al. 2005）。一些研究表明，不同的探针序列作为跨平台的误差来源，也会对结果产生很大的影响。其他的技术问题包括扫描仪的设置、图像分割和定量（Yauk et al. 2004，Repsilber and Ziegler 2005，Yauk et al. 2005）。

即便我们假设测量技术很完美、没有误差，但仍然存在着物种内的生物学差异，譬如不同细胞间或不同个体间的差异。这种差异是由于个体遗传特性和环境条件的差异性造成的。生物学差异性可以通过在相同条件下多次进行实验和收集多个样本（我们称之为重复）来进行估测。通常，对所获结果的显著性评估时，生物学差异似乎要比技术差异更加重要。因此，生物学重复要优于技术重复（Allison et al. 2006）。

生物学重复会受到技术和生物学差异双重影响，从而可用于评估实验的整体差异。如果重复的数目高于可获得的微阵列数目，通常使用混合方法来产生混合样品。

9.4.1.2 重复多少次?

重要的是要认识到任何基因表达测量方法都会有生物学差异，不光是使用微阵列技术。因此，任何分析都需要有生物学重复性。对于微阵列实验，从生物体收集细胞抽提足够量的 RNA 是非常复杂的，尤其是处理真核生物。因此，一个实验必要的重复次数是一个非常重要的问题。

非常不幸的是，这个重复问题没有简单的答案。三次重复是常用的重复次数（Lee et al. 2000，Yang and Speed 2002），可以作为最低数目的经验法则。现在很多学术杂志越来越要求发表的文章至少有三次生物学重复，同时还需要通过其他技术对实验结果进行验证。然而合适的重复次数取决于多个因素。生物学差异性和观察者想观察到的差异影响程度是最重要的。如果生物学差异很大，需要增加重复次数。类似地，检测到的差异影响越小时，就

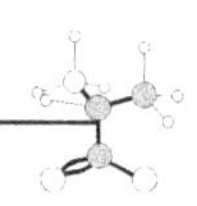

需要更多的重复来检测到微小的差异。研究所需要的精确性是另外一个重要因素，如果实验仅是想找到一些可用于进一步研究的候选基因，减少重复是明智的选择。

所谓的效能分析方法（power analysis methods）可以用来计算大概必须的重复次数。这些方法根据所需的效能（检测大部分差异表达基因的能力）、置信水平和数据的差异性，通过估测重复次数来辅助实验设计。参见 Pan et al.（2002）、Black and Doerge（2002）和 Li et al.（2005），尤其是 Page et al.（2006）已经将 PowerAtlas 软件应用于基于公共数据的效能分析。

9.4.2 基因表达分析

对微阵列实验得到的数据进行分析所采用的方案，取决于实验目的和实验问题。虽然如此，一些常规的基于数据特性的分析步骤是确定的。在文献中经常看到的分析步骤，将在下文中做详细讲解。

9.4.2.1 图像分析

标准的微阵列实验数据分析从分析扫描仪软件产生的图像数据开始。一些新型的微阵列平台通过电化学反应直接读取信号，对这些平台来说，图像分析步骤不是必要的。对于确实需要进行图像分析方法，每个通道生成的图像是分割开的，换句话说，在基板表面的特征位置要确定。大部分图像分析软件用能半自动分割斑点图像的网格管理器进行校正。也存在可以进行全自动分割的算法，但是需要特别小心，分割的错误会导致错误的结果，因为大量特征斑点的正确位置缺失会导致结果赋值到错误的特征位点上。

下一步，图像分析软件计算每个特征点的强度，并且同时进行诸如变异衡量和背景估计的统计分析。大多数软件附加信号质量检测，这些通常被称为标记（flags），可以用来过滤掉低信号强度或不规则形状的斑点。自动产生的标记需要特别小心对待，因为不知道这些测量是怎么进行的，而不同的软件使用的分析方法不同。如有疑问，最好忽略这些标记步骤，在分析后期再进行清除工作。

9.4.2.2 标准化

为了回答实验性问题，图像分析产生的各种测量值需要转换成一个单一值或较少的值，用来描述特征斑点的强度或强度差异。

微阵列数据实现标准化的目的是让一个实验的不同微阵列数据具有可比性。因此，有必要从数据集中移除系统偏差（Quackenbush 2002）。数据中的系统偏差可能来自于不同样本 RNA 浓度的差异、扫描仪设置的差异以及染料标记、脱色和检测等操作过程。从杂交相同标记提取物的技术重复微阵列的检查，发现扫描仪设置也会导致微阵列实验间的很大偏差。花色素苷染料3和5（Cy3和Cy5）是目前双通道微阵列芯片最常用的荧光标记。这些染料发出不同的光强，杂交目标数量和这个关系不是线性的。这可能导致非线性失真，在双通道强度测量值的散点图上可以看出来（见图9.5）。

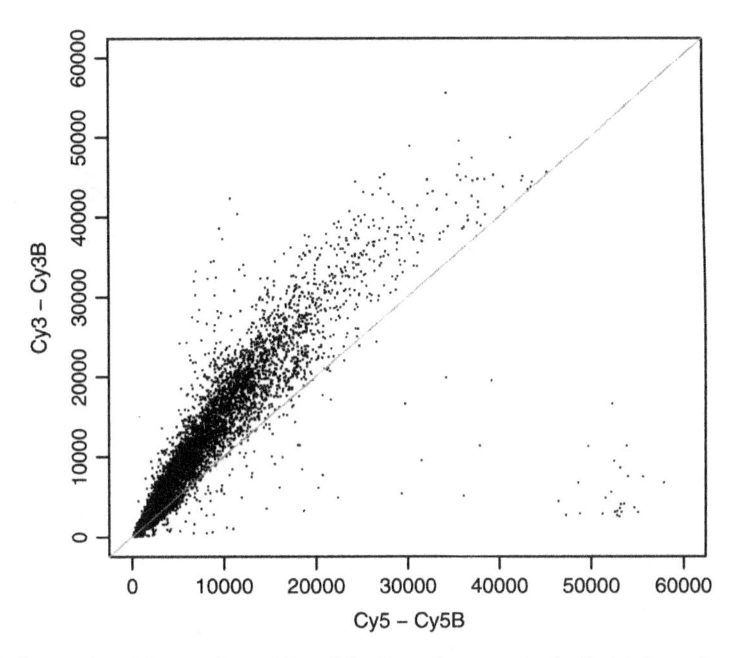

Fig. 9.5 Scatterplot of the raw intensities of the first microarray in the Swirl demo data from the microarray package for R. The raw channel intensities are background adjusted for each channel and plotted for each spot. The main diagonal is plotted as a *grey line*. The data distribution shows a visible deviation from the main diagonal

图9.5　来自R软件微阵列数据包中的漩涡演示数据中的第一个微阵列的原始信号强度散点图。原始通道强度是校正各个通道的背景，被绘制为单个点；主对角线用灰色线绘制，此图的数据分布显示出与主对角线的一个明显偏差。

Yang and Speed（2002）开发出一种名为散点图平滑函数进行标准化的操作方法。他们还提出对微阵列原始信号强度值进行特殊的对数变换，以用于在对数尺度上绘制表达差异图。这种转换将随机背景消除的强度值，转化为对数化的比率（log-ratio，M）和对数化的信号强度（log-intensity，A）：

$$M_i = \log R_i - \log G_i$$

$$A_i = \log R_i - \log G_i$$

其中 R_i和 G_i表示每个通道中第 i 个特征斑点的信号强度。这种表示方法相对于提供差异表达（M）对称测量的正常比率要具有优势，这样上调基因

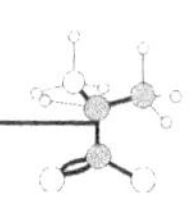

的绝对值和下调基因的绝对值一样。Yang 和 Speed 提出了 MA 散点图（MA-plots），它将两种测量方法结合在一块，以检查系统偏差和染料偏倚。从此，这种散点图成为微阵列数据分析的标准工具（见图 9.6）。

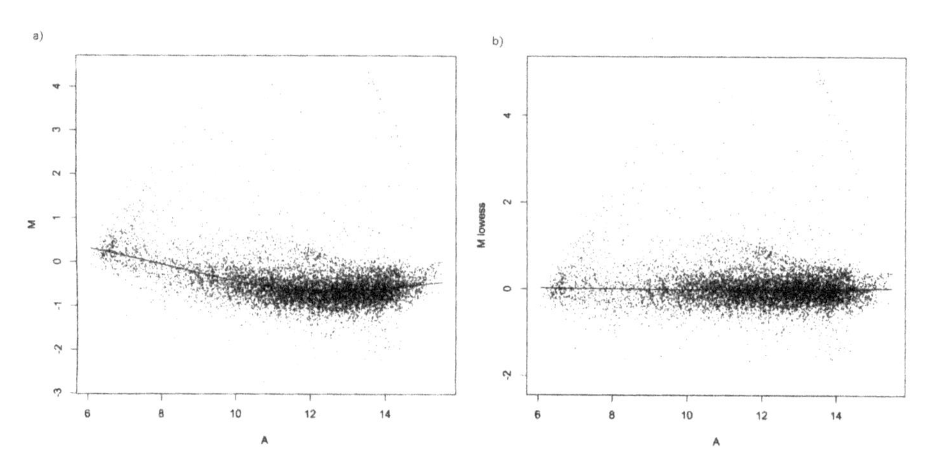

Fig. 9.6 The same data as in Fig. 9.5 after transformation. The y-axis corresponds to the logarithmic differential expression (M-value), the x-axis represents the logarithmic absolute expression measure (A-value). (**a**) depicts the data before and (**b**) after lowess-normalization. The lowess function used for normalizing the data is plotted as a *curve* following the centre of the data distribution

图 9.6　与图 9.5 相同数据转化之后的图。Y 轴表示差异表达的对数值（M 值），X 轴表示绝对表达量的对数值（A 值）。(a) 是用来描述局部加权线性回归标准化之前的数据图 (b) 是用来描述局部加权线性回归标准化之后的数据图；用于数据标准化的局部加权线性回归函数，按照数据分布中心绘制曲线。

英文注释：Lowess-normalization，标准化局部加权回归散点平滑法

9.4.2.3　显著变化的检测

微阵列实验之后最基本的问题是样本中或两个样本间哪些基因显著性（差异地）上调或下调。推断步骤是最重要的，这是许多实验唯一的相关分析步骤（之前的数据采集和处理步骤可以被视为推断步骤的准备工作）。同样，对于机器学习步骤，统计推断在数据简化方面能发挥重要作用。在早期的微阵列研究中，对比率或对数化的比率使用固定的阈值（cut-offs）。然而，选择特定的阈值具有随意性并且很快发现实际效果不好（Quackenbush 2001）。这种被称为倍数变化的方法并不能提供一个可测量的误差估计。没有对可变性进行评估，就无法评估观察到的样本内发生偶然事件的概率（在这种情况下是一个特定的 M 值）。

统计推断或统计检验被用来进行重复样本间的差异表达分析。针对微阵列数据，最近多种统计检验方法被开发出来。这些是在 W. S. Gosset、Fischer

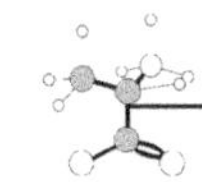

以及其他许多研究人员于20世纪初建立的经典检验理论基础之上开发出来的。假设检验总是基于同样的原则。

存在有两种相反的假设，即原假设或零假设（the null hypothesis，HO；表示转录水平差异不显著）和备择假设（the alternative hypothesis，H1；表示转录水平差异显著）。

假定原假设（HO）是正确的，除非有足够的证据拒绝它而接受H1假设。从样本数据算出统计检验值，来描述数据的经验分布。作为数据的总结，检验统计量可用来决定是否拒绝原假设，需计算出拒绝原假设的统计检验的阈值。阈值设定是通过以下方式：当HO是真，随机观察到该阈值是低概率的。假定检验统计的分布是已知的，如果HO是真，我们可以计算观察到一个至少为极端的统计值的概率，这个概率值就叫P值（p-values）。不幸的是，对P值的解释常常存在一些混乱。以微阵列作为例子，P值可以解读为观察到可以判断基因差异表达的测量值的概率，即使在特定的实验中没有这样的测量值存在。

有数种方法可以用来做微阵列数据的统计检验。经典的t检验（t-test）是其中最常用的方法。当输入的数据是正态分布时，可使用t检验。如果t检验不可用，还有其他统计检验可用，例如威尔科克森秩和检验（Wilcoxon's Rank Sum Statistic）或基因芯片显著性分析（Significance analysis of microarrays，SAM；Tusher et al . 2001）方法。其他的方法诸如CyberT方法（Baldi and Long 2001），是为小数量的重复样本的分析问题提供解决方法。CyberT最主要是借用微阵列上其他具有相似表达值的基因的方差信息。

9.4.2.4 聚类分析

聚类分析（clustering analysis，或简称clustering，聚类），已经成为检测多变量微阵列数据中的潜在结构和共调控基因非常常用的方法。聚类分析流行是可以理解的，因为它不需要或很少需要关于数据的前提假设。聚类分析的应用是出于“连接数法”假设。如果基因具有一个共同的调控机制（例如相同的转录因子），它们也可能是功能相关的。因此聚类分析可以非常有效地用于查找具有相似表达谱的基因群。

聚类算法把基因表达测量值分组成类。对单一分类或多个类加权渐变，聚类分类将非常困难。测量值通过两两间的不同来比较。因此，对差异性或距离进行测量的概念在所有聚类算法中起着关键作用。

分层聚类算法创建一个类似对象的层次结构，形成一个有根二叉树状图。分层聚类存在凝聚层次聚类和分裂层次聚类两种类型。其中凝聚聚类是最通

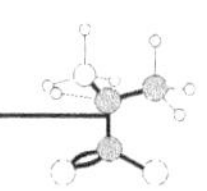

用的方法。

凝聚聚类采用自底向上的策略。首先将每一个对象作为一个类，然后每一步最相似的类被合并成更大的类。分裂聚类采取的策略正好相反，它首先将所有的基因置于一个大类中，然后再逐渐细分为越来越小的类。使用分层聚类方法来分析微阵列数据是由 Eisen et al.（1998）推广而流行起来，他们也开发了微阵列聚类分析软件，在这篇文章中也提出了以热图（heat map）的形式来可视化展示分层聚类结果的方法。表达值通过颜色代码来表示：负对数化的比率值用绿色表示，正值用红色表示，黑色表示数值接近0。

9.4.2.5 分类

有时，在事先知道某个样品来源信息的情况下，微阵列数据可以通过基因表达谱数据来预测样品的类别。分类的过程可被定义为将测量值分配成离散的类标签过程。

所有的分类方法都涉及到训练阶段和分类阶段。用已知来源的数据作为机器学习的训练数据集，然后可将未知来源的样本分类为一个各自来源的类。在分类阶段，训练分类器应用于数据，并对这些数据进行类标签的预测。

Golub et al.（1999）的报道是首先应用这个方法的研究之一，他们通过微阵列数据来分类不同类型的白血病。其中一个最简单但非常有用的方法就是k最近邻（k nearest-neighbour，简称kNN）分类算法。该方法基于k个最相似（即特征空间中最邻近）的样本中的大多数所属类别，来预测未知样本的类别（Cover and Hart 1967）。Wu et al.（2005）使用了一个kNN分类器来比较不同标准化方法的优点。

Vapnik（1999）开发了支持向量机（Support Vector Machine，简称SVM）方法用于分类。SVM适合于寻找一个最佳的分离函数，来确定线性和非线性的边界。经常在文献中可以看到将SVM应用于微阵列数据的例子（例如，Pavlidis et al. 2002，Liu et al. 2005），而且这种应用正变得越来越流行。

9.4.2.6 微阵列数据的分析软件

数种微阵列相关的分析软件在最近几年被开发出来。这些软件应用的领域也常常重叠。因此，在这里我们将专注于几个例子。目前还没有完美的解决方案，因为不同的软件包都具有各自的优点和缺点。尤其是没有一个软件，既能提供高度灵活的分析手段又具有易于操作的图形界面。对于大多数分析任务，这意味着分析将涉及多个软件程序的使用。商业软件可完成大多数任务，但使用许可费昂贵，可以选用开源的免费解决方案，同时也有免费的开源解决方案。软件选择的另一个重要方面是需要考虑软件所需要的硬件和软

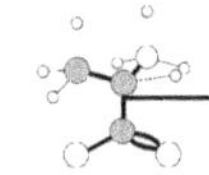

件环境。一些软件包，特别是基于数据库的系统，需要很强的软件安装和专业管理能力。在以下小节中将介绍几种不同的软件。

9.4.2.7　图像分析软件

图像分析软件将扫描器得到的图像结果进行分析。这个过程通常是双重的，需要通过分割化步骤来鉴定特征的位置和它们在图像内的边界，从而与微阵列的斑点对应。根据使用的软件，这一步可以手动操作；如果使用预先定义的网格信息，则可半自动操作或全自动操作。

例如 TIGR Spotfinder 软件是一个学术开源的项目，支持多个操作系统。还存在一个名为 Spot 的软件包，可在统计环境 R 下使用。Affymetrix 微阵列的用户必须使用 Affymetrix 的 MAS5 算法进行图像分析。

9.4.2.8　单一分析系统

单一分析系统进行各种分析算法的收集，但它们通常不进行数据管理。数据通常以文件形式存储，用户互动可通过图形用户界面（GUI）实现。譬如作为第一批开发的 Cluster 和 TreeView 软件（Eisen et al. 1998），专注于聚类算法，只能在 Windows 下工作。此外，还有 Genesis 软件可提供多种聚类方法（Sturn et al. 2002），是用 Java 来实现的，因此使用独立的操作系统。ArrayNorm（Pieler et al. 2004）和 MIDAS（Saeed et al. 2003）是为双色微阵列数据提供标准化和一些统计检验的软件。

应用 Affymetrix 芯片的用户可使用 dChip 软件，为寡核苷酸芯片提供标准化、聚类和分类分析（Li and Wong 2001，Lin et al. 2004）。

总之，中小型实验室推荐使用纯分析系统。通常这种系统对资源需求低，但是不提供数据管理和协作功能。对于有许多参与者的较大项目，这些系统可以作为数据库系统的补充。

9.4.2.9　通用数据库系统

还有一类系统兼具通用数据分析功能和数据管理能力，这些系统使用一个数据库系统来系统地存储和检索数据。此外，通过存储协议可进行实验数据的注释分析。这些系统包括开源系统 BASE（Saal et al. 2002）、MARS（Maurer et al. 2005）、MADAM（Saeed et al. 2003）和 EMMA（Dondrup et al. 2003；Dondrup et al. 2009）。所有系统提供用户和账户管理以及诸如让实验数据共享等协助功能。通常，软件通过网页浏览器进入。这些系统对服务器安装和管理具有相对较高的要求，这让小实验室使用不便。但对于大型研究机构，尤其是对参与国际协作的研究团队来说，设置和维护一个这样的中央安装系统是一个不错的选择。

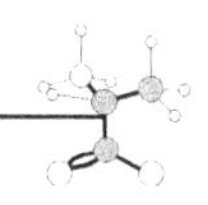

9.4.2.10 R 和 BioConductor

统计学环境 R（Team 2008）在所有分析系统中具有特别重要的作用。它是通用的统计环境，也是功能强大的编程语言。BioConductor 项目为包括芯片和序列分析等很多生物信息学应用程序打包和提供附加包（Gentleman et al. 2004）。与其他分析系统相比，R 具有很高的灵活性并提供最大量的分析算法，其中的常用统计功能也适用于微阵列数据分析，包含各种统计检验方法和图形化显示。在 BioConductor 包中有用于双色微阵列和 Affymetrix 芯片标准化的方法。伴随这种灵活性的不利因素，是该系统需要高水平的专业知识，通过命令行界面与用户进行交互，复杂的操作可能需要编程。不管怎样，我们依然推荐研究人员学习并且试着使用 R 和 BioConductor，因为这个组合在方法选择以及分析过程的全程控制方面提供了最大的灵活性，并且新方法通常最先以 BioConductor 软件包的形式进行应用。我们推荐读者阅读文献（Gentleman et al. 2005）来进行深入的学习。

9.4.3 数据共享和公共数据资源库

基因表达实验涉及复杂的实验流程并产生巨量数据。相对于基因组测序方法，功能注释需要更加精确，因为一个好的功能基因组学功能注释也涉及到环境条件。在数据量增长、数据收集流程和更进一步的实验过程中，样品需要变更数次，并且涉及到复杂的微阵列设计，需要容纳成百上千的特征。

微阵列实验结果需要和这些元信息结合在一起。不然，我们无法解释数据结果，甚至不能独立地重复微阵列实验，虽然重复仍然是困难和不被看重的工作。为了保证科学标准，几乎所有的功能基因组学刊物在发表结果前都要求上传微阵列数据到一个公共数据资源库里（Ball et al. 2004）。

常用的公共数据资源库包括：

- EBI 的 ArrayExpress(http://www.ebi.ac.uk;Parkinson et al. 2007)
- NCBI 的基因表达数据库(GEO; http://www.ncbi.nlm.nih.gov/;Barrett et al. 2007)
- 斯坦福大学的斯坦福微阵列数据库(SMD;http://www.stanford.edu;Demeter et al. 2007)

这三个公共资源库都有 Web 访问界面用于查询数据，并且提供诸如标准化、过滤和聚类等数据分析功能。它们都提供一个基于 Web 界面的提交程序，这对处理中小型数据量的实验非常适合。对数据库的选择存在个人偏好，但推荐每个实验室最好持续使用同一数据库。

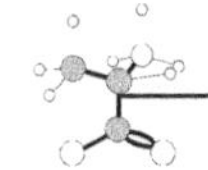

微阵列基因表达数据（Microarray Gene Expression Data，简称 MGED）协会（http://www.mged.org）发展和提升了共享微阵列数据的标准、建议规范和工具。发表微阵列结果最重要的内容是微阵列实验最小信息（Minimal Information About a Microarray Experiment，简称 MIAME）标准（Brazma et al. 2001），其目的是对提交到公共数据库的必要数据内容进行标准化。

MGED 协会基于 XML 开发的微阵列基因表达标志语言（MicroArray Gene Expression Markup Language，简称 MAGE-ML）格式，为微阵列数据交换提供了标准化的文档格式（Spellman et al. 2002）。但是这种文档格式非常复杂，因而只能用于软件应用间的数据交换。后来，一种制表符分隔的基于电子表格的 MAGE-TAB 格式对 MAGE-ML 进行了补充（Rayner et al. 2006）。MAGE-TAB 格式的结构更简化，现作为上传大规模数据的首选格式。

9.4.4 基因表达分析章节的总结

基因表达分析是海洋基因组学中非常灵活和有用的方法，表达数据能极大地方便推断序列的功能。使用微阵列进行转录组的定量分析，然后使用定量 RT-PCR 进行验证。如果只是研究少量已知基因，就没必要使用微阵列方法，而用定量 RT-PCR 即可。

qRT-PCR 和微阵列都需要转录本的序列信息是已知的，而 SAGE 方法在实验过程中产生转录本序列数据，因此可应用于尚未测序的基因组。只能用 SAGE 和微阵列方法来研究基因组范围的表达图谱，其中微阵列技术是二者中更流行的分析方法，因为微阵列技术可以用低成本的方式，平行研究成千上万基因的 mRNA 水平，另外 SAGE 只应用于真核生物。

好的实验设计和适度的重复对基于微阵列的实验至关重要。最佳设置取决于生物学问题。原则上生物性样本重复要优先于技术性重复，并且至少需要三次生物性样本重复。

现在微阵列和相关服务供应商越来越多。对于大规模比较研究，并没有明确的平台偏好，因此对微阵列供应商的选择应基于预算、设计和服务的可用性以及特别是方法的实际应用经验等标准。对于中小规模的实验室，由于高成本和长时投资，不推荐其建立自己的微阵列生产平台。对于大型实验室或联盟，这仍然只是一种选择，当偶尔应用时可考虑选择提供全方位的服务提供商。

任何实验结果都存在一定程度的不可避免的误差，但技术误差可以通过严格遵照实验操作规程得到降低。

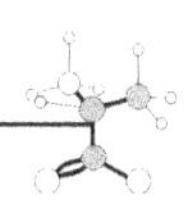

数据分析是一项复杂的工作，需要生物信息学和统计学的专门知识。根据实验设置，简单的分析，诸如统计检验通常是足够的，不过微阵列数据标准化至关重要。首先，建议用成熟和众所周知的统计方法（如 t-检验和 ANOVA）识别具有显著表达差异的基因，同时批判性评估给定的应用新工具的优劣。大多数情况下，数种备选方法的效果都需要进行检验。

对于小实验室或偶尔的应用，低管理要求的独立软件应首先采用。对于大型实验室或国际合作，可使用基于数据库的系统，它们可以提供诸如数据共享的协作功能。为了减少管理要求，提供了数据库远程安装和分析软件，以便减少管理工作量。R 和 BioConductor 操作环境分析比较灵活，可考虑使用或作为补充工具使用。

文章发表前，实验数据和注释信息需先上传到一个公共数据库中。对于大量数据上传，应考虑使用基于 MAGE-TAB 的自动提交方法。

尽管现在微阵列方法很成功，我们预计基于测序的方法会更加流行和有效。对于经典的基因表达研究，测序方法（特别是鸟枪转录组测序）在未来几年，会在费用和准确性方面超越微阵列技术。

参考文献

Adams CP, Kron SJ, Mosaic Technologies USA (1997) Method for performing amplification of nucleic acid with two primers bound to a single solid support. US Patent 5641658.

Alexandersson M, Cawley S, Pachter L (2003) SLAM: cross-species gene finding and alignment with a generalized pair hidden Markov model. Genome Res 13(3): 496 - 502

Allen JE, Salzberg SL (2005) JIGSAW: integration of multiple sources of evidence for gene prediction. Bioinformatics 21(18): 3596 - 3603

Allison DB, Cui X, Page GP et al (2006) Microarray data analysis: from disarray to consolidation and consensus. Nat Rev Genet 7(1): 55 - 65

Altschul SF, Gish W, Miller W et al (1990) Basic local alignment search tool. J Mol Biol 215 (3): 403 - 410

Ashburner M, Ball CA, Blake JA et al (2000) Gene ontology: tool for the unification of biology. The Gene Ontology Consortium. Nat Genet 25(1): 25 - 29

Aziz RK, Bartels D, Best AA et al (2008) The RAST Server: rapid annotations using subsystems technology. BMC Genomics 9: 75

Badger JH, Olsen GJ (1999) CRITICA: coding region identification tool invoking comparative analysis. Mol Biol Evol 16(4): 512 - 524

Baldi P, Long AD (2001) A Bayesian framework for the analysis of microarray expression da-

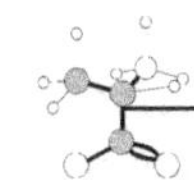

ta:regularized t-test and statistical inferences of gene changes. Bioinformatics 17(6):509 - 519

Ball CA, Brazma A, Causton H et al (2004) Submission of microarray data to public repositories. PLoS Biol 2(9):E317

Bammler T, Beyer RP, Bhattacharya S et al (2005) Standardizing global gene expression analysis between laboratories and across platforms. Nat Methods 2(5):351 - 356

Barrett T, Troup DB, Wilhite SE et al (2007) NCBI GEO: mining tens of millions of expression profiles-database and tools update. Nucleic Acids Res 35(Database issue):D760 - D765

Bartels D, Kespohl S, Albaum S et al (2005) BACCardI - a tool for the validation of genomic assem-blies, assisting genome finishing and intergenome comparison. Bioinformatics 21(7):853 - 859

Bauerle RH, Margolin P (1966) The functional organization of the tryptophan gene cluster in *Salmonella typhimurium.* Proc Natl Acad Sci U S A 56(1):111 - 118

Bekel T, Henckel K, Küster H et al (2009) The sequence analysis and management system-SAMS-2.0: data management and sequence analysis adapted to changing requirements from traditional sanger sequencing to ultrafast sequencing technologies. J Biotechnol 140(1 - 2):3 - 12

Bendtsen JD, Nielsen H, von Heijne G et al (2004) Improved prediction of signal peptides: SignalP 3.0. J Mol Biol 340(4):783 - 795

Benson DA, Karsch-Mizrachi I, Lipman DJ et al (2008) GenBank. Nucleic Acids Res 36: D25 - D30 Berman H, Henrick K, Nakamura H (2003) Announcing the worldwide Protein Data Bank. Nat Struct Biol 10(12):980

Besemer J, Borodovsky M (2005) GeneMark: web software for gene finding in prokaryotes, eukaryotes and viruses. Nucleic Acids Res 33:W451 - W454

Besemer J, Lomsadze A, Borodovsky M (2001) GeneMarkS: a self-training method for prediction of gene starts in microbial genomes. Implications for finding sequence motifs in regulatory regions. Nucleic Acids Res 29(12):2607 - 2618

Birney E, Clamp M, Durbin R (2004) GeneWise and Genomewise. Genome Res 14(5):988 - 995

Black MA, Doerge RW (2002) Calculation of the minimum number of replicate spots required for detection of significant gene expression fold change in microarray experiments. Bioinformatics 18(12):1609 - 1616

Brazma A, Hingamp P, Quackenbush J et al (2001) Minimum information about a microarray experiment (MIAME)-toward standards for microarray data. Nat Genet 29(4):365 - 371

Brejova B, Brown DG, Li M et al (2005) ExonHunter: a comprehensive approach to gene finding. Bioinformatics 21(Suppl 1):i57 - i65

Brent MR (2007) How does eukaryotic gene prediction work? Nat Biotechnol 25(8):883 - 885 Brunak S, Danchin A, Hattori M et al (2002) Nucleotide sequence database policies. Science

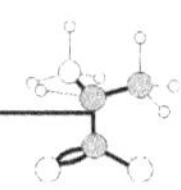

298(5597):1333

Burge C, Karlin S (1997) Prediction of complete gene structures in human genomic DNA. J Mol Biol 268(1):78 – 94

Chen YA, Lin CC, Wang CD et al (2007) An optimized procedure greatly improves EST vector contamination removal. BMC Genomics 8:416

Chothia C, Gough J, Vogel C et al (2003) Evolution of the protein repertoire. Science 300 (5626):1701 – 1703

Cochrane G, Bates K, Apweiler R et al (2006) Evidence standards in experimental and inferential INSDC Third Party Annotation data. Omics 10(2):105 – 113

Cochrane G, Akhtar R, Aldebert P et al (2008) Priorities for nucleotide trace, sequence and anno-tation data capture at the ensembl trace archive and the EMBL nucleotide sequence database. Nucleic Acids Res 36:D5 – D12

Codd EF (1990) The relational model for database management: version 2. Addison-Wesley Longman Publishing Co. , Inc, New York.

Conesa A, Gotz S, Garcia-Gomez JM et al (2005) Blast2GO: a universal tool for anno-tation, visualization and analysis in functional genomics research. Bioinformatics 21(18):3674 – 3676

Consortium U (2008) The universal protein resource (UniProt). Nucleic Acids Res: D190 – D195

Cover T, Hart P (1967) Nearest neighbor pattern classification. IEEE Trans Inf Theory 13 (1):21 – 27

Dandekar T, Snel B, Huynen MA et al (1998) Conservation of gene order: a fingerprint of proteins that physically interact. Trends Biochem Sci 23(9):324 – 328

Datson NA, van der Perk – de Jong J, van den Berg MP et al (1999) MicroSAGE: a modified pro-cedure for serial analysis of gene expression in limited amounts of tissue. Nucleic Acids Res 27 (5):1300 – 1307

Delcher AL, Bratke KA, Powers EC et al (2007) Identifying bacterial genes and endosymbiont DNA with Glimmer. Bioinformatics 23(6):673 – 679

Delcher AL, Harmon D, Kasif S et al (1999) Improved microbial gene identification with GLIMMER. Nucleic Acids Res 27(23):4636 – 4641

Demeter J, Beauheim C, Gollub J et al (2007) The Stanford microarray database: implementation of new analysis tools and open source release of software. Nucleic Acids Res 35:D766 – D770

Djebali S, Delaplace F, Crollius HR (2006) Exogean: a framework for annotating protein-coding genes in eukaryotic genomic DNA. Genome Biol 7(Suppl 1):S7 – S10

Dondrup M, Goesmann A, Bartels D et al (2003) EMMA: a platform for consistent storage and efficient analysis of microarray data. J Biotechnol 106(2 – 3):135 – 146

Dondrup M, Albaum S, Griebel T et al (2009) EMMA 2 – A MAGE-compliant system for the collaborative analysis and integration of microarray data. BMC Bioinformatics 10(1):50

Dressman D, Yan H, Traverso G et al (2003) Transforming single DNA molecules into fluorescent magnetic particles for detection and enumeration of genetic variations. Proc Natl Acad Sci USA 100(15):8817 – 8822

Durbin R, Eddy S, Krogh A et al (1998) Biological sequence analysis. Cambridge University Press, Cambridge.

Edwards RA, Rodriguez-Brito B, Wegley L et al (2006) Using pyrosequencing to shed light on deep mine microbial ecology. BMC Genomics 7:57

Eisen MB, Spellman PT, Brown PO et al (1998) Cluster analysis and display of genome-wide expression patterns. Proc Natl Acad Sci U S A 95(25):14863 – 14868

Elsik CG, Mackey AJ, Reese JT et al (2007) Creating a honey bee consensus gene set. Genome Biol 8(1):R13

Emanuelsson O, Nielsen H, von Heijne G (1999) ChloroP, a neural network-based method for predicting chloroplast transit peptides and their cleavage sites. Protein Sci 8(5):978 – 984

Emanuelsson O, Brunak S, von Heijne G et al (2007) Locating proteins in the cell using TargetP, SignalP and related tools. Nat Protoc 2(4):953 – 971

Ewing B, Hillier L, Wendl MC et al (1998) Base-calling of automated sequencer traces using phred. I. Accuracy assessment. Genome Res 8(3):175 – 185

Fedurco M, Romieu A, Williams S et al (2006) BTA, a novel reagent for DNA attachment on glass and efficient generation of solid-phase amplified DNA colonies. Nucleic Acids Res 34 (3):e22

Fleischmann RD, Adams MD, White O et al (1995) Whole-genome random sequencing and assembly of *Haemophilus influenzae* Rd. Science 269(5223):496 – 512

Flicek P, Aken BL, Beal K et al (2008) Ensembl 2008. Nucleic Acids Res 36:D707 – D714

Florea L, Hartzell G, Zhang Z et al (1998) A computer program for aligning a cDNA sequence with a genomic DNA sequence. Genome Res 8(9):967 – 974

Gaasterland T, Sczyrba A, Thomas E et al (2000) MAGPIE/EGRET annotation of the 2.9 – Mb *Drosophila melanogaster* Adh region. Genome Res 10:502 – 510

Gartemann KH, Abt B, Bekel T et al (2008) The genome sequence of the tomato-pathogenic acti – nomycete *Clavibacter michiganensis* subsp. michiganensis NCPPB382 reveals a large island involved in pathogenicity. J Bacteriol 190(6):2138 – 2149

Gentleman R, Huber W, Carev VJ (eds) (2005) Bioinformatics and computational biology solutions using R and bioconductor. Springer, New York.

Gentleman RC, Carey VJ, Bates DM et al (2004) Bioconductor: open software development for computational biology and bioinformatics. Genome Biol 5(10):R80

Goesmann A, Linke B, Bartels D et al (2005) BRIGEP – the BRIDGE-based genome-tran-

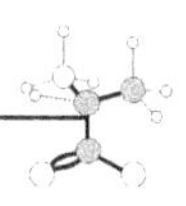

scriptome-proteome browser. Nucleic Acids Res 33:W710 – W716

Goldberg SMD, Johnson J, Busam D et al (2006) A Sanger/pyrosequencing hybrid approach for the generation of high-quality draft assemblies of marine microbial genomes. Proc Natl Acad Sci U S A 103(30):11240 – 11245

Golub TR, Slonim DK, Tamayo P et al (1999) Molecular classification of cancer: class discovery and class prediction by gene expression monitoring. Science 286(5439):531 – 537

Gordon D, Abajian C, Green P (1998) Consed: a graphical tool for sequence finishing. Genome Res 8(3):195 – 202

Gordon D, Desmarais C, Green P (2001) Automated finishing with autofinish. Genome Res 11 (4):614 – 625

Gouy M, Gautier C (1982) Codon usage in bacteria: correlation with gene expressivity. Nucleic Acids Res 10(22):7055 – 7074

Green P (2002) Whole-genome disassembly. Proc Natl Acad Sci U S A 99(7):4143 – 4144

Gresham D, Ruderfer DM, Pratt SC et al (2006) Genome-wide detection of polymorphisms at nucleotide resolution with a single DNA microarray. Science 311(5769):1932 – 1936

Gross SS, Brent MR (2006) Using multiple alignments to improve gene prediction. J Comput Biol 13(2):379 – 393

Guigo R, Reese MG (2005) EGASP: collaboration through competition to find human genes. Nat Methods 2(8):575 – 577

Guigo R, Flicek P, Abril JF et al (2006) EGASP: the human ENCODE Genome Annotation Assessment Project. Genome Biol 7(Suppl 1):S2 – S31

Guo FB, Ou HY, Zhang CT (2003) ZCURVE: a new system for recognizing protein-coding genes in bacterial and archaeal genomes. Nucleic Acids Res 31(6):1780 – 1789

Haas BJ, Salzberg SL, Zhu W et al (2008) Automated eukaryotic gene structure annotation using EVidenceModeler and the program to assemble spliced alignments. Genome Biol 9(1):R7

Henrick K, Feng Z, Bluhm WF et al (2008) Remediation of the protein data bank archive. Nucleic Acids Res 36:D426 – D433

Herring CD, Raghunathan A, Honisch C et al (2006) Comparative genome sequencing of *Escherichia coli* allows observation of bacterial evolution on a laboratory timescale. Nat Genet38 (12):1406 – 1412

Huang X, Madan A (1999) CAP3: a DNA sequence assembly program. Genome Res 9(9): 868 – 877 Huang X, Adams MD, Zhou H et al (1997) A tool for analyzing and annotating genomic sequences. Genomics 46(1):37 – 45

Iizuka M, Yamauchi M, Ando K et al (1994) Quantitative RT-PCR assay detecting the transcrip – tional induction of vascular endothelial growth factor under hypoxia. Biochem Biophys Res Commun 205(2):1474 – 1480

Iseli C, Jongeneel CV, Bucher P (1999) ESTScan: a program for detecting, evaluating, and recon-structing potential coding regions in EST sequences. Proc Int Conf Intell Syst Mol Biol 7:138 – 148

Ju J, Kim DH, Bi L et al (2006) Four-color DNA sequencing by synthesis using cleav – able fluorescent nucleotide reversible terminators. Proc Natl Acad Sci U S A 103(52):19635 – 19640

Kaiser O, Bartels D, Bekel T et al (2003) Whole genome shotgun sequencing guided by bioinformatics pipelines-an optimized approach for an established technique. J Biotechnol 106(2 – 3): 121 – 133

Kall L, Krogh A, Sonnhammer EL (2007) Advantages of combined transmembrane topology and signal peptide prediction-the Phobius web server. Nucleic Acids Res 35: W429 – W432

Kanehisa M, Goto S (2000) KEGG: kyoto encyclopedia of genes and genomes. Nucleic Acids Res 28:27 – 30

Kent WJ (2002) BLAT – the BLAST – like alignment tool. Genome Res 2(4):656 – 664

Korf I (2004) Gene finding in novel genomes. BMC Bioinformatics 5:59

Korf I, Flicek P, Duan D et al (2001) Integrating genomic homology into gene structure prediction. Bioinformatics 17(Suppl 1):S140 – S148

Krause A, Ramakumar A, Bartels D et al (2006) Complete genome of the mutualistic, N2 – fixing grass endophyte *Azoarcus* sp. strain BH72. Nat Biotechnol 24(11):1385 – 1391

Krause L, McHardy AC, Nattkemper TW et al (2007) GISMO – gene identification using a support vector machine for ORF classification. Nucleic Acids Res 35(2):540 – 549

Krogh A, Larsson B, von Heijne G et al (2001) Predicting transmembrane protein topology with a hidden Markov model: application to complete genomes. J Mol Biol 305(3):567 – 580

Küster H, Becker A, Firnhaber C et al (2007) Development of bioinformatic tools to support EST – sequencing, in silico- and microarray-based transcriptome profiling in mycorrhizal symbioses. Phytochemistry 68(1):19 – 32

Lafay B, Lloyd AT, McLean MJ et al (1999) Proteome composition and codon usage in spirochaetes: species-specific and DNA strand-specific mutational biases. Nucleic Acids Res 27(7): 1642 – 1649

Lagesen K, Hallin P, Rodland EA et al (2007) RNAmmer: consistent and rapid annotation of ribosomal RNA genes. Nucleic Acids Res 35(9):3100 – 3108

Lander ES, Waterman MS (1988) Genomic mapping by fingerprinting random clones: a mathematical analysis. Genomics 2(3):231 – 239

Larsen TS, Krogh A (2003) EasyGene-a prokaryotic gene finder that ranks ORFs by statistical significance. BMC Bioinformatics 4:21

Lawrence JG, Roth JR (1996) Selfish Operons: horizontal transfer may drive the evolution of gene clusters. Genetics 143(4):1843 – 1860

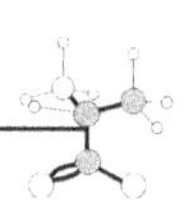

Lee ML, Kuo FC, Whitmore GA et al (2000) Importance of replication in microarray gene expres-sion studies: statistical methods and evidence from repetitive cDNA hybridizations. Proc Natl Acad Sci U S A 97(18):9834 – 9839

Li C, Wong WH (2001) Model-based analysis of oligonucleotide arrays: expression index computation and outlier detection. Proc Natl Acad Sci U S A 98(1):31 – 36

Li SS, Bigler J, Lampe JW et al (2005) FDR – controlling testing procedures and sample size determination for microarrays. Stat Med 24(15):2267 – 2280

Lin M, Wei LJ, Sellers WR et al (2004) dChipSNP: significance curve and clustering of SNP – array-based loss-of-heterozygosity data. Bioinformatics 20(8):1233 – 1240

Linke B, McHardy AC, Neuweger H et al (2006) REGANOR: a gene prediction server for prokaryotic genomes and a database of high quality gene predictions for prokaryotes. Appl Bioinformatics 5(3):193 – 198

Liolios K, Mavromatis K, Tavernarakis N et al (2008) The genomes on line database (GOLD) in 2007: status of genomic and metagenomic projects and their associated metadata. Nucleic Acids Res 36: D475 – D479

Lipshutz RJ, Fodor SP, Gingeras TR et al (1999) High density synthetic oligonucleotide arrays. Nat Genet 21(1 Suppl):20 – 24

Lipshutz RJ, Morris D, Chee M et al (1995) Using oligonucleotide probe arrays to access genetic diversity. Biotechniques 19(3):442 – 447

Liu JJ, Cutler G, Li W et al (2005) Multiclass cancer classification and biomarker discovery using GA-based algorithms. Bioinformatics 21(11):2691 – 2697

Lomsadze A, Ter Hovhannisyan V, Chernoff YO et al (2005) Gene identification in novel eukaryotic genomes by self-training algorithm. Nucleic Acids Res 33(20):6494 – 6506

Lowe TM, Eddy SR (1997) tRNAscan-SE: a program for improved detection of transfer RNA genes in genomic sequence. Nucleic Acids Res 25(5):955 – 964

Lukashin AV, Borodovsky M (1998) GeneMark. hmm: new solutions for gene finding. Nucleic Acids Res 26(4):1107 – 1115

Majoros WH, Pertea M, Salzberg SL (2004) TigrScan and GlimmerHMM: two open source ab initio eukaryotic gene-finders. Bioinformatics 20(16):2878 – 2879

Majoros WH, Pertea M, Salzberg SL (2005) Efficient implementation of a generalized pair hidden Markov model for comparative gene finding. Bioinformatics 21(9):1782 – 1788

Mangalam H (2002) The Bio * toolkits-a brief overview. Brief Bioinform 3(3):296 – 302

Mao X, Cai T, Olyarchuk JG et al (2005) Automated genome annotation and pathway identification using the KEGG Orthology (KO) as a controlled vocabulary. Bioinformatics 21(19):3787 – 3793

Mardis ER (2008) Next-generation DNA sequencing methods. Annu Rev Genomics Hum

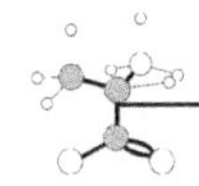

Genet 9:387 – 402

Margulies M, Egholm M, Altman WE et al (2005) Genome sequencing in microfabricated high-density picolitre reactors. Nature 437(7057):376 – 380

Mathe C, Sagot MF, Schiex T et al (2002) Current methods of gene prediction, their strengths and weaknesses. Nucleic Acids Res 30(19):4103 – 4117

Matsumura H, Reich S, Ito A et al (2003) Gene expression analysis of plant host-pathogen interactions by SuperSAGE. Proc Natl Acad Sci U S A 100(26):15718 – 15723

Maurer M, Molidor R, Sturn A et al (2005) MARS: microarray analysis, retrieval, and storage system. BMC Bioinformatics 6:101

McHardy AC, Pühler A, Kalinowski J et al (2004a) Comparing expression level-dependent fea-tures in codon usage with protein abundance: an analysis of predictive proteomics . Proteomics 4 (1):46 – 58

McHardy AC, Goesmann A, Pühler A et al (2004b) Development of joint application strategies for two microbial gene finders. Bioinformatics 20(10):1622 – 1631

Meyer F, Goesmann A, McHardy AC et al (2003) GenDB – an open source genome annotation system for prokaryote genomes. Nucleic Acids Res 31(8):2187 – 2195

Millar CD, Huynen L, Subramanian S et al (2008) New developments in ancient genomics. Trends Ecol Evol 23(7):386 – 393

Miron M, Nadon R (2006) Inferential literacy for experimental high-throughput biology. Trends Genet 22(2):84 – 89

Moore JE, Lake JA (2003) Gene structure prediction in syntenic DNA segments. Nucleic Acids Res 31(24):7271 – 7279

Mott R (1997) EST_GENOME: a program to align spliced DNA sequences to unspliced genomic DNA. Comput Appl Biosci 13(4):477 – 478

Mulder NJ, Apweiler R, Attwood TK et al (2007) New developments in the InterPro database. Nucleic Acids Res 35:D224 – D228

Nagaraj SH, Deshpande N, Gasser RB et al (2007) ESTExplorer: an expressed sequence tag (EST) assembly and annotation platform. Nucleic Acids Res 35:W143 – W147

Nakano M, Komatsu J, Matsuura S-i et al (2003) Single-molecule PCR using water-in-oil emulsion. J Biotechnol 102(2):117 – 124

Needleman SB, Wunsch CD (1970) A general method applicable to the search for similarities in the amino acid sequence of two proteins. J Mol Biol 48(3):443 – 453

Nekrutenko A, Chung WY, Li WH (2003) ETOPE: evolutionary test of predicted exons. Nucleic Acids Res 31(13):3564 – 3567

Ng P, Wei C – L, Sung W – K et al (2005) Gene identification signature (GIS) analysis for transcriptome characterization and genome annotation. Nat Methods 2(2):105 – 111

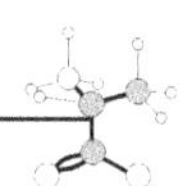

Ng P, Tan JJS, Ooi HS et al (2006) Multiplex sequencing of paired-end ditags (MS-PET): a strat-egy for the ultra-high-throughput analysis of transcriptomes and genomes. Nucleic Acids Res 34(12): e84

Noguchi H, Park J, Takagi T (2006) MetaGene: prokaryotic gene finding from environmental genome shotgun sequences. Nucleic Acids Res 34(19): 5623 – 5630

Ou HY, Guo FB, Zhang CT (2004) GS-Finder: a program to find bacterial gene start sites with a self-training method. Int J Biochem Cell Biol 36(3): 535 – 544

Overbeek R, Disz T, Stevens R (2004) The SEED: a peer-to-peer environment for genome annotation. Commun ACM 47(11): 47 – 51

Overbeek R, Fonstein M, D'Souza M et al (1999) The use of gene clusters to infer functional coupling. Proc Natl Acad Sci U S A 96: 2896 – 2901

Overbeek R, Larsen N, Pusch GD et al (2000) WIT: integrated system for high-throughput genome sequence analysis and metabolic reconstruction. Nucleic Acids Res 28(1): 123 – 125

Overbeek R, Larsen N, Walunas T et al (2003) The ERGO genome analysis and discovery system. Nucleic Acids Res 31: 164 – 171

Overbeek R, Begley T, Butler RM et al (2005) The subsystems approach to genome annotation and its use in the project to annotate 1000 genomes. Nucleic Acids Res 33(17): 5691 – 5702

GP, Edwards JW, Gadbury GL et al (2006) The PowerAtlas: a power and sample size atlas for microarray experimental design and research. BMC Bioinformatics 7: 84

Pan W, Lin J, Le CT (2002) How many replicates of arrays are required to detect gene expression changes in microarray experiments? A mixture model approach. Genome Biol. research0022.

Parkinson H, Kapushesky M, Shojatalab M et al (2007) ArrayExpress-a public database of microarray experiments and gene expression profiles. Nucleic Acids Res 35: D747 – D750

Parra G, Agarwal P, Abril JF et al (2003) Comparative gene prediction in human and mouse. Genome Res 13(1): 108 – 117

Pavlidis P, Weston J, Cai J et al (2002) Learning gene functional classifications from multiple data types. J Comput Biol 9(2): 401 – 411

Pearson WR, Lipman DJ (1988) Improved tools for biological sequence comparison. Proc Natl Acad Sci U S A 85(8): 2444 – 2448

Pertea G, Huang X, Liang F et al (2003) TIGR Gene Indices clustering tools (TGICL): a software system for fast clustering of large EST datasets. Bioinformatics 19(5): 651 – 652

Pieler R, Sanchez-Cabo F, Hackl H et al (2004) ArrayNorm: comprehensive normalization and analysis of microarray data. Bioinformatics 20(12): 1971 – 1973

Prober JM, Trainor GL, Dam RJ et al (1987) A system for rapid DNA sequencing with fluorescent chain-terminating dideoxynucleotides. Science 238(4825): 336 – 341

Pruitt KD, Tatusova T, Maglott DR (2007) NCBI reference sequences (RefSeq): a curated

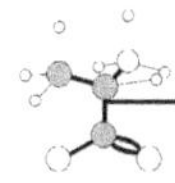

non-redundant sequence database of genomes, transcripts and proteins. Nucleic Acids Res 35: D61 – D65

Quackenbush J (2001) Computational analysis of microarray data. Nat Rev Genet 2(6):418 – 427

Quackenbush J (2002) Microarray data normalization and transformation. Nat Genet 32(Suppl):496 – 501

Quackenbush J (2003) Genomics. Microarrays-guilt by association. Science 302(5643):240 – 241 Quevillon E, Silventoinen V, Pillai S et al (2005) InterProScan: protein domains identifier.

Nucleic Acids Res 33: W116 – W1120

Rayner TF, Rocca-Serra P, Spellman PT et al (2006) A simple spreadsheet-based, MIAME-supportive format for microarray data: MAGE-TAB. BMC Bioinformatics 7:489

Reeck GR, de Haen C, Teller DC et al (1987) Homology in proteins and nucleic acids: a terminology muddle and a way out of it. Cell 50(5):667

Reese MG, Kulp D, Tammana H et al (2000) Genie-gene finding in *Drosophila melanogaster*. Genome Res 10(4):529 – 538

Repsilber D, Ziegler A (2005) Two-color microarray experiments. Technology and sources of variance. Methods Inf Med 44(3):400 – 404

Ronaghi M, Uhlén M, Nyrén P (1998) A sequencing method based on real-time pyrophosphate. Science 281(5375):363 – 365

Rutherford K, Parkhill J, Crook J et al (2000) Artemis: sequence visualization and annotation. Bioinformatics 16(10):944 – 945

Saal LH, Troein C, Vallon-Christersson J et al (2002) BioArray Software Environment (BASE): a platform for comprehensive management and analysis of microarray data. Genome Biol 3 (8): SOFTWARE0003.

Saeed AI, Sharov V, White J et al (2003) TM4: a free, open-source system for microarray data management and analysis. Biotechniques 34(2):374 – 378

Saha S, Sparks AB, Rago C et al (2002) Using the transcriptome to annotate the genome. Nat Biotechnol 20(5):508 – 512

Salamov AA, Solovyev VV (2000) Ab initio gene finding in *Drosophila* genomic DNA. Genome Res 10(4):516 – 522

Sanger F, Nicklen S, Coulson A (1977) DNA sequencing with chain-terminating inhibitors. Proc Natl Acad Sci USA 74:5463 – 5467

Schena M, Shalon D, Davis RW et al (1995) Quantitative monitoring of gene expression patterns with a complementary DNA microarray. Science 270(5235):467 – 470

Schiex T, Moisan A, Rouzé P (2001) Eugène: an eukaryotic gene finder that combines several sources of evidence. In: Computational Biology, selected papers from JOBIM 2000 number 2066 in

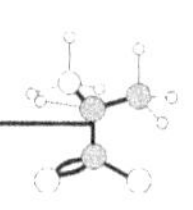

LNCS, Springer Verlag, New York, pp. 111 – 125.

Schneiker S, Martins dos Santos VA, Bartels D et al (2006) Genome sequence of the ubiq – uitous hydrocarbon-degrading marine bacterium *Alcanivorax borkumensis*. Nat Biotechnol 24(8): 997 – 1004

Schneiker S, Perlova O, Kaiser O et al (2007) Complete genome sequence of the myxobacterium Sorangium cellulosum. Nat Biotechnol 25(11): 1281 – 1289

Shendure J, Mitra RD, Varma C et al (2004) Advanced sequencing technologies: methods and goals. Nat Rev Genet 5(5): 335 – 344

Shendure J, Porreca GJ, Reppas NB et al (2005) Accurate multiplex polony sequencing of an evolved bacterial genome. Science 309(5741): 1728 – 1732

Shendure JA, Porreca GJ, Church GM (2008) Overview of DNA sequencing strategies. Curr Protoc Mol Biol Chapter 7: Unit 7: 1

Skovgaard M, Jensen LJ, Brunak S et al (2001) On the total number of genes and their length distribution in complete microbial genomes. Trends Genet 17(8): 425 – 428

Slater GS, Birney E (2005) Automated generation of heuristics for biological sequence compari – son. BMC Bioinformatics 6: 31

Smith MW, Feng DF, Doolittle RF (1992) Evolution by acquisition: the case for horizontal gene transfers. Trends Biochem Sci 17(12): 489 – 493

Smith TF, Waterman MS (1981) Identification of common molecular subsequences. J Mol Biol 147(1): 195 – 197

Sonnhammer EL, von Heijne G, Krogh A (1998) A hidden Markov model for predict-ing transmembrane helices in protein sequences. Proc Int Conf Intell Syst Mol Biol 6: 175 – 182

Spellman PT, Miller M, Stewart J et al (2002) Design and implementation of microarray gene expression markup language (MAGE-ML). Genome Biol 3(9): RESEARCH0046.

Stanke M, Waack S (2003) Gene prediction with a hidden Markov model and a new intron submodel. Bioinformatics 19(Suppl 2): ii215 – ii225

Stanke M, Tzvetkova A, Morgenstern B (2006) AUGUSTUS at EGASP: using EST, protein and genomic alignments for improved gene prediction in the human genome. Genome Biol 7(Suppl 1): S11 – S18

Sturn A, Quackenbush J, Trajanoski Z (2002) Genesis: cluster analysis of microarray data. Bioinformatics 18(1): 207 – 208

Sugawara H, Ogasawara O, Okubo K et al (2008) DDBJ with new system and face. Nucleic Acids Res 36: D22 – D24

Suzek BE, Ermolaeva MD, Schreiber M et al (2001) A probabilistic method for identifying start codons in bacterial genomes. Bioinformatics 17(12): 1123 – 1130

Tamames J, Casari G, Ouzounis C et al (1997) Conserved clusters of functionally related genes in two bacterial genomes. Mol Evol 44:66 – 73

Tatsuov RL, Mushegian AR, Bork P et al (1996) Metabolism and evolution of *Haemophilus influenza* deduced from a whole-genome comparison with *Escherichia coli*. Curr Biol6(3):279 – 291

Tatusov RL, Fedorova ND, Jackson JD et al (2003) The COG database: an updated version includes eukaryotes. BMC Bioinformatics 4:41

Team RDC (2008) R: a language and environment for statistical computing. R Foundation for Statistical Computing, Vienna, Austria.

Tech M, Meinicke P (2006) An unsupervised classification scheme for improving predictions of prokaryotic TIS. BMC Bioinformatics 7:121

Thieme F, Koebnik R, Bekel T et al (2005) Insights into genome plasticity and pathogenicity of the plant pathogenic bacterium *Xanthomonas campestris* pv. vesicatoria revealed by the complete genome sequence. J Bacteriol 187(21):7254 – 7266

Tusher VG, Tibshirani R, Chu G (2001) Significance analysis of microarrays applied to the ionizing radiation response. Proc Natl Acad Sci U S A 98(9):5116 – 5121

Usuka J, Zhu W, Brendel V (2000) Optimal spliced alignment of homologous cDNA to a genomic DNA template. Bioinformatics 16(3):203 – 211

van Baren MJ, Brent MR (2006) Iterative gene prediction and pseudogene removal improves genome annotation. Genome Res 16(5):678 – 685

Vapnik VN (1999) The nature of statistical learning theory. Springer, New York.

Velculescu VE, Zhang L, Vogelstein B et al (1995) Serial analysis of gene expression. Science 270(5235):484 – 487

von Mering C, Jensen LJ, Snel B et al (2005) STRING: known and predicted protein-protein associations, integrated and transferred across organisms. Nucleic Acids Res 33:433 – 437

Vorhölter FJ, Schneiker S, Goesmann A et al (2008) The genome of Xanthomonas campestris pv. campestris B100 and its use for the reconstruction of metabolic pathways involved in xanthan biosynthesis. J Biotechnol 134(1 – 2):33 – 45

Wei C, Brent MR (2006) Using ESTs to improve the accuracy of de novo gene prediction. BMC Bioinformatics 7:327

Wilkinson MD, Links M (2002) BioMOBY: an open source biological web services proposal. Brief Bioinform 3(4):331 – 341

Wu J, Mao X, Cai T et al (2006) KOBAS server: a web-based platform for automated annotation and pathway identification. Nucleic Acids Res 34: W720 – W724

Wu TD, Watanabe CK (2005) GMAP: a genomic mapping and alignment program for mRNA and EST sequences. Bioinformatics 21(9):1859 – 1875

Wu W, Xing EP, Myers C et al (2005) Evaluation of normalization methods for cDNA microarray data by k – NN classification. BMC Bioinformatics 6:191

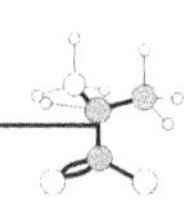

Yang YH, Speed T (2002) Design issues for cDNA microarray experiments. Nat Rev Genet 3(8):579 - 588

Yauk C, Berndt L, Williams A et al (2005) Automation of cDNA microarray hybridization and washing yields improved data quality. J Biochem Biophys Methods 64(1):69 - 75

Yauk CL, Berndt ML, Williams A et al (2004) Comprehensive comparison of six microarray technologies. Nucleic Acids Res 32(15):e124

Zhang MQ (2002) Computational prediction of eukaryotic protein-coding genes. Nat Rev Genet 3(9):698 - 709

Zhang Z, Schwartz S, Wagner L et al (2000) A greedy algorithm for aligning DNA sequences. J Comput Biol 7(1 - 2):203 - 214

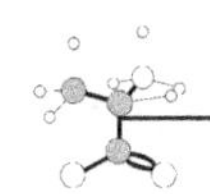

词汇表

2R hypothesis hypothesis suggesting that the genomes of the early vertebrate lineage underwent two rounds of whole-genome duplications, reflected e. g. by the existence of 4 remnants of Hox clusters.

2R 假说. 该假说指出早期脊椎动物在进化过程中经历了两次全基因组复制，4 个 HOX 基因簇的存在就是证据。

Accession number An identifier supplied by the curators of the major biological databases upon submission of a novel entry (usually sequence data) that uniquely identifies that database entry.

登录号 在提交一个新信息时（通常是序列数据），由生物数据库管理员提供的一个标识码，是该数据库中本条信息的独特标识。

AFLP This technique produces DNA fragments that are separated by size (length of sequence) using polyacrylamide gel electrophoresis or capillary sequencers. Bands at a particular site on the gel (equivalent to alleles) are counted as being present or absent for the analysis.

AFLP 随机扩增长度多态性 该技术使用聚丙烯酰胺凝胶电泳或毛细管测序仪将 DNA 片段根据片段大小（序列长度）分离开。以凝胶上特定位点条带出现与否（相当于等位基因）作为分析的依据。

Algae all photosynthetic eukaryotes other than land plants.

藻类 除陆生植物之外所有能进行光合作用的真核生物。

Algal bloom rapid multiplication of one or a small number of algal species. In some cases the algae are toxic.

藻类水华 一种或少数种藻类快速繁殖的现象。在有些情况下，这些藻是有毒的。

Alignment overview of both differences and similarities between two or more DNA, RNA or protein sequences created by aligning residues that are putatively homologous. Gaps may be introduced to optimally align the sequences.

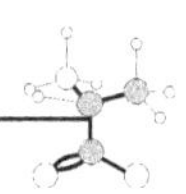

比对 通过比对碱基来比较两种或多种 DNA、RNA 或蛋白序列之间的差异以及相似性，用以推定序列之间的同源性。为获得最佳比对结果，允许存在缺口。

AMD Acid Mine Drainage.

AMD 酸性矿山废水

API Application Programming Interface.

API 应用程序编程接口

BAC bacterial artificial chromosome. a cloning vector carrying up to 300 kbp of insert sequence.

BAC 细菌人工染色体。一种可容纳 300kb 插入片段的克隆载体。

Barcoding A specific piece of DNA (usually part of the mitochondrial cytochrome c oxidase sub unit I gene) is sequenced many times from different species. There are sufficient interspecies differences in this gene, for the sequence of DNA to act as a unique code for a species and therefore be used as a taxonomic aid.

条形码 对不同物种进行多次测序得到的一段特定 DNA 序列（通常是线粒体细胞色素 C 氧化酶亚基 I 的部分序列）。在此段特定的基因上有充分的种间差异，因为这段 DNA 序列对于每个物种都是唯一的，因此可以作为物种分类的辅助手段。

Bioinformatics The application of information technology to the field of molecular biology.

生物信息学 将信息技术应用于分子生物学领域的学科。

BioJava, BioPerl Programming frameworks to handle the most common file formats and tools that are used by bioinformaticians.

BioJava, BioPerl 用于处理最常见的文件格式的编程模块，是生物信息学家处理生物学数据的工具。

BLAST "Basic local alignment search tool", a well-known and established tool for finding homologue sequences in large sequence databases.

BLAST 基于局部比对搜索工具。一种为大家所熟知和使用的工具，广泛用于在大型序列数据库中查找同源序列。

CAMERA Community Cyberinfrastructure for Advanced Marine Microbial Ecology Research and Analysis.

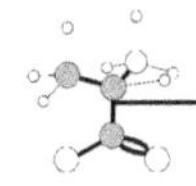

CAMERA 高级海洋微生物生态学研究及分析数据库共享网络基础设施。

Clone the term "clone" can refer to a bacterium carrying a piece of cloned DNA, or to the cloned DNA itself.

克隆 "克隆"是指携带一段克隆 DNA 的细菌，或指克隆的 DNA 本身。

Clustering EST sequences are clustered according to their sequence, so that they can be assembled afterwards. The set of all Clusters form a Clusterset.

聚类 表达序列标签（EST）按照自身的序列聚类在一起，便于后续进行组装。所有聚类形成一个聚类集。

COG/KOG The COG (Cluster of Orthologous Groups of proteins) database consists of 138,458 proteins, which form 4,873 COGs and comprise 75% of the 185 505 (predicted) proteins encoded in the 66 genomes of unicellular organisms that were used for the database. The eukaryotic orthologous groups (KOGs) include proteins from 7 eukaryotic genomes.

COG/KOG 直系同源蛋白质组真核生物直系同源组 直系同源蛋白质组（COG）数据库有 138458 个蛋白，分为 4873 个 COGs，占现有蛋白库的 75%。现有蛋白库包含由 66 个单细胞生物基因组编码的 185，505 个预测蛋白。真核生物直系同源组（KOGs）包含 7 个真核生物基因组的蛋白质。

Cohort Fish in a stock born in the same year.

同生群 同一年出生的鱼类个体所组成的群体。

Contig The term "contig" comes from a shortening of the word "contiguous". It can be used to refer to the final product of a shotgun sequencing project. When individual lanes of sequence information are assembled to infer the sequence of the larger DNA piece, the product consensus sequence is called a "contig".

重叠群 "contig"（重叠群）是"contiguous"的缩写形式。它是指鸟枪法测序项目中产生的最后产物。将每个通道中的序列片段组装成较长的 DNA 大片段，最终产生的共有序列就称为一个重叠群。

Cryptic species A cryptic species is a species that is reproductively isolated (or at least genetically distinct) from a second species but which resembles the second species so closely that both have traditionally been considered a single species.

隐蔽种 隐蔽种是指和另一个物种存在生殖隔离（或至少有遗传差异），但在形态上又十分相似以至于被认为是同一个物种。

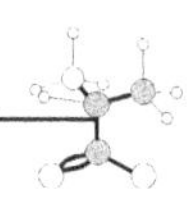

ddNTP dideoxy-nucleotide.

ddNTP 双脱氧核苷酸

DGGE Denaturing gradient gel electrophoresis. Molecular fingerprinting technique that separates PCR products based on sequence differences that result in distinctly different denaturing characteristics of the DNA.

DGGE 变性梯度凝胶电泳。一种可根据序列差异导致 DNA 具有明显不同变性特征来分离 PCR 产物的分子指纹技术。

dNTP deoxy-nucleotide.

dNTP 脱氧核糖核苷三磷酸

Dye A chemical compound used to label biological material such as cDNA allowing it to be detected. Often fluorescent compounds are used.

染料 一种用来标记生物材料（如 cDNA）的化合物，可让生物材料被检测到。其中荧光化合物较为常用。

EGT Environmental Gene Tags.

EGT 环境基因标签

ELISA Enzyme Linked Immuno Sorbent Assay; Biochemical test to detect and quantify an antigen in a sample using a specific antibody.

ELISA 酶联免疫吸附测定；用特异抗体检测及定量样品中抗原的生化检测方法。

Emulsion PCR a technique for generating a clonally amplified piece of DNA in vitro along with a microbead, within a mineral oil emulsion.

微乳液 PCR 在矿物油乳液里，沿着微珠进行 DNA 片段体外扩增的一种技术。

Endosymbiosis An endosymbiont is an organism that lives within the body or cells of another organism. Endosymbiosis played an important role in the evolution of the eukaryotes as mitochondria and plastids were derived from endosymbiotic organisms captured by ancestral eukaryotic cells.

内共生 内共生体是指在一种生物体或其细胞内共生的另一种生物体。线粒体和叶绿体是原始真核细胞捕获的内共生生物演化而成的，因此，内共生现象在真核生物的进化过程中起着重要作用。

Epitope Region of a protein or peptide recognised by an antibody.

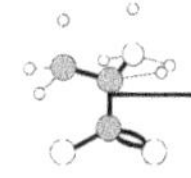

抗原决定簇 蛋白质或多肽上与抗体相结合的特定区域。

eQTL expression Quantitative Trait Loci; the transcriptome is associated with thousands of expression traits (see also QTL).

eQTL 表达数量性状基因座。转录组与成千上万的表达性状有关（参照 QTL）。

EST Expressed Sequence Tag. For a gene expression analysis, it is possible to extract the mRNA from a cell, translate it into cDNA and sequence the cDNA from either one or both ends. The reads are called ESTs. EST data may contain sequencing errors.

EST 表达序列标签 在基因表达分析中，首先从细胞中提取 mRNA，将其反转录为 cDNA，随后从 cDNA 的一端或两端测序。所有读序被称为表达序列标签。EST 数据有可能包含测序错误。

FISH Fluorescence in situ hybridisation.

FISH 原位杂交荧光。

Fosmid A cloning system based on the E. coli F factor. These clones have an average insert size of 40 Kbp, with a very small standard deviation.

Fosmid 基于大肠杆菌 F 因子的一种克隆系统。这种克隆的插入片段平均长度为 40 Kbp，标准偏差很小。

Gametophyte In plants and algae that undergo alternation of generations, a gametophyte is the multicellular generation that produces gametes.

配子体 在植物和藻类进行世代交替时，配子体是多细胞世代产生的配子。

Gene chip Thousands of sequences are individually attached (spotted) onto a small glass slide (typically the size of a microscope slide). Each spot on the slide represents a gene sequence. These can be screened to generate expression profiles of a cell, tissue or organism (s) under different conditions. Also called a microarray.

基因芯片 将成千上万条序列逐一固定在载玻片上（大小通常与显微玻片相近），玻璃板上的每一个点都代表一个基因序列。经过筛查可产生不同条件下细胞、组织或生物体的表达谱。基因芯片也被称为微阵列。

Gene knockdown Refers to techniques such as RNA interference by which the expression of a gene is reduced.

基因敲除 指利用诸如 RNA 干扰等技术降低基因表达活性的方法。

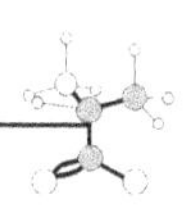

Gene library Collection of pieces of DNA or cDNA cloned into artificial vectors, which can be replicated, sequenced and screened.

基因文库 是克隆至人工载体上的DNA或cDNA集群，它们可被复制、测序和筛选。

Gene repertoire the set of genes (or gene families) encoded in the genome of an organism.

基因库 编码生物体基因组的一整套基因（或基因家族）。

Genetic hitch-hiking A process by which an allele or mutation may spread through the gene pool because it is closely linked to a gene that is being selected for.

遗传搭车 当等位基因或突变与被选择的基因紧密连锁时，会由于受到选择而发生扩散。

Genome All the genetic material of an organism.

基因组 一个生物体的全部遗传物质。

Genomics the study of the structure and function of genomes (see genome).

基因组学 研究基因组结构和功能的学科（参照基因组）。

GOS Global Ocean Sampling.

GOS 全球海洋取样。

GSC Genomic Standards Consortium.

GSC 基因组标准协会。

GUI Graphical user interface. A common means of interaction between humans and computers, which uses graphical representations of data and offers user interaction via pointing devices (e.g. a mouse).

GUI 图形用户界面。一种常见于人与电脑之间互动的方式，它使用图形数据并通过指示设备（如鼠标）提供用户与电脑间的互动。

HMMs Hidden Markov Models (HMMs) allow the integration of diverse sequence features into a coherent, probabilistic framework. For the task of gene identification, HMMs may include states modeling introns, exons, intergenic regions, start and stop codons, splice signals, polyadenyation signals and ribosomal binding sites.

HMMs 隐马尔可夫模型 该模型可以将不同序列的特征整合成一个连贯的随机框架。该模型用于基因识别时，建立包括内含子、外显子、基因间区、启动子和终止密码子、剪

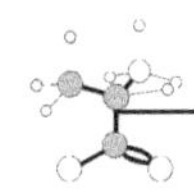

接信号、多聚腺苷化信号和核糖体结合位点在内的状态模型。

Homeodomain proteins a class of proteins characterized by the existence of a specific motif, the homeodomain, which is primarily involved in DNA binding.

同源域蛋白 存在特定基序特征的一类蛋白质。同源域主要参与 DNA 的结合。

Homologue In order to describe the evolutionary relation of proteins, the terms homology, orthology, and paralogy are used. In this context, homology means that two proteins or sequences share a common ancestor. Two principal types of homology can be distinguished (Orthology and Paralogy).

同源性 为了描述蛋白质的进化关系，出现了同源性、直系同源和旁系同源这些术语。在本书中，同源性是指两种蛋白质或两条序列拥有共同的祖先。同源性主要分为两种类型，即直系同源和旁系同源。

Horizontal (or lateral) gene transfer A process in which genes from one organism are integrated into the genome of a second organism (which may be very distantly related to the first) and subsequently inherited with the rest of the genetic material of the cell.

水平基因转移（或侧向基因转移） 基因从一个物种整合至另外一个物种（相互间的亲缘关系有可能较远）的基因组的过程，且能随着细胞的其他遗传物质遗传给下一代。

Hox cluster genomic array of Hox genes, which constitute a subgroup of homeodomain proteins. Spatial Hox gene expression during development tends to correlate with the spatial arrangements of genes in the genomes of a broad panel of animals. Hence, Hox clusters also serve as a paradigm to understand the commonalities and changes during the evolution of animal body plans.

Hox 基因簇 Hox 基因在基因组上的排列，包括一系列同源域蛋白亚群。发育过程中 Hox 基因的空间表达通常与一系列动物基因组中基因的空间排列有关。因此，Hox 基因簇也可用来了解动物体型进化过程中的共性和变化。

HSP high-scoring segment pair.

HSP 高比值片段对。

IMG The Integrated Microbial Genomes system.

IMG 整合的微生物基因组系统。是一个新型数据管理与分析平台，其中的微生物基因组数据皆由联合基因组研究所（Joint Genome Institute，JGT）提供（译者注）。

INSDC International Nucleotide Sequence Database Collaboration. It consists of the DDBJ

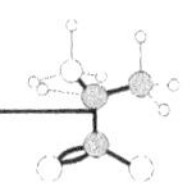

(Japan), EMBL (Europe) and GenBank (USA) Nucleotide Sequence Database. The three databases exchange new and updated data on a daily basis to achieve optimal synchronization.

INSDC 国际核苷酸序列数据库合作组织。由 DDBJ（日本）、EMBL（欧洲）和 GenBank（美国）三个核苷酸序列数据库组成。这三个数据库每天都会进行数据的新增与更新，以期实现最佳同步效果。

IUPAC International Union of Pure and Applied Chemistry. This non-governmental organisation has developed a system of naming chemical elements and their compounds such as amino acids or nucleotides.

IUPAC 国际理论与应用化学联合会，一个开发了化学元素以及化合物（例如氨基酸或核苷酸）命名系统的非政府组织。

JCoast Comparative Analysis and Search Tool.

JCoast 比较分析和搜索工具。

JGI Joint Genome Institute, a sequencing centre in the USA.

JGI 联合基因组研究所，位于美国的一个测序中心。

KEGG The KEGG (Kyoto Encyclopedia of Genes and Genomes) is a bioinformatics resource for linking genomes to life and the environment.

KEGG 京都基因与基因组百科全书，是一个将基因组与生命和环境联系起来的生物信息学资源。

Kb or Kbp kilobase or kilobase pair, region of DNA 1000 nucleotides long.

Kb or Kbp 千碱基或千碱基对，是指长度为 1000 个核苷酸的 DNA 长区域。

Linkage map Genetic map produced using recombination values of genes or other markers to identify the linear order and relative distance of these markers on a chromosome.

连锁图 用基因或其他标记的重组值绘制的遗传图谱，可以鉴定这些标记在染色体上的线性顺序和相对距离。

LPS Lipopolysaccharides; large molecules containing a lipid and a polysaccharide linked by a covalent bond; major components of the outer membrane of Gram-negative bacteria; induce strong immune responses in animals.

LPS 脂多糖，是含有以共价键连接脂类与多糖的大分子；是革兰氏阴性细菌外膜的主要成分；在动物中可以诱导强烈的免疫反应。

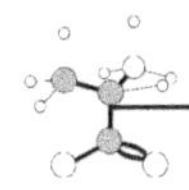

Local alignment Sequence alignment which focuses on matching part of one sequence with part of another.

局部比对 集中用于将一条序列的部分匹配到另一条序列的部分的序列比对。

MAGE-TAB A data format used to transfer microarray data to public repositories (e. g. ArrayExpress). The MAGE-TAB format consists mainly of spread sheets, which can be represented as tabulator- or comma-separated files.

MAGE-TAB 一种用以将微矩阵数据转换成公共存储数据（例如 ArrayExpress）的数据格式。这个格式主要是以制表符或逗号分割文件为代表的电子数据表。

MAS Marker-Assisted Selection; Selection of a genetic determinant of a trait of interest (e. g. productivity, disease resistance and quality) through the use of a marker (morphological, biochemical or one based on DNA/RNA variation).

MAS 标记辅助选择，通过使用形态学、生物化学或基于 DNA/RNA 变异的标记对经济性状（如产量、抗病性以及品质）的遗传决定因素作出选择。

MA-plot A certain kind of scatter-plot commonly used for microarray analysis where the intensity measurements from microarrays are transformed in a special way. The x-axis corresponds to the absolute intensity of the spot (A-value), the y-axis corresponds to a measure of differential expression (M-value).

MA-plot 一种常用于微阵列分析的散点图，以特定方式将芯片测量到的强度转换出来，其中 x 轴对应特定点的绝对强度（A 值），y 轴对应差异表达的测量值（M 值）。

Match position with identical bases or amino acids in a sequence alignment.

配对 在序列比对中，相同碱基或氨基酸对应的位置。

Mate-pair paired reads of the two ends of a cloned DNA molecule. Mate-pairs are often used by genome assembly programs to orient and order contigs, taking into account the distance between the ends of the DNA molecule and the orientation relationship of the sequences.

Mate-pair 是一个克隆 DNA 分子两端的碱基对读取片段（reads）。Mate-pairs 被用来计算 DNA 分子两端的距离和序列的取向关系，经常在基因组组装程序中用来确定重叠群（contig）的方向和排序。

MHC Major Histocompatibility Complex; A large genomic region playing an important role in the immune system, autoimmunity and reproductive success.

MHC 主要组织相容性复合物。在免疫系统、自身免疫和生殖成功中起到重要作用

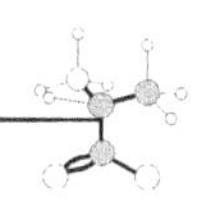

的基因组大区域。

MDA Multiple displacement amplification. Isothermal DNA amplification method using random hexamers and the DNA polymerase of bacteriophage φ29.

MDA 多重置换扩增。采用随机六聚体和噬菌体 φ29DNA 聚合酶进行的等温 DNA 扩增方法。

Metabolomics High throughput analysis of the metabolites present in a cell type, a tissue or an organism and the modifications to metabolite pools under different conditions.

代谢组学 对细胞、组织或生物体中的代谢产物以及不同条件下修饰的代谢产物进行高通量分析。

Metagenomics (or environmental sequencing) Application of genomic analysis, particularly high-throughput sequencing to environmental samples such as uncultured microbial communities.

宏基因组学（或环境测序） 对未经培养的微生物群落等环境样品进行基因组学分析的方法，主要是通过高通量测序方式。

Metatranscriptomics Global analysis, usually by sequencing, of the expressed genetic information (gene transcripts) produced by the collection of organisms in an ecosystem.

宏转录组学 通过测序方式对生态系统中所有生物所表达的遗传信息产物（基因转录本）进行的整体分析。

MIAME Minimal Information About a Microarray Experiment. MIAME is a recommendation that was established by a joint group of microarray experts (the Microarray and Gene Expression Data society or MGED) and describes the minimal requirements for data and information that a submission to a public microarray database should contain. The aim is to make microarray data understandable and reproducible.

MIAME 基因芯片实验的最小限度信息。是由基因芯片专家联合组织（芯片和基因表达数据协会，简称 MGED）制定的规范细则，描述了向公共芯片数据库提交数据或信息所需具备的最低要求，目的是使芯片数据变得容易理解并可被重复。

Microsatellite Small DNA stretches of a repeated core sequence of few base pairs (e.g. GT repeat units) that are highly polymorphic, particularly in terms of the number of repeated units. These are mainly used to create genetic maps.

微卫星 是由少量碱基的重复构成 DNA 片段（例如 GT 重复单元），具有高度多态性，

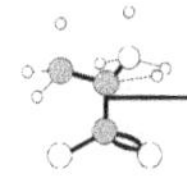

尤其重复单元的数量方面。主要用于遗传图谱的构建。

MIMS Minimum Information about a Metagenome Sequence.

MIMS 宏基因组序列的最低信息量。

Mismatch position with different bases or amino acids in a sequence alignment.

错配 在序列比对中，碱基或氨基酸的位置错误导致的配对。

Morpholino Synthetic molecules of usually 25 bases in length that bind to complementary sequences of RNA, blocking the access of cell components to those sequences. Morpholinos can block translation, splicing, miRNAs or their targets and ribozyme activity.

吗啉 通常为可以与互补的 RNA 序列结合的 25 个碱基长度的合成分子，用以阻止细胞组分靠近这些 RNA 序列。吗啉可以阻止翻译、剪接、miRNA 或它们的靶标以及核酸酶活性。

MPSS Massive Parallel Signature Sequencing; Tool to analyse the level of expression of virtually all genes in a sample by counting the number of individual mRNA molecules produced from each gene.

MPSS 大规模平行测序技术；通过计算每个基因转录产生的 mRNA 分子数量来分析样品中所有基因表达水平的工具。

Northern blot A method for analysing RNA molecules involving separation on an agarose gel and detection of specific RNAs using radioactive or fluorescently labelled probes (also called an RNA gel blot).

Northern 印迹 也被称做 RNA 凝胶印迹。一种用于 RNA 分析的方法，包括用琼脂糖凝胶将 RNA 分子分离，然后用放射性或荧光标记探针对特定的 RNA 进行检测。

Ontology In the context of computer and information sciences, an ontology defines a set of representational primitives with which to model a domain of knowledge or discourse. The Gene Ontology for example provides a controlled vocabulary to describe gene and gene product attributes in any organism.

语义（学） 在计算机和信息科学范畴，语义（学）定义一套有代表性基本元素，它可以模式化说明一个知识或论述的领域。以基因语义（学）为例，其提供了描述任何生物的基因和基因产物特性的数据库。

ORF Open Reading Frame, a series of codons beginning with a start codon and ending with

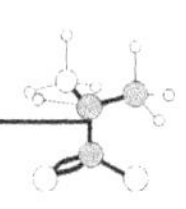

a stop codon, without any internal stop codons.

ORF 开放阅读框，一系列以起始密码子开始和以终止密码子结束的编码区域，其序列内部没有终止密码子。

Organelle A differentiated structure within a cell, such as a chloroplast, mitochondrion or vacuole, which performs a specific function.

细胞器 细胞内执行特定功能的分化结构，例如叶绿体、线粒体和液泡。

Orthologue Orthologous sequences are homologous sequences that have been separated by a speciation event. A gene that exists in a particular species that then diverges into two separate species, will be present in two copies afterwards. These copies are called orthologues and will typically have the same or a similar function.

直系同源（种间同源） 直系同源序列是指由于物种分化而形成的同源序列。一个基因存在于一个特定物种中，由于物种分化就会形成两个拷贝。这两个拷贝就叫做直系同源基因，它们通常具有相同或相似的功能。

PAMP Pathogen-Associated Molecular Patterns; Small molecular motifs conserved within a class of microbes; LPS (see above) is considered to be the prototypical PAMP.

PAMP 病原相关分子模式，一类微生物中保守的小分子基序（motif）。细菌内脂多糖（LPS）被认为是典型的病原相关分子模式。

Pangenome Unique set of proteins for a given species. In bacteria, the pangenome can be twice the size of the genomes of individual members of a species.

泛基因组 一个特定物种的独特全套蛋白。在细菌中，泛基因组大小可能是一个物种个体基因组的两倍。

Paralogue Homologous sequences that have been separated by a gene duplication event in an ancestral genome are called paralogous. Paralogous genes may mutate and acquire new functions because the original selective pressure is reduced.

旁系同源 在祖先基因组中通过基因复制造成分离的同源序列。因为原先的选择性压力降低，旁系同源基因可能发生变异并获得新的功能。

PCR Polymerase chain reaction. A technique for replicating a specific piece of DNA in-vitro. Oligonucleotide primers are added (which initiate the copying of each strand) along with nucleotides and Taq polymerase. By cycling the temperature, the target DNA is repeatedly denatured and copied allowing it to be amplified in an exponential manner.

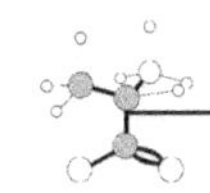

PCR 聚合酶链式反应。一段特定DNA片段的体外复制技术，加入寡核苷酸引物以及核苷酸和Taq聚合酶后启动每一条链的复制。通过温度循环，目标DNA重复变性和复制，从而实现指数扩增。

PDB Protein Data Bank. A single worldwide repository for the processing and distribution of 3-D biological macromolecular structure data.

PDB 蛋白质数据库，一个用于加工处理和共享生物大分子三维结构的世界性数据库。

Phytoplankton The photosynthetic organisms present in the plankton.

浮游植物 可以进行光合作用的浮游生物。

Picoeukaryotes microscopic eukaryote species with a cell size of less than 3 μm.

微型真核生物 细胞直径小于3微米的微小真核物种。

Primary endosymbiosis Used to describe the initial capture of a cyanobacterium by a eukaryotic cell that gave rise to the algae (see also endosymbiosis).

初级内共生 用来描述最初真核细胞捕获蓝藻菌从而形成水藻的过程（详见内共生）。

Primer Small oligonucleotide (anywhere from 6 to 50 nucleotides long) used to prime DNA synthesis.

引物 用来引导DNA合成的短核苷酸（一般长度为6至50个核苷酸）。

Probe/Reporter An oligonucleotide attached to the surface of a microarray.

探针 可以依附在基因芯片表面的寡核苷酸。

Proteomics High-throughput analysis of the proteins present in a cell type, a tissue or an organism.

蛋白质组学 对存在于某一细胞、组织或生物体内的蛋白质进行高通量分析。

Pseudogene A gene that is no longer functional either because its protein-coding sequence is disrupted or because it is no longer expressed.

假基因 一个由于蛋白编码序列被中断或者不再表达的不再具有功能的基因。

PTM Post-transcriptional modification. Alterations made to pre-mRNA before it leaves the nucleus and becomes mature mRNA.

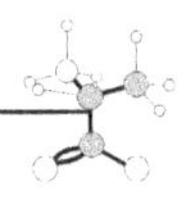

PTM 转录后修饰，在细胞核内将 mRNA 前体转变为成熟 mRNA 的过程。

p-value The result of a statistical test can be summarised using a p-value as a measure of significance. It can be described as the confidence in the rejection of the null-hypothesis. In the case of microarrays, p-values are often used to assess whether a gene is differentially expressed.

p-值 一个可以用来衡量统计结果显著性的数值，其可以被看做拒绝无效假设的可信度指标。对于基因芯片分析而言，p-值经常被用于评估一个基因是否存在差异表达。

Pyrosequencing Massively parallel DNA sequencing technique without the requirement of a prior cloning of the DNA.

焦磷酸测序 一种前期不需要 DNA 克隆的大规模 DNA 平行测序技术。

Q-PCR, qPCR or qRT-PCR Quantitative real-time reverse-transcription PCR. A PCR-based assay, which involves the direct measurment of the incorporation of a fluorescent dye into the PCR product. The level of fluorescence is a direct reflection on the amount of prduct present. By comparing the point in the PCR reaction where the reaction enters the log phase of replication for a control and a treated sample and determining the difference between the two, a measure of the change in relative gene expression caused by the treatment is determined.

Q-PCR、qPCR 或 qRT-PCR 实时定量反转录 PCR。一个基于 PCR 的方法，通过直接检测 PCR 产物中的荧光染料水平来反应 PCR 产物的量。当对照和受处理样品的 PCR 反应进入对数期时通过比较二者的差异，从而确定处理样品中基因的相对表达量。

QTL Quantitative Trait Loci; A region of a chromosome that is associated with a particular measurable trait (see also eQTL).

QTL 数量性状位点；染色体上与特定可测量性状相关联的区域（详见 eQTL）。

RAPD Random Amplification of Polymorphic DNA; DNA fragments randomly amplified by PCR from genomic DNA with short primers (8 – 12 nucleotides) of arbitrary nucleotide sequence.

RAPD 随机扩增多态 DNA；用任意的短引物（8 – 12 核苷酸）序列，对基因组 DNA 进行随机 PCR 扩增而获得的 DNA 片段。

RAST Rapid Annotation using Subsystem Technology.

RAST 基于子系统技术的快速注释。

Replicate A repetition of the same experimental measurement. The replicate can either be technical, where the same biological extract is analyzed twice, e. g. by using the same mRNA for

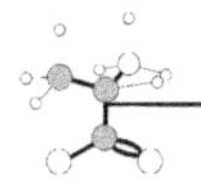

multiple microarrays; or it can be a biological replicate, when for example sample material is harvested from different individuals grown under the same conditions.

重复 相同实验性测量的重复。重复可以是技术性的，即对同一种生物抽提物分析两次，例如在多重微阵列中使用相同的 mRNA；或者是生物性重复，譬如在相同条件下取不同个体获得的重复。

RNA interference (RNAi) a genetic mechanism in which double-stranded RNA is cleaved into small fragments and acts a signal to either initiate the degradation of a complementary messenger RNA or to interfere with its translation.

RNA 干扰（RNAi） 以双链 RNA 被裂解成小片段作为信号，进而启动降解互补或干扰翻译的一种遗传机制。

ROS Reactive Oxygen species; Small molecules or ions formed by the incomplete one-electron reduction of oxygen; ROS contribute to oxidative stress/damage of DNA, lipids (oxidations of polydesaturated fatty acids), proteins (oxidations of amino acids) and inactivate specific enzymes (oxidation of co-factors).

ROS 活性氧簇，由单电子还原氧形成的小分子或离子；活性氧簇促使 DNA、脂类（高度不饱和脂肪酸）和蛋白质（氨基酸）的氧化应激或损伤，以及特殊酶的失活（辅助因子的氧化）。

SAGE Serial Analysis of Gene Expression. A high-throughput method to measure gene-expression by sequencing. Short fragments (tags) are generated from cDNA, which are then ligated to form long concatenated sequences, and sequenced.

SAGE 基因表达的系列分析，一种通过测序方法来测量基因表达量的高通量方法。将 cDNA 生成可拼接成长序列的短片段（标签），并测序。

Seaweed Brown, red or green macroalgae.

海藻 褐藻、红藻或绿藻等大型藻类。

Scatter-plot A two-dimensional plot where each measurement value is depicted by a single dot. Scatter-plots are often used to inspect data distributions.

散点图 一种将每个测量值绘制成一个单点的二维绘图方式。通常用于检查数据的分布情况。

SDS-PAGE Sodium dodecylsulphate-polyacrylamide gel electrophoresis; method used to separate proteins involving electrophoresis on an acrylamide gel.

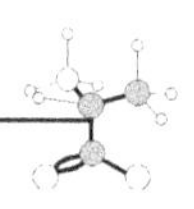

SDS-PAGE 十二烷基硫酸钠-聚丙烯酰胺凝胶电泳。一种应用丙烯酰胺凝胶电泳来分离蛋白质的方法。

Secondary endosymbiosis An event in which a eukaryotic cell enslaves another eukaryotic cell that possesses a plastid derived from a primary endosymbiosis (see also endosymbiosis and primary endosymbiosis).

次级内共生 一个真核细胞捕获另外一个拥有初级共生体的真核细胞的过程（详见内共生和初级内共生）。

Sequence trace file Raw sequence data in the form of a chromatogram from a Sanger sequencing reaction. The NCBI Trace Archive (http://www.ncbi.nlm.nih.gov.gate1.inist.fr/Traces/home/) and the Ensembl Trace Server (http://trace.ensembl.org/) are a public repositories for this sort of data.

序列跟踪文件 Sanger测序反应生成的以色谱峰图表示的原始序列数据。NCBI Trace Archive（http://www.ncbi.nlm.nih.gov.gate1.inist.fr/Traces/home/）和Ensembl Trace Server（http://trace.ensembl.org/）都是存储这类数据的公共存储器。

Short Read Archive (SRA) An NCBI database that stores raw data from sequencing platforms such as the Roche 454 System.

SRA 短片段序列文档 NCBI中存储来自测序平台（比如罗氏454系统）原始数据的数据库。

Singleton A single EST sequence that does not cluster with other ESTs.

单体 不与其他表达序列标签（EST）聚在一起的单一表达序列标签。

SNP Single-nucleotide polymorphism; particular sites (base pairs) in a sequence that are polymorphic and can be used as markers.

SNP 单核苷酸多态性，序列中含有多态性的特定位点（碱基对），可用作物种标记。

SOM Self Organizing Maps.

SOM 自组织映射。

Sporophyte A spore-producing plant.

孢子体 能产生孢子的植物（世代）。

Spot/feature A small area on the surface of a microarray containing a defined probe. Fea-

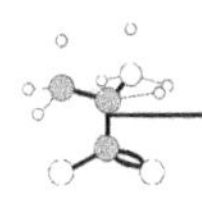

tures can be produced using a spotter or by direct synthesis of oligonucleotides on the substrate.

位点 在基因芯片表面包含有一类特定探针的小区域。这些寡核苷酸位点可通过仪器产生或在基质表面直接合成。

SSR Simple Sequence Repeat; also known as microsatellite (see definition above).

SSR 简单序列重复，也叫微卫星（详见微卫星定义）。

SteN Statistical electronic Northern blot.

SteN 统计性电子 Northern 印迹。

SVM Support Vector Machine, high performance machine learning technique that has been used to improve classification accuracy in biological applications such as gene prediction, detection of protein family members, RNA and DNA binding proteins and the functional classification of gene expression data. SVMs can solve non-linear classification problems by learning an optimally separating hyperplane in a higher-dimensional feature space. By use of non-linear kernel functions such as a Gaussian kernel, complex and non-linear decision functions can be realised.

SVM 支持向量机，一种用于提高分类准确性的高性能机器学习技术。在生物学里，应用于如基因预测、蛋白家族预测、RNA 和 DNA 结合蛋白预测及基因表达数据的功能分类。SVM 通过将向量映射到更高维函数空间找到最佳分离超平面，实现线性可分，解决非线性分类问题。通过非线性核函数（如高斯核函数），可以实现复杂和非线性的决策函数。

Synteny evolutionary preservation of the order of genes in a genomic region that indicates that this order also reflects the arrangement of genes in the last common ancestor of two compared species. Subdivided into macro-synteny (conserved aspects on the chromosomal level) and micro-synteny (local, gene-by-gene synteny).

共线性 在染色体一定区域上存在着进化上保守的基因群序列，这个顺序可以用来反映两个相比较物种的最近共同祖先在这段染色体的基因排列。可细分为宏观共线性（染色体水平的保守层面）和微观共线性（局部的基因共线性）。

Target In a microarray experiment, the target is labelled population of molecules corresponding to the RNA or DNA extracted from mixtures of cells, tissues, cell cultures or other samples. This labelled population of molecules is often referred to erroneously as the probe.

靶标 在基因芯片实验中，靶标是被标记的分子群对应于从细胞混合物、组织、细胞培养物或其他样品中抽提获得的 RNA 或 DNA。这个被标记的分子群经常会被误作“探针”。

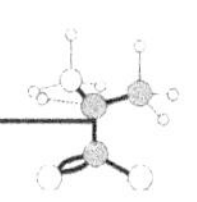

TC Tentative Consensus Sequence. EST sequences can be clustered together using assembly programs to form TCs.

TC 暂定一致性序列。表达序列标签（EST）可以使用组装程序来聚类，从而形成暂定一致性序列。

Tertiary endosymbiosis An event in which a eukaryotic cell enslaves another eukaryotic cell that possesses a plastid derived from a secondary endosymbiosis (see secondary endosymbiosis). This has occurred several times in the dinoflagellates, and the captured plastid is thought to have replaced a pre-existing secondary plastid (see also endosymbiosis).

三级内共生 一个真核细胞捕获另一个拥有次级内共生体的真核细胞的过程（详见次级内共生）。这种情况在海鞭藻中多次发生，而且被吞噬的质体会替代已存在的次级内共生体（详见次级内共生）

TETRA A tool that can be used to calculate, how well tetranucleotide usage patterns in DNA sequences correlate. Such correlations can provide valuable hints about the relatedness of DNA sequences.

TETRA 计算DNA序列四核苷酸多态性的概率，并通过比较不同DNA序列四核苷酸多态性的相关性提供关于DNA序列相关性的有价值的提示。

TGICL TIGR Gene Index Clustering tools.

TGICL 美国基因组研究所（TIGR）开发的用于快速对大量的ESTs和mRNA数据进行基因索引聚类的软件系统。

Third Party annotation (TPA) sequence database A database designed to capture experimental or inferential results that support submitter-provided annotation for sequence data that the submitter did not directly determine but can be derived from DDBJ/EMBL/GenBank primary data.

第三方注释序列数据库 一个数据库中序列的注释信息是由非序列提交者即“第3者”通过实验或自动计算方法推论得到，但其序列数据来源于DDBJ/EMBL/GenBank数据库的原始序列数据。

Transcriptomics High-throughput analysis of the mRNA transcripts present in a cell type, a tissue or an organism.

转录组学 对细胞、组织或生物体中的mRNA转录进行高通量分析。

Transcriptome the full complement of transcripts in a given species.

转录组 一个特定物种的全部转录本。

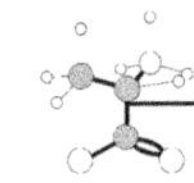

UTR untranslated region of a transcript. The complementary regions to the ORF (see ORF).

UTR 非编码区。一个转录本中的非编码区域，与开放阅读框互补的区域（参见ORF）。

Web services Provide a standardized method to access information or perform computations over a network via the exchange of XML-based messages.

网络服务 提供一个获取信息或通过换成 XML 的数据在网络上进行计算的标准化方法。

Western blot A method for analysing proteins involving separation on by SDS-PAGE (see above) and detection of specific proteins using antibodies.

免疫印迹 一种先通过 SDS-PAGE 胶将蛋白质分离，然后使用抗体来检测特定蛋白的分析方法。

XML Extended Markup Language, a specification to create a common representation of different types of data.

XML 可扩展标记语言。对不同类型的数据创建共有代码的规范。